TEV GAMMA-RAY ASTROPHYSICS

Library of Congress Cataloging-in-Publication Data

TeV gamma-ray astrophysics / Heinrich J. Völk and Felix A. Aharonian.
p. cm.

1. Gamma ray astronomy--Congresses. 2. Astrophysics--Congresses.
I. Völk, Heinrich J. II. Aharonian, Felix A.
QB471.A1T48 1996
522'.6862--dc20 95-47425

ISBN-13: 978-94-010-6561-0 e-ISBN-13: 978-94-009-0171-1
DOI: 10.1007/978-94-009-0171-1

Published by Kluwer Academic Publishers,
P.O. Box 17, 3300 AA Dordrecht, The Netherlands.

Kluwer Academic Publishers incorporates
the publishing programmes of
D. Reidel, Martinus Nijhoff, Dr W. Junk and MTP Press.

Sold and distributed in the U.S.A. and Canada
by Kluwer Academic Publishers,
101 Philip Drive, Norwell, MA 02061, U.S.A.

In all other countries, sold and distributed
by Kluwer Academic Publishers Group,
P.O. Box 322, 3300 AH Dordrecht, The Netherlands.

Printed on acid-free paper

Softcover reprint of the hardcover 1st edition 1996

TEV GAMMA-RAY ASTROPHYSICS

THEORY AND OBSERVATIONS PRESENTED AT THE HEIDELBERG WORKSHOP, OCTOBER 3–7, 1994

Edited by

HEINRICH J. VÖLK
Max-Planck-Institut für Kernphysik, Heidelberg, Germany

and

FELIX A. AHARONIAN
Max-Planck-Institut für Kernphysik, Heidelberg, Germany

Reprinted from *Space Science Reviews*, Vol. 75, Nos. 1–2, 1996.

KLUWER ACADEMIC PUBLISHERS
DORDRECHT / BOSTON / LONDON

Heidelberg Workshop on TeV Gamma-Ray Astrophysics (Participants)

TABLE OF CONTENTS

FOREWORD

When we participated in the second workshop of the series 'Towards a Major Air Cherenkov Detector' in Calgary, in July 1993, we knew that the field of Very High Energy (VHE) Gamma-Ray Astronomy was well on its way technically: the first convincing result of a VHE gamma-ray observation of a celestial source (the Crab Nebula) had been published in 1989 by the Whipple collaboration. This had shown that the air Cherenkov technique could obtain important astronomical results. There was also a scientific community which had begun to organize itself with the aim to develop further the imaging air Cherenkov technique and to hold workshops, started in Palaiseau in 1992, on the possibility to integrate these efforts into one or several larger instruments. That in fact this integration has not taken place but that the effort rather has resulted in the planning and implementation of several independent installations around the world is not a negative development but shows that the field is branching out in different directions: large single telescopes with superior imaging qualities, systems of medium sized telescopes for stereoscopic observations of individual air showers, and finally the idea to use solar power plant-type large light collector arrays or very large fixed-dish telescopes in an effort to decisively reduce the energy threshold by an order of magnitude to bridge the gap between space-borne and ground-based detectors.

The imaging Cherenkov technique, as the presently major tool of TeV gamma-ray astronomy, has the great problem that the telescopes necessarily have a rather small field of view, not exceeding a few degrees. In addition, rather long integration times are required for any given exposure. Therefore, the typical development of an astronomical field, starting with large angle surveys and then going over to pointed observations, is not a possible avenue. This implies not only the need for detailed theoretical investigations of astrophysical source characteristics but also a multiwavelength observational approach. At the same time we believe that the nature of different types of sources would also require different instrumental and data evaluation techniques. Obviously, extended and point-like sources, as well as burst-like and quasi-steady sources need different approaches. For example extended emitters with a typical angular size of about one degree, which presumably constitute important classes of VHE gamma-ray sources, certainly require fields of view exceeding four degrees for imaging cameras. Such large fields of view appear also necessary to improve the observation efficiency for sources whose positions are known, like many EGRET sources, only within about one degree. Thus the astrophysical characteristics of potential sources play an increasing role.

It were these considerations which prompted us during the Calgary workshop to propose a workshop concentrating on the astrophysical aspects, complementary to the plans of Tadashi Kifune to hold a third workshop of the Palaiseau series

Space Science Reviews **75:** VII–IX, 1996.

earlier in 1995 in Tokyo. The idea was well received, with Alan Watson saying that he thought it was a significant development when theoreticians started to develop systematically this side of the field.

The 'Heidelberg Workshop on TeV Gamma-Ray Astrophysics' was held during October 3–7, 1995, at the Max-Planck-Hause in Heidelberg. The idea was to bring together astrophysicists from the various fields which play a role in the formation of high energy gamma-ray emission. To extent that it is relevant, this implies cosmic ray physics and particle physics, the physics and astrophysics of compact objects like pulsars and quasars, particle acceleration in SNRs, galactic winds and AGNs, cascade processes in universal background photon fields, etc. Also, we strongly believed that the different types of present and future detectors, their capabilities and sensitivities needed to be discussed.

The Workshop began with two invited lectures by Trevor Weekes and Michael Hillas that set the scene, also for the sizeable number of students which has been encouraged to participate. The main part of the Workshop was concentrated in review talks on gamma-ray observations starting from the MeV-region, future instrumental capabilities, and models of potential VHE gamma-ray sources as well as gamma-ray propagation through the intergalactic medium. One afternoon/evening was reserved for a *Mini-Workshop* on recent results in TeV astronomy. The contributed papers of this session are not published in this volume. We believe that the many original results presented there should be more appropriately published in the regular journal literature.

Most of the invited speakers were able to put their contributions in the form of review papers. According to the rules of *Space Science Reviews*, all submitted manuscripts were reviewed by different, anonymous referees. We take this opportunity to express our gratitude to the referees for their important work which has often led to substantial improvements and clarifications of the final manuscripts. The ordering of papers here corresponds to the order of the Workshop program. Kluwer Academic Publishers will also print a *Proceedings* version in bookform. One of the papers given (by Hans Bloemen) has been recently published in *Space Science Reviews*. It will therefore only be reprinted in the *Proceedings*. Another, brief and informative paper (given by Marc Lachièze-Rey) had been planned as a tutorial on the subject of the universal 2.7 K background radiation, extensively reviewed in the literature in recent years. It is likewise printed only in the *Proceedings*.

Many individuals contributed to the success of the Workshop, Mitchell Begelman, Catherine Cesarsky, Philippe Goret, Werner Hofmann, Gottfried Kanbach, Tadashi Kifune, John Kirk, German Krymsky, Richard Lamb, Hinrich Meyer, Alan Watson, Trevor Weekes, and Gaurang Yodh agreed to serve on the Scientific Organizing Committee and gave important advice. Many of the Heidelberg students helped in the organization and the later editing of the papers. Most importantly, Waltrud Butsmann and Christian Wiedner ran the local organization. Jacqueline Suppanz-Pirsch and Antje Kohnle organized much of the social program. We express our sincere gratitude to all of these colleagues, without whose efforts the meeting could not have taken place.

HEINZ VÖLK FELIX AHARONIAN

LIST OF PARTICIPANTS

Aharonian, Felix A. (MPI Kernphysik) aharon@boris.mpi-hd.mpg.de / Fax: (49) 6221–516–324
Alexandreas, Dimitris (INFN) dimitris@v2phpd.pd.infn.it / Fax: +39–49–844245
Arons, J. (Calif. Univ.) arons@astroplasma.berkeley.edu / Fax: 510–642–3411
Arqueros, Fernando (Complutense Univ.) arqueros@gae.ucm.es / Fax: +34–1–3945193
Auriemma, Giulio (Basilicta Univ.) 39942::auriemma / Fax: +39–6–8610082
v. Ballmos, P. (Centre d'Etude Spatiale des Rayonnements) 17449::pvb / Fax: (+33) 61 55 67 01
Berezhko, E. (MPI Kernphysik) berozko@Dkosmo.mpi.hd.mpg.de / Fax: (49) 6221–516–540
Bernlöhr, Konrad (MPI Kernphysik) bernlohr@eu1.mpi-hd.mpg.de / Fax: (49) 6221–516–540
Bhuttacharjee, Pijush (Indian Astrophysics Inst.) pijush@iiap.ernet.in
Bloemen, Hans (SRON Utrecht) h.bloemen@sron.ruu.nl / Fax: +30–540860
Bloom, Elliott (Stanford Univ.) elliott@slac.stanford.edu / Fax: 415–926–4335
Boettcher, Markus (MPI Radioastronomie) mboett@mpifr-bonn.mpg.de / Fax: +49+228 525259
Chilingarian, Ashot (Yerenran Phys. Inst.) chilin@crd.erphy.armenia.su / Fax: 007–8852–350030
Chounet, Louis-Marie (L.P.N.H.E.) chounet@frepn11.in2p3.fr / Fax: (33) 169 33 30 02
Coppi, P. (Yale Univ. Astronomy) coppi@blazar.astro.yale.edu / Fax: (01) 203 432 5048
Coyne, Donald (UC Santa Cruz) coyne@sld.slac.stanford.edu / Fax: (408) 459–3043
Cresti, Marcello (Padova Univ.) cresti@padova.infn.it / Fax: 39–49–844245
Daum, Andreas (MPI Kernphysik) daum@eu1.mpi-hd.mpg.de / Fax: (49) 6221–516540
Deckers, Thorsten (Kiel Univ.) deckers@ifkax1.ifkhep.uni-kiel.de / Fax: (49) 431–880–1647
Degrange, Bernard (PNHE) degrange@frcpn11.in2p3.fr / Fax: (33) 169 333002
Dingus, Brenda L. (NASA) bld@egret.gsfc.nasa.gov / Fax: (301) 286–1682
Djannati-Atai, Arache (LPC) djannati@frcpn11.in2p3.fr / Fax: (00331) 43546989
Dorfi, Ernst (Wien Univ. Astronomie) ead@astro.ast.univie.ac.at / Fax: (00431) 470 6800 15
Drury, Luke (Dublin Inst.) ld@cp.dias.ie / Fax: +353–1–6621477
Duffy, Peter (MPI Kernphysik) duffy@boris.mpi-hd.mpg.de / Fax: (49) 6221–516324
Espigat, P. (LPC College de France) / Fax: (33) 1–4354–6989
Eschstruth, Paul (Paris-Sud Univ.) eschstruth@lalcls.in2p3.fr / Fax: (33) 169071526

Space Science Reviews **75:** XI-XV, 1996.

Fegan, David J. (Dublin Univ.) djfegan@ferdia.ucd.ie / Fax: (01) 706 1185
Fernandez, Juan (MPI Physik und Astrophysik) juan@hegral.mppmu.mpg.de / Fax: 089–3226 704
Fleury, Patrick (L.P.N.H.E.) fleury@poly52.in2p3.fr / Fax: (33) 1–69 33 30 02
Fonseca, Victoria (Madrid Univ.) fonseca@eucmmv.gae.ucm.es / Fax: 34 1 3945193
Fontaine, Gerard (Ecole Polytechnique) fontaine@frcpn11.in2p3.fr /Fax: (33) 169 33 3002
Gaidos, James A. (Purdue Univ.) gaidos@purvsc.Physics.purdue.edu / Fax: (317) 494–0706
Gallant, Yves (MPI Kernphysik) gallant@yves.mpi-hd.mpg.de / Fax: (49) 6221 516324
George, Roger (L.P.N.H.E.) george@frcpn11.in2p3.fr / Fax: (33) 14427 4638
Goret, Philippe (DAPNIA Saclay) goret@sapvxg.saclay.cea.fr / Fax: 0033–16908 6577
Grenier, Isabelle (Paris Univ.) grenier@sapvxg.saclay.cea.fr / Fax: (00331) 6908 6577
Gurevich, Alex (P.N. Lebedev Inst.) alex@lpi.ac.ru
Harding, Alice (NASA) harding@1heavx.gsfc.nasa.gov / Fax: (301) 286 1682
Heintze, Joachim (Heidelberg Univ.) heintze@physi;uni-heidelberg.de / Fax: +(49) 6221–516540
Heinzelmann, Götz (Hamburg Univ.) f11hei@dsyibm.desy.de / Fax: 040 8998 2196
Hermann, German (MPI Kernphysik) hermann@enull.mpi-hd.mpg.de Fax: (49) 6221–516540
Hess, Markus (MPI Kernphysik) hess@eu1.mpi-hd.mpd.de / Fax: (49) 6221 516540
Heusler, Andreas (MPI Kernphysik) hsl@vampi.mpi-hd.mpg.de / Fax: (49) 6221 516540
Hillas, Michael (Leeds Univ.) phy6hill@phys-irc.novell.leeds.ac.uk / Fax: (44) 532 333 846
Hofmann, Werner (MPI Kernphysik) wh@enull.mpi-hd.mpg.de / Fax: (49) 6221–516 540
Hurley, K. (California Univ.) khurley@space6.ssl.berkeley.edu / Fax: +510 643 8302
Johnson, Paul A. (Adelaide Univ.) pjohnson@physics.adelaide.edu.au / Fax: +61 8 303 4380
Kanbach, Gottfried (MPI extraterr. Physik) gok@mpe-garching.mpg.de / Fax: 089 3299 3569
Karle, Albrecht (DESY-IfH Zeuthen) karle@hpbail.ifh.de / Fax: 49–33762–77330
Kifune, Tadashi (Tokyo Univ.) tkifune@asun2.icrr.u-tokyo.ac.jp / Fax: 81–424–62–3096
Kirk, John (MPI Kernphysik) kirk@kirk0.mpi-hd.mpg.de / Fax: (49) 6221 516 324
Köhler, Carsten (MPI Kernphysik) koehler@eu1.1mpi-hd.mpg.de / Fax: (49) 6221 516 540
Kohnle, Antje (MPI Kernphysik) kohnle@eu1.mpi-hd.mpg.de / Fax: (49) 6221–516540
Konopelko, Alexander (MPI Kernphysik) sashho@eu1.mpi-hd.mpg.de / Fax: (49) 6221 516 50
Krennrich, Frank
Kronberg, Philipp (Toronto Univ.) kronberg@radio.astro.utoronto.ca / Fax: 001–416–412–0201
Krymsky, G. (Yakutsk Inst. Cosmophys. Res. and Astr.) centr@ikfia.infonet.yakutia.su / Fax: 007 095 230 2919
Krys, Andrzej (Lodz Univ.) akrys@krysia.uni.lod.edu.pl / Fax: (4842) 787087
Krys, Elzbieta (Lodz Univ.) ekrys@krysia.uni.lod.edu.pl / Fax: (4842) 787087

Ksenofontov, Leonid (MPI Kernphysik) ksenofon@boris.mpi-hd.mpg.de / Fax: (49) 6221 516 540
Lachieze-Rey, M. (DAPNIA Saclay) 33589::Lachieze-Rey / Fax: 0033–16908 9266
Lamb, R. C. (IOWA State Univ.) rcl@iastate.edu / Fax: 515 294 6027
Lebohec, S.
Leffers, Georg (MPI Kernphysik) leffers@eu1.mpi-hd.mpg.de / Fax: (49) 6221 516 540
Lemke, D. (MPI Astronomie) lemke@mpiahd.mpi-hd.mpg.de / Fax: (49) 6221 528 246
Lindner, Axel (Hamburg Univ.) lindner@hp-cluster.desy.de / Fax: 040–8998 2950
Lorenz, Eckart (MPI Physik) 14037::ecl / Fax: 089–3226704
Magnussen, Norbert (Wuppertal Univ.) magnussen@wpvx11.physik.uni-wuppental.de / Fax: 0202 439 2662
Mannheim, Karl (Göttingen Univ.) kmannhe@uni-sw.gwdg.de / Fax: (0551) 395043
Mariotti, Mose (Padova Univ.) mariotti@axpd01.pd.infn.it / Fax: +39 49 844 245
Marsella, Giovanni (INFN) marsella@pisa.infn.it / Fax: +39 50 880 451
Mastichiadis, Apostolos (MPI Kernphysik) masti@aposto.mpi-hd.mpg.de / Fax: +49 6221 516 324
Matheis, Volker (MPI Kernphysik) matheis@eul.mpi-ht.mpg.de / Fax: (49) 6221–516540
Mattheus, Robert (MPI Kernphysik) mattheus@eu1.mpi-hd.mpg.de / Fax: (49) 6221–516540
Mattilla, K. (Helsinki Univ.) mattila@cc.Helsinki.FI / Fax: 358–0–1912952
Mayer-Hasselwander, Hans (MPI extraterr. Physik) hrm@mpe-garching.mpg.de / Fax: 089–32993554
Meyer, Hinrich (Wuppertal Univ.) meyer@wpvx11.physik.uni-wuppertal.de / Fax: 0202–4392662
Mirzoyan, Razmik (MPI Physik) ram / Fax: 089–32 267 04
Morse, Robert (Wisconsin Univ.) morse@wishep.physics.wisc.edu / Fax: 608 263 0800
Müller, Dietrich (Chicago Univ.) muller@odysseus.uchicago.edu / Fax: (312) 702–6645
Mukherjee, Reshmi (NASA GSFC) muk@egret.gsfc.nasa.gov / Fax: (301)–286–1611
Orford, Keith J. (Durham Univ.) k.j.orford@durham.ac.uk / Fax: 0044 91 3743 749
Panter, Michael (MPI Kernphysik) panter@enull.mpi-hd.mpg.de / Fax: (49) 6221–516540
Pare, E. (Ecole Polytechnique de Saclay) pare@frcpn11.in2p3.fr / Fax: (33) 169 33 30 02
Perez-Fournon, Ismael (Inst. de Astrof. Canarias) ipf@ll.iac.es / Fax: +34–22–605210
Petry, Dirk (MPI Physik) petry@hegra1.mppsu.mpg.de / Fax: 089–3226704
Plaga, Rainer (MPI Physik) plaga@hegra1.mppmu.mpg.de / Fax: 089–3226704
Pohl, Martin (MPE Garching) mkp@mpesun8.mpe-garching.mpg.de / Fax: 089 3299–3569
Porter, N. A. (Dublin Univ.) djfegan@ferdia.ucd.ie / Fax: (00353 1) 706 1185
Prahl, Jürgen (Hamburg Univ.) prahl@ips106.desy.de / Fax: 040–8998–2950

Primack, Joel (California Univ.) joel@lick.ucsc.edu / Fax: 408 459 3043
Procureur, Jacques (Bordeaux Univ.) jproc@frcpn11.in2p3.fr / Fax: (0033) 56 89 8759
Prosch, Christian (MPI Physik) chris@hegral.mppmu.mpg.de / Fax: 089–3226704
Punch
Quebert, Jean (CENBG) quebert@frcpn11.in2p3.fr / Fax: 33 56751180
Raubenheimer, Christo (Potchefstroom Univ.) fskbcr@puknet.puk.ac.za / Fax: 27–148–99–2421
Rauterberg, Gerhard (Kiel Univ.) grauterberg@kernphysik.uni-kiel.d400.de / Fax: (49) 431–880–1647
Resvanis, L. (Athen Univ.) 22718::resvanis / Fax: 301–3633413
Rhode, Wolfgang (Wuppertal Univ.) rhode@wpvx11.physik.uni-wuppertal.de / Fax: 0202 4392662
Ribel (KFZ Karlsruhe) / Fax: O7247–824075
Rivoal, Monique (Paris Univ.) rivoal@frcpn11.in2p3.fr / Fax: (0033 1) 4427 4638
Rose, H. J. (Leeds Univ.) phy6jr@sun.leeds.ac.uk / Fax: (44) 532 33 3846
Rozanzka, M. (Krakow Univ.) rozanska@hegral.mppmu.mpg.de
Saggion, Antonio (Padova Univ.) saggion@padova.infn.it / Fax: +39–49–844245
Samorski, Manfred (Kiel Univ.) msamorski@kernphysik.uni-kiel.d400.de / Fax: (49) 431–880–1647
Sartori, Giorgio (Padova Univ.) gsartori@padova.infn.it / Fax: 39–49 844245
Schatz, Gerd (KFS Karlsruhe) joe@ik3dc1.kfk.de / Fax: 07247–824075
Schlickeiser, R. (MPI Radioastronomie) p337sch@mpifr-bonn.mpg.de / Fax: (0228) 5252
Schmele, Dirk (Hamburg Univ.) schmele@hp-cluster.desy.de / Fax: 040–89982950
Schmitz, Norbert (MPI Physik) nos / Fax: 089–32 308–436
Sigl, Günter (Chicago Univ.) sigl@oddjob.uchicago.edu / Fax: 001–312–702–8212
Stamenov, J. W. (Bordeaux Univ.) jproc@frcpn11.in2p3.fr / Fax: (33) 56 89 8759
Stamm, Wilhelm (Kiel Univ.) wstamm@kernphysik.uni-kiel.d400.de / Fax: (49) 431–880–1647
Stecker, F. (Goddard Space Flight Center) stecker@lheavx.gsfc.nasa.gov / Fax: 301–286–1682
Thornton, Greg (Adelaide Univ.) gthornto@physics.adelaide.edu.au / Fax: 301–286–1682
Trunk, Ulrich (MPI Kernphysik) trunk@eu1.mpi-hd.mpg.de / Fax: (49) 6221–516 540
Tuffs, Richard (MPI Kernphysik) rjt@taurus.mpi-hd.mpg.de Fax: (49) 6221–516 324
Turver, T. (Durham Univ.) k.e.turver@durham.ac.uk / Fax: (44) 91 3744745
Ulrich, Mathias (MPI Kernphysik) ulrich@eu1.mpi-hd.mpg.de / Fax: (49) 6221 516540
Völk, Heinrich (MPI Kernphysik) vlk@vampi.mpi-hd.mpg.de / Fax: (49) 6221–516549
Wahl, Heinrich (CERN) wahl@cernvm.cern.ch / Fax: 004122–7830672
Watson, A. (Leeds Univ.) a.a.watson@leeds.ac.ul / Fax: 0113–2–333-846
Weekes, Trevor C. (Smithsonian Inst.) weekes@egret.sao.arizona.edu Fax: 602–670–5739

Westerhoff, Stefan (Wuppertal Univ.) westerhoff@wpvx11.physik.uni-wuppertal.de / Fax: 0202 4392662
Wiedner, Christian A. (MPI Kernphysik) wiedner@vampi.mpi-hd.mpg.de / Fax: (49) 6221–516540
Wieland, Christian (MPI Kenphysik) wieland@eu1.mpi-hd.mpg.de / Fax: (49) 6221–516540
Willmer, Martin (Kiel Univ.) willmer@kernphysik.uni-kiel.d400.de / Fax: (49) 431–880–1647
Wirth, Hermann (MPI Kernphysik) wirth@vampi.mpi-hd.mpg.de Fax: (49) 6221–516540
Yodh, Gaurang B. (Irvine Univ.) yodh@master.ps.uci.edu / Fax: 714 856 7478
Zink, Ralf (MPI Kernphysik) zink@eu1.mpi-hd.mpg.de / Fax: (49) 6221–516540

Westerhoff, Stefan (Wuppertal Univ.) westerhoff@wpvs11.physik.uni-wuppertal.de
/ Fax: 0202 4392662

[illegible]

[illegible]

[illegible]

[illegible]

[illegible]

THE ATMOSPHERIC CHERENKOV TECHNIQUE IN VERY HIGH ENERGY GAMMA-RAY ASTRONOMY

TREVOR C. WEEKES
Whipple Observatory,
Harvard-Smithsonian Center for Astrophysics,
P.O. Box 97,
Amado AZ 85645-0097 U.S.A.

Abstract. The role of the atmospheric Cherenkov technique in the development of Very High Energy Gamma Ray Astronomy is described; it is shown that the Crab Nebula has played a critical role in this development. The ability of a TeV telescope to see a 5 sigma signal from a source in an hour of observation is now a reality. Sources as weak as 1/20th the strength of the Crab are detectable in principle now at TeV energies.

1. Introduction

In the past decade a new astronomical window (from 100 GeV to 10 TeV) has opened; the atmospheric imaging Cherenkov technique has played a major role in this exciting development. The sensitivity achieved to date with atmospheric Cherenkov imaging telescopes effectively complements the Compton GRO (where the EGRET experiment cuts off at 30 GeV); with this sensitivity important detections have been made of gamma-ray sources. However only a small fraction of the sky has been surveyed and there are reasons to believe that important scientific advances can be made with some extension of existing ground-based detection techniques. With a relatively minor investment both the flux sensitivity and energy threshold of these telescopes can be reduced and these scientific issues can be addressed.

1.1. History

Just as gamma-ray astronomy in the 100 MeV region can be said to date from the historic paper of Morrison (1957), astronomy at TeV energies seems to date from the prescient prediction by Cocconi (1959) of the detectability of TeV gamma- ray from the Crab Nebula. Although his model for the gamma-ray emission was not correct (he overestimated the eventual detected flux by a factor of 1000) he sowed the seeds for the first serious atmospheric Cherenkov experiments to detect very high energy gamma rays from cosmic sources. His model of the Crab Nebula assumed that relativistic protons were produced in the initial explosion and trapped in the Nebula; collisions with ambient gas produced charged pions which subsequently produced the relativistic electrons which caused the observed synchrotron radiation. Gam-

Space Science Reviews **75**: 1–15, 1996.

ma rays would come from the decay of the neutral pions produced in the same way.

Cocconi's prediction was based on the detection of air showers from the gamma rays by existing particle arrays; (as of this date (1994) the Crab Nebula has still not been detected by a particle air shower array). He was apparently unaware of the work of Bill Galbraith and John Jelley, who, in the pioneering experiment of ground-based gamma-ray astronomy, made the daring assumption that air showers, which up until then had been detected by the surviving particles at ground level, might be detectable optically by their atmospheric Cherenkov radiation; they made a series of simple, but elegant, experiments to show that this was so (Galbraith and Jelley, 1953). The descriptions of the early experiments (Galbraith and Jelley, 1955, Jelley and Galbraith, 1955) should be mandatory reading for all newcomers to the subject; even today it is important that to remember that fundamental experiments can be made with simple equipment and that every measurement does not require a CAMAC crate and a workstation.

In these early phenomenological experiments it was realized that this new technique offered possibilities for discrete source astronomy at TeV energies. These experiments relied on the detection of a directional anisotropy amongst the arrival directions of cosmic ray air showers, the assumption being made that the interstellar and interplanetary magnetic fields rendered the charged component isotropic and that there were no other neutral particles that could come from a discrete source. It was also generally assumed that the sources would be steady emitters or, after the discovery of pulsars, steady periodic sources.

The first really serious attempt to use this technique to make gamma-ray observations was in the Crimea by the group from the Lebedev Institute in Moscow (Chudakov et al., 1965); they began their work in 1960 and were inspired to do so by the Crab prediction. Their detector consisted of an array of 12 telescopes, each of 1.5 m aperture, mounted in sets of three on railway cars and directed in parallel. They surveyed the most important sources suspected to be sites of high energy particle activity for possible gamma-ray emission; these included supernova remnants and radio galaxies. The sensitivity achieved in this experiment was about two orders of magnitude below the flux predicted by Cocconi; it is a matter of conjecture now as to how rapidly this whole field might have developed if the Crab flux had exceeded this threshold sensitivity.

Although the results were negative, they were the first real limits in this new energy regime and others were encouraged by the simplicity to pursue the technique. It was particularly attractive to groups with small resources; hence in the early sixties it was also pursued in Ireland and in India.

1.2. The Atmospheric Cherenkov technique

The atmospheric Cherenkov technique was particularly suited for gamma-ray astronomy for a number of reasons including: (i) the forward momentum of the shower is so great, and the Cherenkov angle of emission in the atmosphere so small, that the Cherenkov light retains the original direction of the primary photon; hence the inherent angular resolution of the technique is high; (ii) at the same time the light does spread out appreciably so that the light pool that reaches ground level has dimensions of several hundreds of meters; hence even a small optical detector has a collection area of > 5×10^8 cm^2; (iii) the amount of light radiated is proportional to the total number of particles in the shower and is not strongly absorbed by the atmosphere; hence the Cherenkov light is a calorimetric component of the shower and can be used as a good estimator of the primary energy; (iv) the very short duration of the light pulses is well-matched to fast pulse counting electronics and enables the shower to be detected against the ambient light of the night-sky.

The atmospheric Cherenkov telescope is a field instrument in which the detector is at a remote location, is open to the night- sky and uses the earth's atmosphere as its detection medium. This has the advantage that the gas does not need to be replenished and the detector need never be lifted into orbit, but it also means that the experimenter is totally at the mercy of the environment. Temperature, pressure, and transmissivity, may vary in a haphazard way and the optical experiment is constantly limited by the optical background. The sun, the moon, the background stars, are obvious sources of interference as is the earth's varying airglow. Clouds are a constant limitation and drive experiments to desert sites. Meteors and distant lightning are annoying sources of pulsed radiation; recently the background due to city lights, beacons, aircraft, and satellites has become increasingly important. But the most troublesome background is, of course, the charged hadronic component of the cosmic radiation which may be more than one thousand times more numerous than the expected flux of gamma-rays at these energies. In this respect the ground-based gamma-ray astronomer faces the same limitation as his/her space-borne lower energy counterpart; the latter has the luxury of an anti-coincidence shield and the unique interaction (pair production) to reduce this background to practically nothing.

The atmospheric Cherenkov telescope is not a wide field survey-type instrument; instead, like a conventional optical or radio telescope, it requires a predetermined target which for some reason is supposed to be a source of TeV gamma radiation. Candidate objects come from observations at other wavelengths, with some modelling required to extrapolate the emission to these high energies.

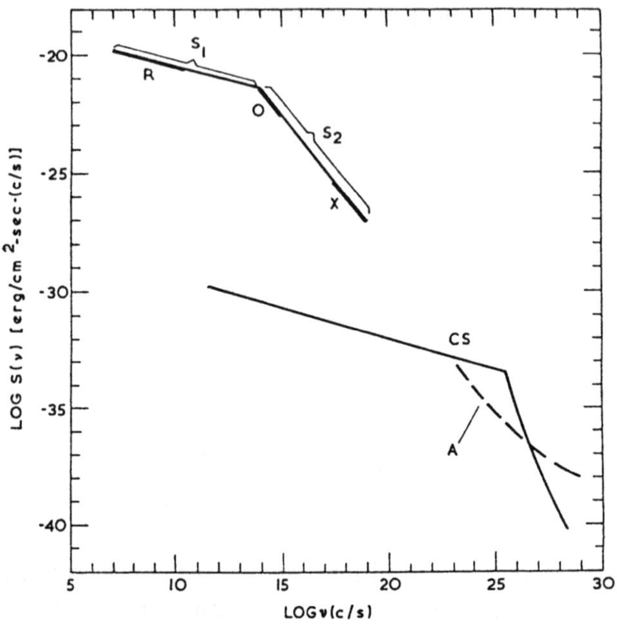

Fig. 1. The Compton-synchrotron spectrum (CS) calculated by Gould (1965) for the Crab Nebula; A is the spectrum expected from interaction with the 2.7° field (Jelley 1971). The only free parameter in this model was the magnetic field which was assumed to be near the equipartition value.

2. Development

2.1. The Crab Revisited

The development of a new model for the Crab Nebula was the catalyst for a new stage of development of atmospheric Cherenkov telescopes. Gould (1965) showed that if the amorphous radiation from radio to x-rays from the Nebula was due to synchrotron radiation by relativistic electrons then these same electrons should Compton-scatter the photons, boosting them to gamma-ray energies. The resultant gamma-ray spectrum would be most easily detectable at 100-1000 GeV energies, dipping sharply thereafter because of the Klein-Nishina cross-section (Figure 1).

Encouraged by this prediction the Smithsonian Astrophysical Observatory built a 10 m optical reflector on Mount Hopkins in Arizona (Figure 2); this was for many years the largest reflector built solely to do gamma-ray astronomy. Three years of observation led to a three sigma detection but this was rightly treated as an upper limit (Fazio et al. 1972). With hindsight one can now identify this observation as the first indication of the detection of the Compton-synchrotron spectrum of the Crab Nebula; if the detection had been a little more significant, the history of VHE astronomy might have been quite different. During this period there were active experiments in

Fig. 2. The 10 m Optical Reflector at the Whipple Observatory in southern Arizona. In the focal plane there is an array of 109 photomultipliers.

Ireland, the United States, the U.S.S.R., Australia, and India. There were several other sources that gave tentative evidence of detectability (PSR0531, Centaurus A, Cygnus X-3) but it would be true to say that, by the end of that decade, the credibility of the VHE detections, amongst gamma-ray astronomers was not high.

2.2. Second Generation Telescopes

As Monte Carlo techniques of shower simulation were developed, it became apparent that the initial assumption, that gamma-ray and hadronic showers were identical, was simplistic and that it might be possible to discriminate between the two types of shower, at least in a general way. Because of the smaller transverse momentum in electromagnetic interactions, the electromagnetic cascade is a much more compact and tidier entity than its hadronic counterpart (Figure 3a,b).

The principal qualitative differences between gamma-ray and hadron showers are explored in a quantitative fashion in Prof. Hillas's talk at this workshop. Experiments which measure the lateral or angular distribution of the Cherenkov light have the added advantage that they increase the angular resolution of the telescope and hence reduce the background for geometrical reasons.

It was apparent that if progress was to be made in gamma-ray astronomy in the TeV energy region it would be necessary to exploit the basic differences

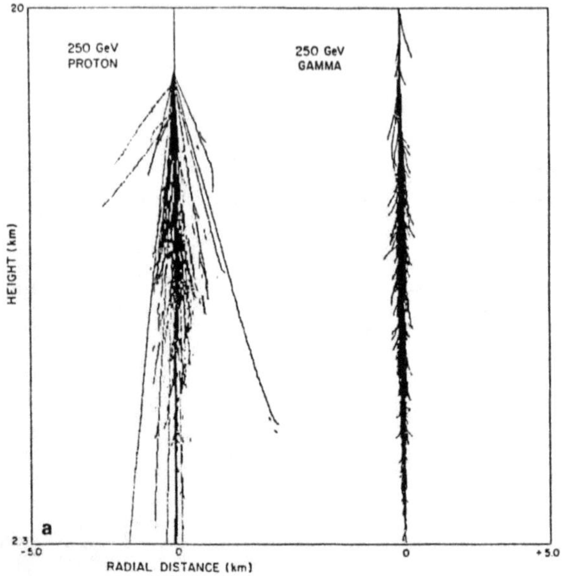

Fig. 3. Development of proton and gamma-ray initiated air showers in the atmosphere. Not only does the gamma-ray develop earlier but it has a more regular lateral distribution (at detector level) and does not have penetrating particles (mostly muons) which tend to dominate (or at least play a major role in the appearance of) the Cherenkov light image of a hadron shower.

between the gamma-ray showers and the hadronic background. It was not clear which of these differences would be the most effective discriminant or whether they would be best used in combination.

In 1977 at a gamma-ray workshop a method of improving the atmospheric Cherenkov technique was described (Weekes and Turver, 1977). This paper pointed out that hadronic showers would become increasingly less efficient at radiating Cherenkov light than gamma-ray showers of the same energy as the energy decreased; a combination of large reflectors, all equipped with multiple detector elements, was advocated so that the angular spread (image) of the light could be recorded and its arrival direction determined. In practice the imaging was to be achieved by an array of phototubes in the focal plane of a large optical reflector such as the Whipple Observatory 10 m optical reflector. It was the availability of this reflector with its good angular resolution that led to consideration of imaging as a technique with improved sensitivity. This method was considered primarily as a means of increasing the angular resolution; it was only later that the possibility of using the images to identify the primary was considered (Weekes, 1981; Hillas, 1985).

The atmospheric Cherenkov imaging technique, as this method came to be called, was actually based on two earlier developments of the Cherenkov technique. (1) As early as 1960, image intensifier pictures of the Cherenkov light from air showers had been recorded (Hill and Porter, 1961) and the potential of this method for gamma-ray astronomy was discussed (Jelley and Porter, 1963). However, as this technology was then very difficult and the energy threshold with image intensifier systems in excess of 100 TeV, it did not result in a practical gamma-ray telescope. (2) In the early seventies a simple stereoscopic system (the "double beam technique") was used to reduce the hadron background (Grindlay, 1972). In this approach two reflectors, with two "pixels", separated by 100 m, were used to give a crude stereoscopic image of showers. Although this method produced some promising results (Grindlay et al., 1975a,b; Grindlay et al., 1976), it was not developed any further.

3. Discovery

3.1. The Imaging Technique

As discussed above, there are basic differences in the appearance of the Cherenkov light image of a gamma-ray shower (centered in the field of view) and that of a typical background hadron shower coming from a random direction. The differences arise for two reasons: a) physics: the smaller transverse momentum in electromagnetic interactions compared with hadronic interactions means that the Cherenkov radiating particles in the gamma-ray shower are, on average, closer to the direction of the primary; also since there is no penetrating component the local contribution to the light is small and the fluctuations in the shower image are far less; b) geometry: the orientation of the roughly elliptical image of a shower depends on the angle it makes with the optic axis of the telescope; a shower whose axis coincides with the optic axis (the trivial case) gives a round image that is centered on the axis; showers which are parallel to the axis but fall up to 120 m away have elliptical images whose major axis intersect the optic axis (Figure 4); this holds true for gamma-ray and hadron showers but it is easier to characterize the axis of a gamma-ray shower because of the tighter image.

Although the image could be recorded on any of the imaging devices currently used in two-dimensional astronomical photometry, an array of small phototubes is the best match to the crude optics of the collectors, the short duration of the light flash, the high data rate, and low contrast images expected from air showers. Cameras developed at the Crimean Astrophysical Observatory (Stepanian et al. 1983) and the early camera at the Whipple Observatory (Weekes et al. 1989) consisted of 37 pixels (5 cm diameter phototubes) spaced 0.5° apart with a full-field of 3.5°.

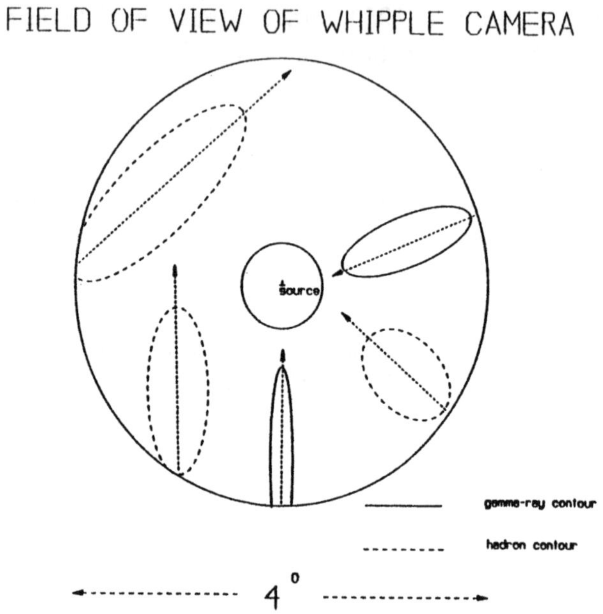

Fig. 4. The contours of typical Cherenkov light images as seen in the focal plane of the Whipple Observatory Camera (diameter 3.5°). Images of gamma-ray showers coming from a source along the optic axis (cross) are shown by full lines; they are narrow and point towards the center. The images from background cosmic rays are broader and have no preferred pointing direction.

3.2. A Source at last: the Crab Nebula.

The first clear evidence that the Crab Nebula was to play the same pivotal role in VHE astronomy as it played in radio, x-ray and 100 MeV gamma-ray astronomy came in 1989 with the publication of the results of observations with the imaging detector at the Whipple Observatory (Weekes et al., 1989); this publication was based on 60 hours of observation in the epoch 1986-88 with the 37 pixel camera. Selection of showers, based on the predicted properties of gamma-ray showers (Hillas, 1985), led to rejection of 97% of the background; a steady signal at the 9 sigma level was detected. This strong detection of a weak steady signal (about 0.2% of the observed cosmic ray background) was the strongest evidence for the existence of TeV gamma-ray sources.

The Crab has been detected by at least seven groups using a variety of versions of the atmospheric Cherenkov technique (including one group in the southern hemisphere!) . The most significant detections were reported using the imaging technique but is most important that the source was detected also by other versions of the ACT (Goret et al. 1993; Baillon et al. 1993).

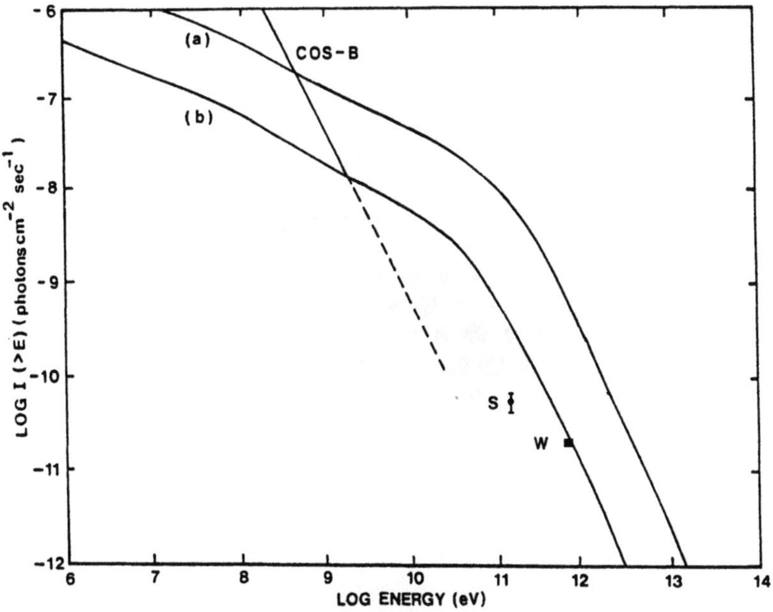

Fig. 5. **First VHE energy detection of the Crab Nebula compared with the Compton-synchrotron model (Weekes et al. 1989). The initial Compton-synchrotron model had been updated by several authors to take account of a better understanding of the nebula (Rieke and Weekes, 1969; Grindlay and Hoffman, 1971); the measured flux was in good agreement with the predicted flux for a value of magnetic field close to the equipartition value.**

4. Current Status

4.1. Imaging Technique

The imaging technique has now become established as the technique of choice in the 100 GeV-10 TeV energy range as can be seen from Table 1 which lists the current experiments in this range. At high end of the range the Cherenkov light wavefront timing approach is competitive as is the MILAGRO water Cherenkov experiment (particularly as a survey instrument). At lower energies where there is a strong impetus to bridge the gap with space experiments which cut off above 30 GeV there is not yet a clear technique. The very feeble light signal demands very large mirror areas and the background is dominated by the light from single muons. Some of the approaches favored have been discussed in recent workshops (Pare 1993; Ong 1994). For the next few years the imaging technique will probably provide the most useful results; this will be described in more detail with the present Whipple camera (Figure 6) taken as a prototype and the Alpha analysis method described in some detail.

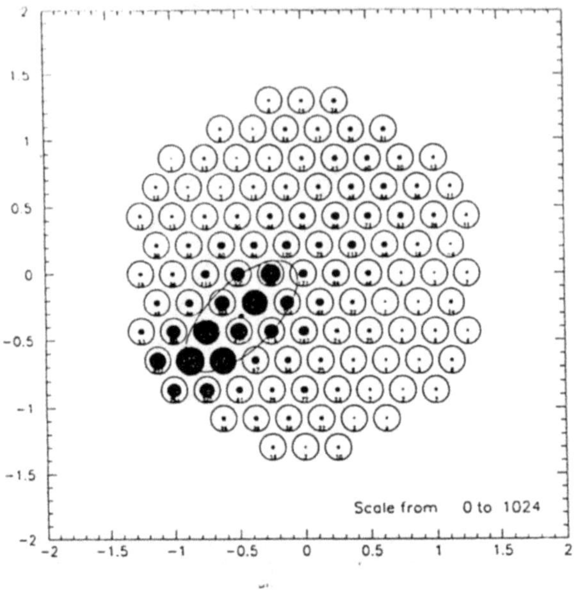

Fig. 6. The pixel layout in the high resolution camera of the Whipple Observatory with typical image superimposed; this image would be classified as a gamma-ray because it points to the center (the position of the putative source) and because of its small angular size. The numbers are photoelectrons. The current high resolution version of the Whipple camera on the 10m optical reflector (Cawley et al. 1991) has 109 pixels with typical spacing 0.25°. The camera is triggered when two pmt's exceed a preset threshold. Upon command the light level at each pixel is digitized and record.

Table 1.

Atmospheric Cherenkov Gamma Ray Observatories c. 1994.

Organization/Location	Long./Lat.	Elev. (km)	Threshold (TeV)	Technique
SPACE, South Pole	90S	2.5	0.5	Duration
Potchefstroom, South Africa	27E, 27S	1.4	1.0	Conventional
CANGAROO Australia	137E,31S	0.0	1.0	Imaging
Durham,Narrabri, Aust.	145E,31S	0.2	0.1	Imaging
CAO,Crimea,Ukraine	34E, 45N	2.1	1.0	Imaging
Lebedev, Tien Shan, C.I.S.	75E, 42N	3.3	1.0	Imaging
HEGRA La Palma, Spain	18W, 29N	2.2	0.5	Imaging
Whipple Coll., Arizona	111W,32N	2.3	0.2	Imaging
IHEP, Beijing, China	117E,40N	1.0	1.0	Conventional
ASGAT, Pyrenees, France	1W, 43N	1.5	0.5	Timing
Themistocle, Pyrenees, France	1W, 43N	1.5	3.0	Timing
CLUE (Pisa), La Palma, Spain	156W,21N	2.2	??	UV
Tata, Pachmari, India	78E, 22N	1.1	??	Lateral
Tactic, Mt. Abu, India	73E, 24N	1.7	0.2	Imaging

There are many ways in which the gamma-ray events that constitute the signal can be differentiated from the hadron background off-line; the method currently favored is the so- called Alpha analysis (Punch et al. 1992;

Reynolds et al. 1993). Events are selected on the basis of their parameterized images (simple moment fitting after some image cleaning); only those whose Length and Width (Figure 7a) fall within the "gamma-ray domain" (predicted from simulations) are accepted. The orientation of these candidate gamma-ray events within the field of view is then analyzed; those coming from the direction of the source (center of the field of view) should have their major axis aligned so that it passes close to the center. This pointing is characterized by the angle, Alpha; if Alpha is less than 15°, the event has the characteristics of a gamma-ray event from the source direction (Figure 7b).

In principle this "a priori" selection requires no control observations (no OFF observations) but, in practice, in the developmental phase of this technique it is safer to make an equal number of ON and OFF observations and to see the signal as a difference in the populations of the two sets in parameter space. Once the technique is fully understood, this will not be necessary; since the orientation will be peculiar to the source direction it is possible to search the field for the presence of a source using the other directions as the control (Akerlof et al., 1991; Fomin et al., 1991). The advantage of this method is that it separates out the differentiation on the basis of image angular size (Width and Length) and orientation (Alpha). Examples of detected sources are shown in the next section.

Although this introduction to the subject has stressed the role of the Whipple telescope in the development of the discipline it should be emphasized that there are exciting new developments in the imaging technique at many observatories worldwide; these are described in detail in other papers at this workshop. Particular mention should, however, be made of the CANGAROO project (Tanimori et al. 1994). This stereoscopic telescope is currently the most sensitive TeV system in the Southern Hemisphere; CANGAROO is an acronym for the Collaboration of Australia and Nippon for a Gamma-Ray Observatory in the Outback and, as such, is more descriptive than most acronyms. The observatory site is at Woomera in central Australia. It is made up of two independent telescopes, one developed by the University of Adelaide and the other by a consortium of Japanese institutions; the two telescopes can be operated independently or in coincidence. The Australian telescope, BIGRAT, has been in operation since November, 1988; it consists of three reflectors each of diameter 4m. This hybrid telescope has recently had a 37 pixel camera mounted on its central reflector.

The most exciting results have come from the advanced imaging camera on the Japanese telescope (Tanimori et al. 1994; Kifune et al. 1994); this telescope has been in operation since February, 1992. The basic optical element is a high quality 3.8 m diameter mirror (previously used in lunar laser ranging) which is mounted on an accurate alt-azimuth mount. The camera

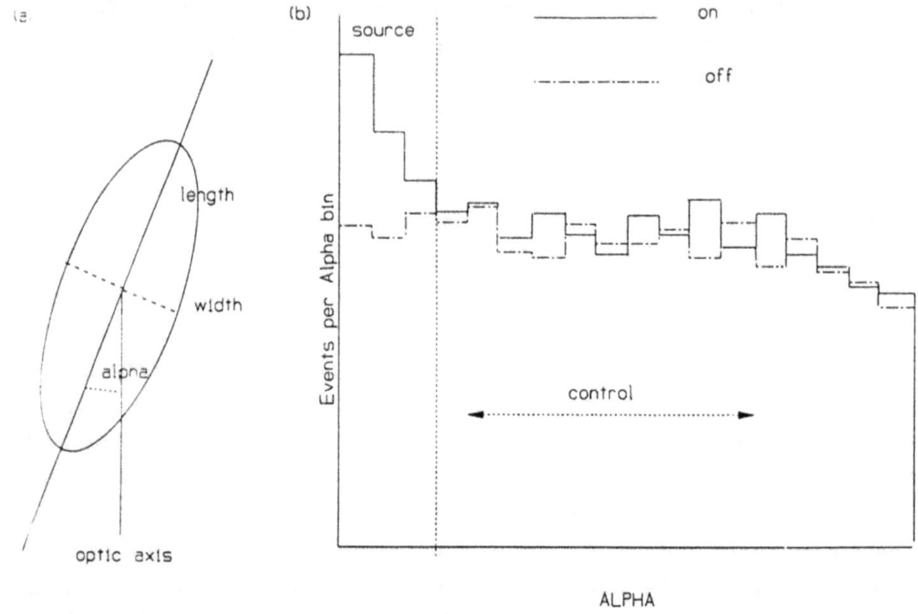

Fig. 7. (a) The parameter definitions used to characterize images; (b) After selection using the Width and Length parameters (Shape), the surviving events are sorted according to their Alpha parameter (Orientation). The values of the parameters used in the selection are derived from simulations and optimized on the Crab Nebula which is regarded as a standard source.

has 220 pixels (but may be expanded) with a full field of 3°; each square pixel is 0.14° (8 mm).

The University of Durham (U.K.) is also developing an imaging telescope in Narrabri, Australia; this will have a mirror collection area of 120 m^2 and an energy threshold of 70 GeV. There are major imaging telescopes being developed in France (CAT) and in Germany (HEGRA) as well as in India and Russia. The French collaboration (CAT) will have a camera with more than 500 pixels.

Other methods are now under study for extending the Atmospheric Cherenkov technique to lower energies i.e. to 20 GeV or lower; these include the construction of large arrays of imaging telescopes, the use of large solar power plants during the night-time hours and the construction of a large "Arecibo- type" dish on a high mountain plateau in Tibet.

An alternative approach to Cherenkov imaging is provided by the MILAGRO experiment which utilizes a giant water Cherenkov detector to measure the arrival directions and energies of air showers at mountain altitudes. With high sensitivity to gamma-ray showers of energy 1 to 100 TeV over a broad range of elevation angles this telescope is eminently suited for survey work (to detect bright transients) and as a burst monitor. When completed (in

1998) this telescope will complement the atmospheric Cherenkov telescopes and extend their energy coverage (Barwick et al. 1993).

4.2. Sources

The TeV region of the electromagnetic spectrum has had more than its share of controversy with the detections of more than 20 sources claimed but with a much smaller number verified; these have been the subject of a number of reviews (Weekes 1991; Chadwick et al. 1991; Cronin et al. 1994) and will be reviewed further at this workshop. Here we will note only that there are five sources now that are given strong credence; the status of these is listed in Table 2. .

Table 2.

Status of TeV Sources

Source	Variability	Imaging	Detections
Crab	No	Yes	Whipple, Sandia, Crimea, ASGAT, HEGRA, CANGAROO, Themistocle
PSR1706-44	No	Yes	CANGAROO
AE Aqr	Yes, periodic	No	Durham, Potchefstroom
Vela X-1	Yes, periodic	No	Durham, Potchefstroom
Markarian 421	Yes	Yes	Whipple

The Alpha plots for the two galactic sources that have been detected using the imaging technique are shown in Figure 8.

As it has in many other bands of the electromagnetic spectrum the Crab Nebula has become the standard candle for TeV astronomy. In a recent report from the EGRET experiment on the Compton GRO (Nolan et al. 1993), the predicted turnup in the spectrum of the Crab above 1 GeV has been observed; to this author's knowledge this is the first instance where a gamma-ray feature that was first seen in ground-based observations has been verified in a gamma-ray space experiment. It is worth noting also that the magnetic field assumed in the De Jager and Harding model is in agreement with estimates by other methods.

5. Outlook

Gamma-ray astronomy at TeV energies is now a viable science but it is in its infancy. With the sensitivity that the new telescopes provide, it becomes possible to make a serious investigation of TeV emission from such classical sources as the galactic plane, the molecular clouds, the radio pulsars and the supernova remnants. Extragalactic sources, in particular the superluminal quasars, are strong candidates for emission. The atmospheric Cherenkov

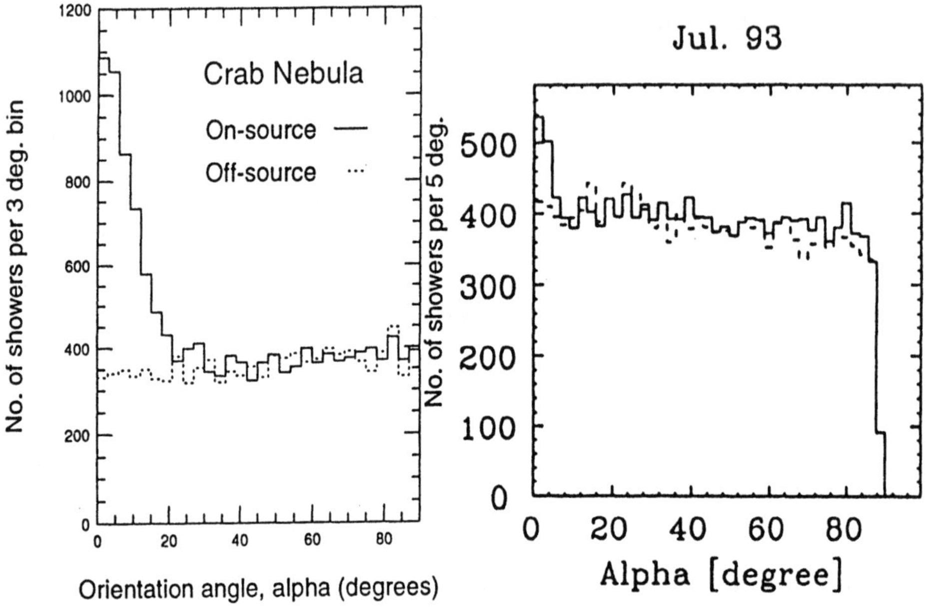

Fig. 8. **The observed Alpha distributions from the Crab Nebula and from one season of observations on PSR1706-44**

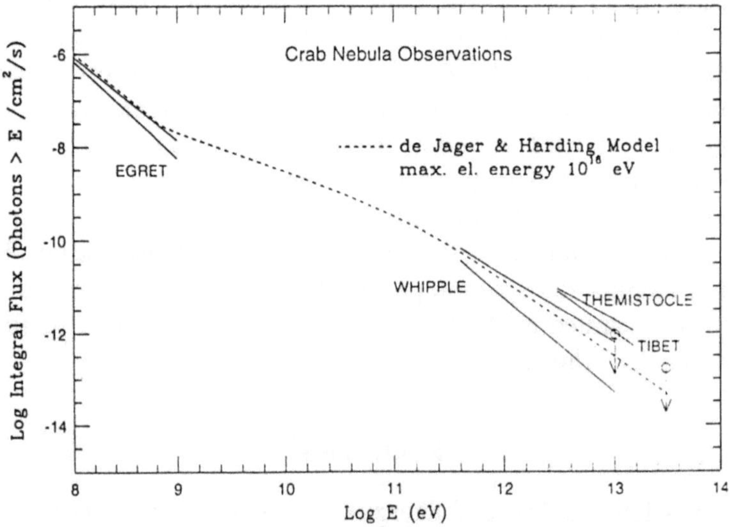

Fig. 9. **High energy gamma-ray spectrum of the Crab Nebula. There is no evidence for any significant variation of this signal with time at TeV energies; the Compton-synchrotron model would not predict short-term variations although there might be a long-term secular decline.**

technique has justified its early promise and has become a sophisticated tool for the investigation of cosmic ray sources.

Acknowledgements: Research in TeV gamma-ray astronomy at the Whipple Observatory is supported by the U.S. Department of Energy and the Smithsonian Scholarly Studies Fund.

References

Akerlof, C.W. et al.: 1989, Proc. *GRO Science* Workshop (Goddard), 4-49.
Akerlof, C.W. *et al.*: 1991, *ApJ Lett* **377**, L97
Baillon , P. *et al.*: 1993, *Astroparticle Phys* **1**, 341
Barwick, S. *et al.*: 1993, Proceedings of the 23rd ICRC (Calgary), **1**, 436
Cawley, M.F. *et al.*: 1990, *Exp. Astron.* **1**, 173
Chadwick, P.M. *et al.*: 1990, *J. Phys. G: Nucl. Part. Phys.* **16**, 1773
Chudakov, A.E. *et al.*: 1965, *Transl. Consultants Bureau, P.N.Lebedev Phys. Inst.* **26**, 99
Cocconi, G.: 1959, Proceedings of the 7th ICRC, (Moscow), **2**, 309
Cronin, J.W., Gibbs, K.G. & Weekes, T.C.: 1994, *Ann. Rev. Nucl. Part. Sci.* **43**, 883
De Jager, O.C. & Harding, A.K.: 1992, *ApJ* **396**, 161
Fazio, G.G. *et al.*: 1972, *ApJ Lett* **175**, L117
Fomin, V.P. *et al.*: 1994, *Astropart. Phys* **2**, 151
Galbraith, W. & Jelley, J.V.: 1953, *Nature* **171**, 351
Galbraith, W. & Jelley, J.V.: 1955, *J.Atmos. Phys.* **6**, 250
Goret, P. *et al.*: 1993, *A & A* **270**, 401
Gould, R.J.: 1965, *Phys. Rev. Lett.* **15**, 511
Grindlay, J.E.: 1972, *ApJ Lett.* **174**, L9
Grindlay, J.E. & Hoffman, J.A.: 1971, *Astrophys. Lett.* **8**, 209
Grindlay *et al.*: 1975a, *ApJ* **201**, 82
Grindlay *et al.*: 1975b, *ApJ Lett* **197**, L9
Grindlay *et al.*: 1976, *ApJ* **209**, 592
Hill, D.A. & Porter, N.A.: 1961, *Nature* **191**, 690
Hillas, A.M.: 1985, Proceedings of 19th Inter. Cosmic Ray Conf.(La Jolla), **3**, 445.
Jelley, J.V.: 1971, IAU Symposium 46, *The Crab Nebula*, Eds. R.D.Davies, F.G.Smith, Publ. D.Reidel Publ., 32.
Jelley, J.V. & Galbraith, W.: 1955, *J. Atmos. Phys.* **6**, 304
Jelley, J.V. & Porter, N.A.: 1963, *Quart. J. Roy. Ast. Soc.* **4**, 275
Kifune, T. *et al.*: 1994 ApJ. (in press)
Morrison, P.: 1958, *Nuovo Cimento* **7**, 858
Pare, E.: 1993, Proceedings of Workshop *"Towards a Major Atmospheric Cherenkov Detector-II*, Calgary, Canada Publ. Iowa State Univ. editor: R.C.Lamb, 250.
Punch, M. *et al.*: 1992, *Nature* **358**, 477
Ong,R. *et al.* : 1994, *Atmospheric Cherenkov Detector-III* Tokyo, Japan (in press).
Rieke, G.H. & Weekes, T.C.: 1969, *ApJ* **155**, 429
Reynolds, P.T. *et al.*: 1993, *ApJ* **404**, 206
Schubnell, M.S. *et al.*: 1992, Compton GRO Symposium, St. Louis, AIP Conference Proceedings, **280**, 1171.
Stepanian, A.A. *et al.*: 1983, *Izv.Krym.Astrofiz.Obs.* **66**, 234
Tanimori, T. *et al.*: 1994, *ApJ Lett* **429**, L61
Vacanti, G. *et al.*: 1991, *ApJ* **377**, 467
Weekes, T.C.: 1981, Proceedings of 17th. Inter. Cosmic Ray Conf. (Paris), **8**, 34.
Weekes, T.C.: 1991, *Space Sc. Rev.* **59**, 315
Weekes, T.C. & Turver, K.E.: 1977, Proceedings of 12th Eslab Symp. (Frascati), ESA SP-124, 279.
Weekes, T.C. *et al.*: 1989, *ApJ* **342**, 379

DIFFERENCES BETWEEN GAMMA-RAY AND HADRONIC SHOWERS

A.M. HILLAS
Physics Department, University of Leeds, Leeds LS2 9JT, UK

Abstract. Cosmic gamma-ray ray sources may be sought if high-energy gamma-rays may be detected without confusion from the very intense isotropic background of hadronic cosmic rays. Ground-based methods are needed at energies above tens of GeV, using air showers, and at energies below tens of TeV, the detection of muons in showers is not the most efficient way to reject hadronic showers. The shape and orientation of Čerenkov images can reject far more than 99% of the background. The way in which Čerenkov radiation is distributed in showers is discussed, and the possibilities of using image shape, distribution of light on the ground, time profile, spectrum and polarization of the light are briefly discussed. Imaging alone appears to be the most powerful. Simulations suggest that the uv content of the light should not be a useful diagnostic.

Key words:

Čerenkov radiation – Cosmic gamma-rays – Extensive Air Showers

1. Introduction

Amongst the high-energy particles reaching the Earth, which are mostly protons and stripped atomic nuclei, there is a small flux of gamma-rays, clearly detected at energies up to 10^{13} eV. The gamma rays travel undeflected from their sources and are capable of locating sources of high energy particles, so their detection is of great interest. Spacecraft provide the only method for detecting these gamma rays at energies up to a few GeV, but ground-based techniques take over at energies above about 100 GeV, where one can detect the electron-photon cascades which the incoming gamma rays generate in the atmosphere. At energies below 10 TeV (i.e. 10^{13} eV) the cascades die out in the upper atmosphere and are detected by the Čerenkov radiation which the cascade particles emit, and which is bright enough to be picked up by large focusing mirrors on the ground. At higher energies some shower particles reach the ground and can be detected by scintillators, first at mountain altitudes, but beyond a few hundred TeV detection is possible even at sea level. However, almost all such air showers are caused by an isotropic flux of hadronic cosmic rays, and the study of gamma rays requires an efficient technique for recognizing and rejecting the hadronic showers, and accepting only pure electron-photon showers.

Accurate location of the direction from which the primary particle is coming has been an important feature of successful observations so far: a sky map of arrival directions will show a uniform distribution due to hadronic showers, with a narrow cluster around the direction of a gamma-ray source, if

Space Science Reviews **75**: 17–30, 1996.

there is one. But such sources are so weak that one needs additional methods to reject the majority of hadronic showers from the records.

In hadronic showers, several percent (at least) of the particles reaching the ground are muons, whilst the number of muons in photon-induced showers of the same energy should be lower by about a factor 30. So in studies of showers reaching the ground muon detectors may be incorporated, to reject hadrons. However, the muons are spread over a very large area (averaging hundreds of metres from the shower axis), so the muon signature may be missed. A much more efficient rejection has been possible at TeV energies, where the greater transverse spread of hadronic showers is a distinguishing feature, and also their (usually) greater and more ragged penetration into the atmosphere. Čerenkov light emitted by the particles can be used to form a low-resolution picture of the shower, to estimate its width, length and alignment with an interesting astronomical object which may be a source. Surprisingly, with even crude imaging, it has proved possible to reject 99.7% of hadronic shower images (over a field of 3°) at a few hundred GeV, whilst still accepting most of the gamma showers, and at energies of many TeV the hadron rejection can be 99.99% (unpublished). It has been suggested that other characteristics of the image such as the proportion of ultra-violet light might also be diagnostic (if gamma showers die out earlier, less of their uv will penetrate to the ground), or that the time structure of the light may be a useful pointer. And the lateral spread of Čerenkov light on the ground has also been used as an alternative or adjunct to imaging. The question addressed here is which of these tests should offer a very high selectivity. The emphasis will be on the TeV regime.

Most of the diagrams shown below are obtained from Monte-Carlo simulations using the author's MOCCA simulation program, which has proved very successful in predicting detailed features of Čerenkov radiation from air showers and of particle distributions in showers.

2. Structure of air showers

Figure 1 illustrates typical showers generated by (a) a 0.32 TeV gamma ray, and (b) a 1 TeV proton, showing the much greater width of the latter, due to transverse momentum given to pions in strong interactions. To make this clearer, the lateral scale has been expanded by a factor 5. Only a fraction of the particle track length radiating Čerenkov light has been shown, to avoid saturation. The more collimated gamma ray shower results in a pool of Čerenkov light on the ground with a radius about 120m (at typical mountain altitude), beyond which the light intensity falls rapidly, as shown in figure 2. The more widely spread tracks in the proton shower produce a different light distribution, also shown in this diagram, often with prominent local peaks. The longitudinal spread of the shower – or more specifically the distribution

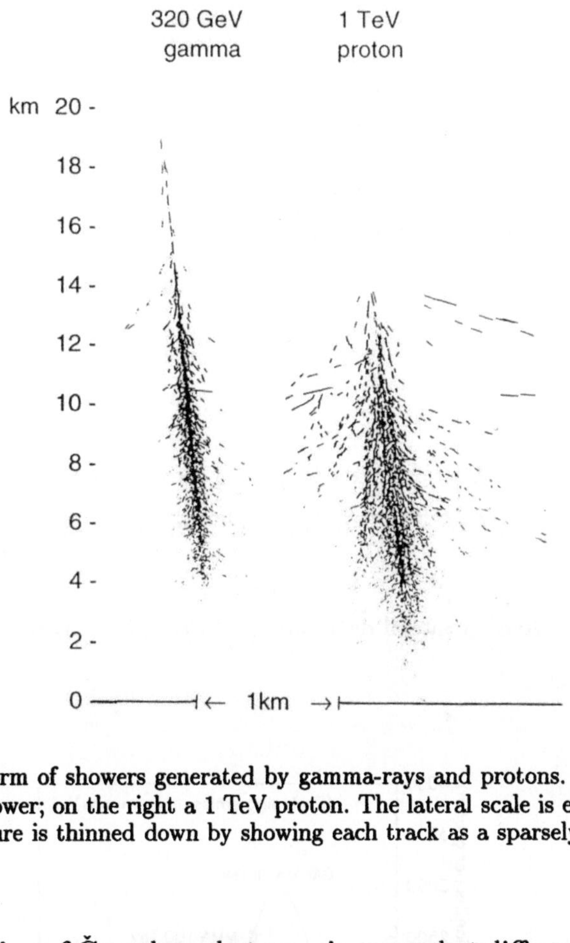

Fig. 1. Different form of showers generated by gamma-rays and protons. On the left, a 320 GeV gamma shower; on the right a 1 TeV proton. The lateral scale is exaggerated by a factor 5. The picture is thinned down by showing each track as a sparsely dashed line.

in points of emission of Čerenkov photons – is somewhat different in the two cases: proton showers have a longer penetrating tail, some particles reaching the ground and giving a more intense Čerenkov radiation near the shower axis. Figure 3 shows the average distribution in atmospheric depth of the emitted light, but omitting light emitted directly by muons. The muons, in proton showers, do penetrate to the ground over a wide area, and an intense local peak of Čerenkov light is seen around each muon impact point. Proton showers show much greater shower-to-shower fluctuations in longitudinal profile than do gamma showers. Figure 4 illustrates typical cases (again omitting the more widely spread muon "tail", to make the point that even the core electron-photon cascades differ for the two types of shower). These differences in longitudinal profile will not, however, identify every proton shower. Showers due to heavy nuclei of the cosmic radiation are even more well distinguished from gamma showers. For the same energy, they are even more laterally diffuse, and contain more muon tracks, contributing to patchy muon light (at TeV energies).

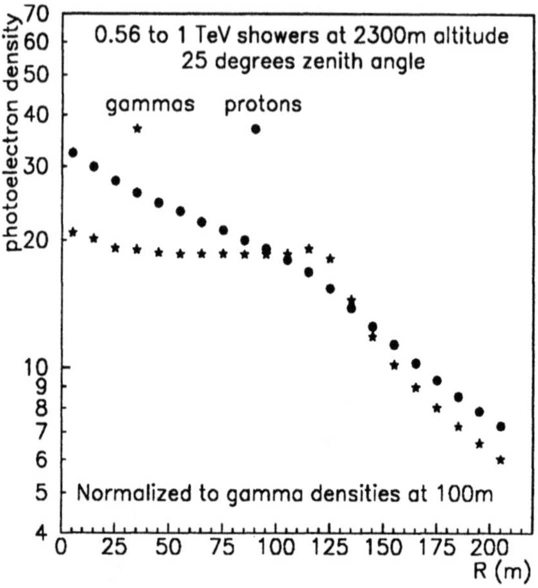

Fig. 2. Average lateral distributions of Čerenkov light at about 700 GeV

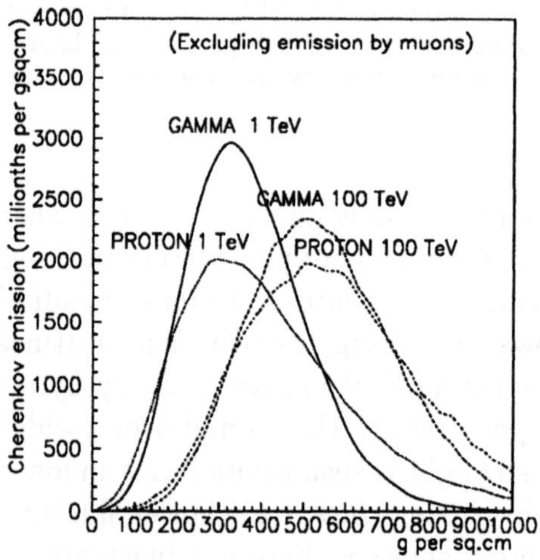

Fig. 3. Average longitudinal distribution of Čerenkov emission

2.1. Main characteristics of Čerenkov light from showers

Figure 5 shows schematically the anatomy of Čerenkov light in a shower.

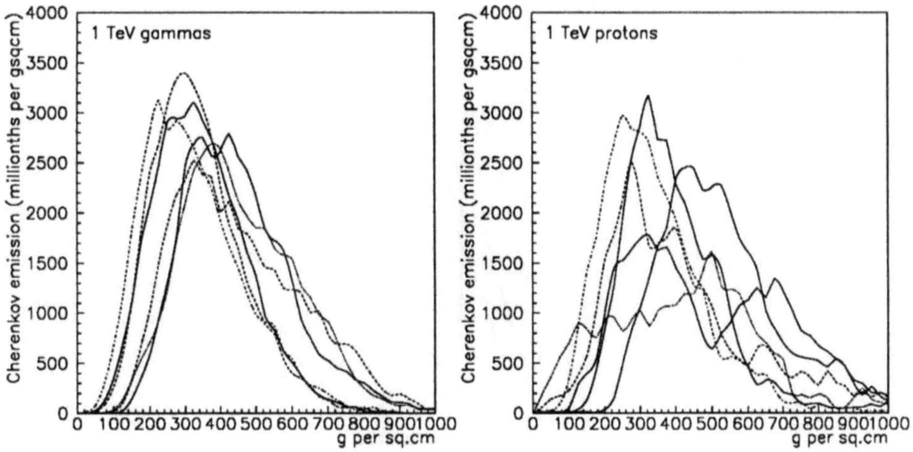

Fig. 4. Longitudinal structure of individual showers, showing greater variation amongst proton showers.

If all the charged particles moved along the shower axis, with no lateral displacements, the light would be emitted at the characteristic Čerenkov angle, which is 1.0^o for very fast particles at around 8km above sea level, near maximum development of the shower. At the highest altitudes, due to lower air density, this angle decreases, whilst at sea level it is around 1.5^o. The result of this changing angle is that light from ultra-relativistic particles reaches the ground at 120m from the shower axis if it is emitted anywhere from 20 to 10km above the ground, forming a ring of intense light of this radius. Light emitted closer to the ground has a shorter distance to travel, outweighing the somewhat larger Čerekov emission angle, and so it strikes the ground closer to the axis. When one allows for the angular and lateral spread of the emitting particles, this "Čerenkov ring" becomes more blurred.

A simplification which is very useful in interpreting Čerenkov signals from showers is to regard the light reaching the ground as made up of three parts. The main bulk of particles, near shower maximum, produce the light seen at 100-130m from the shower axis: here the intensity of the light is a good measure of the calorimetric energy deposition in the atmosphere. Close to the shower axis (within 40m, say) one sees the light from local particles – the tail of the shower – so the intensity here fluctuates very much from shower to shower, and should be closely correlated with the signal in particle detectors on the ground. And particles actually reaching the ground are surrounded by a $1/r$ distribution of light intensity, so there are localized peaks due to penetrating muons, in hadronic showers, and due to electrons reaching the

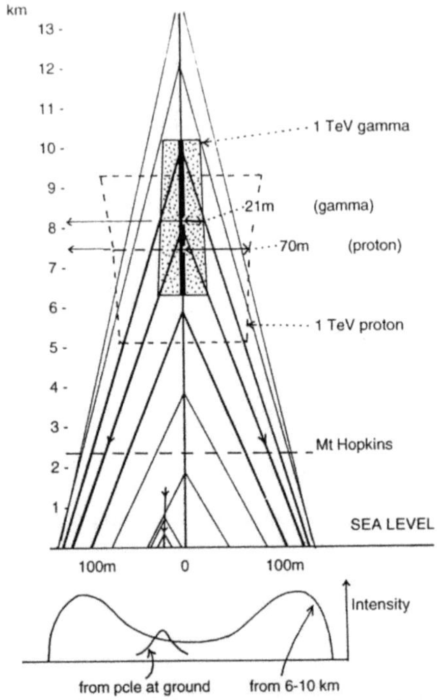

Fig. 5. Geometry of Čerenkov rays in air showers. The stippled box contains the main region of emission of light in gamma-ray showers: 25% is emitted above the top of the box and 25% below the bottom. The median altitude is 8.1 km. At these heights, half the emission occurs at radial distances within the box. The dashed-line box indicates the corresponding main reagion of emission in proton showers of the same energy. Čerenkov light is emitted from ultra-fast particles travelling along the axis at the angles shown (with exaggerated horizontal scale), resulting in a peak of intensity on the ground just beyond 100m from the centre.

ground within about 30m of the axis even in the more energetic gamma showers. (A light imaging system shows arcs, of around 1.3^o radius, if such particles land very close by.)

A focusing mirror on the ground, at, say, 60 to 130m from the shower axis, will see what looks like a glowing column in the air, where the shower is at its maximum development, the median altitude of emission in 1 TeV gamma showers being about 8km above sea level for vertical incidence (and rather more for inclined showers). Half of the light emission from a gamma shower of this energy arises within 21m of the axis, so if the glowing column is viewed from a distance of 6km (e.g. from the Whipple Observatory, at 2.3km altitude), it is seen to have a radius of about 0.2^o. The main electron-photon cascade of a 1 TeV proton shower, however, emits half its light within 70m of the axis, and the median height is 0.5km closer. One expects to see a

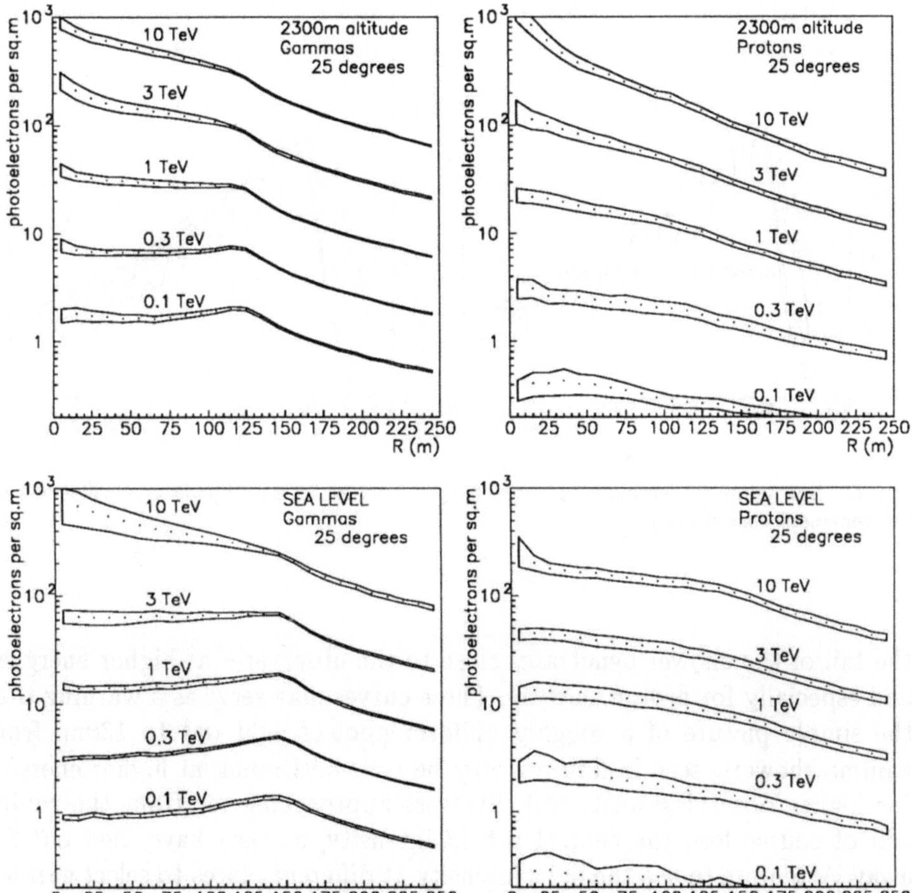

Fig. 6. Lateral distribution of Čerenkov signal at two altitudes. The width of each band indicates the uncertainty in the average: the actual fluctuation amongst different showers is very much larger.

radius of around 0.75^o for the image in this case. (See the "median cylinders" of emission in figure 5. Half the vertical extent of Čerenkov emission occurs within the cylinders indicated, and half of the emission from this zone is emitted within the radial distance indicated by the cylinder.)

These effects are illustrated in Figure 6 , showing the lateral distribution of Čerenkov photons on the ground. The "Čerenkov ring" is most prominent (though underemphasized in this highly compressed logarithmic scale) when the shower dies out far above the observer, as in low-energy gamma showers, especially seen from sea level. The intensity near the shower axis rises when

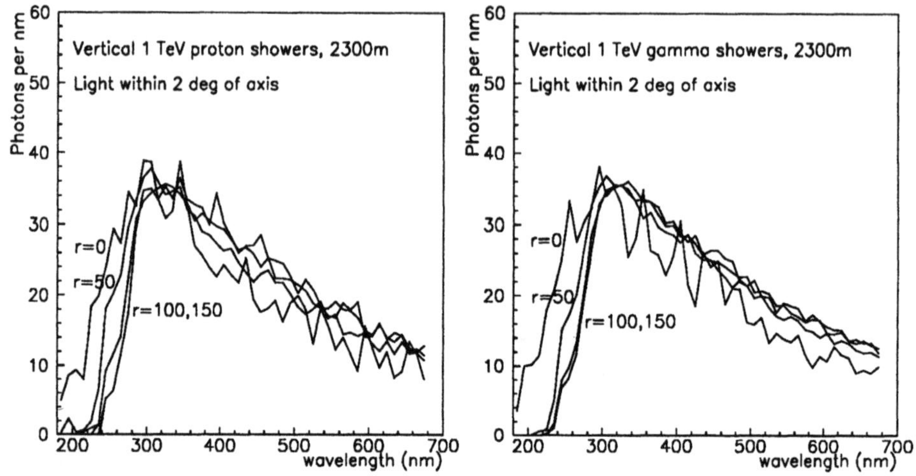

Fig. 7. Spectrum of Čerenkov light reaching ground at 2300m. The large oscillations do not represent real effects.

the tail of the shower penetrates close to the observer – at higher energies, and especially for proton showers. These curves may serve as a warning that the simple picture of a roughly uniform pool of light out to 130m, from gamma showers, seen in figure 2 may be very misleading at higher energies (or higher observing altitudes). Showers approaching far from the zenith will of course lose the central rise in intensity, as they have died out far away. Attempts to use the light intensity at different places to select gamma showers should take into account these variations in expected profile with different energies or zenith angles.

Light from muons passing close to the detector will suffer less attenuation of the uv component. However, calculation of the spectrum of light reaching the observer (here at 2.3 km altitude) does not reveal any notable difference in the spectrum for gamma and proton showers. Figure 7 shows the spectrum at distances 0, 50, 100 and 150m from the shower axis. (The large fluctuations seen in the r=0 cases are associated with low statistics in the sampling.) The calculations assumed that the fraction of light transmitted vertically to an altitude of 2.3km from various initial altitudes was as shown in figure 8, from a Whipple Observatory internal report.

2.2. Čerenkov images

As mentioned above, a hadronic shower seen from the ground usually looks much wider than does a gamma shower, and may also have sporadic patches of light due to muons passing near by. If one wishes to detect gamma rays

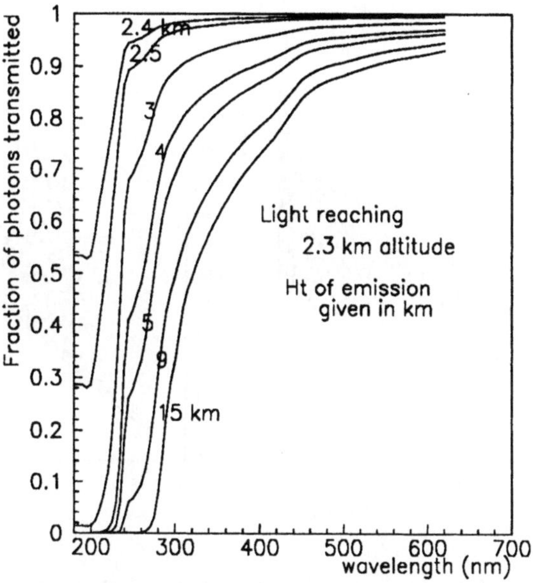

Fig. 8. Vertical transmission of Čerenkov light to 2300m, from various initial altitudes.

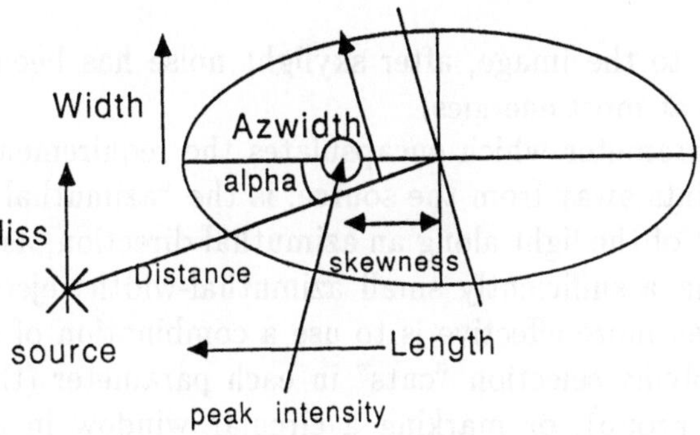

Fig. 9. Parameters used to describe a shower image

coming from a specific point-like object (such as the Crab nebula), these showers will be aligned with this point in the sky. It has been found in practice at the Whipple Observatory (in excellent agreement with Monte-Carlo predictions) that if shower images are characterized by parameters describing their width, length, orientation, angular distance from the object in question and possibly other features (compactness), most of the background images not due to gamma rays from this source can be identified and rejected. Figure 9 illustrates the parameters of an image, if the "best

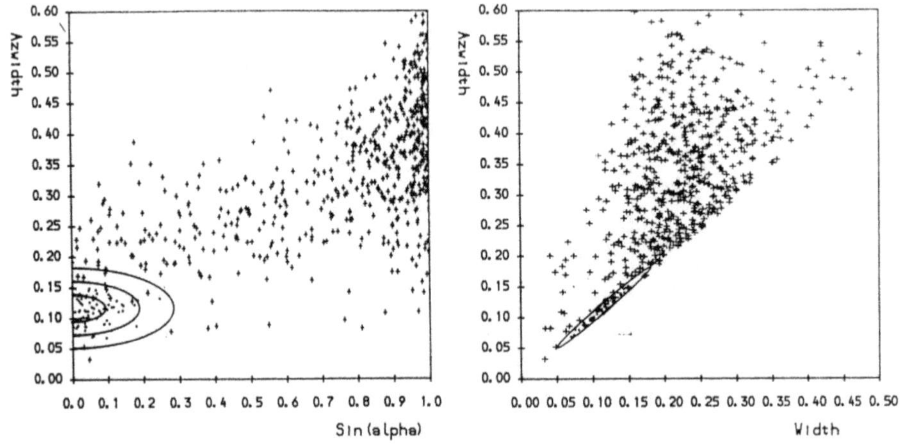

Fig. 10. Separation of gamma-ray and proton showers using pairs of image parameters. Most of the gamma images lie within the (central) elliptical "windows", the parameters for proton showers being spread widely. In practice more than two parameters are used, so one looks at a hyper-ellipsoid.

ellipse" is fitted to the image, after skylight noise has been cleaned away (very important at most energies).

One single parameter which encapsulates the requirement for a narrow image which points away from the source, is the "azimuthal width", which is the rms extent of the light along an azimuthal direction, as centred on the source. Requiring a sufficiently small azimuthal-width rejects most of the background. Even more effective is to use a combination of several parameters, either applying selection "cuts" in each parameter (the "Supercuts" of the Whipple group), or marking a circular window in an appropriate multi-parameter space (the "cluster" or "hypersphere" method). The latter is especially well adapted to being defined a priori from shower simulations. Figure10 illustrates the separation of gamma images from background in the simplest case of two-parameter plots.

This method gives much the most effective background rejection to date, the rejection levels quoted in the introduction being those actually found in Whipple observations of the Crab nebula. (Further discussion of imaging may be found in the paper by Fegan at this symposium.)

Would it be improved still further by adding information about the lateral distribution of Čerenkov light on the ground, or about the time distribution of the Čerenkov photons received? The first question has been tested by calculating the lateral distribution of light in gamma showers, and in many

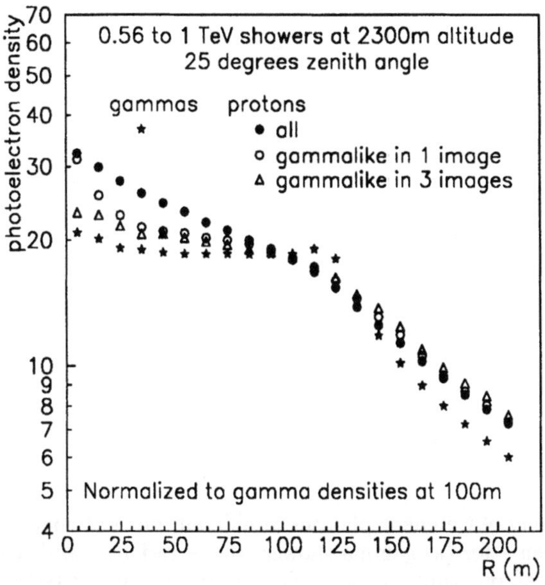

Fig. 11. Lateral distribution of Čerenkov light from proton showers: effect of selecting showers with "gamma-like" images.

proton showers, and picking out those which produced "gamma-like" images, passing the tests mentioned above. The average distribution of light for the gamma showers, shown in figure 11, with the characteristic sharp bend near the "Čerenkov ring", is very different from the average for "all protons". But if one concentrates on showers which gave a gamma-like image, or those which looked gamma-like from three points on the ground (about 130m apart in this test), the proton showers lose much of their characteristic rise near the axis. They still do not have such a sharp shoulder to the distribution, but this may not be easy to test in practice. Hence, the tests are not independent, and the lateral distribution is less helpful than might be expected if one already has a good imaging system.

The average Čerenkov density as a function of axial distance does not tell the whole story, though. In hadronic showers there are irregular azimuthal variations in density, due to the local peaks surrounding muons (or indeed any charged particles) reaching the ground: the rms fluctuation in Čerenkov photon density (measured over a $10m^2$ area) is calculated to be 16% for gamma showers, and 36% for proton showers (at around 1 TeV), at 100m from the axis; but if three separated detectors see a "gamma-like" image of a nucleon shower, this fluctuation is somewhat reduced, to 27%. Thus gamma showers give more uniform light intensities over several detectors (but not

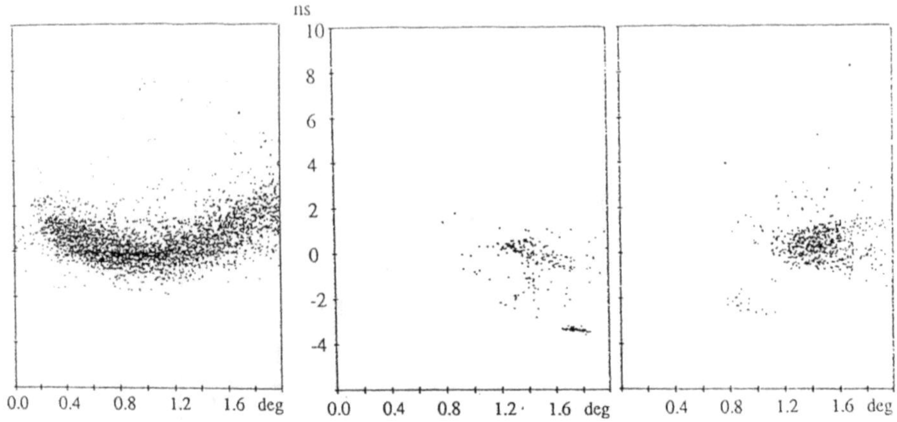

Fig. 12. Arrival times of Čerenkov photons at different parts of image – different distances from source direction – for (a) gamma showers (little variance) and (b) and (c) individual proton showers. Small outlying groups of photons are seen in the latter, due to single local muons.

necessarily "flat" – independent of radial distance: this depends on being well beyond shower maximum: see Figure 6).

The time structure of the Čerenkov pulse has also attracted interest as a possible discriminant between primary particles. One physical cause of a difference is that muons can travel for a long distance faster than light and so their local Čerenkov emission arrives before the light from the main part of the shower, giving precursor pulses. Figure 12 shows the simulated spread in times of arrival of photons at different phototubes along an image (with optics better than the Whipple 10m telescope, but a similar aperture), for an average gamma shower (there being small fluctuation) and for two individual proton showers, the latter showing some distinct precursor islands of light, 2 to 3 ns before the main body of photons. It should be noted, however, that in many hadronic showers these precursors are not evident, and that they are presumably associated with distinct muon (or local-particle) arcs, so that imaging would have given the same information.

One characteristic feature of Čerenkov radiation which has not been exploited is its intrinsic polarization. Is there any useful information to be gained here? If one could measure the intensity of light seen through linear polarizers aligned radially and azimuthally in the shower the ratio of these intensities is calculated to be as shown in Figure 13. If all charged particles had been travelling along the shower axis, this ratio should be zero, but in fact the projected ground impact points of radiating particles are scattered very widely on the ground, so that individual contributions have widely

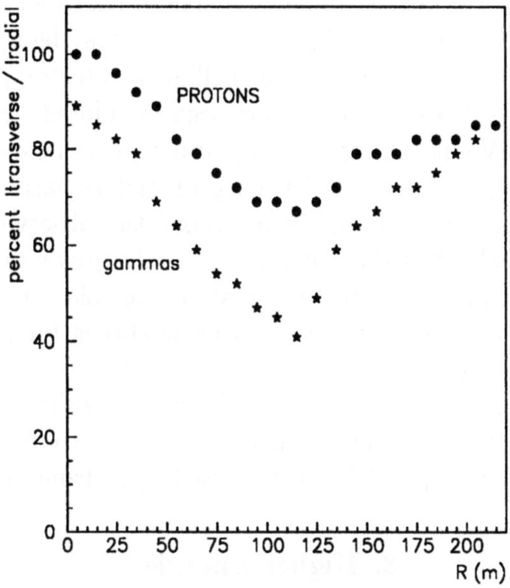

Fig. 13. Polarization of Čerenkov light

varying polarization vectors. The modest difference expected would not be of much help (and more than two polarizers would be needed as the radial direction is not known in advance).

2.3. Applications of stereoscopic Čerenkov imaging of showers

Using Čerenkov light we see a glowing column of light along the axis of the shower in the upper atmosphere, and it should be aligned with the source of gamma-rays. This provides one important tool for the elimination of most of the isotropic background radiation: we select primary particles from source directions along (part of) a line defined by this axis. If two views are available, from detectors spaced 60 to 150m apart, say, the second view provides another line on which the source lies, reducing the target area of sky, and further reducing the hadronic background. The detectors must have a low triggering threshold if their collection areas are to nearly coincide.

Stereoscopic systems may become important also at very low energies. Firstly, they could eliminate muon arcs as a source of small images that could be confused with small air showers: if they are over 120m apart they should not both be triggered by a single muon, and in any case muons impacting between them would produce images about 3° apart in the two views. And at very low energies – say below 150 GeV – there may be too little Čerenkov light collected to measure an image shape. Orford and Turver suggest that in this case the parallactic displacement seen in the centroid of the light emission viewed from different locations may be a useful diagnostic tool. As an example, a system of three collinear detectors, two 146m apart and a third midway between them, has been simulated. Images of showers produced by 60 GeV gamma rays usually appeared to be displaced by about 0.4^o and in the direction expected from geometrical parallax when looking at a single source of light about 10km above the observer. But the three images of nucleon showers showed much more irregular displacements, the angular separation and the direction of shift both showing much variation. Hence few of the hadrons would have been mistaken for gamma rays. The main reason for this is that a few pions at notably divergent angles create sub-showers in the hadronic case, and different subshowers will dominate the intensity seen from different viewpoints on the ground, so one does not reconstruct a consistent point by tracing back light from different detectors.

3. Higher Energies

Muons are characteristic of hadronic cascades – surely they provide the most obvious test? This is true, but there are rather few of them, so their non-detection does not achieve a 99% hadron rejection efficiency. In a 1 TeV proton shower, there is only about one muon within the central 30m radius (at 2.3km altitude), and one must go to 30 TeV or more for reasonably high efficiency, using a very large sensitive area, such as is planned for MILAGRO.

Muons can also be detected over a large area with Čerenkov imaging: a muon passing within about 40m of a 10m telescope will produce a short arc on the image (part of a circle of 1.3^o radius, becoming a complete arc if the muon strikes the light collecting mirror). These separate arcs may typically appear about a degree away from the image of the main cascade, and add to the "width" of the image, causing it to be rejected as "not gamma-like". However, there is on average less than one muon arc fragment per image at the energies most commonly studied, somewhat below 1 TeV, so again, muons do not play the main part in hadron rejection.

Although hadronic showers should get narrower at high energies, and presumably less different from gamma showers, except for an increasingly reliable contribution from muons, it is found in practice that (using pixels 15 arc minutes apart, with unfilled spaces between the phototube sensitive areas) if showers are characterized by several image parameters (typically width, length, distance from source and angular deviation of image axis from line pointing to source) discrimination against hadrons improves up to at least 10 TeV.

PRESENT STATUS
OF VERY HIGH ENERGY GAMMA RAY OBSERVATIONS
on Southern Sky Objects

T. KIFUNE
Institute for Cosmic Ray Research, University of Tokyo, Tokyo 188, Japan

Abstract. Recent results from observations of the southern sky objects are summarized. The unpulsed, persistent very high energy (VHE) emission from the gamma ray pulsars, the Crab and PSR1706-44, is discussed. A process of energetic electrons ejection may take place from a variety of other objects such as from X-ray binaries, similarly to the pulsars. Such an effect may be seen also in "pair halos" around extragalactic VHE gamma ray emitters, the observational study of which is still in a preliminary stage in the southern hemisphere.

Key words: very high energy gamma ray astronomy, Čerenkov telescopes, pulsars, X-ray binaries, supernova remnants, active galactic nuclei

1. Introduction

VHE Gamma Ray Astronomy has, after 30 years' efforts, reached sensitivity sufficient to detect objects bright in VHE gamma rays, enabled by the second generation of Čerenkov telescopes designed for imaging gamma ray showers. The launching of Compton Gamma Ray Observatory (CGRO) satellite in 1991 was very timely, as it was made when the imaging technique had become available. Many gamma ray sources have been discovered from 100 MeV to 10 GeV region by the Energetic Gamma Ray Experiment Telescope (EGRET) of CGRO (Fichtel et al. 1994). It is expected that the detection of VHE gamma rays from these objects will help VHE gamma ray astronomy develop closer links with GeV gamma ray astronomy and with other energy bands. This paper, presented at the Heidelberg Workshop for VHE gamma rays (October 1994), summarizes the southern sky observations and attempts to understand the implications and limitations of current VHE results and to understand better about the future prospects for VHE gamma ray astronomy.

Evidence of VHE emission has been reported for various kinds of objects; enigmatic remnants from supernova explosions such as the Crab and Vela pulsars, X-ray binaries (XRB), radio pulsars and extra-galactic objects. Some of the positive results, if true, are surprising as they imply an unexpectedly intense energy output in the VHE band above ~100 GeV without detectable emission in the lower energy region at ~1 GeV.

The southern hemisphere, which contains the galactic center, has the advantage of a number of putative VHE sources along the galactic disc. The

Space Science Reviews **75**: 31–41, 1996.

role of southern observations, if such a 'role' for each hemisphere make any sense, would be in obtaining a better understanding on the various types of galactic sources such as pulsars, XRBs and supernova remnants (SNR), by compiling a list with greater number of such sources than northern observations. The CANGAROO telescopes at Woomera in Australia have until recently been the only imaging systems in the southern hemisphere. Recently, the Durham group in Narrabri, Australia, has constructed a new telescope with imaging capability, scheduled for operation early in 1995 (Turver, this workshop).

2. Observational Results

2.1. Extra-galactic objects

The search is in the southern hemisphere still in its early days for VHE emission from active galactic nuclei (AGN) detected by EGRET. The upper limits so far obtained are too large to argue about their implications. The CANGAROO group has started observing sources such as PKS 0521-365, which has a redshift of 0.055, the second lowest after Markarian 421. The current upper limit on this object is found in the Durham data (Bowden et al. 1993a) as 2.5×10^{-10} cm^{-2} s^{-1} for > 200 GeV, which is considerably higher than the flux from Markarian 421 (Punch et al. 1991). It will take some more time before the accumulation of data becomes sufficient to detect any VHE signal or set meaningful upper limits.

The best studied extra-galactic object so far in the southern sky is Centaurus A, the nearest active galaxy at a distance of $\sim$ 5 Mpc. An early observation by Grindlay et al. (1975) encouraged further efforts with a 4.5 σ detection of a flux of $(4.4 \pm 1.1) \times 10^{-11}$ cm^{-1} s^{-1} at 300 GeV. No convincing positive detection has been made since then. There have been claims of transient time-variable signals in data from particle arrays (Allen et al. 1993; Clay et al. 1994) which have the advantage of being able to observe with a high on-time efficiency. The EGRET group has reported two unidentified sources near Centaurus A, J1317-44 (CGRO News No. 102, January 1993) and J1314-42 (Fichtel et al. 1994). This complex emission region may contain 2$\sim$3 sources, one of which might be Centaurus A (Hartman 1994).

2.2. Gamma Ray Pulsars

Vela: A persistent flux, modulated with the 89 ms pulsar period, was first reported by Grindlay et al. (1975) and a possible detection of $\sim 1 \times 10^{-12}$ cm^{-2} s^{-1} at > 5 TeV with 4σ was reported by the Tata group (Bhat et al. 1987). However, this pulsed signal has not been confirmed by other groups: the upper limits are 5×10^{-11} cm^{-2} s^{-1} above 300 GeV (Bowden et al. 1993a), 6×10^{-12} cm^{-2} s^{-1} above 2.3 TeV (Nel et al. 1993) and 7×10^{-11} cm^{-2} s^{-1} above 800 GeV (Edwards et al. 1994). An unpulsed signal not modulated with the pulsar rotation might be more likely in VHE band as is observed in the Crab case, and would be detectable with an imaging telescope. Since the Vela pulsar is as close as 500 pc, the radio emission from the SNR covers the field of view of Čerenkov telescope with an embedded, comparatively compact but not point-like, X-ray synchrotron nebula. A shift from the pulsar position and/or extended emission may be the case, and the data need a careful analysis. Haines (1993) mentioned the preliminary CANGAROO result on the Vela pulsar. Analysis on data accumulated since then is in progress in the CANGAROO experiment.

PSR1706-44: A flux of $\sim 8 \times 10^{-12}$ cm^{-2} s^{-1} was detected by the CANGAROO group (Kifune et al. 1995) at 1 TeV energy with a statistical significance at the level of 10 σ. It is unlikely that the total of this observed flux is pulsed. A major portion at least is unpulsed emission. The Durham group have examined their data set for emission (without period analysis), and find an excess which currently stands at $\sim$3 σ (Turver's talk in this workshop). Cheng and Ding (1994) studied parameters which describe the pulsar magnetosphere based on "outer gap model", and concluded that PSR 1706-44 is the most likely object among the gamma ray pulsars in which the pulsed emission would extend to the VHE region. The Durham and the Potchefstroom groups have set upper limits on the pulsed flux, 8×10^{-11} cm^{-2} s^{-1} at 250 GeV (Bowden et al. 1993a) and 5.8×10^{-12} cm^{-2} s^{-1} at 2.6 TeV (Nel et al. 1993), respectively.

PSR1055-52: The Durham group's upper limit on the pulsed flux is 5×10^{-11} cm^{-2} s^{-1} at 300 GeV (Bowden et al. 1993a). A preliminary analysis of the CANGAROO data seems to show no positive excess from the object, precluding a flux larger than 1×10^{-11} cm^{-2} s^{-1} at 1 TeV. The very flat spectrum up to 10 GeV, as observed by EGRET, does not continue to VHE region.

Other short period pulsars: EGRET has made searches for GeV gamma rays from a number of rapidly rotating pulsars, setting upper limits of a few times 10^{-7} cm^{-2} s^{-1}. There are so far four milli-second pulsars in the northern sky which have been claimed as VHE gamma ray emitters, all with comparatively larger fluxes than an extrapolation of the EGRET upper limits to the VHE domain: PSR 1937+21, PSR 1953+29 and PSR 1855+09 (by the Durham group), and PSR 1957+20 (Potchefstroom group).

In the southern sky, eleven pulsars with period from 1.8 to 5.8 msec are found within one minute of arc in the globular cluster 47 Tucanae (NGC104), though EGRET failed so far to detect gamma rays from this direction (Michelson et al. 1994). The total flux of VHE gamma rays – the sum of contributions from all pulsars – is estimated to be as large as 3×10^{-11} cm^{-2} s^{-1} for > 400 GeV (Smith 1993), though period analysis can not be applied to the total flux. The current upper limit is, however, 3×10^{-10} cm^{-2}

s^{-1} (Bowden et al. 1993a). Imaging telescopes have considerably better sensitivity which can examine the predicted flux.

2.3. X-RAY BINARIES

EGRET has so far failed in detecting gamma rays from XRBs, which suggests XRBs are not encouraging candidates for VHE gamma ray observations. Alternatively, it may mean that XRBs show their most enigmatic face at very high energies.

TABLE I
Southern X-ray binaries

object	spin period (s)	orbital period (days)	spent time (hrs)	flux (cm^{-2} s^{-1}) and/or fraction of cosmic ray intensity	Rayleigh probability	group
SMC X-1	0.714	3.89	333	1.8 %	$\sim 10^{-5}$	D
Cen X-3	4.84	2.09	207	6×10^{-10}, 4.7 %	$\sim 10^{-6}$	D
			71	1.3×10^{-10}	2×10^{-4}	P
LMC X-4	13.5	1.41	107	2×10^{-10}, 3.8 %	2.2×10^{-4}	D
Vela X-1	283	8.96	114	2×10^{-11}	7×10^{-3}	P
			103	7.4×10^{-11}	1.1×10^{-4}	D
4U1145-619	292	188	45	1.3×10^{-10}, 1.7 %	1.5×10^{-4}	D
AE Aquarii	33.1	9.98	278	2.3×10^{-11}	5×10^{-3}	P
			99	$\sim 1 \times 10^{-10}$	2×10^{-3}	D

D: Durham group and P: Potchefstroom group
The threshold energy is about 300 GeV and 1 TeV for D and P, respectively.

A summary of VHE results on southern XRBs including one cataclysmic variable AE Aquarii (Meintjes et al. 1992; Bowden et al. 1992c) is given in Table I (compiled mostly from the review by Chadwick et al. 1990). Most of evidence for XRBs is based on the detection of a signal modulated with the neutron star/white dwarf rotation and/or orbital period of binary system. Orbital motion affects the frequency of spin period modulation, and, if gamma ray emission occur preferably at some particular orbital phase, observed spin frequency can be shifted from the known period of X-ray emission. The frequency may also depend on uncertainty of emission region in the binary system (Bowden et al. 1993b), as well as on reprocessed emission mechanisms (e.g. Lamb et al. 1988 about Her X-1). A long time more than 100 hours duration is necessary per object and have been spent on observations for Rayleigh powers typically of 10^{-4} to 10^{-5}, which corresponds to roughly 4σ significance.

Both the Durham and Potchefstroom groups have reported persistent emission from Vela X-1 (modulated with the 283 sec spin period) at all orbital phases: $(7.4 \pm 1.5) \times 10^{-11}$ cm^{-2} s^{-1} for $E_\gamma > 300$ GeV (Carraminana et al. 1989) and $(4.7\pm0.6) \times 10^{-11}$ cm^{-2} s^{-1} $E_\gamma > 1200$–1900 GeV (Raubenheimer et al. 1994), respectively. From these two results, the VHE emission must have a very flat energy spectrum. Rather than arguing about this feature, let us point out that the persistent VHE emission, which both results suggest, is irrespective of orbital phase and will be detectable with better accuracy with imaging telescope. The results were from 103 and 114.5 hours observation, respectively. If we use the imaging technique that has about a factor of 100 of the figure of merit for rejecting background, 10 hrs or 1 hr observation (respectively, for the Durham and Potchefstroom result) may confirm the results with a significance similar to the reported ones.

Examples of enhanced Rayleigh power were shown (Turver, this workshop), when an imaging and "stereo" cut was applied to preliminary data on AE Aquarii and Vela X-1. If event rate is plotted as a function of orientation angle α of the Čerenkov light image, a peak of events at zero values of α, would provide evidence free from the complexities of periodic analyses (which are particularly important for Vela X-1 with its long spin period). Such an analysis enables us to obtain interesting information on unpulsed emission from XRB. Unpulsed VHE gamma rays might be detected from XRB, and detection of such a signal is no less important than claims of pulsed emission.

2.4. Other objects

The number of potential sources is now fairly large. EGRET unidentified sources amount to 37 and 43 respectively for low and high galactic latitudes. Thus, it will take years before VHE observation can cover many of the objects. From one of these, 2CG006-00/J1758-23, the CANGAROO group noted an interesting hint of signal in 1992 data (Kifune et al. 1993), which needs confirmation by the accumulation of more data.

3. Discussions

3.1. Supernova remnants and cosmic ray protons

The VHE gamma rays from the Crab (Vacanti et al. 1991; Lewis et al. 1993; Baillon et al. 1993; Tanimori et al. 1994), PSR1706-44 (Kifune et al. 1995) and AGN Markarian 421 (Punch et al. 1991) probably result from the interactions of electrons. In contrast, energetic cosmic ray protons are believed to be accelerated in the shock at SNR. VHE gamma rays are expected from protons with a flat spectrum of a power index 2.0–2.2 in SNR (Drury et al. 1994; Naito and Takahara 1994). Above several hundred GeV, the estimated flux dominates the emission from the galactic disk, making the detection most promising at sub-TeV energies with imaging Čerenkov telescopes, if the SNR is closer than $\sim$ 2 kpc.

The VHE gamma rays from SNR can be identified by a spectral index of 2.0–2.2, which will in turn provide strong evidence for the shock acceleration theory. In the error circle of several unidentified EGRET sources, radio pulsars and SNR are found. Associations of SNRs and pulsars now number $\sim$10 (e.g. Caraveo 1993), most of which are in the southern sky. The energy spectrum would be useful to distinguish whether the SNR and/or pulsar contributes to the VHE gamma ray emission. The Crab is so young that the pulsar's activity dominates the SNR, and its spectral index of 2.7 (Lewis et al. 1993) can be evidence that the VHE energy spectrum from the pulsar (or from pulsar nebula) differs from that expected from SNR. In addition, there must be an enhancement of gamma rays from molecular clouds. The energy spectrum from the clouds can be different from SNR. In fact, the case of ρ Ophiuchus shows a steeper spectrum at GeV energies (Hunter et al. 1993). Galactic VHE sources must exhibit a variety of energy spectra with power indices in the range 2 – 3, possibly with spatially extended emission.

3.2. Unpulsed emission and electrons

Gamma-ray pulsars: The Crab has an X-ray synchrotron nebula that is observationally well established, and the unpulsed VHE gamma ray emission is explained naturally as arising in the nebula (e.g. De Jager and Harding 1992). Other gamma ray pulsars, with the exception of Vela, are not associated with (at least well established) nebulae. In the Crab case, most of spin-down energy is spent in the pulsar wind going eventually into the nebula. Generally, the intensity of unpulsed VHE gamma rays must be related to the spin-down energy used for energetic electrons which are then transported into outer space well beyond the light cylinder. Different mechanisms may allow VHE emission for different pulsars. According to Kwok et al. (1991), unpulsed VHE gamma rays are emitted in a region at a few times the light cylinder radius from the neutron star.

The luminosity in X-rays and VHE gamma rays is compared for three gamma ray pulsars in Table II. An estimate is given with a very simplified view of approximating the density of ambient photons only by the 2.7 K microwave background. The photon density in a nebula with a given observed luminosity decreases with increasing nebula size. The size of the Crab nebula is about 1 pc, and with this size the total photon density is of a similar order of magnitude as the 2.7 K photon density. Estimates based on this view will not be very far from reality unless the nebula size is much smaller than 1 pc and/or the synchrotron luminosity is much intenser than the Crab's. The relative intensity of synchrotron X-rays to TeV gamma rays through the inverse-Compton effect on 2.7 K photons depends on magnetic field strength: roughly estimated to be $L_{VHE} \sim (6 \times 10^{-12}/B^2) \cdot L_X$, where

TABLE II
nebula X-ray emission

object	B (μG)	X-ray luminosity observed (erg s^{-1})	X-ray luminosity estimated from TeV luminosity (erg s^{-1})	typical energy of synchrotron X-rays (keV)
Crab	250	2×10^{37}	4×10^{37}	1.2
Vela	15	3×10^{33}	9×10^{32} (*)	0.05
1706-44	> 3	1×10^{32}	$> 4 \times 10^{33}$	> 0.01

* by assuming a steady flux of $10^{-11}\ cm^{-2}\ s^{-1}$ flux at 1 TeV.

B is in gauss. The Crab and Vela pulsars (with an assumed steady VHE emission for the latter) is quite consistent in this relation of X and VHE luminosity. However, for PSR 1706-44, the X-ray luminosity detected by the ROSAT satellite (Becker et al. 1992) is smaller than expected from VHE gamma ray luminosity even when we minimize the expected X-ray luminosity with a magnetic field as low as the interstellar field of 3 μ gauss. A softer X-ray spectrum within the uncertainty of the ROSAT data may allow TeV emission compatible with the X-ray luminosity of ROSAT (Harding, this workshop). The minimized strength of magnetic field implies that the X-ray synchrotron radiation peaks at longer wavelengths, in the optical to ultraviolet, to which ROSAT is not sensitive. The typical energy k_s (in keV) of synchrotron radiation and gamma ray energy k_{ic} (in TeV) from inverse Compton scattering on 2.7 K photons are related to each other as $k_s \sim 3 \cdot 10^3 B \cdot k_{ic}$ from the same progenitor electrons, where the magnetic field B is in gauss. If the pulsar nebula similar to the Crab's does not exist or the nebula magnetic field is much weaker, the energetic electrons can rather efficiently Compton scatter ambient photons into VHE gamma rays. The importance of 2.7 K photons for explaining the Crab spectrum was indicated by Aharonian (this workshop).

X-ray halos and XRB: If XRBs emit VHE gamma rays through protons colliding with the atmosphere of companion star, abundant electrons and positrons are expected from $\pi^{\pm}$ decay, and an extended emission of X and gamma rays is expected. When electrons as energetic as $\sim$ 100 TeV are emitted into interstellar space with a $\sim 3\mu$ gauss magnetic field, the synchrotron radiation would be in the keV X-ray band. This effect should appear as an X-ray halo and can be seen for Cyg X-3 in EXOSAT data (Kifune et al. 1987) and also in the GINGA data (Kafuku et al. 1994). The high energy electrons would also form a gamma ray halo by inverse Compton scattering 2.7 K photons. Table III compares VHE gamma ray with X-ray luminosity for the objects which have been claimed as VHE sources. The X and VHE luminosities from the halo are considered not much smaller than L_{VHE} in Table III.

TABLE III

Luminosities from Southern X-ray binaries

object	L_X ($erg\,s^{-1}$)	L_{VHE} (erg s^{-1})
SMC X-1	5×10^{38}	7.5×10^{37}
Cen X-3	1×10^{38}	2×10^{35}
LMC X-4	4×10^{38}	1×10^{37}
Vela X-1	5×10^{36}	3.6×10^{34} (Durham)
		7.8×10^{33} (Potchefstroom)
4U1145-619	1×10^{35}	1.8×10^{34}
AE Aquarii	$\sim 10^{31}$	1.5×10^{32} - 10^{34} (burst)

Extragalactic GeV gamma ray halos: A gamma ray halo is expected also from AGN gamma rays, through the process in which energetic gamma rays create electron-positron pairs while colliding with infrared background photons in intergalactic space, and the secondary electrons (and positrons) inverse Compton scatter 2.7 K micro-wave photons into gamma rays (e.g. Aharonian et al. 1994). Thus, the gamma ray halos indicate higher energy emission than observed. The "direct" gamma rays at 1–10 TeV correspond to 1–100 GeV gamma rays in the halo, thus, at GeV energies in the EGRET domain. It is interesting to note that a halo at GeV energies from distant VHE emitters might be found in the EGRET data and, if so, it is likely to dominate the time-variable "direct" emission after adding a number of observation periods (Hartman et al. 1993). The apparent angular size of the halo would be in proportion to the distance to the source. A bump at ~ 1 GeV of energy spectrum can be caused by ~ 1 TeV primary gamma rays, and would be relevant to the intensity of extragalactic infrared photons.

3.3. Time variability and outbursts

The emissions from the EGRET AGN are time-variable with occasional outbursts. Similarly, this feature characterizes the claimed VHE emission from XRBs, which has caused some controversy as confirmation of burst detections is not easy. Episodic bursts of XRB were observed to take place at certain orbital phases, for instances near eclipse of Vela X-1. However, a complete coverage of orbital phase takes a long time because the orbital period is typically of the order of days (e.g. Bowden et al. 1992b).

The Whipple group has detected a burst from Markarian 421 in May 1994 with 6.5σ significance (Kerrick et al. 1995). However, note that either if the threshold energy were not 250 GeV but as high as 1 TeV or if the imaging technique were not in use, the significance would have been as small as $1-2\sigma$. With the $\sim$ 100 GeV threshold of the telescopes as planned by various groups (see e.g. Goret 1995), high enough significances will be obtained from one nights data. The prospects of time-variability should encourage continual monitoring with imaging telescopes, which may be rewarded with the detection of similar VHE outbursts.

By detecting the halo emission around VHE sources, the time averaged activity over past years can be observed for violently variable objects. The accretion life time of AGN with a black hole of 10^8 solar mass is about 10^8 years (e.g. Rees 1984) which is as long as the propagation and life time of pair electrons created by energetic gamma rays in extragalactic space.

3.4. Comments on the Imaging Technique

The counting rate of detecting Čerenkov light is affected by a factor as large as several percent when the direction of the geomagnetic field varies relative to the source direction (Bowden et al. 1992a). Bowden et al. suggested that the imaging technique might not be useful when the geomagnetic field is transverse to the source direction. Recently, Lang et al. (1994) examined the Whipple data on the Crab and were unable to detect any such effects. The gamma ray signal from PSR 1706-44 gives the strongest evidence against the effect of the transverse magnetic field; the angle between the geomagnetic field and object direction at culmination is 17° and 34° for Crab and Markarian 421 (in the Whipple observation), and 25° and 72°, respectively, for Crab and PSR1706-44 when observed by CANGAROO.

VHE emission from many interesting objects is expected to be spatially extended. The angular accuracy is as good as $\sim 0.1°$ for imaging telescopes (Akerlof et al. 1991). It is necessary with this accuracy to identify the direction of the most intense gamma rays within the error circle of EGRET detections, and also to develop techniques for measuring emission extending over $\sim 1°$. A lower energy threshold enables a wider energy range for observations, which is necessary to determining VHE energy spectra, such as from SNR.

4. Conclusion

The unpulsed VHE emission from gamma ray pulsars is probably caused by energetic electrons. Various processes, possibly different for each particular object, take place for the ejection of high energy electrons, and X- and gamma rays are emitted through synchrotron and inverse Compton radiation. If electrons are transported far enough, X- and gamma ray halos may be formed around the object and "pair halos" around distant AGN at far distances. The VHE emission from distant AGN may be evidenced by GeV gamma ray halos. On the other hand, the interaction of energetic protons are believed to produce gamma rays in such objects as SNR. A variety of emission mechanisms, of electron or proton progenitor, would result in different energy spectra of VHE gamma rays. Much remains to be done in VHE gamma ray astronomy in years to come!

Imaging Čerenkov telescopes are believed to be able to detect VHE gamma rays from the mechanisms discussed in this paper. However, for better understanding, further improvement is desirable in obtaining a lower threshold energy and for a better capability of resolving extended emission.

Acknowledgements

The author thanks F.A. Aharonian and H.J. Völk for inviting him to the Heidelberg Workshop and giving the opportunity to present this paper. The author is indebted to P.G. Edwards for reading the manuscript and comments.

References

Aharonian, F.A., Coppi, P.S., and Völk, H.J., 1994, *Astrophys. J. Lett.*, **423**, L5
Akerlof, C.W., et al., 1991, *Astrophys. J. Lett.*, **377**, L97
Allen, W.H., et al., 1993, *Astroparticle Phys.*, 1, 269
Baillon, P., et al., 1993, *Proc. of 23rd Int. Cosmic Ray Conf. (Calgary)*, 1, 271
Becker, W. et al., 1992, *IAU Circular*, **5554.2**
Bhat, P.N., et al., 1987, *Astron. Astrophys.*, **178**, 242
Bowden, C.C.G., et al. 1992a, *J. Phys. G; Nucl. Part. Phys.*, **18**, L55
Bowden, C.C.G., et al. 1992b, *J. Phys. G; Nucl. Part. Phys.*, **18**, L127
Bowden, C.C.G., et al. 1992c, *Astroparticle Phys.*, **1**, 47
Bowden, C.C.G., et al. 1993a, *Proc. of 23rd Int. Cosmic Ray Conf. (Calgary)*, **1**, 294
Bowden, C.C.G., et al. 1993b, *Proc. of 23rd Int. Cosmic Ray Conf. (Calgary)*, **1**, 298
Carraminana, A. et al. 1989, *Astrophys. J.*, **346**, 967
Caraveo, P.A., 1993, *Astrophys. J. Lett.*, **415**, L111
Chadwick, P, McComb, and Turver, K.E., 1990, *J. Phys. G.*, **16**, 1773
Cheng, K.S., and Ding, K.Y., 1994, Astrophys. J., **431**, 724
Clay, R.W., Dawson, B.R., and Meyhandan, R., 1994, *Astroparticle Phys.*, in press
De Jager, O.C., and Harding, A.K., 1992, *Astrophys. J.*, **396**, 161
Drury, L.O'C., Aharonian, F.A., and Völk, H.J., 1994, *Astron. Astrophys.*, **287**, 959
Edwards, P.G., Thornton, G.J., et al., 1994, *Astron. Astrophys.*, **291**, 468
Fichtel, C.E., et al., 1994, *Astrophys. J. Suppl.*, **94**, 551
Goret, P., 1995, *Proc. of Int. Workshop, "Towards a Major Atmospheric Čerenkov Detector III" (Eds. by T. Kifune; Universal Academy Press Inc., Tokyo)*, p359
Grindlay, J.E., Helmken, H.F., Hambury Brown, R., Davis, J., and Allen, L.R., 1975, *Astrophys. J.*, **201**, 82
Haines, T.J., 1993, *Proc. of 23rd Int. Cosmic Ray Conf. (Calgary)*, Volume for Invited, Rapporteur and Highlight Papers, 341
Hartman, R.C. et al. 1993, *Proc. of 23rd Int. Cosmic Ray Conf. (Calgary)*, **1**, 164
Hartman, R.C. 1994, personal communication
Hunter, S.D. et al. 1993, *Proc. of 23rd Int. Cosmic Ray Conf. (Calgary)*, **1**, 140

Kafuku, S., Yamauchi, S., Yamauchi, M., Kawai, N., Matsuoka, M., Koyama, K., and Kifune, T., 1994, *Mon. N. R. Astron. Soc.* **268**, 437
Kerrick, A.D. et al., 1995, *Astrophys. J. Lett.*, **438**, L59
Kifune, T., Sadzinska, M., Wdowczyk, J., Wolfendale, A.W., Norton, A.J., and Warwick, R.S., 1987, *Mon. N. R. Astron. Soc.* **228**, 243
Kifune, T., Ogio, S., et al., 1993, *Proc. of 23rd Int. Cosmic Ray Conf. (Calgary)*, **1**, 444
Kifune, T., Tanimori, T., Ogio, S., Tamura, T., et al., 1995, *Astrophys. J. Lett.*, **438**, L91
Kwok, P.W., Cheng, K.S., and Lau, M.M., 1991, *Astrophys. J.*, **379**, 653
Lamb, R.C., et al., 1988, *Astrophys. J. Lett.*, **328**, L13
Lang, M.J., et al. 1994, *J. Phys. G; Nucl. Part. Phys.*, **20**, 1841
Lewis, D.A., et al. 1993, *Proc. of 23rd Int. Cosmic Ray Conf. (Calgary)*, **1**, 279
Meintjes, P.J., Raubenheimer, B.C., De Jager, O.C., Brink, C., Nel, H.I. and North, A.R., 1992, Astrophys. J., **401**, 325
Michelson, P.F., et al., 1994, Astrophys. J., **435**, 218
Naito, T., and Takahara, F., 1994, *J. Phys. G; Nucl. Part. Phys.*, **20**, 477
Nel, H.I. et al.,, 1993, *Astrophys. J.*, **418**, 836
Punch, M. et al., 1991, *Nature*, **358**, 477
Raubenheimer, B.C., North, A.R., et al. 1994, *Astrophys. J.*, **428**, 777
Rees, M.J., 1984, *Ann. Rev. Astron. Astrophys.*, **22**, 471
Smith, I.A., 1993, *Astrophys. J.*, **408**, 468
Tanimori, T., Tsukagoshi, T., Kifune, T., et al., 1994, *Astrophys. J. Letters*, **429**, L61
Vacanti, G. et al., 1991, *Astrophys. J.*, **377**, 467

GAMMA-RAY BURSTS AND VERY HIGH ENERGY GAMMA-RAY ASTRONOMY

K. HURLEY
University of California, Berkeley, Space Sciences Laboratory

Abstract. The current status of gamma-ray burst astronomy is reviewed briefly, with emphasis on experiments which might determine the burster distance scale. Recent observations by the EGRET experiment aboard the Compton Observatory, indicating the presence of delayed, high energy emission, are summarized. It is shown that searches for TeV emission from burst sources are feasible, and may be fruitful. Detection of such emission might resolve the question of whether burst sources are at cosmological distances.

1. Introduction

Can VHE gamma-ray astronomy help resolve the mystery of cosmic gamma-ray bursts? Recent observations suggest that there is indeed a chance that it may. One of the biggest mysteries surrounding this phenomenon is their distance scale. It is now well known that the spatial distribution of burst sources is apparently isotropic, while their brightness distribution is inhomogeneous. Based on an analysis of 743 events detected by the Burst and Transient Source Experiment (BATSE) aboard the Compton Observatory, Meegan *et al.* (1994) have determined the dipole and quadrupole moments in galactic coordinates to be $\langle cos\theta \rangle = 0.031 \pm 0.021$ and $\langle sin^2 b - 1/3 \rangle = -0.007 \pm 0.011$ where θ is the angle between the burst location and the galactic center, and b is the galactic latitude. The brightness distribution is characterized using the V/V_{max} test: $\langle V/V_{max} \rangle = 0.321 \pm 0.013$. Burst sources could therefore be very close (within the heliosphere), in an extended galactic halo, or at cosmological distances. Apart from theoretical arguments favoring one distribution or another, there is still little or no observational evidence which points clearly to the correct distance scale. For comprehensive reviews of the observations and their possible interpretations, see Harding (1994), Higdon and Lingenfelter (1990), Hurley (1993), and Paczynski (1993).

2. Near or Far?

Heliocentric theories of gamma-ray bursts often invoke comets (*e.g.*, White 1993). Oort cloud distributions are compatible with both the BATSE $\langle V/V_{max} \rangle$ and cumulative number-intensity distribution (Horack et al. 1994). Models of galactic bursts usually involve lone neutron stars undergoing sudden or gradual accretion (*e.g.* Blaes *et al.* 1992, Mitrofanov and Sagdeev, 1990) or glitches (Ruderman, 1991a,b,c). Here the BATSE data taken as a

Space Science Reviews **75**: 43–52, 1996.

whole effectively rule out most galactic arm models (Smith 1994a), although some subsets of it appear to be consistent with spiral arm distributions (Quashnock and Lamb 1993). Pure disk models, too, are incompatible with the data (Hakkila *et al.* 1994), but disk plus halo models can be accomodated (Higdon and Lingenfelter 1994). Neutron stars in an extended halo, possibly ejected from the disk at high velocities (Podsiadlowski, Rees, and Ruderman 1995), are also consistent with the data. Cosmological models are often built around neutron star-neutron star (Eichler *et al.* 1989) or neutron star-black hole (Paczynski 1991) mergers; such models naturally satisfy both the isotropy and inhomogeneity constraints imposed by the data. From this rapid survey of models, it should be clear that the simple answer to the question "near or far?" *i.e.* cosmological or non-cosmological, would not solve the GRB mystery, since it would not, by itself, tell us exactly what produces bursts. But it is a necessary and sufficient condition for serious, detailed modeling, and in addition, it would give observers a clear indication of how to proceed with multi-wavelength counterpart searches.

3. Distance Indicators

Table 1 presents a number of distance indicators which might be used to answer the "near or far" question. Some are rather unusual by conventional astronomical standards, but each has the potential of revealing crucial information about gamma-ray burst sources. Column 1 describes the indicator. Columns 2 and 3 indicate whether a null or a positive result would actually give definitive information about the distance. Column 4 gives an assessment of whether the measurement is likely to take place and yield interesting information. We discuss each in turn briefly.

The detection of a quiescent counterpart to a burst source in practically any wavelength range would almost certainly lead not only to the distance scale, but also to the ultimate resolution of the GRB puzzle. No unambiguous identifications exist yet, despite many searches (Schaefer 1994). Still, the high level of activity, coupled with the great sensitivity of available instrumentation, makes the likelihood high that progress will be made in this area. A null result, however, leads to no clear indication of the distance.

Two types of spectral features, interpreted as cyclotron resonance absorption in teragauss fields, and redshifted positron-electron annihilation, have been reported for many bursts (e.g., Mazets *et al.* 1981, Murakami *et al.* 1988) and were long considered to be a prime piece of evidence that gamma-ray bursts were generated on galactic neutron stars. The BATSE data have not confirmed these observations, although this is not necessarily evidence against their existence (Band *et al.* 1994). It is quite likely that the germanium spectrometer on the recently launched WIND spacecraft (Owens *et al.* 1991) will observe enough bursts with good statistics to confirm the previous observations. This would add considerable credence to the galactic neutron star hypothesis, although cosmological models would not be completely ruled out. Non-detection of lines, however, might still leave the question of their existence open.

A frequency-dispersed radio signal is expected if radio emission accompanies bursts at large distances (Paczynski and Rhoads, 1993; Palmer 1993). Detection of a dispersed radio burst would give a clear indication of distance. The delays are relatively short for galactic bursts, making it less likely that they would be detected as radio sources, but the longer delays for cosmological events, and the numerous searches now being carried out, make detection rather likely if radio emission is indeed a property of GRB sources. A null result, however, can always be explained by assuming that such emission is below the threshold of detectability.

Gravitational lensing is predicted for cosmological bursts (Paczynski 1987), but the number of lensed events is likely to be quite small, and current searches have not revealed any clear candidates. Detection of lensed events would be an unambiguous indication of cosmological origin, but non-detection is inconclusive.

If GRBs are due to merging neutron stars, neutrino emission is expected. The strongest bursts might produce neutrino fluxes which are detectable by DUMAND II (Paczynski and Xu, 1994). The rate of such bursts is low, however, and uncertainties in the models would cause a null result to be inconclusive. Gravitational radiation is similarly predicted for mergers, but the detection rate for LIGO is predicted to be of the order of only one per decade (Kochanek and Piran 1993).

It is not known whether GRB emission extends to TeV energies. As discussed below, there are indications that it may, and futhermore, that such emission may be delayed with respect to the lower energy emission. Searches for TeV emission are therefore practical, and detection of it could provide an important distance limit.

Measurement of the energy spectra of bursts around 1 keV might reveal the signature of absorption, and lead to an estimate of the column depth to the source (Owens and Schaefer 1994). Since absorption around the source itself is expected to be negligible, a positive detection would be a fairly clear indication of the distance. Non-detection, however, could be interpreted as the signature of a source either in the halo of our own galaxy, or that of an external galaxy, and could therefore be ambiguous. The likelihood of a chance observation of a burst in one of the current low energy x-ray telescopes is negligible; however, several proposals for wide field monitors have been submitted. If any is accepted, the chances of measuring absorption are good.

Paczynski (1991) pointed out that GRB sources should be surrounded by transient X-ray halos due to scattering on dust, if they are not nearby.

Klose (1994) has studied this possibility in detail, and concludes that the very strongest bursts might have halos which are detectable around 1 keV by current instruments. Uncertainties in the dust distribution, the relative infrequency of strong bursts, and the need for rapid, lengthy followup observations make it unlikely that this measurement can actually be achieved, however.

The low energy spectra of cosmological bursts might display redshifted absorption edges or atomic lines (E. Fenimore, private communication, 1994). Detection of features due to oxygen or iron could give the source redshift and distance. This measurement requires excellent energy resolution and is beyond the capability of detectors now flying, but is under consideration for future missions.

Detection of an excess of bursts from the direction of a nearby galaxy such as M31 or the LMC would confirm a galactic halo origin for GRBs (Atteia and Hurley 1986; Liang 1991). No such excess has yet been identified in the BATSE data (Meegan et al. 1994), however, and the lack of such an excess does not appear to set significant constraints on the halo model (Smith 1994b; but see also Hakkila *et al.* 1994).

If bursts are generated in the Oort cloud, nearby stars like α Cen might be expected to display an excess of events. Horack *et al.* (1994) report no detectable dipole moment in this direction; a negative result rules out some distance scales, but does not rule out a heliospheric origin.

Parallax, *i.e.* a significant displacement between the positions of a burst determined by a single detector and multiple detectors (such as triangulation by the Interplanetary Network) would be an indication that bursts are quite nearby (Connors *et al.* 1993). The fact that no such discrepancies have been found is probably a good indication that burst sources are not closer than 500 AU, but does not rule out a heliospheric origin.

The unambiguous detection of a repeating classical GRB source would rule out most cosmological models. Several suggestions have been made that there is indeed evidence in the BATSE catalog for recurrence (Wang and Lingenfelter 1993; Quashnock and Lamb 1993), but as the BATSE error circles are large, these suggestions are not yet completely convincing. Reduction of the error circle size is now in progress, and may resolve this difficulty soon.

Evidence for a redshift-luminosity relation for burst time histories (Norris *et al.* 1994) and energy spectra (Nemiroff *et al.* 1994) has been presented. Whether this is truly a cosmological effect or one produced by the intrinsic properties of bursts has been questioned (Band 1994; Atteia *et al.* 1994). Future analysis of a larger data base may help to resolve some of these issues.

Afterglows due to interactions of gamma radiation with the medium surrounding the source have been suggested as distance indicators (Jennings 1983; Fencl, Boyd, and Hartmann 1993). Their detection relies on the rapid availability of precise burst positions. Within the next year, the HETE mission will make such positions available (Ricker *et al.* 1992).

4. Time Histories and Energy Spectra

As measured in the tens to hundreds of keV energy range, the durations of gamma ray bursts present a broad distribution from about 10 ms to 1000 s with an average value of several tens of seconds (Hurley 1992; Kouveliotou *et al.* 1993). As new instruments covering higher energy ranges have come into operation, the energy spectra of bursts have been found to extend to higher and higher energies. An interesting relationship between the energy spectrum of a burst, its most rapid time history fluctuations, and its distance has been pointed out (Schmidt 1978). Estimating the source size from the rapid fluctuations leads to a relation between the distance at which the source becomes optically thick to photon-photon interactions and the maximum photon energy in the spectrum, beyond which a spectral break should occur. Schmidt concluded that the observation of MeV photons in a burst implied a galactic origin, unless the emission was beamed. Spectral breaks have indeed been observed in BATSE bursts (Schaefer *et al.* 1992). However, the EGRET spark chamber aboard the Compton Observatory has detected five bursts at energies up to tens of GeV with no apparent spectral softening. Table 2 presents the characteristics of these bursts. Column 1 gives the date of the burst and references. Columns 2 and 3 give the maximum energy and the duration as observed by EGRET. Column 4 gives the energy spectrum in the MeV region averaged over the event. Columns 5 and 6 give the rank of the event, out of 793 bursts, in terms of 50-300 keV flux and fluence, as observed by BATSE (C. Kouveliotou, private communication). It is clear from the last two columns that all events detected by EGRET have been exceptionally bright. The first four events displayed extended or delayed emission, *i.e.* the high energy emission observed by EGRET continued after the lower energy emission had died out. The February 17 event was the most dramatic in this respect (Hurley *et al.* 1994). As observed in the several hundred keV range, the burst lasted 180 s. However, EGRET detected emission for a total of about 1.5 h, including an 18 GeV photon.

5. VHE Astronomy and the Distance Scale

Two models attempt to explain extended, high energy emission by the interaction of an expanding fireball, following a merger, with surrounding matter (Meszaros and Rees, 1994; Katz, 1994). If this picture is correct, such emission could be a common feature of bursts. If so, searches for TeV emission might be fruitful. Figure 1 shows the integral flux as a function of energy for the first four bursts of Table 2. Figure 2 shows the integral fluence of

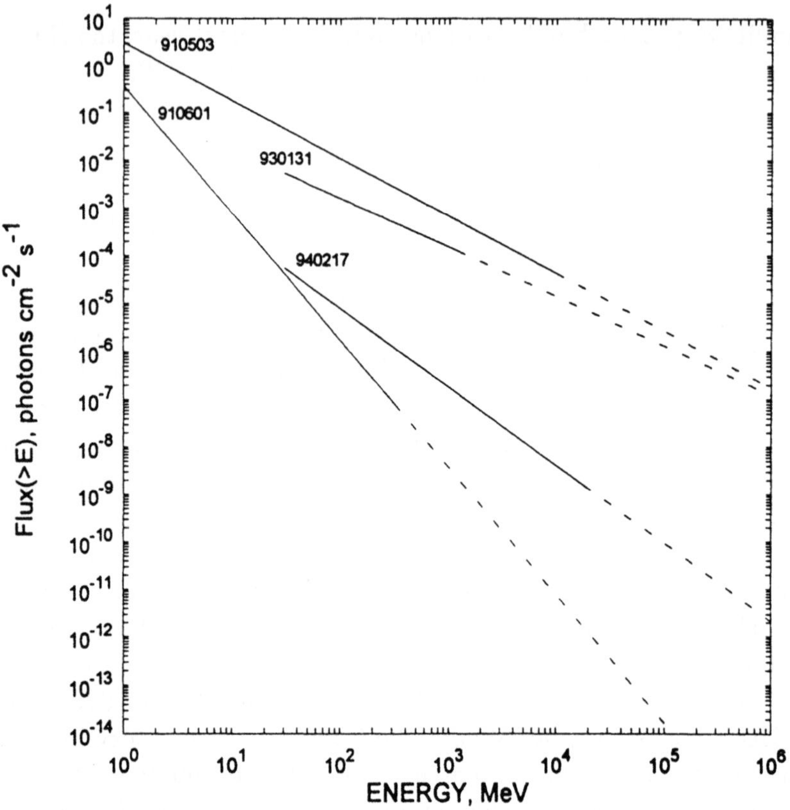

Fig. 1. Integral flux as a function of energy for four of the bursts in Table 2. The solid line indicates the energy region where EGRET measured delayed, high energy emission. The dashed line represents an extrapolation to 1 TeV.

these events. Due to Earth occultation, the true high energy fluence of the February 17 burst could not be determined, and the minimum possible value is shown; the true value could be about three times greater. Some of the fluxes and fluences appear to be at the threshold of detectability for current and future VHE experiments.

The detection of a burst at TeV energies would in itself be an astonishing achievement: a clear indication that an entire component of the output from bursters has been neglected in the energetics of these events. However, the significance of such measurements might go far beyond this. Just as the detection of Markarian 421 (Punch *et al.* 1992) at TeV energies appears to be consistent with suggestions that TeV photons suffer strong attenuation by extragalactic starlight (Stecker, deJager, and Salamon 1992), so the clear detection of an unattenuated power law TeV component from bursts would indicate that they cannot be cosmological. As with all other distance indicators, however (Table 1), the non-detection of such a component, or even the detection of an attenuated component, could be attributed to intrinsic

TABLE I

Distance Indicators for Gamma-Ray Bursts

Indicator	Null	Positive	Likelihood
Counterpart	no	yes	high
Cyclotron lines	no	maybe	high
Emission lines	no	maybe	high
Dispersed radio signal	no	yes	moderate
Gravitationally lensed GRB	no	yes	low
Coincident neutrino burst	no	yes	low
Coincident gravitational radiation	no	yes	low
TeV emission from a GRB	no	yes	moderate
Low energy absorption	no	maybe	moderate
Soft x-ray scattering halo	no	maybe	low
Absorption edges	no	yes	moderate
Excess of bursts from nearby galaxy	no	yes	moderate
Excess of bursts from nearby star	no	yes	moderate
Parallax	no	yes	moderate
Repeating classical GRB source	no	yes	moderate
Redshift-luminosity relation	no	yes	moderate
Afterglow	no	yes	moderate

TABLE II

Gamma-Ray Bursts Detected by EGRET

Date	Maximum Energy (GeV)	Duration of High Energy Emission	Event-Averaged Energy Spectrum	Flux Rank	Fluence Rank
1991 May 3[1,2]	10	84 s	$E^{-2.2}$	3	9
1991 June 1[1,3]	0.314	200 s	$E^{-3.7}$	7	7
1993 Jan 31[1,4]	1.2	100 s	$E^{-2.0}$	5	68
1994 Feb 17[5]	18	1.5 h	$E^{-2.6}$	4	2
1994 Mar 1[6]	0.16	30 s	$E^{-2.5}$	36	16

1. Dingus *et al.* 1994 2. Schneid *et al.* 1992 3. Kwok *et al.* 1994 4. Sommer *et al.* 1994
5. Schneid *et al.* 1994 (in preparation)

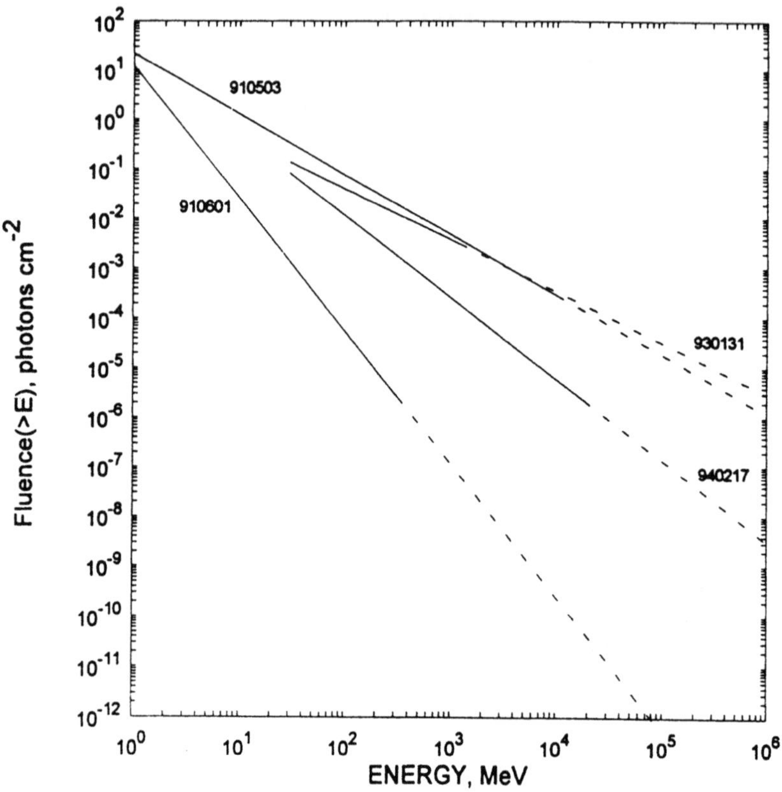

Fig. 2. Integral fluence as a function of energy for four of the bursts in Table 2. As in Figure 1, the solid line represents the energy region measured by EGRET; the dashed line is an extrapolation to 1 TeV. The February 17 fluence could actually be about three times greater.

properties, and would be ambiguous. See, for example, the CYGNUS results at energies above 100 TeV (Alexandreas *et al.* 1994). Still, the rapid availability of coarse burst positions from the BATSE Coordinate Distribution Network (BACODINE: Barthelmy *et al.* 1994) and the future availability of arcsecond to arcminute positions with HETE (Ricker *et al.* 1992) make this search feasible.

Acknowledgements

This work was supported under JPL Contract 958056 and NASA Grant NAG5-1560. The author is also indebted to the organizers of this meeting for travel support.

References

Alexandreas, D. *et al.*: 1994, *Ap. J.* **426**, L1

Atteia, J.-L., and Hurley, K.: 1986, *Adv. Space Res.* **6(4)**, 39
Atteia, J.-L. *et al.*: 1994, *Astron. Astrophys.* **288**, 213
Band, D.: 1994, *Ap. J.* **432**, L23
Band, D. *et al.*: 1994, 'Cyclotron Line Search: BATSE-Ginga Consistency' in G. Fishman, J. Brainerd, and K. Hurley, ed(s)., *Gamma Ray Bursts*, AIP Press: New York, 256
Barthelmy, S. *et al.*: 1994, 'BACODINE: The Real-Time BATSE Gamma-Ray Burst Coordinates Distribution Network' in G. Fishman, J. Brainerd, and K. Hurley, ed(s)., *Gamma Ray Bursts*, AIP Press: New York, 643
Blaes, O., Blandford, R., Madau, P., and Yan, L.: 1992 *Ap. J.* **399**, 634
Connors, A. *et al.*: 1993, *Astron. Astrophys. Suppl. Ser.* **97**, 75
Dingus, B. *et al.*: 1994, 'EGRET Observations of Three Gamma-Ray Bursts at Energies > 30 MeV' in G. Fishman, J. Brainerd, and K. Hurley, ed(s)., *Gamma Ray Bursts*, AIP Press: New York, 22
Eichler, D., Livio, M., Piran, T., and Schramm, D.: 1989, *Nature* **340**, 126
Fencl, H., Boyd, R., and Hartmann, D.: 1993, *Ap. J.* **407**, L21
Hakkila, J., *et al.*: 1994, *Ap. J.* **422**, 659
Harding, A.: 1994, 'Gamma-Ray Burst Models: Present Problems and Possible Solutions' in C. Fichtel, N. Gehrels, and J. Norris, ed(s)., *The Second Compton Symposium*, AIP Press: New York, 30
Higdon, J., and Lingenfelter, R.: 1990, *Ann. Rev. Astron. Astrophys.* **28**, 401
Higdon, J., and Lingenfelter, R.: 1994, *Ap. J.* **434**, 552
Horack, J., Koshut, T., Mallozzi, R., Storey, S., and Emslie, G.: 1994, *Ap. J.* **429**, 319
Hurley, K.: 1992, 'Gamma-Ray Burst Observations: Past and Future' in W. Paciesas and G. Fishman, ed(s)., *Gamma-Ray Bursts*, AIP: New York, 3
Hurley, K.: 1993, *Ann. N.Y. Acad. Sci.* **688**, 588
Hurley, K. *et al.*: 1994, *Nature* **372**, 652
Jennings, M.: 1983, *Ap. J.* **273**, 309
Katz, J.: 1994, *Ap. J.* **432**, L27
Klose, S.: 1994, *Ap. J.* **423**, L23
Kochanek, C., and Piran, T.: 1993, *Ap. J.* **417**, L17
Kouveliotou, C. *et al.*: 1993, *Ap. J.* **413**, L101
Kwok, P. *et al.*: 1994, 'EGRET Observations of Gamma-Ray Bursts on June 1, 1991 and August 14, 1991' in M. Friedlander, N. Gehrels, and D. Macomb, ed(s)., *Compton Gamma Ray Observatory*, AIP Press: New York, 855
Liang, E.: 1991, *Ap. J.* **380**, L55
Mazets, E., *et al.*: 1981, *Nature* **290**, 378
Meegan, C. *et al.*: 1994, 'Two and a Half Years of BATSE' in G. Fishman, J. Brainerd, and K. Hurley, ed(s)., *Gamma-Ray Bursts*, AIP: New York, 3
Meszaros, P., and Rees, M.: 1994, *Mon. Not. R. Astron. Soc.* **269**, L41
Mitrofanov, I., and Sagdeev, R.: 1990, *Nature* **344**, 313
Murakami, T. *et al.*: 1988, *Nature* **335**, 234
Nemiroff, R. *et al.*: 1994, *Ap. J.* **435**, L133
Norris, J. *et al.*: 1994, *Ap. J.* **424**, 540
Owens, A. *et al.*: 1991, *IEEE Trans. Nuc. Sci.* **38(2)**, 559
Owens, A. and Schaefer, B.: 1994, *Comm. Astrophys.* **17(2-3)**, 119
Paczynski, B.: 1987, *Ap. J.* **317**, L51
Paczynski, B.: 1991, *Acta Astronomica* **41**, 257
Paczynski, B.: 1993, *Ann. N.Y. Acad. Sci.* **688**, 321
Paczynski, B., and Rhoads, J.: 1993, *Ap. J.* **418**, L5
Paczynski, B., and Xu, G.: 1994, *Ap. J.* **427**, 708
Palmer, D.: 1993, *Ap. J.* **417**, L25
Podsiadlowski, P., Rees, M., and Ruderman, M.: 1995, *Mon. Not. R. Astron. Soc.* , in press
Punch, M. *et al.*: 1992, *Nature* **358**, 477
Quashnock, J., and Lamb, D.: 1993, *Mon. Not. R. Astron. Soc.* **265**, L59

Ricker, G. *et al.*: 1992, 'The High-Energy Transient Experiment (HETE): An International Multiwavelength Mission' in C. Ho, R. Epstein, and E. Fenimore, ed(s)., *Gamma Ray Bursts: Observations, Analyses, and Theories*, Cambridge University Press: Cambridge, U.K., 288
Ruderman, M.: 1991a, *Ap. J.* **366**, 261
Ruderman, M.: 1991b, *Ap. J.* **382**, 576
Ruderman, M.: 1991c, *Ap. J.* **382**, 587
Schaefer, B.: 1994, 'Search for Gamma Ray Burst Counterparts' in G. Fishman, J. Brainerd, and K. Hurley, ed(s)., *Gamma-Ray Bursts*, AIP: New York, 382
Schaefer, B. *et al.*: 1992, *Ap. J.* **393**, L51
Schmidt, W. K. H.: 1978, *Nature* **271**, 526
Schneid, E. *et al.*: 1992, *Astron. Astrophys.* **255**, L13
Smith, I.: 1994a, *Ap. J.* **429**, L65
Smith, I.: 1994b, *Ap. J.* , in press
Sommer, M. *et al.*: 1994, *Ap. J.* **422**, L63
Stecker, F., deJager, O., and Salamon, M.: 1992, *Ap. J.* **390**, L49
Wang, V., and Lingenfelter, R.: 1993, *Ap. J.* **416**, L13
White, R. S.: 1993, *Astrophys. Space Sci.* **208**, 301

OVERVIEW OF TEV GAMMA RAY OBSERVATIONS

R. C. LAMB
Physics and Astronomy Department
Iowa State University
Ames, Iowa 50011, USA

Abstract. TeV emission has been observed with high signal-to-noise from: two pulsar-driven supernova remnants, the Crab Nebula and PSR B1706-44 and a blazar class AGN, Markarian 421. Other sources, observed without the benefit of major background reduction techniques, are some of the compact object binaries. These discoveries plus the discoveries of EGRET of more than 130 sources at lower energies point to a growing number of scientific questions capable of being addressed by this field (particularly for distant sources via intergalactic absorption processes). Rapid further development will come as new instruments employing a variety of background reduction techniques come on-line.

1. Introduction

In this paper we will try to step back a bit and survey where TeV γ-ray astronomy stands in regard to its sources. We will also remind ourselves what are its motivations, its accomplishments and its prospects. For comprehensive, up-to-date reviews of the field the reader is referred to articles contained in the proceedings of the 1993 Calgary workshop (Raubenheimer, Fontaine, Goret), the proceedings of the 1994 Tokyo workshop (Goret, Aharonian), as well as the article by Kifune (1995) at this workshop.

2. Scientific Issues

In many branches of science, knowledge increases curiosity. The answer to one question leads to many more, a pattern that represents a sort of "fractal" side of science. However, there are some questions that may be irreducible. And when they are answered then one can stand back and say, "Yes, that area of science is now understood. Let's move on." Whether or not TeV γ-ray astronomy can make such truly fundamental discoveries remains to be seen. What we can say is that TeV γ-ray astronomy has entered the main stream of astrophysical research with established sources, the promise of important astrophysical measurements, and a number of proven techniques that can be significantly improved.

There are two principal driving motivations for ground-based high energy γ-ray ray astronomy. The first is that the sky above 10 GeV is still a largely unexplored frontier, *terra incognita*, representing one of the last regions of the electromagnetic spectrum to be explored with sensitive instruments. Although there is yet to be a sensitive survey of the sky at TeV energies,

Space Science Reviews **75**: 53–65, 1996.

recent high-signal-to-background Cherenkov telescopes, pointing at specific objects of promise, have shown that there are at least 3 solidly established sources of TeV γ-rays. Two of these, the Crab Nebula and PSR 1706-44, are pulsar-driven supernova remnants, with the third, Markarian 421, an active galactic nucleus of the *blazar* class. These sources taken together with the $\sim$ 130 sources established by the EGRET detector of the *Compton Gamma Ray Observatory* (Fichtel *et al.* 1994) at energies >100 MeV forecast a rich harvest of scientific results when higher energy portions of the electromagnetic spectrum are fully explored.

A second major motivation for high energy astronomy is the desire to locate the sources of the cosmic rays. Where and how does nature accelerate particles to energies which extend to 10^{20} eV and beyond? Particle energies of this magnitude imply that near their acceleration sites a range of elementary particle physics phenomena is present which is beyond the ability of present-day accelerators to explore. Ground-based γ-ray astronomy may catch a glimpse of such phenomena.

Discoveries from the *Compton Gamma Ray Observatory* and detections at TeV energies make it clear that a combination of space-based and ground-based instruments covering the γ-ray range from 1 MeV to 100 TeV will be able to address a number of significant scientific questions.

A partial list of such questions includes the following:

- What parameters determine the cut-off energy for pulsed emission of γ-rays from pulsars?
- What is the role of shell-type supernovae in the production of cosmic rays?
- At what energies do AGN blazar spectra cut-off?
- Are γ-ray blazar spectral cut-offs intrinsic to the source or due to intergalactic absorption?
- Is the dominant particle species in AGN jets leptonic or hadronic?
- Can intergalactic absorption of the TeV emission of AGN's be a tool to calibrate the epoch of galaxy formation, the Hubble parameter, and the distance to γ-ray bursts?
- Are there sources of γ-rays which are "loud" at TeV energies, but "quiet" at lower photon energies?
- What parameters govern the episodes of apparent emission from the binary sources?

3. Sources

3.1. Active Galactic Nuclei

One of the major discoveries of the EGRET instrument on the *Compton Gamma Ray Observatory* been its detection of more than 40 active galactic nuclei (AGN) (von Montigny *et al.* 1995). All are *blazars*. In the simplest, unified picture of the AGN phenomenon (Antonucci 1993), blazars are those AGN which have associated jets and the orientation of the jet axis is toward the observer.

Although the percentage of AGN with jets is only a few percent, understanding them is important since they are the favored candidates for the acceleration of the highest energy cosmic rays.

As stressed by Bregman (1992), Marscher (1993), and Blandford and Levinson (1995) multiwaveband observations of flares in blazars hold great promise for exploring the region where the relativistic plasma is generated and the jet is focused and accelerated.

The origin of blazar γ-ray emission is currently the subject of much theoretical debate. In one school of thought (Bloom & Marscher 1993; Dermer, Schlickeiser, & Mastichiadis 1992; Sikora, Begelman, & Rees 1994), the γ-ray emission involves inverse- Compton scattering of lower frequency seed photons by high-energy electrons in the inner regions of a relativistically flowing jet aimed almost directly along the line-of-sight. Some of these models predict a cut-off to the spectrum in the range 10–100 GeV. If, on the other hand, the γ-rays result from secondary pion production (Mannheim 1993) from a jet whose composition is mainly hadronic then one expects a hardening of the spectrum above 10 GeV. This distinction between a leptonic and a dominantly hadronic beam presents the high energy γ-ray community with an opportunity for decisive observation if the energy threshold of ground-based instruments can be lowered sufficiently to overlap with those in space.

As yet only one, Markarian 421, has been observed at TeV energies (Punch 1992). Quite recently this source was caught in an outburst which was seen nearly simultaneously by the *ASCA* X-ray satellite (Takahashi *et al.* 1994) and by the Whipple collaboration (Kerrick *et al.* 1995). Observations of this source during the spring of 1994 are shown in Figure 1. Particularly striking is the factor of 10 flare on day 135 (May 15, 1994).

Since the emission is undoubtedly associated in some way with the jet whose orientation is generally towards the observer, the time scale of variation cannot readily be converted into a characteristic size for the emission region. There are at least three different effects which spoil a simple $r = c\Delta t$ calculation. The first is that time differences are foreshortened by $(1 - \beta \cos\theta)$, where the βc is the speed of the emitting region and θ is the angle it makes with respect to the observer. The second is the Lorentz contraction of the emitting region by a factor of $\sim \gamma$. Finally, the variation may have nothing to do with a spatial variation, but rather be due to a variation in θ. Since the luminosity varies as the third or fourth power of the Doppler factor, $[\gamma(1 - \beta cos\theta)]^{-1}$ (Lind & Blandford 1985) small variations in θ are highly leveraged. If we take for the purpose of illustration a value of $\gamma = 10$

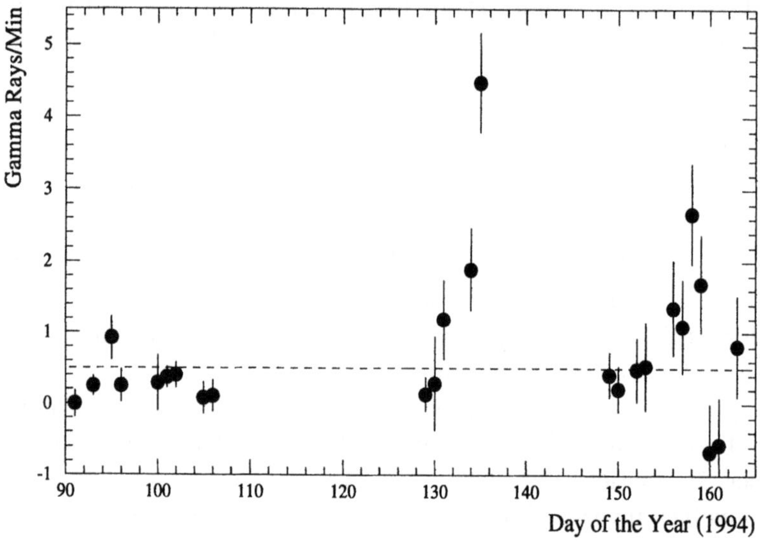

Fig. 1. The variation in net γ-rays/minute for the Markarian 421 observations from 1 April, 1994 (day 91) through 12 June, 1994 (day 163), as observed by the Whipple Observatory (Schubnell *et al.* 1994, Kerrick *et al.* 1995). The peak intensity of 15.25 May (day 135) occurs 27 hours prior to a reported X-ray high state (Takahashi *et al.* 1994). Data from 16-28 May (days 136-148) were not obtained due to bright moon conditions. The dashed line shows the 1994 quiescent level (Schubnell *et al.* 1995)

(e. g. Padovani & Urry 1992), a variation of two degrees about a viewing angle of 10° produces a variation of ~ 3, apart from any spatial variation.

3.2. Pulsars, Pulsar Nebulae, and Supernova Remnants

There are two broad categories of supernova remnants, pulsar-driven and shell-type. In the former case, the remnant has a generally filled center,*i.e.* it is plerionic, and emission may be both pulsed at the spin frequency of the neutron star as well as unpulsed. For the shell-type SNR the energy source is the supernova explosion itself, and high energy γ-ray emission, if present, will be unpulsed.

Six pulsars/nebulae have been observed by the *Compton Observatory*: the Crab, Vela, Geminga, PSR B1055-52, PSR B1509-58, and PSR B1706-44. At these energies the emission is dominantly pulsed. Two of the these sources, the Crab (cf. Fontaine 1993, for an extensive set of references) and PSR B1706-44 (Kifune et al. 1995) have been seen at TeV energies. At TeV energies the emission is dominantly unpulsed.

The observations of unpulsed emission from the Crab Nebula at photon energies above 1 MeV provide an illustration of how observations over an extended range of energies are important, and in particular how space and ground-based observations may complement each other. Several authors (de

Jager et al. 1995, Aharonian 1995, Aharonian & Atoyan 1995) have successfully fit the spectrum using a synchrotron self-Compton model. From these fits one extracts a value of the maximum electron energy accelerated by the pulsar of 2–3$\times 10^{15}$ eV.

One of the most interesting questions concerning the Crab and its role as the prototypical pulsar-driven nebula is the cut-off energy for pulsation. Previous reports of pulsations by the Durham group (Dowthwaite et al. 1984) above 1 TeV, but only upper limits by the Whipple group (Vacanti et al. 1991) above 400 GeV, suggest that the limiting energy for pulsation is time dependent. For those cases where TeV pulsations are observed, then one can say that the pulsed photons must be produced in a region relatively far from the surface of the neutron star, where the magnetic field is weak. Otherwise the high energy photons will be absorbed as they interact with the magnetic field. The condition on absorption (Harding 1991) is that pair production prevents the escape of the γ-ray when the following inequality is satisfied: $E_\gamma/2mc^2 > 0.1B_{cr}/B_\perp$, where $B_\perp$ is the magnetic field perpendicular to the direction of the γ-ray and B_{cr} is the critical field: $m^2c^3/e\hbar = 4.4 \times 10^{13}$ G.

As for the shell-type SNR, recent calculations by Drury, Aharonian, and Völk (1994) indicate that the luminosity of nearby supernova remnants (SNR) should be sufficient for detection by the current generation Cherenkov telescopes. Six or seven of the unidentified EGRET sources (cf. the paper by Kanbach in the proceedings of this workshop) are sufficiently near to catalogued shell-type SNR that, at least some of them, may be causally linked. Statistical studies by Kanbach et al. (1994), Eposito et al. (1995), and Sturner and Dermer (1994) estimate that the chance association of these SNR and the unidentified EGRET sources is $<10^{-3}$.

3.3. Compact Object Binaries

An outstanding observational issue in the field of TeV γ-ray astronomy is the question of the reality of the binary sources which have been reported at these and higher energies. There is the well- documented failure of Cygnus X-3 to shine brightly since 1985; Hercules X-1 has not been observed convincingly since 1986 in spite of efforts with improved equipment. Furthermore, none of the binaries has yet been reported by EGRET. If they are truly sources of γ-ray emission, then their low duty factor will make a search for them in the EGRET database very difficult, although BATSE will provide a useful activity indicator.

Part of the difficulty in accepting the binaries as sources is due to their highly episodic and generally unpredictable behavior. Also they have yet to be observed with good signal-to-noise except in very rare, momentary outbursts lasting only a few minutes, at which time they are exceedingly bright. During these bright flares (cf. Bowden et al. 1992) the raw shower counting rate in a Cherenkov telescope is apparently dominantly gamma rays from the source. I think that it is fair to say that the binaries remain on a list of possible sources but that they are not regarded as secure by many astrophysicists.

Perhaps the most convincing recent observations of a binary are those of the cataclysmic variable white dwarf, AE Aquarii. This object is very close, 50-100 pc, and is the fastest known rotating binary white dwarf with a spin period of 33 seconds. Observations by the Potchefstroom group (Meintjes et al. 1994) and the Durham group (Bowden et al. 1992) are in rough agreement regarding the episodic nature of the emission and the fact that when pulsations are observed they occur at the second harmonic. However the nature of the intense flares associated with the object is unclear since Bowden et al. (1992) report that the flares are pulsed however the flares observed by the Potchefstroom group (Meintjes et al. 1994) show no such periodicity.

One of the lessons that EGRET teaches us is that variability (in some cases of more than an order of magnitude, e.g. 3C279) is a prevalent characteristic of the accreting, massive black-holes represented by the blazars. It is reasonable to assume that variability from accreting compact object binaries might also be expected. Such variability could provide useful constraints on beam-target geometry and even possible acceleration mechanisms. This makes detection at TeV energies of the binaries not only a challenge experimentally but potentially a very rewarding scientific enterprise.

The possible similarity between accreting galactic objects and blazars has been strengthened by the recent observation of jets and superluminal motion in two galactic sources, GRS1915+105 (Mirabel & Rodriguez 1994) and GRO J1655-40 (Reynolds & Jauncey 1994).

4. Intergalactic Absorption of High Energy Photons

As described in the section on scientific issues, intergalactic absorption of high energy photons may become a very useful diagnostic for settling a number of scientific questions such as the epoch of galaxy formation, the distance to the γ-ray bursts, and even the Hubble parameter. It is therefore well to review the physics and history of this process.

The dominant absorption process for high energy photons in intergalactic space is pair creation through collisions with low energy photons. The threshold energy for this process occurs when the product of the energies of the two interacting photons equals $(mc^2)^2$, and the peak in the cross section occurs when this product is $\sim$ twice this value. Thus for 1 TeV photons, target photons of $\sim$ 0.5 eV (i.e. infrared photons) have the maximal absorbing power.

Nikishov (1962) first considered this process for photons in the TeV region interacting with intergalactic starlight. Shortly after the discovery of the microwave background radiation Jelley (1966) and Gould and Schréder

(1966, 1967) independently calculated its effect on photons with PeV energies.

Absorption calculations using the microwave background show that those photons which encounter the largest cross section in colliding with it (photon energies of $\sim$ 1.7 PeV) have a mean free path of only $\sim$ 8 kpc, and even 0.1 PeV photons are strongly absorbed for distances beyond twice the distance to the Virgo Cluster (20 Mpc). The microwave background thus presents an effective barrier to the direct study of cosmologically distant γ-ray sources above this energy.

At lower energies, relevant to the atmospheric Cherenkov technique, a number of authors (Stecker, de Jager, and Salamon 1992; Dwek and Slavin 1994; MacMinn and Primack 1995) have recently examined absorption effects of starlight radiation of TeV photons. Some of the incentive to refine such calculations has come from the Whipple Observatory's report of TeV photons from Markarian 421 at a red-shift of $z = 0.03$, at which distance absorption effects begin to play a role.

In the most detailed calculations, MacMinn and Primack (1995) have sought to model realistically the formation of the galaxies and their evolution. Their calculations basically consider two alternate models for galaxy formation: a cold dark matter model which, in their words, "represents a moderate era of galaxy formation" and a cold plus hot dark matter (CHDM) model representing a late era of galaxy formation. Their conclusion is that the amount of the absorption depends on the model and that, within a given model, the variation of absorption with energy is determined by the epoch at which galaxies began to form.

In Figure 2, the absorption they calculate for a particular choice of CHDM model for sources at redshift distances of $z = 0.2$ and 0.5 is shown. One sees that absorption begins gently for photons with energies as low as 10–30 GeV with a sharp cut-off above 10 TeV. However, there is a modulation to the spectrum. One could therefore envisage a situation in which the epoch of galaxy formation would be determined (or at least strongly constrained) by precise spectral measurements of many such AGN's. It is clear from the figure that the range over which the spectra need to be measured is quite large, and would ideally involve both satellite instrumentation at the lower energies and ground-based detectors at higher energies.

In addition to the absorption's dependence on the details of galaxy formation, the expression for the optical depth depends inversely on the Hubble parameter, H_0. Salamon, Stecker, and de Jager (1994) have pointed out this dependence and have suggested that from observations of many AGN at a suitable ensemble of red-shifts H_0 may be determined.

For both the Hubble constant determination and the determination of the epoch of galaxy formation, one must have a reasonably large number of AGN whose intrinsic spectra can be confidently extended to multi-TeV

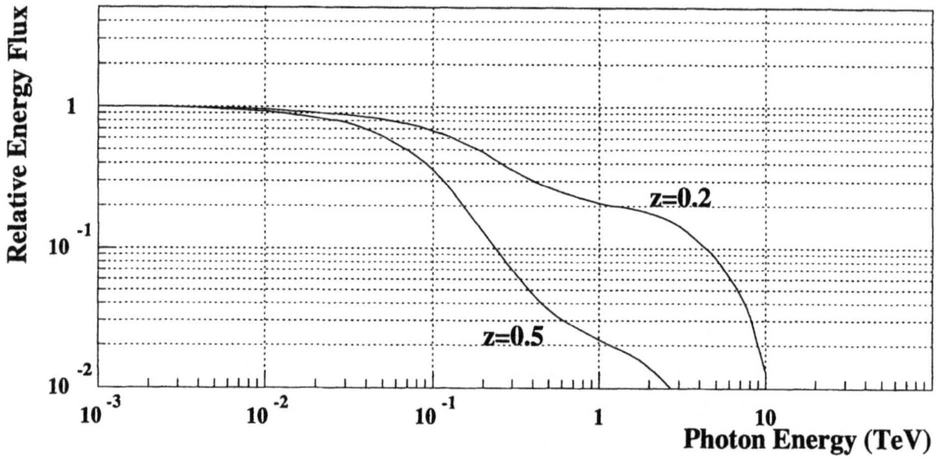

Fig. 2. Modulation to an assumed power-law spectrum from γ-ray sources at redshifts $z = 0.2$ and 0.5 (one of the cold plus hot dark matter models calculations of MacMinn and Primack 1995). The dip is due to the absorption of the very high energy photons by intergalactic photons red-shifted from the visible to the infrared. The depth and shape of the amount of absorption are dependent on the precise epoch at which galaxies first formed.

energies. In view of the fact that there is as yet only one AGN which has been found at TeV energies, Markarian 421, the extrapolation to many seems premature.

A less problematic use of the intergalactic absorption would involve the observation (or not) of γ-ray bursts (GRB's) at energies above 100 GeV. One of the major questions confronting high energy astrophysics today is the distance scale for GRB's. The present ensemble of GRB's from BATSE is consistent with isotropy on the plane of the sky but with a strong indication of a cut-off in the line-of-sight direction. Only two possibilities have the spatial characteristics which satisfy the constraints of the observations. Either the bursts are from a Galactic population corresponding to a very extended, roughly spherically symmetric halo with a scale height greater than 70 kpc. Or they are at cosmological distances.

From the detection of some of the BATSE bursts by the EGRET detector at GeV energies an extrapolation to TeV energies may be made. For example, the so-called Super-Bowl burst of January 29, 1993 (Sommer et al. 1994) has an observed spectrum:

$$dN/dE \approx 10^{-5}(E/E_o)^{-2} \text{ photons/cm}^2\text{/s/MeV}.$$

The duration of this burst is observed to be 25 seconds. If one extrapolates this spectrum to TeV energies, one finds:

$$N(\text{photons} > 100\text{GeV}) = 10^{-6} \text{ photons/cm}^2\text{/s}.$$

If we assume, conservatively, a collection area of only 10^8 cm^2, then 2500 photons would be detected in a Cherenkov telescope with a threshold energy of 100 GeV. Therefore significantly fainter GRB's could easily be detected. Thus if the GRB's are from a galactic population one would expect no absorption features in the energy regime covered by the Cherenkov detectors. On the other hand if they are cosmological then only the nearer will be detectable above 100 GeV and characteristic absorption features as illustrated in Figure 2 will be present.

5. Experimental Techniques for Background Reduction

Without some means of reducing the enormous background from cosmic ray air showers, putative γ-ray signals are condemned to be of order 1% or less of that background, except in those rare cases where the source flares strongly. Operating in such a regime invites trouble due to the possible presence of low-level systematic errors.

One of the most exciting developments of recent years has been the recognition and verification that the cosmic ray background for ground-based Cherenkov experiments can be drastically reduced. Recent workshops devoted to the developments in the air Cherenkov technique (Palaiseau-1992, Calgary-1993, and Tokyo-1994) have highlighted several promising avenues for the reduction of the cosmic ray background. All such techniques are founded on the demonstrable fact that there are substantial differences between the characteristics of γ-ray induced air showers and cosmic-ray air showers (cf. Hillas 1995 these proceedings).

One of the most promising techniques is that of imaging, which has been used by both the Whipple Observatory (Vacanti et al. 1991) and the Tokyo University group (Tanimori et al. 1994) to detect the Crab Nebula. This technique has also been employed to detect PSR 1706-44 (Kifune et al. 1995). With this technique it is possible to reject more than 99% of the cosmic ray background while keeping a substantial fraction of the γ-rays. To illustrate this fact, the Whipple Observatory collaboration's observations of the Crab Nebula (Vacanti et al. 1991) and Markarian 421 (Kerrick et al. 1995) during its May 15, 1994 outburst are shown in Figure 3. In this figure the events have been pre-selected to have the shape expected for γ-ray showers. The peak near zero degrees (superimposed on a relatively flat residual cosmic background) contains the sources' γ-rays. As one can see, near zero degrees, γ-rays outnumber the residual cosmic ray background by a substantial margin.

An alternative technique for background reduction is that of the fast timing array (Ballon et al. 1993). Here one relies on sampling the Cherenkov wave-front over ~100 m to reconstruct the best estimate of the shower's arrival direction. Reconstruction accuracies of order 0.1° are possible. The

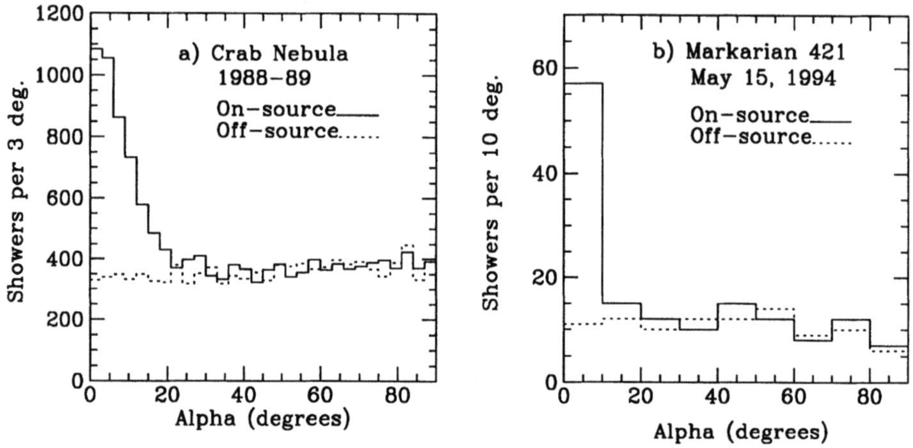

Fig. 3. a) On and off-source orientation angle ("alpha") distributions for the Whipple Observatory 30 hours of on-source observations of the Crab Nebula (Vacanti et al. 1991). b) On and off-source alpha distributions for the May 15, 1994 outburst from Markarian 421. On source time 30 minutes. (Kerrick et al. 1995).

background then is that contained in the associated solid angle. With an array it is also possible to exploit differences in the characteristics of the lateral distributions of γ-rays and cosmic rays (Hillas 1995, these proceedings)

A full Monte Carlo simulation of an array of 10 m class Cherenkov imagers has yet to be done, but it appears from some preliminary work (Krennrich 1995) that the lateral distribution differences and the image differences are somewhat independent. Thus a large array of Cherenkov imagers that measures accurately each of these properties should be capable of substantial gains in sensitivity over what has already been achieved.

5.1. The Cosmic Ray and Electron Backgrounds

As one can see from Figure 3, present detectors are capable of detecting sources as bright as the Crab Nebula with signal-to-backgrounds exceeding unity, and with relatively high instantaneous γ-ray rates, e.g. a few per minute. In order to place this in perspective it is useful to compare the Crab Nebula spectrum with the background cosmic ray spectrum. This comparison is shown in Figure 4 for two different background solid angles, 10^{-3} and 10^{-5} steradians, corresponding to half-angle fields-of-view of 1.0° and 0.1° respectively.

Using Figure 4 and knowledge about the relative Cherenkov light yields of γ-ray and proton showers one can easily calculate what effective angular resolution is required in order to achieve a particular signal-to-background ratio. For example, the 1988/89 detection of the Crab (Vacanti et al. 1991)

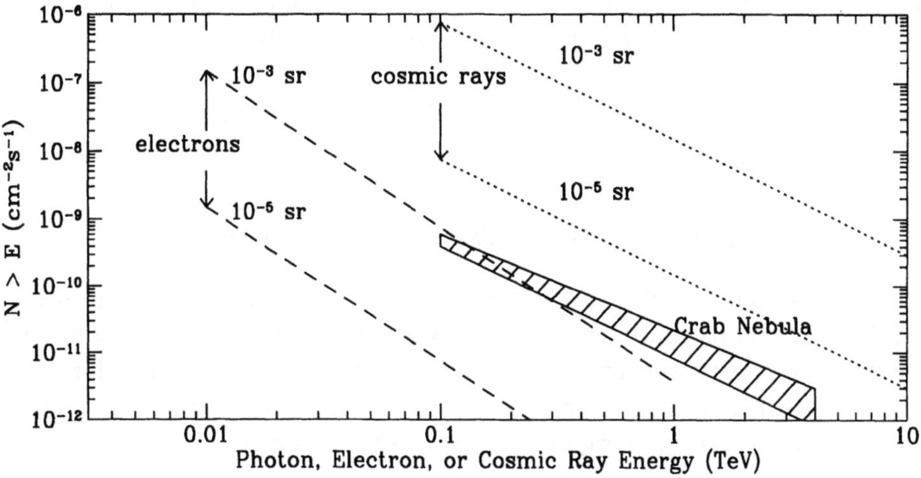

Fig. 4. Backgrounds from cosmic ray nuclei and electrons compared with an approximate representation of the spectrum of the Crab Nebula. The backgrounds are calculated for two values of the solid angle: 10^{-5} and 10^{-3} steradians (0.1° and 1° half-angle fields of view). The electron spectrum is taken from Longair (1992).

with an approximate signal-to-background of unity had an effective threshold energy for γ-rays of 400 GeV. This is equivalent to a proton threshold energy of ~1 TeV. From the figure therefore, one can see that the effective solid angle for this detection is $\sim 5 \times 10^{-6}$ sr, equivalent to a half-angle field-of-view of 0.07°.

With full development of background reduction techniques referred to above, it is conceivable that the cosmic ray background for a point-like γ-ray source can be made virtually negligible for a source as strong as the Crab. However, under these conditions, one must consider the background due to cosmic ray electron showers. Electron showers will not be distinguishable from γ-ray showers, except that they will be isotropically distributed on the plane of the sky in contrast to possible localized γ-ray sources.

In Figure 4, the electron spectrum (taken from Longair 1992) is shown for the two solid angles. As the energy is lowered this electron background becomes increasingly important for two reasons. In the first place, the electron spectrum rises more rapidly than the cosmic ray spectrum ($E^{-2.3}$ compared to $E^{-1.7}$). More important however, is the fact that the Cherenkov light yield from electron showers is virtually the same as the yield from γ-ray showers at a given energy. Whereas for cosmic ray showers the ratio of their Cherenkov light yield to the light yield of γ-ray showers goes to zero as the energy is decreased.

Because of the existence of the EGRET sources and the desire to track their behavior in the region below 100 GeV and the desire to see the most distant blazars (or other sources) unhindered by intergalactic absorption effects it is clear that one of the dominant themes of future Cherenkov detectors will be lower thresholds. Thus the electron background will become an increasingly important limiting factor in the drive for increased sensitivity.

6. Prospects for TeV γ-ray Astronomy

By many standards, TeV γ-ray astronomy is healthy and vigorously growing. Established groups continue to upgrade their detectors, new projects are off the drawing boards and under construction, and there are dreams for even better instruments (cf. the proceedings of the 1994 Tokyo workshop) some of which may reach to energies as low as 10 GeV and thereby overlap with the satellite observations.

The reasons for the health of the field are easy to see: the increasing awareness of significant scientific questions that may be addressed by the field, the discoveries of EGRET and BATSE at lower energies, and the excitement that comes from the opening and flowering of a new field. In this context it is well to remember that the health of TeV astronomy may be dependent on the health of the lower energy γ-ray astronomies conducted in space. We have clearly benefited from discoveries from MeV and GeV γ-ray astronomies; we will return the favor.

In addition to these developments in the atmospheric Cherenkov technique, a new project, MILAGRO (Barwick et al. 1993), promises to provide an alternative and possible competitive technique. The wide-angle, high-duty factor character of MILAGRO lends itself to exploration of episodic phenomena. A major question is whether or not this type of detector can approach the sensitivity already achieved by Cherenkov detectors. If so, then such detectors may eventually provide the all-sky survey that TeV γ-ray astronomy needs.

I would like to thank Paul Ray and Tom Prince for their help with this manuscript.

References

Aharonian, F.: 1994, Proceedings of the workshop, *Towards a Major Atmospheric Cherenkov Detector – III* (Tokyo), in press
Aharonian, F.: 1995, these proceedings
Aharonian, F. A. & Atoyan, A. M.: 1995, *Astroparticle Physics*, in press
Antonucci, R.: 1993, *Ann Rev Astr & Ap* **31**, 473
Ballon, P. *et al.*: 1993, Proceedings of the workshop, *Towards a Major Atmospheric Cherenkov Detector – II* (Calgary), 121
Barwick, S. *et al.*: 1993, Proceedings of the 23rd ICRC (Calgary) **1**, 436

Blandford, R. & Levinson, A.: 1995,*ApJ*, in press
Bloom, S. D. & Marscher, A. P.: 1993, AIP Conf. Proc. 280 (Compton Symposium), ed. N. Gehrels, 578
Bowden, C. C. G. *et al.*: 1992, *Astroparticle Physics* **1**, 47
Bregman, J.N.: 1992, in Physics of AGN, ed. W. Duschl & S. Wagner (Springer-Verlag), 699
Dermer, C. D., Schlickeiser, R., &Mastichiasdis, A.: 1992, *A&A* **256**, L27
de Jager, O. C. *et al.*: 1995, in preparation.
Dowthwaite, J. C. *et al.*: 1984, *ApJ* **286**, L35
Drury, L. O'C., Aharonian, F. A. & Völk, H. J.: 1994, *A&A* **287**, 959
Dwek, E. & Slavin, J.: 1994, *ApJ* **426**, 696
Esposito, J. *et al.*: 1995, in preparation
Fichtel, C. A. *et al.*: 1994, *ApJ Suppl* **94**, 551
Fontaine, G. R.: 1993, Proceedings of the workshop, *Towards a Major Atmospheric Cherenkov Detector – II* (Calgary), 12
Goret, Ph.: 1993, Proceedings of the workshop, *Towards a Major Atmospheric Cherenkov Detector – II* (Calgary), 25
Goret, Ph.: 1994, Proceedings of the workshop, *Towards a Major Atmospheric Cherenkov Detector – III* (Tokyo), in press
Gould, R. J. & Shréder, G.: 1966, *Phys Rev Lett* **16**, 252
Gould, R. J. & Shréder, G.: 1967, *Phys Rev* **155**, 1404
Harding, A. K.: 1991, *Science* **251**, 1033
Hillas, A. M.: 1995, these proceedings
Jelley, J. V.: 1966, *Phys Rev Lett* **16**, 479
Kanbach, G.: 1994, private communication
Kanbach, G.: 1995, these proceedings
Kerrick, A. D.: 1995, *ApJ* **438**, L59
Kifune, T. *et al.*: 1995, *ApJ* **438**, L91
Kifune, T.: 1995, these proceedings
Krennrich, F.: 1995, private communication
Lind, K. R. & Blandford, R. D.: 1985, *ApJ* **295**, 358
Longair, M. S.: 1992, *High Energy Astrophysics, Vol 2, 2nd Ed.*, Cambridge, England, 275
MacMinn, D. & Primack, J: 1995, *ApJ* , submitted
Mannheim, K.: 1993, *A&A* **269**, 67
Marscher, A.P.: 1993, in Astrophysical Jets, STScI Symposium Series, 6, ed. D. Burgarella, M. Livio, & C. O'Dea (Cambridge U. Press), 73.
Meintjes, P. J. *et al.*: 1994, *ApJ* **434**, 292
Mirabel, I. F. & Rodriguez, L. F.: 1994, *Nature* **371**, 46
Nikishov, A. I.: 1962, *Soviet Physics JEPT* **14**, 393
Padovani, P. & Urry, C. M.: 1992, *ApJ* **387**, 449
Punch, M. *et al.*: 1992, *Nature* **160**, 477
Raubenheimer, B. C. 1993, Proceedings of the workshop, *Towards a Major Atmospheric Cherenkov Detector – II* (Calgary), 3.
Reynolds, J. & Jauncey, D.: 1994, *IAU Circ.* 6063
Sikora, M., Begelman, M. C., & Rees, M. J.: 1994, *ApJ* **421**, 153
Salamon, M. H., Stecker, F. W.,& de Jager, O. C.: 1994, *ApJ* **423**, L1
Sommer, M. *et al.*: 1994, *ApJ* **422**, L63
Sturner, J. A. & Dermer, C. D.: 1994, *A&A* in press
Stecker, F. W., de Jager, O. C., & Salamon, M. H.: 1992, *ApJ* **415**, L71
Takashashi, T. *et al.*: 1994, IAU circ. No. 5995
Tanimori, T. *et al.*: 1994, *ApJ* **429**, L61
Vacanti, G. *et al.*: 1991, *ApJ* **377**, 467
von Montigny, C. *et al.*: 1995, *ApJ* in press

THE ATMOSPHERIC ČERENKOV TECHNIQUE IN THE SEARCH FOR PBH

N. A. PORTER
University College Dublin, Belfield, Dublin 4, Ireland

Abstract. Searches for primordial black holes are reviewed.

1. Introduction

A search for exploding Primordial Black Holes was suggested by Page and Hawking (Page and Hawking, 1976). A detection, or even a restrictive upper limit, could give information on the missing mass problem, chaos in the early Universe, or the quantum theory of gravity.

Two different models of the explosion were suggested by Page and Hawking: the composite particle, or Hagedorn model, in which 10^{34} ergs of gamma rays at about 200 MeV are released in the final 0.1 microseconds; and the elementary particle model based on quarks, in which 10^{30} gamma rays at 5×10^{12} eV are emitted in 0.1 seconds. This latter model has been modified recently by Halzen et al (Halzen et al., 1991), who have calculated a detailed spectrum and time distribution. Smaller numbers of gamma rays are found than in the earlier work: for a time interval of 0.1 seconds, 7×10^{28} gamma rays above 100 GeV, with a flat spectrum extending upwards in energy, are predicted. The mean energy is in the region of 2 TeV.

2. Observations

a) Hagedorn model. The Čerenkov technique can be used with this model, even though the individual gamma rays are only about 200 MeV. A disc of them will impinge on the atmosphere and form small electron-photon cascades by pair production. These in turn will produce Čerenkov light with an angular spread of about 15 degrees (full width), about the direction of the primary gamma rays. This disc of light will hit the ground over a period of about 0.1 microseconds, so that a system with this integration time will collect light originating from different gamma rays in a single pulse. The disc of particles is spread over a very large area, so that to distinguish the signal from that caused by TeV cosmic rays, two or more detectors separated by hundreds of kilometers, can be used in coincidence. This will also considerably reduce effects due to lightning or satellites.

Such an experiment was carried out in 1976 at the Whipple Observatory in Arizona and the White Sands missile site in New Mexico (Porter and Weekes, 1978). The separation was 400 kilometers, and the detectors, the 10

Space Science Reviews **75**: 67–69, 1996.

meter reflector at Mount Hopkins and the 9 meter solar furnace at White Sands. Each had clusters of photomultipliers at the focus to give full fields of view 5 degrees, and the detectors were connected by a telephone line. Local high precision clocks were also used at each station to register coincidences. No coincidences were found. A simple calculation led to an upper limit for PBH explosions of 0.04 /cubic parsec year. The Egret group has recently found a limit of 0.05 explosions /cubic parsec year (Fichtel et al., 1995). Since these numbers are upper limits, the close agreement is presumably coincidental. It would appear that the Čerenkov technique is competitive with other techniques for the Hagedorn model, at least for determining upper limits.

b) Elementary Particle model. This can also be tested with the Čerenkov technique or with observations of particles at ground level. The Halzen et al modification of the original model predicts gamma rays ranging from 100 GeV to about 100 TeV, on time scales of seconds. The lower energy region is accessible with the Čerenkov technique. Original limits of 4.9×10^7 explosions/cubic parsec year (Porter and Weekes, 1979) have been improved several times, the best value being 1.8×10^6 (Nolan et al., 1989).

At higher gamma ray energies, in the region 100 TeV, separated particle detectors can be used to detect small air showers (Fegan et al., 1978). The lowest upper limit obtained is due to the Cygnus Collaboration, (Alexandreas et al., 1993) and is 6.7×10^5 explosions/cubic parsec year. Assuming this limit, and accepting Page and Hawking's suggestion that PBH are clustered in galaxies by a factor of 10^6, the total maximum number of PBH in the Universe can be calculated. A value of about 10^{40} is found. If each has a mass of 10^{15}g, the upper limit to the mass in PBH in the Universe is 10^{55}g. This is comparable with the total mass of the Universe, so the present upper limits with the Elementary Particle model do not constrain the amount of missing matter in the form of PBH. A reduction in the limits is required, preferably by a factor of about 1000. This may be attainable with a dedicated detector and an intelligent triggering system, but will not be easy. If the Hagedorn model were valid, then existing limits would be sufficient to restrict the amount of missing matter in the form of PBH, to about 10^{-7} of the total mass of the Universe. This model is not currently favoured by particle physicists. An intermediate model between the two extremes may eventually be accepted.

References

Alexandreas, D.E. et al. In *Proc. 23nd I.C.R.C. (Calgary)*, page 428, 1993.
Fegan, D.J. et al. *Nature*, 271:731, 1978.
Fichtel, C.E. et al. *Astrophys. J.*, 1995. In press.
Halzen, F. et al. *Nature*, 353:807, 1991.
Nolan, K. et al. In *Proc. 21st I.C.R.C. (Adelaide)*, volume 2, page 150, 1989.

Page, D.N. and Hawking, S.W. *Astrophys. J.*, 206:1, 1976.
Porter, N.A. and Weekes, T.C.. *Mon. Not. Roy. Astr. Soc.*, 183:205, 1978.
Porter, N.A. and Weekes, T.C.. *Nature*, 277:199, 1979.

GAMMA-HADRON SEPARATION AT HIGH ENERGIES IN VIEW OF RECENT CROSS SECTION MEASUREMENTS

GERD SCHATZ

Forschungszentrum Karlsruhe P. O. Box 3640, D-76021 Karlsruhe, Germany

Abstract. Recent measurements of the cross section for hadron photoproduction at HERA are reviewed and their importance for gamma ray astronomy is outlined. Various methods of distinguishing gamma and hadron primaries among cosmic rays in the air shower range are then discussed in view of our (insufficient) knowledge of the relevant hadron hadron cross sections. Theoretical model calculations indicate that the fraction of proton primaries mimicking gamma rays may be as high as one in a thousand.

Key words: Gamma rays, extensive air showers, cross sections

1. Introduction

The most important recent experimental cross section result is the measurement of photoproduction in gamma proton interactions at HERA. The result and its implications for gamma ray astronomy will be discussed in section 2. Hadron hadron cross sections are of relevance to the field because they influence the discrimination of gamma ray from the large excess of primary hadrons. Unfortunately, not much progress has been made in the experimental determination of the relevant data. The situation is described in section 3 where various procedures to discriminate hadrons from gamma rays are investigated.

2. Hadron Photoproduction

Photoproduction of hadrons on protons have been measured by the two experiments at the HERA electron proton collider, H1 (Ahmed et al. 1993) and ZEUS (Derrick et al. 1992). Both results are in excellent agreement and displayed in figure 1 (from DESY 1993). The figure also shows some theoretical predictions compatible with the new results. The enigmatic results of the Kiel group (Samorski and Stamm 1983a) who had observed point source events from Cygnus X-3 with a muon content compatible with primary hadrons (Samorski and Stamm 1983b) had triggered theoretical attempts to achieve a much steeper increase of photoproduction with energy. This possibility now seems to be ruled out by the new data. Figure 2 shows the gamma air cross section calculated from these new results versus the labo-

Space Science Reviews **75**: 71–82, 1996.

ratory energy E_{lab} of the gamma which is related to the cms energy $W_{\gamma p}$ of figure 1 by

$$E_{lab} = \frac{W_{\gamma p}^2}{2mc^2} \tag{1}$$

Here m is the proton rest mass. So up to a laboratory energy of c. 20 TeV we can now be sure that the photoproduction cross section does not change dramatically. Figure 2 also includes the Bethe-Heitler pair production cross section on air and extrapolations of two of the theoretical curves in figure 1. The one labelled ALLM is based on Abramowicz et al. (1991) who have given an explicit formula for the cross section. The other one is a simple-minded geometrical extrapolation of the corresponding curve in figure 1. The latter procedure is questionable, at best (Werner 1994). It is included in figure 2 in order to show that even a very bold extrapolation of the experimental values available today yields photoproduction cross sections well below conventional pair production way up into the PeV region. So the attempts to explain the observations of the Kiel group (Samorski and Stamm 1983a,b) by a steep increase of hadron photoproduction now seem to be ruled out. On the other hand, we are on firm ground now in the energy range where gamma rays have been identified today, i. e. up to c. 15 TeV (Baillon et al. 1993).

3. Hadronic Cross Sections and Gamma Hadron Discrimination

Among the most difficult problems in very high energy gamma ray astronomy is the suppression of the hadron 'background' which is larger than the signal by at least a factor of 10^3. (I wonder whether something which represents more than 99.9 total should not be called foreground rather than background.) In the energy range of air Cherenkov telescopes, i. e. up to few tens of TeV, this can be achieved by analyzing the distribution of light in the image plane. The procedures employed are described by Hillas (1994; cf. also Weekes et al. 1989 and Vacanti et al. 1991) in these proceedings and will not be discussed here. At higher energies the small fields of view and the limitations of duty cycle render the application of Cherenkov telescopes difficult. We will therefore concentrate on the prospects of gamma hadron discrimination in the range of extensive air showers.

Showers initiated by gamma rays and by hadrons in the atmosphere differ considerably in average, especially concerning their muon and hadron content. The difficulties of discrimination lie in the large intensity ratio and the enormous fluctuations of all observable quantities in extensive air showers, especially those initiated by primary protons. In any space of measurable shower parameters gamma showers will sit on the tail of a huge hadron (mainly proton) peak. The possibilities of discrimination therefore depend

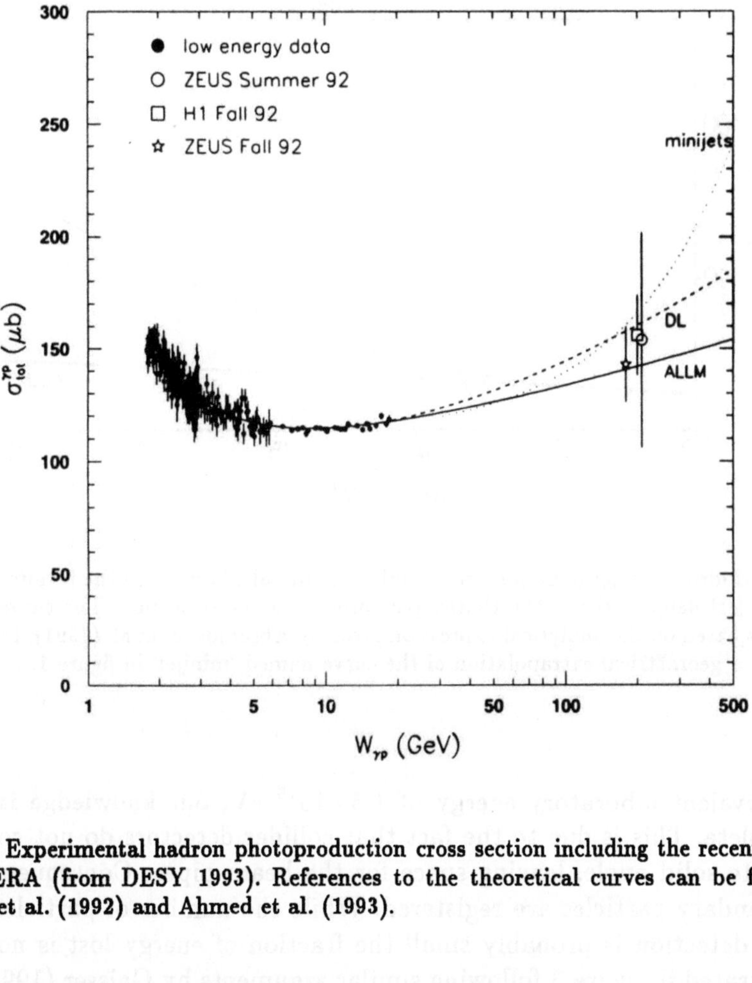

Fig. 1. Experimental hadron photoproduction cross section including the recent results from HERA (from DESY 1993). References to the theoretical curves can be found in Derrick et al. (1992) and Ahmed et al. (1993).

on the frequency and behaviour of very untypical hadron events and hence on a very small fraction of the total hadron cross section. In this respect protons are the most important primary hadron species to be considered because the fluctuations of showers initiated by heavier nuclei are expected to be smaller (Schatz et al. 1994a). In the following sections we will first discuss the (deplorable) status of the relevant experimental data and then study various procedures employed to discriminate between gamma and hadron showers.

3.1. Hadronic Cross Sections of Relevance to Gamma Ray Astronomy

Unfortunately, little progress has been made recently in our knowledge of cross sections required for a comprehensive understanding of extensive air showers. Even in the energy range accessible to operating colliders, up to

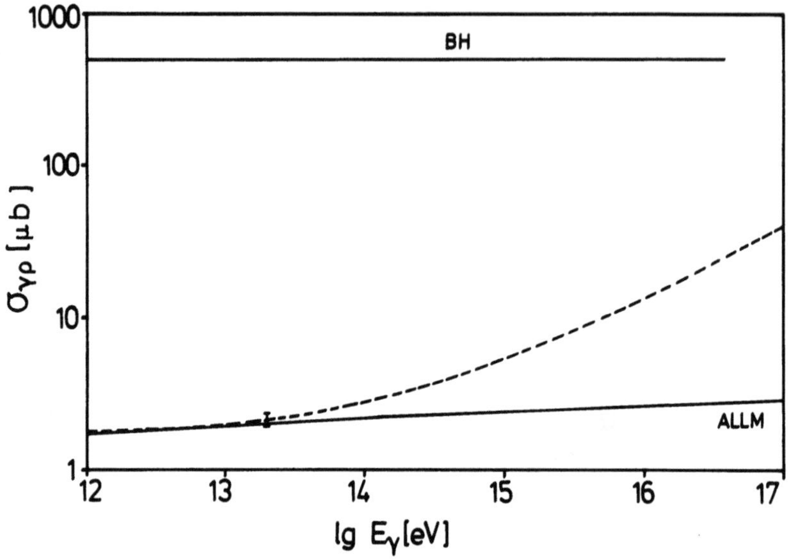

Fig. 2. Hadron photoproduction cross section on air calculated from the results shown in figure 1. BH denotes the Bethe-Heitler pair production cross section. The curve labelled ALLM is based on the analytical expression given by Abramowicz et al. (1991), the dashed curve on a geometrical extrapolation of the curve named 'minijet' in figure 1.

an equivalent laboratory energy of $1.6 \cdot 10^{15}$ eV, our knowledge is rather incomplete. This is due to the fact that collider detectors do not cover the complete solid angle, leaving space for the beam pipes. Consequently, not all secondary particles are registered. While the number of particles which escape detection is probably small the fraction of energy lost is not. This is illustrated in figure 3 following similar arguments by Gaisser (1992). The experimental points, taken from Alner et al. (1986), represent the pseudorapidity distribution of all secondary particles. (Pseudorapidity is defined by $\eta = \ln \tan \frac{\theta}{2}$ and hence equivalent to angle, so the data give essentially an angular distribution.) When we assume for simplicity that all secondary particles are pions and have the average transverse momentum we can calculate their energy in the c.m.s. and laboratory systems. The lower curve, referred to the right hand scale, then gives the fraction of the primary energy carried by all secondary particles below the corresponding value of pseudorapidity. While the assumptions made do not hold strictly the procedure should nevertheless give a reasonable estimate. As can be seen from figure 3 only ca. 30 % of the laboratory energy is carried by the measured particles. So we have to conclude that the bulk of the laboratory energy is in (probably few) particles not seen by the collider detectors and these particles determine the development of extensive air showers and hence the observable quantities to

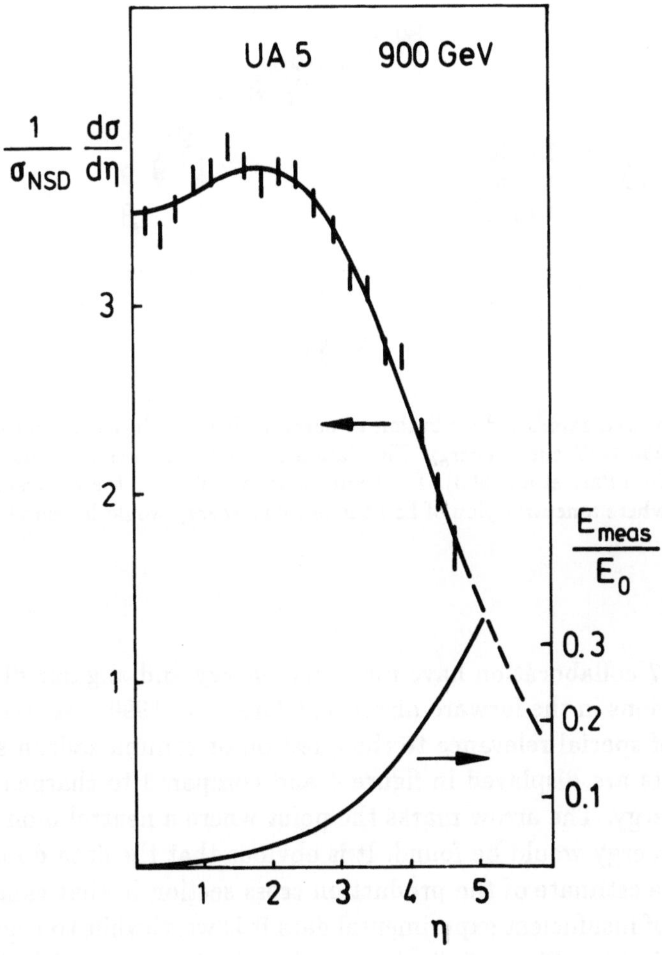

Fig. 3. Upper curve and experimental points: pseudorapidity distribution of all charged secondary particles in proton- antiproton collisions at 900 GeV c.m.s. energy (from Alner et al. 1986). The smooth curve through the data points represents an analytical interpolation. Lower curve: fraction of laboratory energy carried by all observed particles of lower pseudorapidity.

a large extent. More complete measurements only exist from the ISR which corresponds to a laboratory energy of 2 TeV.

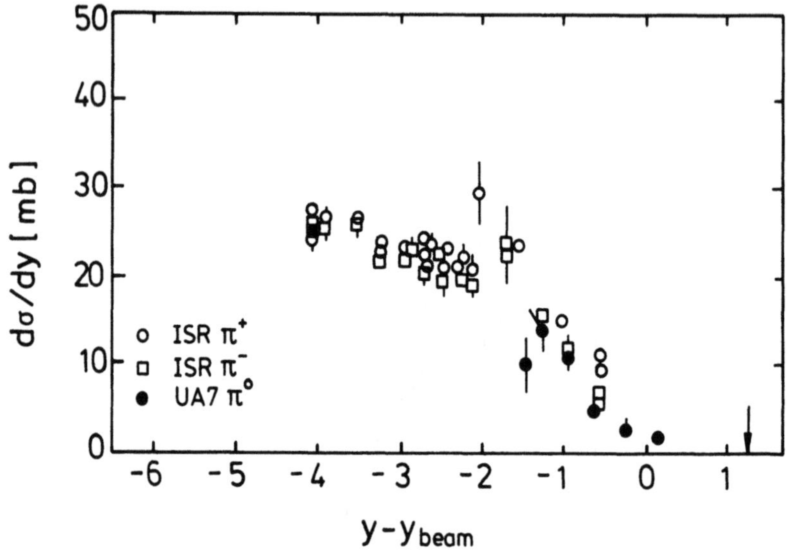

Fig. 4. Measured rapidity distribution of neutral pions produced in proton-antiproton collisions at 630 GeV c.m.s. energy. The data are compared to charged pion production at the ISR (from Pare et al. 1990). The beam rapidity is 6.5 and the vertical arrow points to the value where a neutral pion of half the primary energy would be found if emitted at zero degree.

The UA7 collaboration have measured energy and angular distributions of neutral pions in the forward direction (Pare et al. 1990). Neutral pions are obviously of special relevance to the question of gamma hadron separation. Their results are displayed in figure 4 and compared to charged pion data at lower energy. The arrow marks the point where a neutral pion of half the primary's energy would be found. It is obvious that the data do not allow a quantitative estimate of the production cross section in that range.

In view of insufficient experimental data it is worthwhile to consider theoretical predictions. Figure 5 displays results of calculations with the VENUS model. This model allows to calculate collisions of hadrons with nuclei and yields, for each event, a list of secondary particles with their energies and directions. The model has been described in detail elsewhere (Werner 1993). For a large number of collisions of 17.56 TeV protons and of 10 TeV charged pions with nitrogen nuclei the energies of electromagnetic secondary particles were summed up. Figure 5 shows the cumulative distribution of the number of events versus the electromagnetic energy fraction. According to this model more than three quarters of the primary energy are converted into electromagnetic particles for one event in a thousand. For pions the electromagnetic energy fraction is considerably larger. This can be understood as a consequence of charge exchange from charged into neutral pions

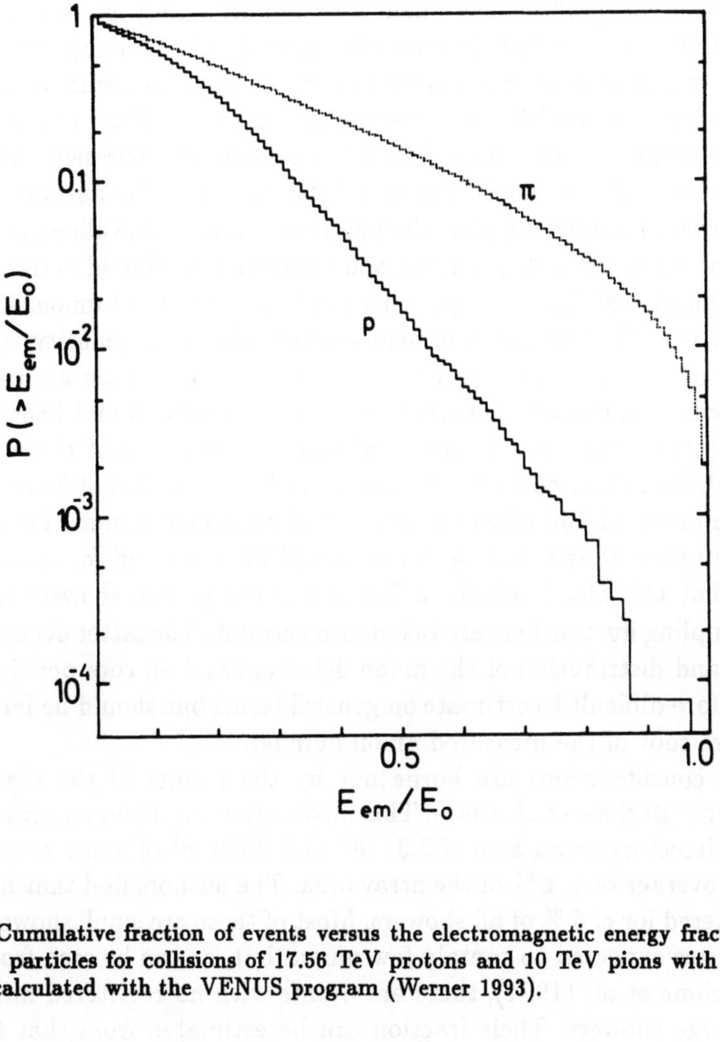

Fig. 5. Cumulative fraction of events versus the electromagnetic energy fraction of all secondary particles for collisions of 17.56 TeV protons and 10 TeV pions with 14N. The data are calculated with the VENUS program (Werner 1993).

(Schatz et al. 1994b). Hence one might expect one primary proton among a thousand to mimic a gamma ray according this model.

3.2. Discrimination via Muons

The ratio of muons to electrons are expected to be higher in extensive air showers initiated by hadrons than in gamma ray showers by more than an order of magnitude. This is a direct consequence of the small cross section for hadron photoproduction as compared to pair production (cf. figure 2). In principle, the measurement of the electron to muon ratio is straightforward. The limitation of this procedure lies in the low density of muons which requires large detector areas, and in the large fluctuation of the ratio for proton induced showers.

Gaisser et al. (1991) have made a systematic study of this possibility. An example of their results is shown in figure 6. The two uppermost curves show the muon number distributions for showers of the same range of electron numbers initiated by protons and gamma rays. They are normalized to the same number of events. The other two curves represent the gamma distributions scaled down by factors of 103 and 104. The authors conclude that, at least at higher energies, the muon to electron ratio allows to discriminate between primary gamma rays and hadrons. One should notice, though, that the abcissa of figure 6 represents the total number of muons above the threshold of 1 GeV in the complete shower. This is a quantity which has never been measured because the muons are spread out on an area several hundreds of meters in diameter. The measurement would hence require huge detector areas. In a large experiment at most 1 % of the muons are registered. So the numbers on the abscissa of figure 6 should be reduced by at least a factor of 100 when the number of registered muons is considered. The figure then shows that no muon would be measured for most gamma showers but this also holds for a fraction of the proton showers especially when sampling fluctuations are taken into account. The latter depend on the position and distribution of the muon detectors and on core position. They are therefore difficult to estimate on general terms but should be larger than the square root of the measured muon number.

These considerations are borne out by the results of the CASA-MIA experiment (Borione et al. 1994). This installation combines an array of 1024 electron detectors on an area of $2.3 \cdot 10^5$ with 2560 m^2 of muon counters, i.e. a muon coverage of c. 1 % of the array area. The authors find that no muons are registered for c. 5 % of all showers. Most of these are small showers which are of course expected to contain few muons but, as can be seen from figure 28 of Borione et al. (1994), there are events with no registered muons also among large showers. Their fraction can be estimated from that figure to approximately one event in 2000.

3.3. Discrimination via Muons and Hadrons

The low muon content of gamma showers reflects the fact that few hadrons are produced by primary gamma rays. So these showers are also expected to contain few hadrons on observation level. Nikolsky et al. (1987) have exploited this feature to discriminate between primary gammas and hadrons in the results of the Tien Shan experimental installations (Erlykin et al. 1965). They selected showers poor in muons as well as in hadrons. It is interesting to recall their figures: A total of 2445 showers centered on the Tien Shan hadron calorimeter and with an electron number above $3.2 \cdot 10^5$ were analysed in more detail. Of these, 80 turned out to be muon poor,

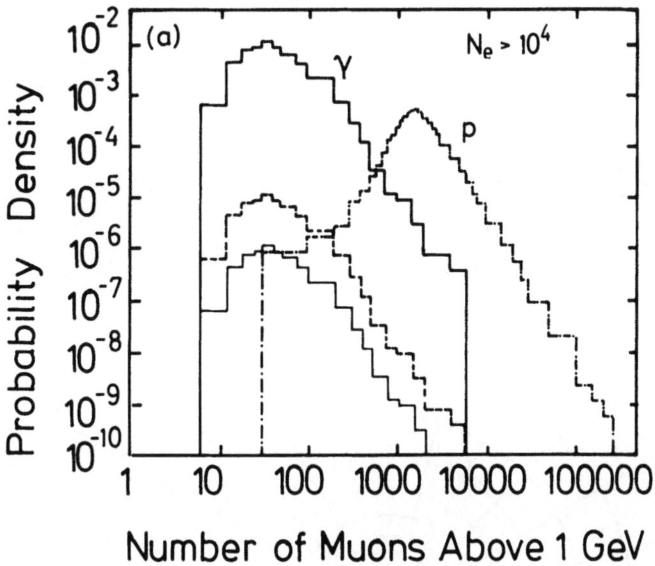

Fig. 6. Distributions of the number of muons for showers initiated by primary gamma rays and protons (from Gaisser et al., 1991). The two uppermost curves are normalized to the same number of events. The dashed and thin lines represent the distributions for gamma events scaled down by factors of 103 and 104, respectively.

according to their definition. But only 7 of these were also poor in hadrons. The authors conclude eventually that one primary cosmic ray particle in a thousand is a gamma quantum above 1 PeV.

This value is larger than most theories of gamma ray production and propagation in the Galaxy would predict. In view of the above discussion of the results displayed in figure 5 this conclusion does not seem compelling. There it was shown that, according to the VENUS model, a very large fraction of the energy of a primary proton is converted into electromagnetic particles for ca. one event in a thousand. Interestingly, this is precisely the level of the showers identified as gamma induced by the Tien Shan group.

3.4. Discrimination via Spatial Distribution of Particles in the Shower Core

An interesting new approach to discriminate between hadron and gamma primaries has recently been proposed by Kempa (1994a). He has studied the distribution of individual particles (mostly electrons) in the cores of extensive air showers, comparing simulations with data obtained by the Kiel neon

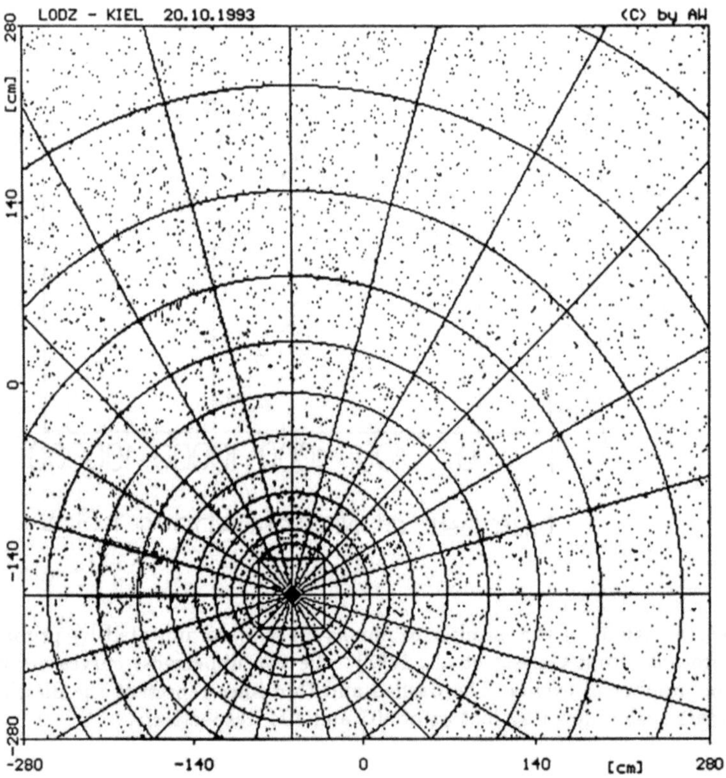

Fig. 7. Distribution of particles (mainly electrons) in the core of an extensive air shower measured with the Kiel neon hodoscope (Bagge et al., 1979). The lines show the grid used for studying the shape of the lateral distribution.

hodoscope (Bagge et al. 1979). Figure 7 gives an example of a measured distribution (Kempa 1994b). The method of analysis uses fractal and multifactorial moments and essentially analyses the azimuthal symmetry at various distances from the shower axis which can be determined to a high precision. Simulations show that gamma, proton and heavy primaries exhibit different distributions which find their counterparts in experiment.

Kempa (1994b; cf. also Dudarewicz et al., 1994) finds that one of 1044 showers analyzed displays features of a primary gamma ray. Incidentally, this is of the same order of magnitude as the rate of events attributed to gamma rays by the Tien Shan group (Nikolsky et al. 1987).

4. Summary and Conclusions

The cross section for the production of hadrons in gamma proton collisions has recently been measured at the HERA electron proton collider (Derrick et al., 1992; Ahmed et al., 1993). The results correspond to an equivalent laboratory energy of 20 TeV and exhibit a very slow increase with energy as expected by conventional theory (Abramowicz et al. 1991). There is no indication of a dramatic increase of hadron photoproduction up to this limit. All reasonable extrapolations to higher energy point to cross section values much smaller than conventional pair production well into the PeV region.

The identification of gammas in primary cosmic rays is hampered by the large excess of hadrons. In the energy range of the air Cherenkov technique very efficient methods of hadron suppression have been developed (Weekes et al. 1989; Vacanti et al. 1991; Hillas 1994). The same cannot be said of the region of extensive air showers (above a few hundred TeV). Distinguishing gamma from hadron primaries will be ultimately limited by hadron induced events where a large fraction of the primary energy is converted into electromagnetic particles during the first few collisions in the atmosphere. Our present knowledge of the relevant hadron nucleus cross sections is insufficient for a reliable estimate of this limit from experimental data. This holds even in the energy range accessible to colliders presently in operation. Theoretical models predict that the fraction of primary protons simulating gamma rays may be of order one in a thousand. This is also the level at which some experiments seem to have found gamma-like extensive air showers. An empirical way of distinguishing gammas from hadrons in the energy range of extensive air showers could possibly be developed if the distribution of gamma ray sources in the sky is nonuniform (in a similar way as the Crab nebula served as a standard candle to develop and refine the methods of gamma hadron discrimination in the Cherenkov range).

References

Abramowicz, H. *et al.*: 1991, *Phys. Lett. B* **269**, 465

Ahmed, T. *et al.*: 1993, *Phys. Lett. B* **299**, 374

Alner, G. J. *et al.*: 1986, *Z. Physik C* **33**, 1

Bagge, E. R., Samorski, M., and Stamm, W.: 1979, *Proceedings 16th Internat. Cosmic Ray Conf.*, Kyoto 1979, vol. 13, p. 260

Baillon, P., *et al.*: 1993, *Astropart. Phys.* **1**, 341

Borione, A., *et al.*: 1994, *Nucl. Instr. Meth. A* **346**, 329

Derrick, M. *et al.*: 1992, *Phys. Lett. B* **293**, 465

DESY 1993: *Wissenschaftl. Jahresbericht 1993* , p. 64

Dudarewicz, A., *et al.*: 1994, *Proceedings VIIIth ISVHECRI*, Tokyo 1994; Waseda University, Y. Fujimoto, S. Hasegawa *et al.* Eds, p. 616

Erlykin, A. D., *et al.*: 1965, *Proceedings 9th Internat. Cosmic Ray Conf.*, London 1965, vol. 2, 731

Gaisser, T. K. *et al.*: 1991, *Phys. Rev. D* **43**, 314

Gaisser, T. K.: 1992, *Proceedings VIIth ISVHECRI*, Ann Arbor 1992; AIP Conf. Proceedings 276, L. W. Jones Ed., p. 397
Hillas, A. M.: 1994, these proceedings
Kempa, J.: 1994a, *J. Phys. G: Nucl. Part. Phys.* **20**, 215
Kempa, J.: 1994b, private communication
Nikolsky, S. I., J. N. Stamenov, S. Z. Ushev: 1987, *J. Phys. G: Nucl. Phys.* **13**, 883
Pare, E., *et al.*: 1990, *Phys. Lett. B* **242**, 531
Samorski, M., W. Stamm: 1983a, *Astrophys. J.* **268**, L17
Samorski, M., W. Stamm: 1983b, *Proceedings 18th Internat. Cosmic Ray Conf.*, Bangalore 1983, vol. 11, p. 244
Schatz, G., *et al.*: 1994a, *J. Phys. G: Nucl. Part. Phys.* **20**, 1267
Schatz, G., *et al.*: 1994b, *Proceedings VIIIth ISVHECRI*, Tokyo 1994; Waseda University, Y. Fujimoto, S. Hasegawa *et al.* Eds, p. 525
Vacanti, G., *et al.*: 1991, *Astrophys. J.* **377**, 467
Weekes, T. C., *et al.*: 1989, *Astrophys. J.* **342**, 379
Werner, K.: 1993, *Phys. Rep.* **232**, 87
Werner, K.: 1994, private communication

FUTURE INSTRUMENTAL CAPABILITIES IN THE ENERGY RANGE OF NUCLEAR TRANSITIONS

P. von BALLMOOS
Centre d'Etude Spatiale des Rayonnements, Toulouse, France

Abstract. Gamma-ray lines are the fingerprints of nuclear transitions, carrying the memory of high energy processes in the universe. Setting out from what is presently known about line emission in gamma-ray astronomy, requirements for future telescopes are outlined. The inventory of observed line features shows that sources with a wide range of angular and spectral extent have to be handled: the scientific objectives for gamma-ray spectroscopy are spanning from compact objects as broad class annihilators, over long-lived galactic radioisotopes with hotspots in the degree-range to the extremely extended galactic disk and bulge emission of the narrow e^-e^+ line.
The instrumental categories which can be identified in the energy range of nuclear astrophysics have their origins in the different concepts of light itself: geometrical optics is the base of coded aperture systems - these methods will continue to yield adequate performances in the near future. Beyond this, focusing telescopes and Compton telescopes, based on wave- and quantum- optics respectively, may be capable to further push the limits of resolution and sensitivity.

Key words: gamma-ray astronomy – instrumentation

1. Introduction

With the Compton Gamma-Ray Observatory and the GRANAT/SIGMA telescope in orbit, high energy astrophysics has for the first time ever the opportunity to study all the facets of celestial gamma-rays side by side. In the energy range of nuclear transitions, the gamma-ray sky has been mapped on various angular scales and a large number of new gamma-ray sources has been discovered (see reviews by Gehrels et al. 1994 and Paul 1995). Maybe the most important scientific potential of the present generation high energy telescopes is their extremely broad common energy coverage. Together with the operating X-ray and high energy gamma-ray telescopes, a quasi-continuous coverage has opened the possibility for multi-wavelength studies of continuum spectra spanning from the keV- to the GeV-range. Figure 1 indicates that the present area once might be called the "golden age of gamma-ray astronomy": never before has the high-energy sky been examined so thoroughly and over such a broad energy range. For many of the high energy sources, multi-wavelength studies may be the only way that leads to an understanding of their complex source mechanisms. A model case is the spectrum of the quasar 3C273 (Lichti et al. 1995) that has been observed - partly simultaneously - from radio to gamma-ray energies. Nevertheless, the gamma-ray telescopes on the Compton Gamma Ray Observatory and on GRANAT also have raised new astrophysical questions and highlighted those which remain unanswered. The future goals of gamma-ray astronomy

Space Science Reviews **75**: 83–96, 1996.

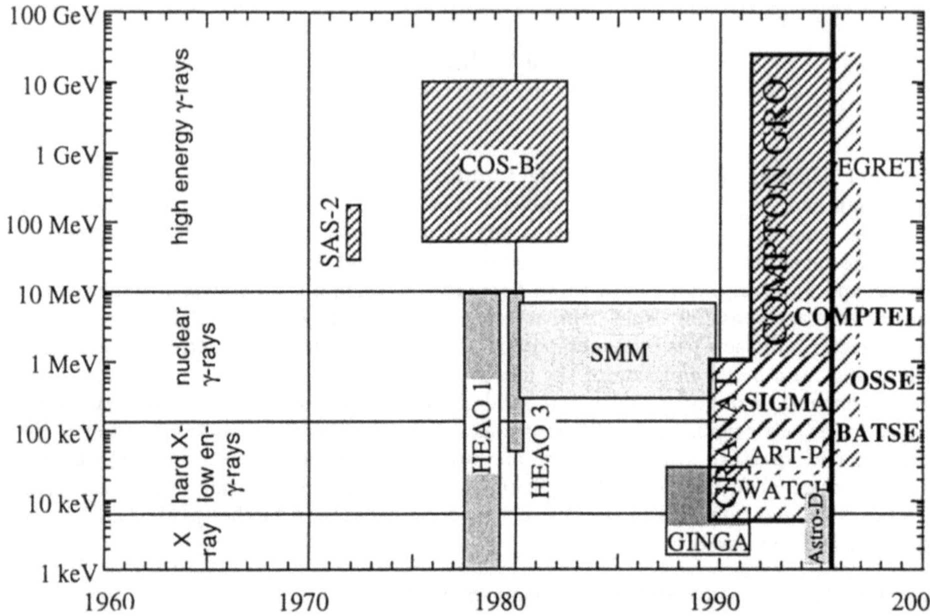

Fig. 1. The "golden age of gamma-ray astronomy" ? Never before the high-energy sky has been examined so thoroughly and over such a broad energy range.

must be defined in this context. The progress of SIGMA, BATSE, OSSE and COMPTEL is based primarily on skymaps, excellent timing analysis, and moderate to fair spectral resolution. The observations have revealed certain aspects of the morphology of celestial gamma-ray emitters, yet, the physical processes at work are often only poorly understood. Frequently, the observed spectra do not sufficiently constrain the emission mechanisms: explaining a relatively simple, featureless continuum with a complex multiparameter model can be ambiguous, moreover, different components may blend into one another, each of them can depend on various physical parameters in the emitting region.

In many ways, the present situation resembles the situation of optical astronomy in the beginning of the last century: Back then, the available observational data mainly consisted in images, starcounts, variability's, and colors. Astrophysics was born when G. Kirchhof and R. Bunsen (here at Heidelberg in 1859) developed the spectral analysis and explained the Frauenhofer-lines in the spectrum of the sun. The exploration of atomic and molecular lines has since turned out to be the most powerful tool for the study of the physical conditions in celestial sources.

Up to today, only little advantage has been taken from the fundamental astrophysical information contained in gamma-ray lines. The reason for this is the modest energy resolution of the existing instruments (typically

$\Delta E/E \sim 10\%$). High resolution spectroscopy will therefore be one of the major goals of the next generation of space borne gamma-ray telescopes.

The scientific potential of nuclear gamma-ray astronomy is outlined in section 2. In section 3, different categories of instruments for spectroscopy in the low and medium gamma-ray channel are presented: coded aperture systems, Compton telescopes and diffraction lens telescopes. The three techniques actually reflect our current perception of light itself - they are based on the principles of geometrical optics, quantum optics and wave optics, respectively.

2. The promise of nuclear gamma-ray astronomy

Ultraviolet, visible and infrared-spectroscopy has long become an extremely valuable technique in astronomy. While optical lines reflect structural changes in the electron shell of atoms, caused by collisions with energies of the order of 10 eV ($T \sim 10^3$K), transition between discrete nuclear energy levels imply MeV energies ($T \sim 10^7$ *to* 10^9K), corresponding to the binding energy of nucleons. Collision energies of this order are characteristic for the temperatures inside stars, particles accelerated by electromagnetic fields in solar flares, or interactions of cosmic ray particles with the interstellar medium. Gamma-ray lines are the fingerprints of nuclear transitions, carrying the memory of high energy processes in the universe. High resolution gamma-ray spectroscopy is a unique tool to identify the presence of exited nuclei, quantitatively determine their abundance's, and provide insight into the physical conditions of the source regions (temperature, density, gravitational or cosmological energy shifts).

The scientific potential of gamma-ray spectroscopy includes better understanding of the chemical evolution of the Galaxy by mapping the sites of recent nucleosynthetic activity (novae, SN, massive stars). The observation of supernovae in the galaxies of our local cluster and closeby superclusters (Virgo cluster) will verify our models of nucleosynthesis in massive stars. Spectral features in solar flares, cyclotron lines in gamma-ray bursts and pulsars are among the further challenges of gamma-ray spectroscopy.

An inventory of observed gamma-ray lines is presented in table I, it reflects the current status of nuclear gamma-ray astronomy. Note that most of the knowledge on gamma-ray lines is based on observations made with scintillation detectors that typically feature energy resolutions of 10%. Nevertheless, these elementary spectroscopic measurements already indicate the tremendous potential of gamma-ray line emission. Performance requirements for future spectroscopy missions become conspicuous when the measured/-anticipated line fluxes are compared to the expected angular scales. Figure 2 indicates that emissions with a wide range of angular and spectral extent are expected, varying in intensity by several orders of magnitude. The sci-

TABLE I
Observed celestial gamma-ray line features

Physical Process	Energy [keV]	Source	Flux ph $cm^{-2}s^{-1}$	Instrument, Detector
Nuclear deexcitation				
$^{56}Fe(p,p',\gamma)$	847	Solar flares	≤ 0.05	SMM, NaI
$^{24}Mg\ (p,p',\gamma)$	1369	Solar flares	≤ 0.08	SMM, NaI
$^{20}Ne\ (p,p',\gamma)$	1634	Solar flares	≤ 0.1	SMM, NaI
$^{28}Si\ (p,p',\gamma)$	1779	Solar flares	≤ 0.08	SMM, NaI
$^{20}12C(p,p',\gamma)$	4439	Solar flares	≤ 0.1	SMM, NaI
	4439	Orion Comp.	$\leq 5{\cdot}10^{-5}$	COMPTEL, sc)
$^{16}O(p,p',\gamma)$	6129	Solar flares	≤ 0.1	SMM, NaI
	6129	Orion Comp.	$\leq 5{\cdot}10^{-5}$	COMPTEL, sc)
Radioactive decay				
$^{56}Co(EC,\gamma)^{56}Fe$	847, 1238	SN 1987A	$\approx 10^{-3}$ b	Ge det's and
	2598			var. scint's,
	847,1238	SN 1991T	$5{\cdot}10^{-5}$ b	COMPTEL, sc)
$^{57}Co(EC,\gamma)^{57}Fe$	122, 136	SN 1987A	$\approx 10^{-4}$	OSSE, NaI-CsI
$^{44}Ti(EC)^{44}Sc(\beta^{+}\gamma)$	1157	Cas A SNR	$7{\cdot}10^{-5}$	COMPTEL, sc)
$^{26}Al(\beta^{+}\gamma)^{26}Mg$	1809	gal. plane	$4{\cdot}10^{-4}$	COMPTEL, sc)
	1809	Vela SNR	$1\text{-}6{\cdot}10^{-5}$	COMPTEL, sc)
$e^{-}e^{+}$ Annihilation				
	511	Gal. bulge	$1.7{\cdot}10^{-3}$	OSSE, NaI-CsI
	511	Gal. disk	$4.5{\cdot}10^{-4}$	OSSE, NaI-CsI
	480±120 a)	1E 1740-29 c)	$1.3{\cdot}10^{-2}$	SIGMA, NaI
	511	Solar Flares	≤ 0.1	SMM, NaI scint.
	479±18 a)	Nova Muscae	$6.3{\cdot}10^{-3}$	SIGMA, NaI
	400-500 a)	Bursts ? c)	≤ 70	various scint.
	440±10 a)	Crab PSR ? c)	$3{\cdot}10^{-4}$	FIGARO, NaI
Neutron Capture				
$^{1}H(n,\gamma)^{2}H$	2223	Solar flares	≤ 1	SMM, NaI
$^{56}Fe(n,\gamma)^{57}Fe$	5947 a)	6/10/1974 tr.	$1.5\ 10^{-2}$	SMM, NaI
Cyclotron Lines				
	20-58	Hercules X-1	$\leq 3{\cdot}10^{-3}$	various scint.

legend: sc) liquid scintillator/NaI scintillator; a) Redshifted line; b) Maximum emission, c) detection uncertain, the feature has not been confirmed yet

entific objectives for gamma-ray spectroscopy are spanning from compact sources as broad class annihilators, over long-lived galactic radioisotopes with hotspots in the degree-range to the extremely extended galactic disk and bulge emission of the narrow $e^{-}e^{+}$ line.

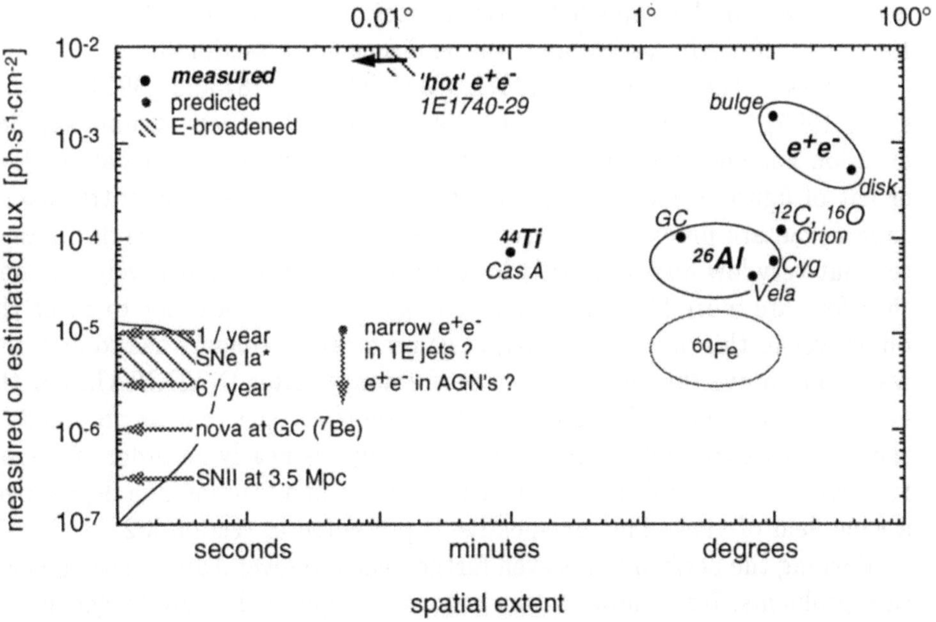

Fig. 2. Future spectroscopy missions have to face emissions with a wide range of angular extent, and with intensities different by several orders of magnitude. The anticipated flux for extragalactic SNe of type 1 has been deduced from the COMPTEL detection of SN1991T (Morris et al. 1995) and by scaling its ^{56}Co 847 keV gamma-ray flux with the optical peak magnitude of observed SNIa.

According to our present view of celestial gamma-ray sources in the energy range of nuclear transitions, narrow lines seem to be generally emitted from extended distributions while broad lines tend to be radiated by compact sources.

3. Instruments for gamma-ray spectroscopy

A natural first objective for a future gamma-ray spectroscopy mission is the mapping of the relatively intense sources (on the upper right of figure 2) which are typically emitting 10^{-4} ph cm^{-2}s^{-1} to a few 10^{-6} ph cm^{-2}s^{-1}. Candidate sources of this intensity are mostly galactic and include the sites of recent nucleosynthesis, regions of e^-e^+ annihilation and clouds where nuclear de-excitation by energetic particles takes place. Some of them might appear as extended structures: either because of their apparently diffuse origin - such as the narrow 511 keV line (Purcell et al. 1994) - or because they are relatively closeby as the nucleosynthesis sites in the local spiral arm (del Rio et al. 1994). An instrument that is adequate for this kind of objectives should provide a sensitivity of several 10^{-6} ph cm^{-2}s^{-1}, a wide field of view and an angular resolution in the degree range. Such a profile corresponds to the expected performances of the SPECTROMETER on ESAs INTEGRAL mission and will be discussed in section 3.1.

Besides of the supernovae 1987A (Matz et al. 1988) and 1991T (Morris et al. 1995) the evidence for point like sources of narrow gamma-ray line emission has been mostly implicit at this point. Yet, in the area at the lower left of figure 2 various objects like e.g. galactic novae and extragalactic supernovae are predicted. These sources should have small angular diameters but very low fluxes - mostly because such objects are relatively rare and therefore are more likely to occur at large distances. In order to cover the objectives in this area, experimental gamma-ray astronomy has to find new ways to improve the observational performances after INTEGRAL. For the energies above 500 keV, solid state Compton telescopes (e.g. as proposed by the ATHENA concept) could achieve sensitivities nearly an order of magnitude lower, while providing angular resolutions of a fraction of a degree over a wide field of view. This perspective is presented in section 3.2.

Pushing the performances even further leads to several apparently unsolvable problems. Up to now, better instruments generally were bigger instruments. Yet, physical limits for the size of spaceborne instruments are being attained today. Moreover, achieving sensitivities better than 10^{-6} ph cm^{-2}s^{-1} and resolutions better than fractions of a degree seems to be principally impossible with the presently practiced instrumental concepts: even larger collection areas are synonymous with larger detectors and thus higher background noise. The ensuing mass/sensitivity dilemma can ultimately only be overcome by a radically new type of gamma-ray telescope. In section 3.3, a possible way out of this impasse is presented: a focusing gamma-ray telescope using a tunable crystal diffraction lens.

3.1. Spectroscopy with coded aperture systems

The underlying concept of coded aperture instruments is geometrical optics since the source photons are considered as traveling on rectilinear paths only. Light is passing through open elements of an otherwise opaque aperture. This aperture system - consisting of masks or collimators - modulates the signal which then reaches the detection plane designed to discern shadow patterns of some kind*. Since the photoelectric effect is the predominant mode of gamma-ray interaction in medium- and high-Z materials, these systems are well adapted to the low energy gamma-ray channel. Pinhole cam-

* The basic phenomenon of coded aperture images has first been described by Aristotle in 4th century BC. Noticing that specks of sunlight under a large tree always are rounded, he uses the concept of light traveling along straigt lines through 'peep-holes' to interpret the images (Aristotle, προβλήματα). Aristotle also describes and interprets the observation of the crescent-shaped images of the sun during an eclipses produced under different 'masks': eg. using the small spaces between the leaves of plane trees, the small holes of a sieve or the holes made by crossing the fingers of one hand over the other.

eras, rastering collimators, coded masks and modulation collimators belong to this category of instruments.

In gamma-ray astronomy, two main classes of coded aperture systems are used, according to whether the signal is encoded by spatial modulation (coded masks telescopes) or by temporal modulation (modulation collimators). These two types stand for a whole spectrum of devices mixing the basic concepts of spatial and temporal modulation.

Typical coded mask telescopes consist of a planar array of opaque and transparent elements located in front of a position sensitive detection plane (PSD). Scanning collimators or rotating modulating collimators use moving aperture parts in order to time modulate the intensity at a detection plane that does not need to provide spatial resolution. Many of these instruments modulate the signal from an extended region of the sky and thus are able to observe a number of sources simultaneously or quasi simultaneously. The telescopes featuring this multiplex advantage over rastering collimators (such as pure 'on-off' techniques) are sometimes called multiplexing devices.

Our present understanding of the low energy (<1 MeV) gamma-ray sky has been acquired mainly by coded aperture systems. Operational satellite instruments of this type include SIGMA, OSSE and BATSE (Paul et al. 1991, Johnson et al. 1993, and Fishman et al. 1989). While the detection plane of all three of them is based on scintillators, the aperture systems used are a coded mask, a collimator, and the earth, respectively.

High resolution spectroscopy with a coded mask instrument will be available by the beginning of the next decade with ESA's INTEGRAL mission. The INTEGRAL SPECTROMETER described below corresponds to the ESA proposal by a consortium between the CESR (Toulouse), MPE (Garching), CEA (Saclay), IFCTR (Milan), IEEC (Barcelona), Univ. of Louvain, Univ. of Valencia, Univ.of Birmingham and US Institutes. Figure 3a) gives a cutaway view of the SPECTROMETER. The instrument features a compact array of high purity Germanium detectors entirely shielded by a BGO/CSI anticoincidence system (for details see e.g. von Ballmoos et al. 1995).

The Ge detector assembly consists of 19 high purity n-type germanium detectors cooled to 85K via pulsed tubes supplied by two cryogenic compressors. With a total geometric detection area of $\approx$ 500 cm^2 the detection plane is the largest one that could be fitted into the overall constraints of the INTEGRAL mission. The aperture subsystem features a tungsten coded mask (HURA with 127 elements) 171 cm above the detector. The coded mask together with the detector plane defines a fully coded field of view of $16° \times 16°$.

The narrow line sensitivity shown in Fig. 4 has been obtained by completely modeling the SPECTROMETER for the radiation environment conditions expected outside the magnetosphere. The sensitivity is based on models for the instrument efficiency (GEANT) and background in ^{70}Ge detectors

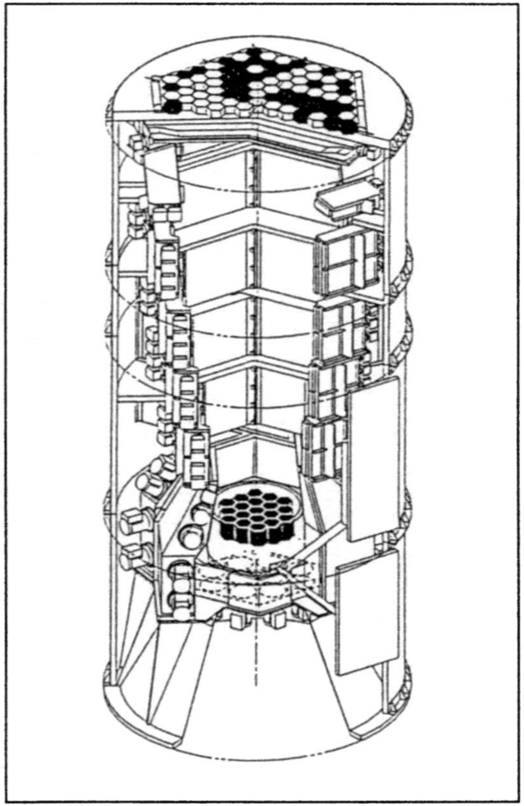

Fig. 3 a) Cutaway view of the proposed INTEGRAL SPECTROMETER.

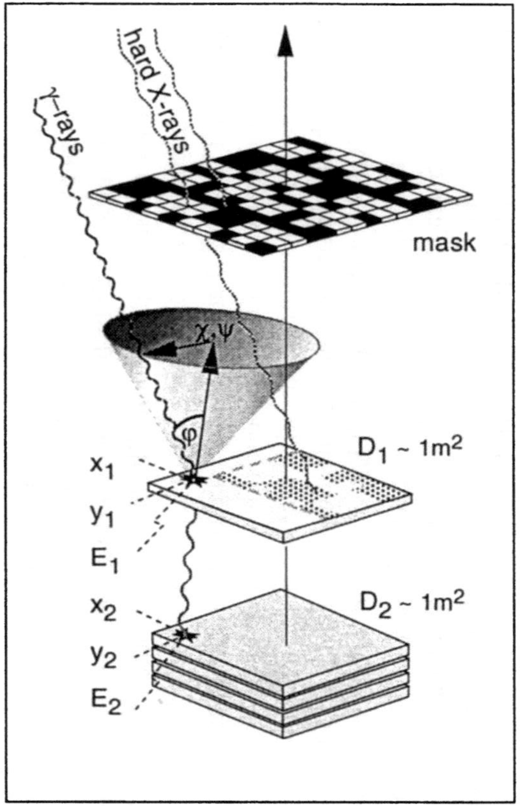

Fig. 3 a) Schematic view of a solid state Compton telescope (ATHENA type) with a coded aperture mask added for hard X-ray imaging.

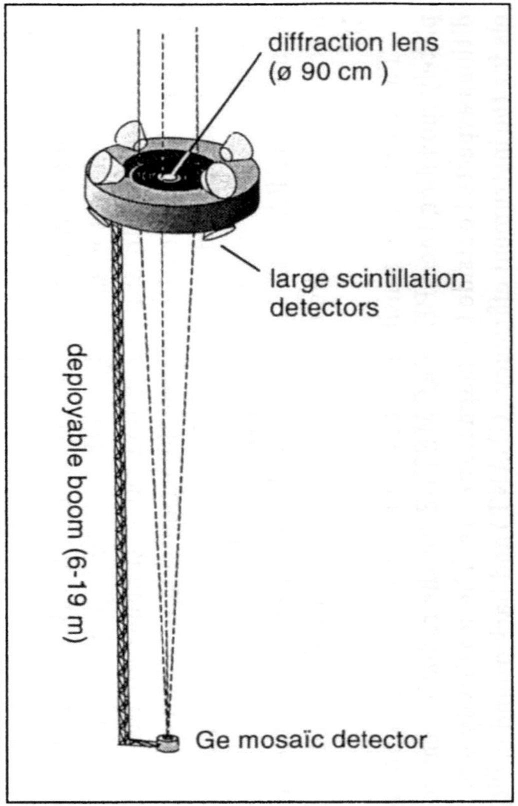

Fig. 3 c) "artists view" of a space borne crystal diffraction telescope. As a counterpart to the extremely directional lens telescope, a full sky monitor (large area scintillators) could complement the payload.

(Naya et al. 1995). The anticipated performance for narrow line spectroscopy is characterized by an energy resolution in the parts per thousand range (typically 2 keV at 1 MeV), an angular resolution of order 2°, and a sensitivity 2-5·10^{-6} ph $cm^{-2}s^{-1}$ in the energy range relevant for nuclear astrophysics. The 511 keV line sensitivity is of the order of 2·10^{-5} ph $cm^{-2}s^{-1}$.

3.2. Spectroscopy with Compton telescopes

While the total cross section has its minimum at a few MeV, the energy range between several hundred keV and about 30 MeV is dominated by the Compton effect. At these energies coded aperture systems run into several problems: the efficiency of the modulation decreases and various background problems related to heavy shielding become increasingly prohibiting. Hence, making use of the quantum nature of the photon interactions is an inevitable choice for instruments at medium and high energy gamma-rays. The idea to make use of the Compton effect instead of fighting it with thicker shielding and modulators has stimulated several groups and resulted in a distinct class of imaging instruments. With COMPTEL on the Gamma-ray Observatory, the concept has definitely proven its unique potential for MeV gamma-ray astronomy (Schönfelder et al. 1993).

In a Compton telescope the incident gamma-ray is identified by successive interactions in the two detector layers D_1 and D_2. Compton scattering in D_1 is favored when low Z material is chosen. Total absorption of the scattered photon in D_2 can be expected when high Z materials are used. Since the $D_1 \wedge D_2$ coincidence condition discriminates against most of the internal nβ background, a Comptom telescope has an extremely low background. At the same time, this coincidence condition causes the relatively low detection efficiency. In order to further reduce the background due to upward scattered events, the time-of-flight between the two detectors is usually measured in Compton telescopes. Neutron interactions can be identified by measuring the pulse shape of the scintillation pulse.

From the interaction location in D_1 and D_2, the direction χ, ψ of the scattered gamma-ray is obtained; the energy deposits in D_1 and D_2 are E_1 and E_2 respectively. The scatter direction χ, ψ, together with the amounts of energy deposited in the two interactions can be used to reconstruct the arrival direction of the gamma-ray. The Compton equation allows to express the scatter angle φ as a function of the energy-deposits E_1 and E_2:

$$cos\overline{\varphi} \quad = \quad 1 - \frac{m_e c^2}{E_2} + \frac{m_e c^2}{E_1 + E_2} \qquad \text{with } m_e c^2 = 511\text{keV}$$

If E_1 and E_2 are measured without systematic errors ($E_{tot} = E_1 + E_2 = E_\gamma$), the derived scatter angle $\overline{\varphi}$ equals the true Compton scatter angle φ. The arrival direction of the incident gamma-ray can then be confined to lie on a cone-mantle with axis χ, ψ and opening angle $\overline{\varphi}$ (see fig 3b). The projection of this cone results in a circle on the sky that is generally called the 'event circle'. No information on the azimuth of the incident photon on the circle can be deduced from the measured parameters. The azimuthal information is lost since state-of-the-art Compton telescopes can not measure the direction of the recoil electron in D_1. Consequently, genuine direct imaging is impossible for present Compton telescopes: the image reconstruction process is handicapped by missing information on the scatter angle of the incident photon.

The ATHENA concept (Kurfess et al. 1994, Kröger et al. 1995, Johnson et al. 1995) is based on D_1 and D_2 layers using Germanium planar strip detectors providing 2-3 keV spectral resolution and spatial resolution of $\sim$2 mm. Such detectors, typically 5 cm $\times$ 5 cm $\times$ 1 cm, are available today and might be integrated into large panels in the future. The ATHENA concept foresees a 1 m^2 D_1 layer consisting of one panel, and a 1 m^2 D_2 layer of four panels, each panel containing 400 Ge strip detectors. Figure 3b) shows a schematic diagram of a solid state Compton telescope combined with a coded mask for low energy gamma-rays. In the Compton mode (300 keV-10 MeV) such an instrument can achieve angular resolutions of 0.2°- 0.3° within a field of view of typically one steradian, and a narrow line sensitivity of several times 10^{-7} ph $cm^{-2}s^{-1}$ above 1 MeV (see fig. 4).

3.3. Spectroscopy with diffraction lens telescopes

Being used to accept that it is 'impossible to reflect or refract gamma rays', high energy astronomy has never considered to make use of the wave-quality of gamma-ray photons. Since focusing was out of question, the detector area of gamma-ray telescopes has always been identical with the photon-collecting area. Consequently, gamma-ray astronomy has come to a mass-sensitivity impasse, since achieving a fair signal to noise ratio is a conflict of trading in signal ($\sim$ detector-surface) against background ($\sim$ detector-volume).

A new type of gamma-ray telescope featuring a Laue-diffraction lens makes use of the periodic nature of light and its interference with the periodic structure of a crystal. The focusing gamma-ray telescope (Smither et al. 1995, von Ballmoos et al. 1995) is designed to collect gamma-ray line photons on a large effective area and focus them onto a Germanium detector matrix with a small equivalent volume for background noise.

The instrument has first been proposed as a balloon-borne telescope with the lens tuned to diffract 511 keV photons only (von Ballmoos and Smither 1994). Such a configuration makes possible the study of galactic 'micro-quasars' and other broad class annihilators in the light of e^-e^+ annihilation during a balloon flight. As a balloon-borne instrument it can provide high energy- and high angular resolution (2 keV, 15 arc sec, respectively) combined with an excellent sensitivity ($\sim 3 \cdot 10^{-5}$ ph $cm^{-2}s^{-1}$ at 511 keV). The

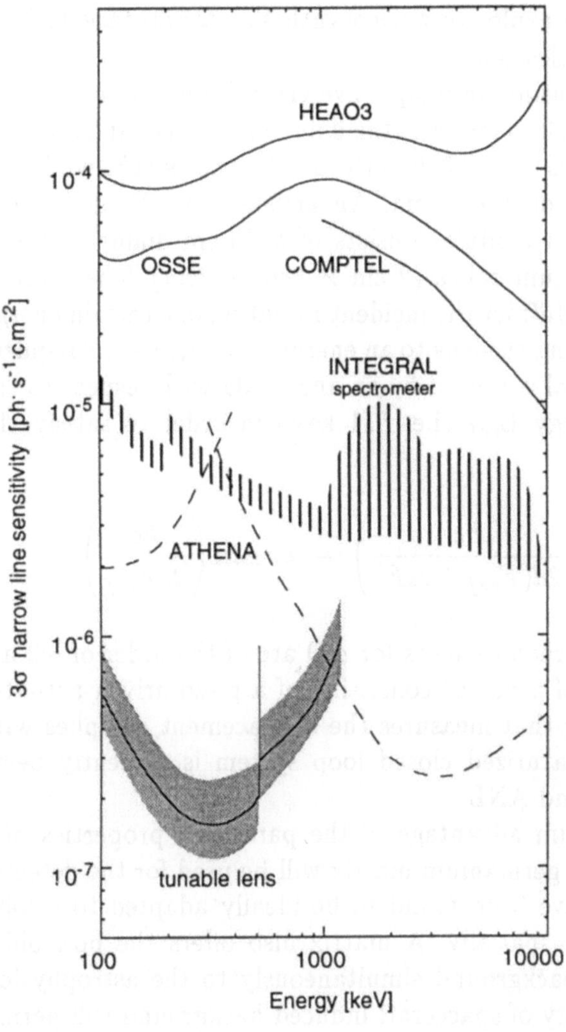

Fig. 4. Comparison of 3 σ narrow line sensitivities: HEAO3 (achieved sensitivity). For all other instruments $T_{obs}=10^6$ sec. The hatched area outlines the sensitivity of the INTEGRAL SPECTROMETER. Lower and upper limits are for an optimistic(o)/pessimistic(p) model: o) including the multiplexing advantage (i.e. sensitivity for any galactcic source during a 10^7 sec survey of the inner galaxy); p) including a β^+ background component. The shaded area indicates different background estimates for the diffraction lens telescope.

performances of this Ge-lens/Ge-matrix system have been verified in summer 1994 during laboratory measurements with a ground based prototype (Naya et al. 1995).

Ultimately however, the concept should be put to use in space where longer exposures and steady pointing would result in outstanding sensitivities. Yet, as a satellite instrument, a monochromatic lens would clearly be a handicap since its scientific objectives are too exclusive - already e.g. the possible annihilation line of most extragalactic sources (AGNs, quasars) would be inaccessible because of cosmological redshift.

The need of a energy tuning has lead to a 'tunable crystal diffraction lens' (von Ballmoos et al. 1995) that permits observation of any identified source at any selected line-energy in the cardinal range of nuclear transitions (200 keV - 1300 keV). The 'sequential' operation mode resulting from such a concept makes the sites of explosive nucleosynthesis natural targets for a tunable crystal diffraction telescope.

A space borne telescope using an adaptative crystal diffraction lens will consist of three modules: the lens module, the detector module, and a boom. Optimally the lens module is located directly on the spacecraft, while the detector module is perched on the boom. An artists view of the concept is shown in Fig 3c. The lens module consists of a 90 cm diameter frame accommodating 700 germanium cubes (2 cm $\times$ 2cm $\times$ 2cm). The crystals are oriented so that they all diffract the incident radiation of a certain energy to the same focal point. Tuning the lens to an energy $E = E_{ref} + \Delta E$ requires that each of the single crystal is rocked by an angle $\Delta\Theta$ with respect to the position of a reference energy E_{ref} (i.e. 511 keV) in order to satisfy the Bragg condition anew.

$$\Delta\Theta = arcsin\left(\frac{hc}{2d(E_{ref} + \Delta E)}\right) - arcsin\left(\frac{hc}{2dE_{ref}}\right)$$

The precision/repeatability requirements for $\Delta\Theta$ are of the order of $\sim$2 arc seconds. The performance of a device consisting of a piezo-driven actuator and an Eddy-current sensor that measures the displacement complies with these requirements. A miniaturized closed loop system is presently being built and tested at CESR and ANL.

In order to take maximum advantage of the particular properties of a focused gamma ray beam, a germanium matrix will be used for the detector module. These detectors have been found to be ideally adapted to resolve the beam energetically and spatially. A matrix also offers the possibility to monitor the remaining background simultaneously to the astrophysical observation. The low intensity of spacecraft induced background will permit the use of a very light anticoincidence shield.

Since the focal length of the lens is increasing with energy, a retractable boom will be used to the vary the distance between the lens and the detector module. An energy of 200 keV and 1240 keV respectively corresponds to a change in focal length of 3 m to 20 m - for the 511 keV positron annihilation line the distance lens-detector is 8.3 m. Booms have been used for space-borne gamma-ray spectroscopy on Apollo 15 (NaI detector) and on Mars-Observer (Ge detector). Deploying the detector on a boom instead of the diffraction lens has several striking advantages: The mechanical requirements on the mast rigidity are less severe since a Germanium detector array is small and lightweight and thus easy to handle on a boom; moreover, twists and bends of even up to a few cms are tolerable, as the focal spot can wander around on the detector array without significant loss of sensitivity. On the other hand, the stringent requirements for the pointing of the lens (typically $\sim 5''$) can be satisfied on board the pointed and stabilized spacecraft. Finally, moving the detector away from the spacecraft should reduce the background by up to an order of magnitude.

The imaging capabilities of a crystal diffraction telescope are defined by its beamwidth which is identical to the field of view of the lens: for compact sources discrete pointings of the object will be an appropriate observation mode, while extended structures as for example the jets of galactic micro-quasars will be scanned with the narrow beam. The field of view depends on the angular width of the mosaic structure of the diffracting crystals - it will typically be $\sim 15''$ FWHM. Below 1 MeV, the narrow line sensitivity is expected to be a few 10^{-7} ph $cm^{-2}s^{-1}$ (see figure 4).

4. Conclusion

With the spectrometer on ESA's INTEGRAL mission, a coded mask instrument for gamma-ray spectroscopy will be available by the beginning of the next decade. The foremost objectives of the INTEGRAL spectrometer will be the mapping of gamma-ray line sources emitting 10^{-4} ph $cm^{-2}s^{-1}$ to a few times 10^{-6} ph $cm^{-2}s^{-1}$. Many of these potential sources will be galactic. Some of them might appear as extended structures - either because of their truly diffuse origin or because they are relatively closeby as the nucleosynthesis sites in the local spiral arm. An energy resolution of $E/\Delta E \approx 500$, a wide field of view and a mid-scale angular resolution make the INTEGRAL spectrometer adequate for such objectives.

On a more distant horizon, experimental gamma-ray astronomy has to find ways to further push the limits of resolution and sensitivity: At energies above one MeV, Compton telescopes can provide angular resolutions of fractions of degrees and sensitivities of a several 10^{-7} ph $cm^{-2}s^{-1}$. At energies below the MeV, tunable crystal diffraction telescopes can achieve sensitivities of a few times 10^{-7} ph $cm^{-2}s^{-1}$ and angular resolutions of the order of $15''$.

References

Aristotle, 'Problemata Physica' book XV, problems 6 and 11, 4th cent. BC, The Loeb Classical Library, (W. Heineman:London), 1936
del Rio E. et al., 199, AIP Conf. Proc. 304, p. 171, (ed. Fichtel C. et al., New York)
Fishman G. et al., 1989 in Johnson, W.N. (Ed), Proc GRO Science Workshop, p.239
Gehrels N. et al., 1994, ApJ Sup. S., 92, 351
INTEGRAL ESA SCI(93)1, ESA phase A study report, April 1993
Johnson W.N. et al., 1993, ApJ Sup. S., 86, 693
Johnson, W.N. et al., 1995, proc. of workshop 'Imaging in High En. Astron.', Capri 1994
Kurfess J.D. et al., NASA proposal for new mission concepts in Astrophysics, NRA 94-OSS-15, 1994
Kröger R.A. et al., 1995, proc. of workshop 'Imaging in High En. Astron.', Capri 1994
Lichti G. et al., Astron Astrophys 1995, in press
Matz S.M. et al., 1988, Nature 331, 416

Morris D. et al. 1995, proc. 17th Texas Symposium on Relativistic Astrophysics
Murphy R.J. et al. 1990, ApJ ,358, 290
Naya J., et al., 1995a, submitted to NIM
Naya J. et al. , 1995b, proc. of workshop 'Imaging in High En. Astron.', Capri 1994
Paul J. et al., 1991, Adv.Space Res., 11 (6), 289
Paul J. et al., 1995, in Signore M. et al (Ed), The Gamma Ray Sky with Compton GRO and SIGMA, p. 15
Purcell W.R et al., 1994, AIP Conf. Proc. 304, p. 403, (ed. Fichtel C. et al., New York)
Schönfelder, V., Hirner, A. and Schneider, K.: 1973, Nucl.Instrum. Meth., 107, 385.
Smither R.K. et al., 1995, proc. of workshop 'Imaging in High En. Astron.', Capri 1994
von Ballmoos P. and Smither R.K. 1994, ApJ Sup. S., 92, 663
von Ballmoos P. et al, 1995, proc. of workshop 'Imaging in High En. Astron.', Capri 1994
von Ballmoos P., 1995, proc. 17th Texas Symposium on Relativistic Astrophysics

AGATE: A HIGH ENERGY GAMMA-RAY TELESCOPE USING DRIFT CHAMBERS

R. MUKHERJEE, B.L. DINGUS and J.A. ESPOSITO
Universities Space Research Association
NASA/GSFC, Code 662, Greenbelt, MD 20771

and

D.L. BERTSCH, R. CUDDAPAH, C.E. FICHTEL, R.C. HARTMAN, S.D. HUNTER and D.J. THOMPSON
NASA, Goddard Space Flight Center, Code 662
Greenbelt, MD 20771

Abstract. The exciting results from the highly successful Energetic Gamma-Ray Experiment Telescope (EGRET) instrument on the Compton Gamma-Ray Observatory (CGRO) has contributed significantly to increasing our understanding of high energy gamma-ray astronomy. A follow-on mission to EGRET is needed to continue these scientific advances as well as to address the several new scientific questions raised by EGRET. Here we describe the work being done on the development of the Advanced Gamma-Ray Astronomy Telescope Experiment (AGATE), visualized as the successor to EGRET. In order to achieve the scientific goals, AGATE will have higher sensitivity than EGRET in the energy range 30 MeV to 30 GeV, larger effective area, better angular resolution, and an extended low and high energy range. In its design, AGATE will follow the tradition of the earlier gamma-ray telescopes, SAS-2, COS B, and EGRET, and will have the same four basic components of an anticoincidence system, directional coincidence system, track imaging, and energy measurement systems. However, due to its much larger size, AGATE will use drift chambers as its track imaging system rather than the spark chambers used by EGRET. Drift chambers are an obvious choice as they have less deadtime per event, better spatial resolution, and are relatively easy and inexpensive to build. Drift chambers have low power requirements, so that many layers of drift chambers can be included. To test the feasibility of using drift chambers, we have constructed a prototype instrument consisting of a stack of sixteen 1/2m $\times$ 1/2m drift chambers and have measured the spatial resolution using atmospheric muons. The results on the drift chamber performance in the laboratory are presented here.

Key words:
gamma-ray detectors – pair production – drift chambers

1. Introduction

The field of high energy gamma-ray astronomy has benefited vastly from the broad set of important new results obtained from the Energetic Gamma-Ray Experiment Telescope (EGRET) instrument on the Compton Gamma-Ray Observatory (CGRO). Prior to the launch of CGRO only a handful of astrophysical sources were known to emit photons of energy > 100 MeV. However, in the first 18 months (Phase 1) of its operation, EGRET has detected 124 sources above 100 MeV (Fichtel et al. 1994). The detected sources include solar flares, the Large Magellanic Cloud, several active galactic nuclei

Space Science Reviews **75**: 97–108, 1996.

(AGN), pulsars, and gamma-ray bursts. The EGRET observations, in the energy range $\sim$ 30 MeV to 30 GeV, have led to a better understanding of many high-energy astrophysical phenomena and significantly changed our ideas of others. We understand better the nature of the mysterious gamma-ray bursts, the collimation of jets of relativistic plasma in AGN, the radiation and plasma processes in the super-strong magnetic fields of pulsars as well as in the magnetic loops of the sun, and the physics of relativistic particle acceleration in these and other exotic astrophysical environments such as the remnants of supernova explosions.

The EGRET instrument is expected to be operational for another 2 or 3 years. The results from EGRET are the primary motivation for continuing the development of instruments for the study of the high energy gamma-ray sky. While EGRET has enhanced our understanding of several astrophysical phenomena, it has at the same time raised numerous questions. A large number of sources in the first EGRET catalog (Fichtel et al. 1994), particularly within 10° of the Galactic Plane, are unidentified with sources at lower frequencies. These sources could be identified with gamma-ray emitting blazars. However, because of the density of confusing Galactic radio sources, it is desirable to have improved positions for the unidentified gamma-ray sources. This can be achieved only with an instrument with better angular resolution and effective area than EGRET. Such improvement is also required in the study of AGN to determine why so many blazars prominent at radio frequencies and with flat radio spectra are not detected by EGRET, and also to enable the detection of at least the closer blazars over the full range of their variations. Some other questions raised by EGRET are the frequency of AGN flares, the rapidity of variation of the gamma-ray flux and whether the gamma-ray variability is correlated with flares at lower energies.

Gamma-ray bursts have been another surprising addition to the detections made by EGRET. EGRET has imaged about five bursts and gamma-rays have been observed with energies well above a GeV, more than two orders of magnitude higher than previously observed. In the burst of 17 February 1994 (Hurley et al. 1995) the highest energy gamma-ray detected was 18 GeV and the emission extended for over an hour or more beyond the sub-MeV emission. This raises the question of what fraction of gamma-ray bursts emit GeV photons, and what is the light curve of the high energy radiation. A detector with greater sensitivity will enable the detection of more bursts and more photons per burst. It will be possible to learn if the high-energy photons seen in some bursts are a property of only the strongest (which are probably the closest) bursts, or an indication of a separate class of bursts. Similarly, there are many unanswered questions in the field of gamma-ray pulsars, solar flares, Galactic and extragalactic diffuse radiation, molecular clouds, and normal galaxies.

In order to achieve the scientific advances mentioned above, several aspects of the design of a future high-energy gamma-ray instrument need to be improved. These aspects are: effective area, angular resolution (these two aspects determine the sensitivity), extended low and high-energy range, and field-of-view (FOV). Table 1 summarizes the improvements needed for the new instrument from the point of view of the science goals (Dingus et al. 1995).

TABLE I
Science driven characteristics of the successor to EGRET.

REQUIREMENT	WHY	HOW
Increased Sensitivity	•Weaker Sources •Faster Time Variability •Source Localization •Transient Sources	•Larger Area •Low Energy Sensitivity •Better Angular Resolution •Larger Field of View
Improved Angular Resolution	•Multiple Source Confusion •Source Localization	•Reduce Multiple Scattering (Low Energies) •Improved Position Resolution (High Energies)
Broad Energy Range	•Low and High E Spectral Breaks or Continuations •Broad Pion Bump at 70 MeV	•Reduce Radiation Lengths (Low Energies) •Increase Area and Prevent Self Veto (High Energies)
Low Background	•10,000 to 100,000 Particles per Gamma Ray •Particles Masquerade as Gamma Rays •Albedo Rejection	•Active Anticoincidence Shield •Directional Veto (e.g. time of flight)

2. High Energy Gamma-Ray Detector Design

The Advanced Gamma-Ray Astronomy Telescope Experiment (AGATE) concept takes into account the aspects of the design of the future high-energy gamma-ray instrument mentioned above. AGATE follows the tradition of the three successful gamma-ray satellites SAS-2, COS B, and EGRET by retaining the four principal components of their design: active anticoincidence, directional coincidence, energy measurement, and track imaging system. However, to meet the scientific requirements mentioned above, AGATE needs a much a larger effective area of approximately 2×2 m^2. Achieving a low instrumental background is one of the most crucial requirements for a gamma-ray instrument. The unambiguous detection of the rare gamma-rays in the high background of energetic charged particles in a space environment is one of the greatest challenges of space instrumentation. The previous successful gamma-ray telescopes used a highly efficient plastic scintillator anticoincidence dome and a directional coincidence system (Kanbach et al. 1989). The use of a single piece plastic scintillator dome for AGATE would be unsuitable because the path length of the scintillation light from the top of the dome would be very long ($\sim$4m), precluding fast time. Also, the sensitivity would be very non-uniform and the "self-veto" which limits the EGRET sensitivity above $\sim$ 10 GeV would be difficult to eliminate. Instead, a segmented dome will be used with several independent overlapping scintillator counters. This design has a maximum path length of $\sim$ 1.3m, permitting fast timing, and when combined with the signal from the coincidence system, it provides coarse directional information for charged particles, which will minimize the self-veto of high-energy gamma-rays.

The coincidence and energy measurement systems for AGATE will be adapted from those of EGRET, with slight modifications. Based on the EGRET results, the energy resolution of AGATE will not be required to be significantly better than $\sim$ 25% FWHM. One idea would be to use a segmented calorimeter ($\sim$ 8 radiation length deep) consisting of many CsI(Tl) crystals coupled to silicon photodiodes. Such a configuration would also enable one to get crude position information. For a more detailed description of the AGATE concept, and the anticoincidence, directional coincidence, and energy measurement sub-systems the reader is referred to Cuddapah et al. (1993) and Hunter et al. (1993). The remainder of this article concentrates on the track imaging system for AGATE.

The main modification in the AGATE design in comparison to SAS-2, COS B, and EGRET will be in its track imaging system. The imaging system of a high energy gamma-ray telescope works on the principle that a gamma-ray pair produces in one of the several layers of high-Z material placed in the active volume. The tracks of the electron-ion pair are imaged to obtain the direction and energy of the incident gamma-ray. The amount of high-Z material placed between measurement layers needs to be small enough to prevent multiple scattering, yet large enough for there to be a sufficiently high probability of the photon converting, and is chosen to be of the order of 0.5 radiation lengths. In EGRET the tracking detector is composed of spark chambers interleaved with 0.02 radiation lengths of tantalum foil. Simply scaling up the EGRET spark chambers to the size needed for AGATE is not practical in order to achieve the improvement in sensitivity needed. In addition, spark chambers have prohibitively long dead times (100 msec), and improving the spatial resolution to substantially better than the 0.8 mm achieved with EGRET, would be difficult. Thus the needed improvement in angular resolution would be hard to achieve. Drift chambers, on the other hand, prove to be an excellent choice. They have low power requirements, small dead times (~ few microseconds), and can achieve a spatial resolution of ~ 100 μm. Large area drift chambers have also been successfully built and operated at accelerators (Charpak, et al. 1968; Sauli 1987; Christian, D.C., et al. 1994) and are relatively simple and inexpensive to build.

In order to test the feasibility of using drift chambers for the track imaging system of AGATE, a laboratory prototype consisting of a stack of sixteen 1/2m × 1/2m drift chambers was constructed and tested. Both argon and xenon gas mixtures were used for the tests. Tracks of atmospheric muons were imaged using the drift chambers and the position resolution was measured. The results obtained with the prototype instrument are described in the following.

3. Drift Chamber Prototype

3.1. Experimental Setup

Figure 1 shows the schematic of the track imaging system used for the laboratory tests. It consists of a stack of 16 drift chamber frames of dimension 1/2m × 1/2m. Each drift chamber frame is made up of rectangular drift cells formed by three parallel grids of wires. An individual drift cell consists of an anode bounded on either side by two cathodes at a distance of 2 cm. A set of 21 uniformly spaced field shaping wires are placed 0.5 cm above and below the anode-cathode plane. There are 11 adjacent drift cells in each frame. The diameter of the anode wire is chosen to be 0.02 mm in order to facilitate gas amplification. The wires are soldered on to a G-10 frame with an accuracy better than 0.05 mm. Positive high voltage (HV) is applied to the anode, while the cathodes are kept at negative HV. Seven of the sixteen drift chamber frames are constructed with an offset of half a drift cell to resolve the "left/right" ambiguity caused by not knowing whether the electrons drift towards the anode from the left or right part of the drift cell (Charpak et al. 1968; Sauli 1987).

The frames are stacked vertically into 3 groups of 4, 8 and 4 frames, as shown in Fig. 1. A plane of aluminized mylar, stretched on an aluminium frame, is placed between each pair of frames, at a distance of 4.7 mm. This plane acts as an equal potential plane and isolates each frame electrically from its neighbours above and below. The drift chambers provide 13 x and 3 y coordinate measurements. The height of the total stack is about 50 cm. The stack is enclosed in a stainless steel pressure vessel with a hemispherical dome.

Experiments with the drift chamber stack were carried out using both argon and xenon, mixed with different molecular gases. Initial tests were performed with 85% argon and 15% isobutane at 1 atm. The results obtained

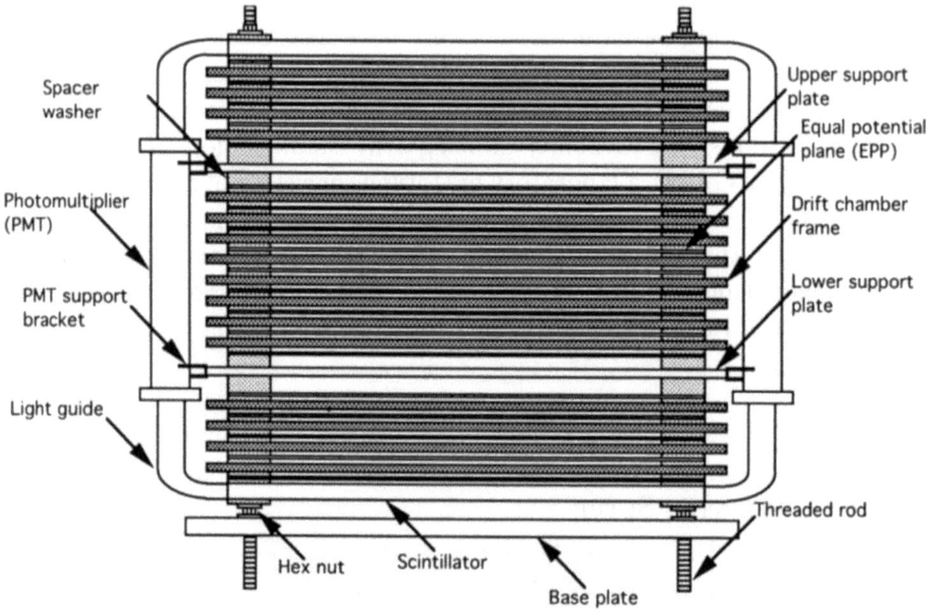

Fig. 1. Schematic drawing of the 1/2m × 1/2m track imaging system. The sixteen drift chambers are stacked in three groups, containing, respectively, four, eight, and four drift chambers. The overall height of the stack is about 50 cm. Scintillator layers are positioned at the top and bottom of the stack of drift chambers.

were compared with tests made with 90% xenon and 10% methane at 1 atm. The drift velocity of electrons in the argon-isobutane and xenon-methane mixtures are 4.2 cm μs^{-1} atm^{-1} (Sauli 1987) and 2.4 cm μs^{-1} atm^{-1} (Christophorou et al. 1980), respectively. A purification system was built for the xenon gas (Mukherjee et al. 1994). No purification was necessary for the other gases used for the tests.

Atmospheric muons were used to test the drift chambers. Ionization tracks of the muons in the drift gas were imaged using the setup described above. The 13 frames providing the x coordinate measurement were used for the tests. The trigger system used two plastic scintillators coupled to photomultiplier tubes (PMT) placed above and below the stack in order to select those tracks traversing the entire drift volume. The drift chamber stack assembly provided two coordinates for the position of each point of interaction, one coordinate being determined by the drift time, and the other by the height of the drift chamber anode wire layer on the stack.

LeCroy drift chamber amplifier and discriminator cards (2735DC) were used for the front end electronics to read out the anode signals. The signals were then fed to LeCroy time-to-digital (TDC) converters (2277) from which they were read out using a CAMAC system. The coincidence signal of the scintillators provided a common stop to the TDCs. The drift chambers were operated with the anode voltage between 1300 V and 1500 V and the cathode voltage at −1400 V.

3.2. Drift Chamber Results

When an atmospheric muon passes through the active volume of the drift chamber it ionizes the drift gas and leaves a track of electrons and ions along its path. The charged particle track is sampled by the 13 drift chambers capable of making the x coordinate measurement. The electrons drift towards the nearest anode wires under the influence of an applied electric field. Near the anode wires the field is very high and charge amplification by a factor of $\sim 10^3$ occurs. The drift time of the electrons to the anode wires is measured with TDCs, and using an average drift velocity, the x coordinate of the trajectory sample can be deduced. A non-linear correction is later made to the drift distance to take into account the non-uniform drift velocity of the electrons near the anode wires. The z coordinate is obtained from the position of the anode wire on the stack. Thus the 2-dimensional information of the track is obtained and the $x - z$ projection of the muon track can be reconstructed.

Figure 2 is an example of a reconstructed track of an atmospheric muon recorded with the drift chambers running with a xenon-methane gas mixture. Of the 13 'x' frames only 10 were used for analysis. The drift field for this run was 700 V cm^{-1} atm^{-1}. The best track direction was determined by doing a least square fit to a straight line through the points. For each reconstructed track the residual, that is, the measured track position minus the best fit track position, was estimated. Several runs were made both with xenon-methane and argon-isobutane gas mixtures to obtain a large event statistics.

In order to estimate the spatial resolution, a distribution of the residuals was made for the data collected in the runs made with both xenon-methane and argon-isobutane gas mixtures. Figures 3a and 3b show the results. A Gaussian fit was made to the data, shown as continuous solid lines in the figures. The standard deviation of $\sigma(x) = 0.141$ mm for argon-isobutane and $\sigma(x) = 0.234$ mm for xenon-methane are a measure of the spatial resolution attained with the drift chambers. The better result obtained with the argon-isobutane runs is probably partly due to the fact that the drift velocity variations have not been completely compensated for in the case of the xenon-methane mixture. The effect of the velocity variations is seen more clearly in Fig. 4, which shows the deviation of the measured position from the fit line, assuming a constant drift velocity, plotted as a function of the

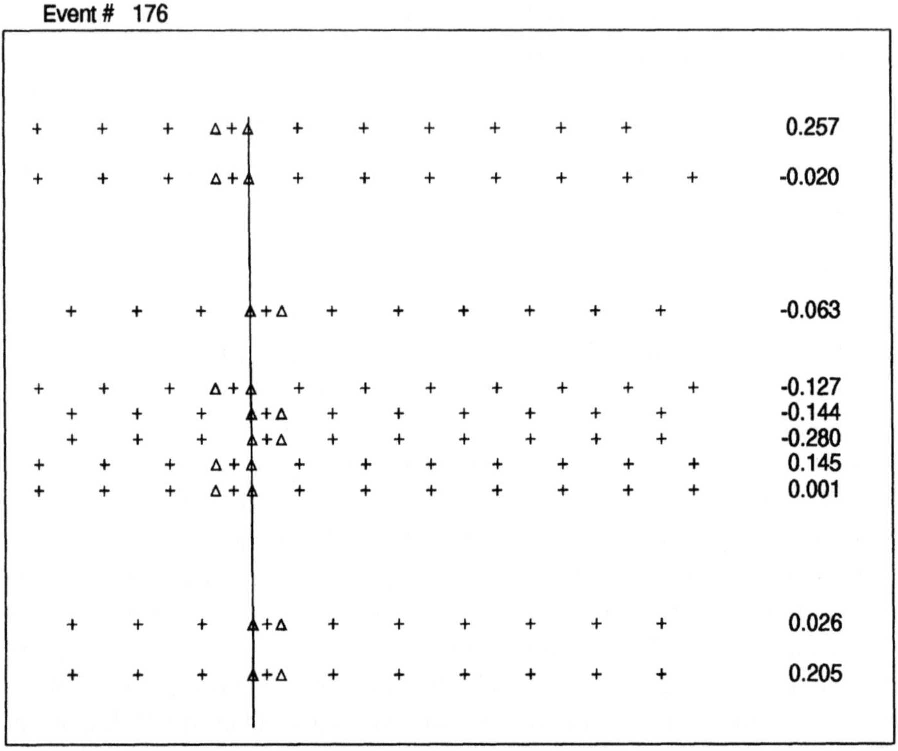

Fig. 2. Reconstructed track of an atmospheric muon recorded with the 1/2 m x 1/2 m prototype drift chambers with 90% Xe and 10 % methane at 1 atm. The positions of the anodes are indicated with crosses. Each horizontal row of anodes corresponds to a frame. Half of the frames are offset by half a drift cell to eliminate the "left/right" ambiguity. The triangles are the hit positions corresponding to drift from the left or right directions towards the anode. The numbers on the right hand side are the residuals in units of mm for each point used in the reconstruction.

distance from the anode. The deviation is more pronounced in the case of xenon-methane, and this has an adverse effect on the spatial resolution.

Figure 5a shows the dependence of the spatial resolution on the incident angle of the cosmic ray track for the runs with xenon-methane and argon-isobutane gas mixtures. The spatial resolution is seen to degrade at large angles. An algorithm to improve the large angle resolution has not yet been fully optimized. Figure 5b shows the spatial resolution as a function of the drift distance. While the spatial resolution measured for the argon-isobutane runs is nearly constant over the drift distance, the xenon-methane data shows some degradation towards the edges of the drift cell. This degradation could be due to an incomplete correction of the drift velocity over the entire drift cell.

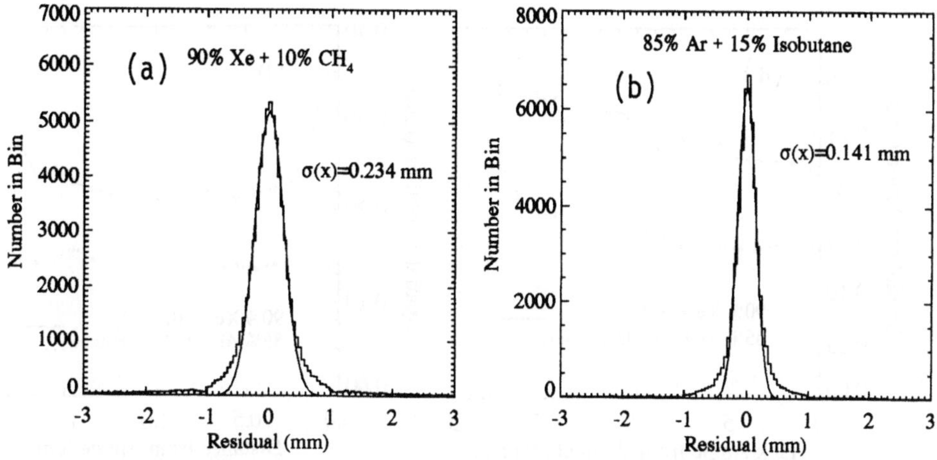

Fig. 3. Distribution of the residuals for the drift chamber runs with (a) xenon-methane and (b) argon-isobutane gas mixtures. The least square fit to a Gaussian yields the standard deviation which is a measure of the spatial resolution attainable with the drift chambers.

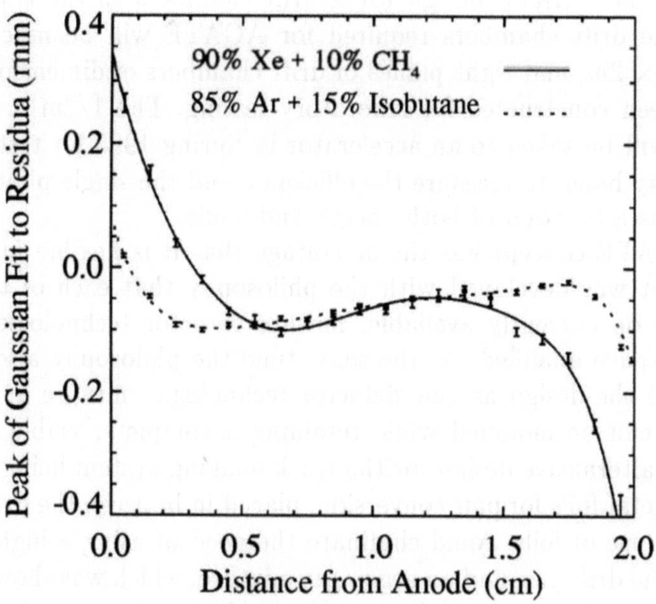

Fig. 4. Deviation of the measured position from the line fit to the muon track, assuming a constant drift velocity for argon-isobutane and xenon-methane

4. Future Plans

The results obtained with the 1/2m × 1/2m drift chambers are very encouraging. The successful operation of the drift chambers is a major step towards

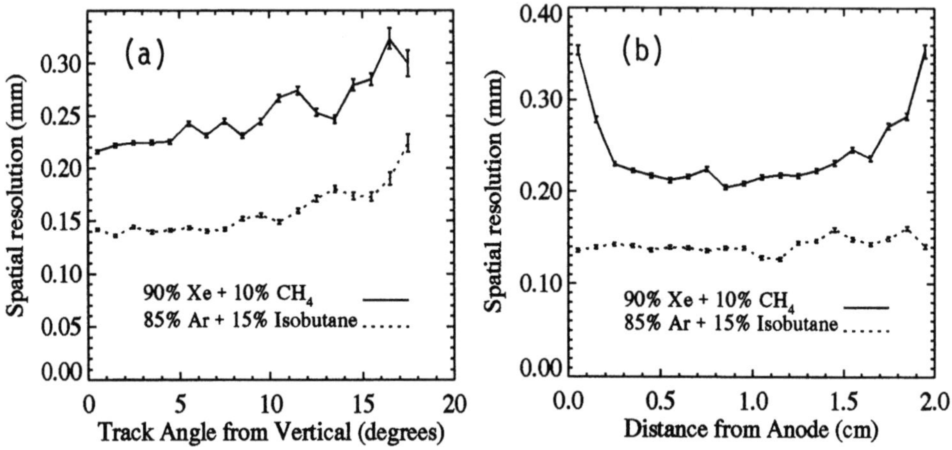

Fig. 5. Spatial resolution plotted as a function of (a) the angle of the muon track and (b) the drift distance for the xenon-methane and argon-isobutane runs.

completing the AGATE design using drift chambers as the track imaging system. The drift chambers required for AGATE will be much larger in size, $\sim$ 2m $\times$ 2m, and eight planes of drift chambers of dimensions 1/2m $\times$ 2m have been constructed for laboratory testing. The 1/2m $\times$ 1/2m drift chambers will be taken to an accelerator in Spring 1995 for testing it with a gamma-ray beam to measure the efficiency and the single photon angular resolution as a function of both energy and angle.

The AGATE concept has the advantage that it is flexible in its design. The concept was developed with the philosophy that each of the systems should rely on currently available, mature detector technologies that can be readily space qualified. At the same time the philosophy also allows for evolution of the design as new detector technologies mature and individual systems can be modified while retaining a complete, viable instrument design. An alternative design for the track imaging system being considered uses thin metal foils for pair conversion, placed in between the drift chamber layers. The use of foils would eliminate the need of using a high-Z gas (eg. xenon) as the drift gas, and an argon gas mixture, which was shown to be the better drift gas in the the 1/2m $\times$ 1/2m drift chamber experiments, could be used. One approach would be to use foils of thickness $\leq$ 0.01 radiation lengths distributed between measurement planes, giving a total depth of 0.5 radiation length, and retaining the large ($\sim$ 200) number of layers of drift chambers.

The number of drift chamber measurement layers is flexible and is determined by power and cost requirements, and the scientific benefits desired. Figure 6 is an example of three different telescope configurations that can be

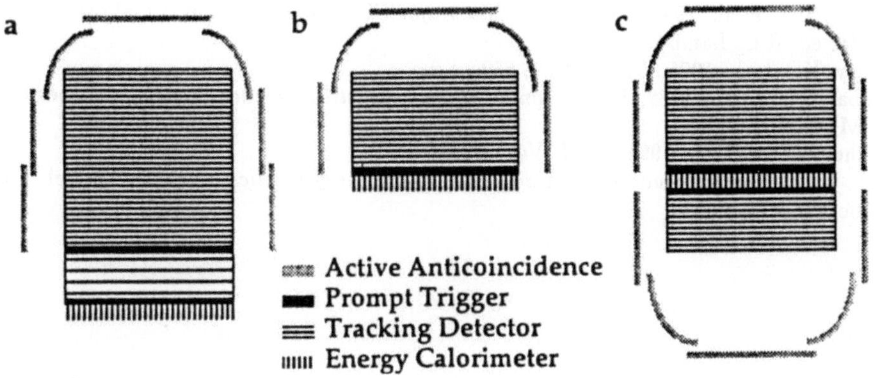

Fig. 6. High energy, gamma-ray telescope options using drift chambers as the tracking detector. (a) High stack with pair conversion material spread out over large number of measurement planes. Best angular resolution is achieved with this configuration. (b) Short stack giving an instrument with a large field-of- view. (c) Drift chambers on both sides of the calorimeter. Such a configuration is possible only because drift chambers have such low power requirements. The livetime of a telescope in this mode is doubled since one side can observe while the other is occulted by the earth. Background rejection is also improved.

achieved with drift chambers. The telescope in Fig. 6a has a large number of measurement layers with the pair conversion material spread out over a large region. In this configuration the telescope has the best angular resolution, although field-of-view and low energy sensitivity are partially compromised. Such a telescope would be ideal as a point source instrument. Fig. 6b shows a shortened stack and hence an instrument with a much larger field-of-view. In this configuration the telescope will be best suited as an all-sky survey type instrument. The configuration in Fig. 6c takes advantage of the low power requirements of drift chambers, enabling the use of a large number of drift chamber layers. It shows drift chambers on both sides of the calorimeter. This has the great advantage of doubling the livetime of a pointed telescope in low Earth orbit, as one side can observe while the other is occulted by the earth. Background rejection is also considerably improved. Several such options are currently under consideration.

References

Charpak, G., et al.: 1968, *Nucl. Instr. & Meth.* **62**, 262
Christian, D.C., et al.: 1994, *Nucl. Instr. & Meth.* **345**, 62
Christophorou, D.V., et al.: 1980, *Nucl. Instr. & Meth.* **171**, 491
Cuddapah, R., et al.: 1993, *1992 IEEE Nuclear Science Symposium* **Vol. 1**, 643

Dingus, B.L., et al.: 1995, *"AGATE: Expanding on the success of EGRET"* **Proc. Workshop on Imaging in High Energy Astrophysics, Capri**, 26–30 September 1994
Fichtel, C.E., et al.: 1994, *ApJS* **94**, 551
Hunter, S.D., et al.: 1993, *Proc. Towards a Major Cherenkov Detector-II* **Calgary, Canada**, ed. R.C. Lamb
Hurley, K., et al.: 1995, *Nature* **372**, 652
Kanbach, G., et al.: 1989, *Proc. Gamma-Ray Observatory Science Workshop* **Greenbelt, MD**, 10-12 April
Mukherjee, R., et al.: 1994, *SPIE* **Vol. 2305**, 2
Sauli, F.: 1987, *Experimental Techniques in High Energy Physics* **ed. T.J. Ferbel**, Frontiers in Physics

GLAST

ELLIOTT D. BLOOM
Stanford Linear Accelerator Center, Stanford University, Stanford, CA 94309

Abstract. A proposal for a new high energy gamma-ray space based detector using silicon strip technology is discussed. The detector is called GLAST for Gamma-ray Large Area Space Telescope. Results from Monte Carlo simulations are presented. The projected physics capabilities of GLAST are discussed in comparison to EGRET, currently in orbit on CGRO. Connections to the ground based gamma-ray observation program are explored.

1. Introduction

Recent results from the Energetic Gamma-Ray Experiment Telescope (EGRET) aboard the Compton Observatory have generated strong interest in space based high-energy ($E_\gamma > 10$ MeV) gamma ray astronomy. This science has wetted our curiosity of what might be observed with an instrument having considerably more capability than EGRET, if such a device were practical in our fiscally difficult times. Advances in silicon technology over the past decade, and the resulting rapid drop in costs, encourage the development of a dramatically new type of high-energy gamma ray space telescope based on silicon strip technology. The GLAST team (GLAST stands for Gamma-ray Large Area Space Telescope)* has been working for the past two years on the design of such an instrument, and the development of the silicon strip hardware and readout electronics needed to realize our design. Figure 1 shows an artist's concept of our current instrument design, including the spacecraft.

As in previous high-energy instruments, GLAST is a pair spectrometer backed by a total absorption electro-magnetic shower counter. Measurement of the energy and direction of the induced electro-magnetic shower provides

* The GLAST Team currently consists of the following institutions and individuals, in alphabetical order by institution: University of Chicago—Corbin Couvault, Rene Ong, Mark Oreglia; Kanagawa University—Katsuaki Kasahara; Lockheed Research Laboratory, Palo Alto, CA—David Chenette; Los Alamos National Laboratory—Geofrey Mills; University of Maryland—John Mattox; Max Planck Institut für Extraterrestriche Physik—Hans Mayer-Hasselwander; NASA Ames Reasearch Center—Jeffrey Scargle; Naval Research Laboratory—Eric Grove, Paul Hertz, W. Neil Johnson, Michael Lovellette, Gerald Share, Kent Wood; University of California at Santa Cruz—Robert Johnson; Sonoma State University—Lynn Cominsky; Stanford University—(Physics Department and HEPL) Yin-Chi Lin, Peter Michelson, Patrick Nolan, Arthur Walker, Jr., (Stanford Linear Accelerator Center) William Atwood, Elliott Bloom, Michael Chen, Gary Godfrey, Alex Luebke, (Aero-Astro Department) Robert Twiggs; University of Tokyo—Tuneyoshi Kamae; Instituto Nazionale di Fisica Nucleare, Trieste—Guido Barbiellini, Alberto Colavita; University of Washington—Thompson Burnett.

Space Science Reviews **75**: 109–125, 1996.

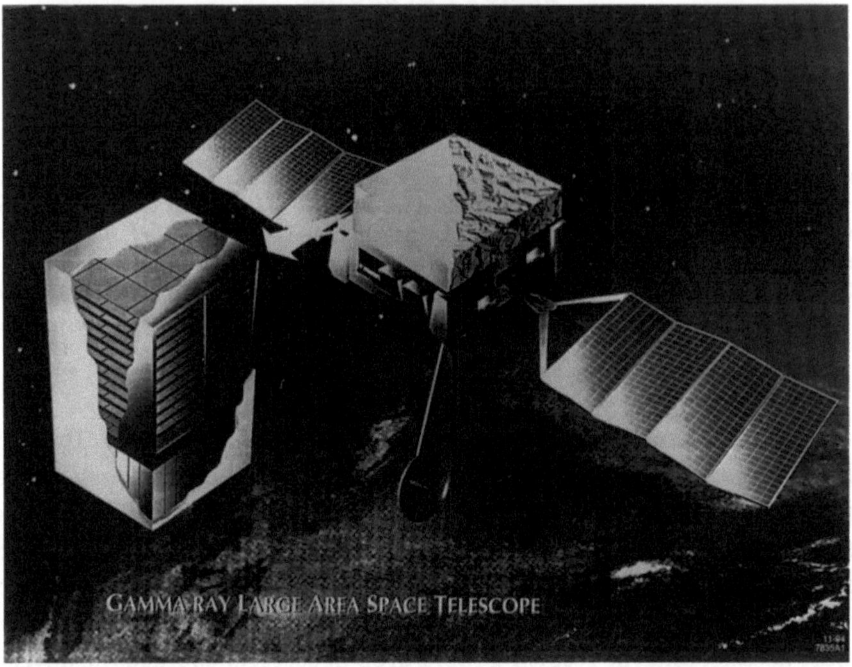

Fig. 1. An artist's conception of the GLAST detector and spacecraft in orbit. The picture also shows an exploded view of one tower of the 49 in the current design.

information about the energy and direction of the incident gamma-ray. However, due to the flexibility and relatively low cost of the silicon strip technology, the telescope has about a factor of 10 increase in effective area over EGRET, and about a factor of 5 increase in field of view. At the same time, the GLAST design is calculated to have much better point source sensitivity, and to have an energy range of 10 MeV $< E_\gamma <$ 300 GeV. Due to the economies of silicon technology, along with weight, and size savings compared to gas based detector technology, we estimate that this instrument can be built and flown as a Delta II mission. Thus, GLAST would easily fit into the NASA intermediate mission category with an estimated total cost of about $200 million.

2. The GLAST Design

Figure 2 shows some details of the current GLAST design. The telescope consists of three elements: a segmented charged particle anticoincidence shield; a gamma-ray tracker/converter, consisting of thin sheets of high-Z converter material interspersed with silicon strip detectors for particle tracking; and a segmented CsI calorimeter to provide good energy resolution at high

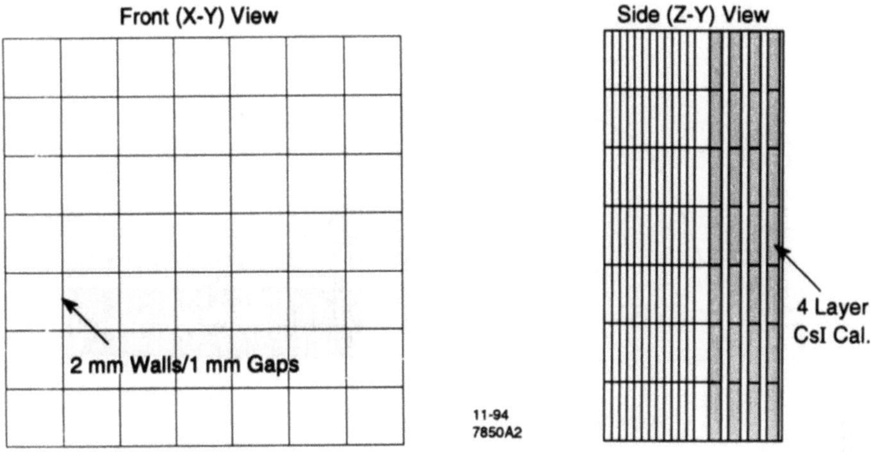

Fig. 2. A detailed view of the GLAST detector geometry in plan or front (left part of figure) and side (right part of figure) views.

energies. As the left part of Figure 2 shows, and by the arrow in Figure 1, the GLAST detector is modular, consisting of a 7×7 array of towers, with each tower containing elements of the anticoincidence shield, the tracker/converter stack, and the calorimeter. The right part of Figure 2 shows, along with the tracker/converter, a longitudinally segmented calorimeter design that can dramatically increase the effective area of the detector at high energy by using the photon angle measurement from the calorimeter alone. Such design details of the calorimeter, as well as other aspects of the detector are still in development.

Note that the detector technology we are using naturally allows GLAST to be wide (W) compared to its depth (D). This results from having no separate time of flight system in the GLAST design. The powerful self-triggering and tracking capabilities of the silicon strip technology obviate the need for TOF.

Figure 3(a) shows the instrument acceptance vs aspect ratio (W/D), where W and D are defined in part (b) of the figure. The large aspect ratio of GLAST greatly increases its angular acceptance, i.e., fraction of the sky seen at one time, compared to previous instruments (EGRET is shown for example in the figure). Part (c) of the figure shows the relative angular acceptance of GLAST vs the cosine of the angle to normal incidence to the detector. The detector cutoff angle, $\cos \Theta_c$, is about 0.1. The large angular acceptance of GLAST, combined with normal spacecraft operation in orbit that is anti-nadir pointing, allows great sensitivity to transient events. This

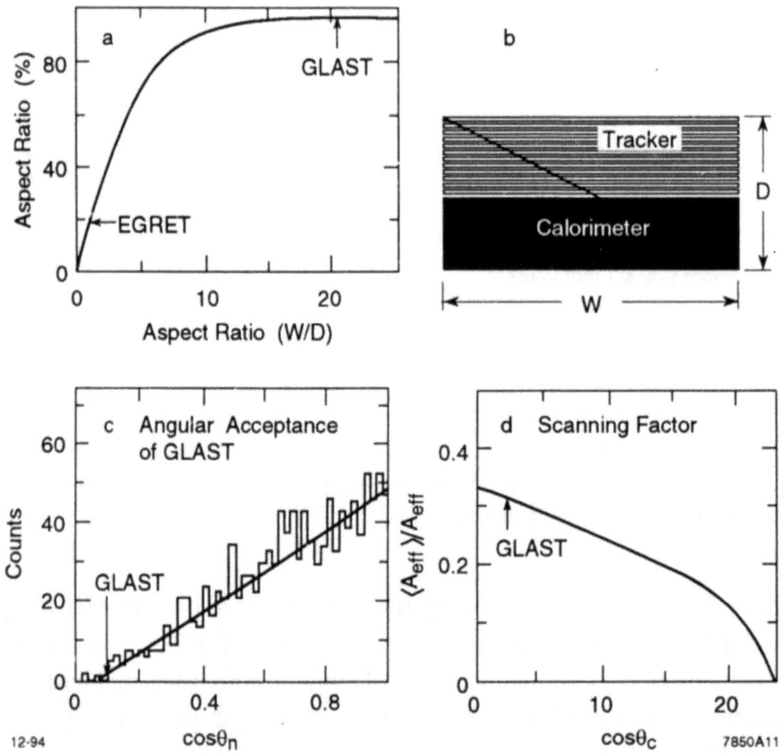

Fig. 3. Part (a) shows the instrument acceptance vs aspect ratio (W/D), where W and D are defined in part (b) of the figure. Part (c) shows the relative angular acceptance of GLAST vs the cosine of the angle to normal incidence to the detector, $\cos\Theta_n$. The GLAST cutoff angle, $\cos\Theta_c$, is about 0.1. Part (d) shows an effective "scanning factor" vs $\cos\Theta_c$. The scanning factor gives the normalized effective area of the instrument, when the spacecraft is operating in anti-nadir pointing mode, over one full orbit (see text).

is quantified in part (d) of Figure 3, which shows an effective "scanning factor" vs $\cos\Theta_c$. The scanning factor gives the normalized effective area of the instrument, when the spacecraft is operating in anti-nadir pointing mode, over one full orbit. Thus, any object in the sky that is seen with A_{eff} when it is at maximum acceptance has a scanning factor of, $< A_{eff} > /A_{eff}$, where

$$\frac{\langle A_{eff}\rangle}{A_{eff}} = \frac{1}{2\pi}\int_{-\Theta_c}^{\Theta_c} \frac{\cos\Theta_{orb} - \cos\Theta_c}{1-\cos\Theta_c} d\Theta_{orb}. \tag{1}$$

GLAST views any object in the sky, within its acceptance, with about 1/3 of the objects maximum acceptance in GLAST averaged over one orbit ($\sim$ 90 minutes). Of course, depending on the orbit, some parts of the sky may never be seen in this mode.

2.1. The Tracker

The gamma-ray tracker/converter, in the current GLAST design, consists of a tracking layer of silicon strips that provides a charged particle anti-coincidence veto, followed by 10 radiator/tracking layers of 0.05 radiation lengths each, spaced 3 cm apart, and finally an additional 2 tracking layers, also spaced 3 cm apart. Behind each radiator is a tracking layer consisting of a pair of single-sided silicon strip detectors that provide (x, y) position determination. Downstream of these 10 layers are two planes of (x, y) single-sided silicon strip detectors. These last two layers without radiator are used to track the previously converted photons. As three points are a minimum needed to establish a track, radiator in these layers will just degrade tracking performance, with no gain in acceptance.

The 240 μm strip pitch will result in a rms position resolution of about 70 μm in x and y. The e^- and e^+ from a just converted photon will make one hit in the silicon immediately following the radiator, thus precisely locating the conversion point. The multiple scattering occurring in the radiator and the intrinsic opening angle of the pair will typically cause two hits in the next layer of silicon (which is after the next radiator). Multiple scattering in the second radiator has a minor effect on the determination of the direction of the tracks since it is located directly on top of the second layer of silicon. The intrinsic two-hit resolution of 480 μm allows for more accurate tracking of the initial pair than in gaseous tracking devices in which the two-hit resolution is typically 2–3 mm.

For very high energy photons, the rms projected angular resolution of this configuration is limited to about 0.03° by the strip pitch and spacing between planes. For the total of 0.56-radiation-length thickness of the tracker/converter, the probability of a photon converting is 0.35 (above 1 GeV).

Large arrays (smaller than GLAST) of silicon strip detectors have been built and successfully operated as elements of particle physics detectors. These devices typically operate in radiation fields that are larger than those expected to be seen by the GLAST instrument. Figure 4 shows a multi-layer silicon strip detector array built and flown in a balloon by the Wizard collaboration*. This device operated successfully.

* Picture and information provided courtesy of G.Barbiellini, A. Colavita, INFN, Trieste, members of the Wizard Collaboration.

Fig. 4. A picture of the Wizard Collaboration silicon strip multilayer detector flown on a balloon. Each layer is a 50 cm by 50 cm array.

2.2. The Calorimeter

The photon energy is estimated by measuring the $e^{\pm}$ shower energy deposited in the calorimeter and correcting for energy losses in the tracker/converter (currently estimated by hit counting). The calorimeter must provide good energy resolution in the energy range from 10 MeV to about 100 GeV, and reasonable energy resolution to about 300 GeV. It must be coarsely pixelated in order to aid in the rejection of backgrounds. Current design calls for a 10-radiation-length calorimeter, with a longitudinal segmentation by 4 and a transverse size of3×3 cm. This results in an array of 64×4 crystals for each GLAST tower, or 12,544 for the entire detector. Thallium-doped cesium iodide is the material of choice in the current design. CsI(Tl) provides excellent energy resolution at modest cost, has a short radiation length (1.86 cm), and Moliere radius (3.8 cm), and provides a signal that is fast for our purposes. This material is also reasonably radiation hard. In addition, larger (compared to GLAST) CsI detectors have been built and operated as elements of particle physics detectors. The practicality of longitudinal segmentation is currently under review.

In summary, the GLAST design has many technical benefits including no consumables, all relatively low voltages, modularity, modern, low cost, robust and long lived technology. It also has its design challenges including a large channel count of about 10^6 channels, and on-board computing requirements. We believe that we currently have viable designs that meet these challenges.

3. Estimated GLAST Performance

High-energy gamma rays from space are excellent probes of the most energetic phenomena that occur in nature. They are emitted over a wide range of angular scales from a diverse population of astrophysical sources: stellar mass objects, in particular neutron stars and black hole candidates; the nuclei of active galaxies (AGN) which are likely to contain massive black holes; interstellar gas in the galaxy that interacts with high energy cosmic rays; the diffuse extragalactic background; supernovae that may be sites of cosmic ray acceleration; and gamma-ray bursts. Many of these sources exhibit transient phenomena.

3.1. Comparison of EGRET to GLAST

Interest in these areas has received a large boost from the EGRET instrument over the past 2 years, with more yet to come. A sample of EGRET results includes:

- Pulsed gamma-ray emission above 100 MeV detected from 5 pulsars [D.J. Thompson et *al.*, 1994].
- More than 30 sources detected in the galactic plane, including high energy gamma emission from the galactic center region [C.E. Fichtel et *al.*, 1994].
- Diffuse emission detected from the Large Magellanic Cloud. This observation along with nondetection of the Small Magellanic Cloud, directly shows that cosmic rays are not universal but are galactic in origin [P. Sreekumar et *al.*, 1992].
- Gamma-ray emission from more than 40 AGN [C. von Montigny et *al.*, 1995]. These have variability observed on time scales of days to months. Ground based observations from 0.5 TeV on up of Mkn 421 by the Whipple observatory have been coordinated with observation by EGRET. Recently a flare of Mkn 421 was simultaneously observed by ASCA, CGRO and Whipple [A.D. Kerrick et *al.*, 1994].
- More than 25 high-latitude unidentified sources [C.E. Fichtel et *al.*, 1994].
- Several high-energy gamma-ray bursts detected. One burst lasted more than 90 minutes with emission to at least 18 GeV [K. Hurley et *al.*, 1994].
- Detection of a solar flare of several hours duration and emission to at least 2 GeV [E.J. Schneid et *al.*, 1994].

Table I. Comparison of GLAST and EGRET γ-ray Telescope Parameters.

	EGRET[1]	GLAST Conceptual Design
Energy range	35 MeV–30 GeV[2]	$\sim$ 10 MeV–300 GeV
Energy resolution[3]		
10 MeV	...	24%
50 MeV	14%	12%
100 MeV	12%	8%
1 GeV	9%	4.4%
10 GeV	12%	6%
100 GeV	...	18%
Effective Area[4]		
10 MeV	...	300 cm^2
50 MeV	250 cm^2	4000 cm^2
1 GeV	1200 cm^2	8000 cm^2
10 GeV	700 cm^2	8000 cm^2
100 GeV	...	8000 cm^2
Solid Angle Acceptance[5]	$0.15 \times \pi$ sr	$0.82 \times \pi$ sr
Full Field of View	$0.2 \times \pi$ sr	$1.8 \times \pi$ sr
Point Source Sensitivity[6]		
E > 100 MeV	5.4×10^{-8} ph cm^{-2} s^{-1}	1.5×10^{-9} ph cm^{-2} s^{-1}
E > 1 GeV	1.2×10^{-8} ph cm^{-2} s^{-1}	1.5×10^{-10} ph cm^{-2} s^{-1}
E > 10 GeV	2.1×10^{-8} ph cm^{-2} s^{-1}	9.5×10^{-10} ph cm^{-2} s^{-1}
Single photon position[7]		
50 MeV	8.4 °	5.2 °
100 MeV	5.6 °	2.5 °
1 GeV	1.5 °	0.42 °
10 GeV	0.5 °	0.1 °
Point Source Location	5–10 arcmin	0.1–1 arcmin
Volume[(8)]	4.8 m^3	$\sim$ 2 m^3
Mass	1830 kg	$\sim$ 2700 kg
Power	190 W	$\sim$ 1 kW
Lifetime	$\sim$4 yr	> 4 yr

1. Parameters for the EGRET instrument are from Refs.[D.J. Thompson et *al.*, 1993] [A.H Walker, 1990].
2. This is the energy range when the EGRET trigger system is operated with the NaI calorimeter (TASC) in coincidence. This is the normal operating mode. With TASC not in coincidence, the energy range is 20 MeV to about 30 GeV. This latter mode is not normally used because of the higher rate of spark chamber gas consumption.
3. Equivalent Gaussian σ. For EGRET, both the e^- and e^+ enter the calorimeter for the resolutions shown.
4. Area $\times$ γ-ray detection efficiency. The effective areas given for GLAST include γ-ray detection inefficiencies due to background rejection analysis cuts. The geometric area of GLAST is 28,200 cm^2. For comparison, the geometric area of the EGRET spark chamber is 6,400 cm^2.
5. FWHM. The effective area at the edge of the field is 50% of its on-axis value.
6. This is the flux sensitivity at a 5σ detection threshold. The background assumed is 2×10^{-5} photons cm^{-2} s^{-1} sr^{-1} $(100\ MeV/E)^{1.1}$, typical of the background seen by EGRET at high galactic latitudes. In each case, the exposure assumed is that obtained from a 1 year all-sky survey.
7. Angle for 68% containment from a point source. The high energy angular resolution of GLAST is determined by the spacing between converter planes, the radiation lengths per converter, and the pitch of the Si strips. For the numbers shown here, the spacing between planes is 3.0 cm, the converters are each 0.05 radiation lengths thick, and the readout pitch is 240 μm ($\sigma = 69\mu$m).

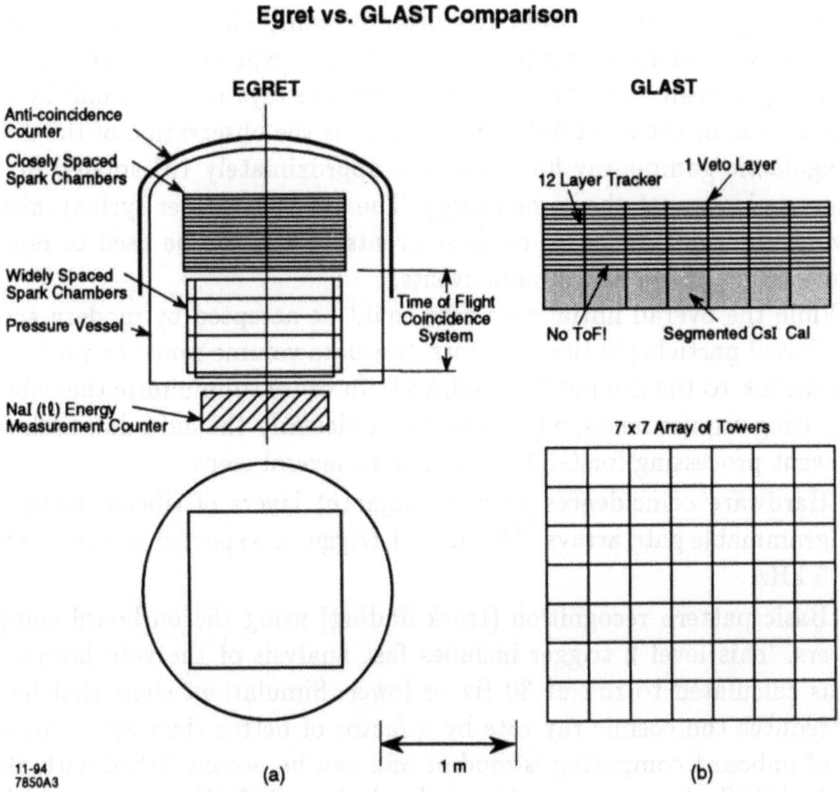

Fig. 5. A comparison of the physical sizes of EGRET and GLAST to the same scale. A meter is shown in the figure for comparison. The major subsystems of each detector are also enumerated in the figure. Given its small size and relatively light weight, GLAST can be launched on a Delta II rocket.

GLAST can make a dramatic improvement on EGRET observations. Due to its larger effective area and field of view, lower energy threshold, and deadtimeless operation GLAST will be much more sensitive to transient events. (The use of spark chambers in EGRET introduces a 0.1 second deadtime per event.) Figure 5 and table 1 compare EGRET to our current GLAST design. The figure shows the relative sizes of the two instruments. The weight of GLAST allows the instrument to be launched on a Delta II class rocket. The table details a comparison of a number of important observational and other parameters for the two instruments.

3.2. GLAST PERFORMANCE

One of the main challenges in operating a gamma-ray detector in space is control of the backgrounds in the detector. One expects that many charged and neutral particles (cosmic-ray protons, ions, electrons, Earth albedo gamma-rays, neutrons, trapped radiation, etc.) will interact with GLAST and its spacecraft. The flux of cosmic gamma-rays is very small in comparison. One of the most difficult problems is the observation of the diffuse extragalactic gamma-ray flux, which is approximately 10^4 smaller in rate than cosmic rays at the same energy. The GLAST trigger system initially provides information for all of these events, which can be used to reject a large fraction of the background events.

While the overall initial data rate could be accepted by modern accelerator based particle physics detectors, this data volume would be prohibitive to downlink to the ground from GLAST. In order to minimize the onboard computing requirements, while limiting sufficiently the data downlink rate, the event processing for GLAST is done in several steps:

- Hardware coincidences between adjacent layers of silicon, using programmable gate arrays. This level 1 trigger is expected to run at about 5 kHz.
- Basic pattern recognition (track finding) using the on board computers. This level 2 trigger includes fast analysis of the veto layers, and is calculated to run at 30 Hz or lower. Simulations show that level 2 reduces the cosmic ray rate by a factor of better than 200. This level of onboard computing is modest and can be accomplished with about 15 MIPS of computing. At this level about 100 Kbits/sec would have to be downlinked to the ground.
- The use of additional event information, including the pattern of energy deposition in the calorimeter. This level 3 trigger requires more onboard computing power. Simulations indicate that event filtering at this level reduces the downlink rate to 25 Kbits/sec; a very comfortable rate.
- The rest of the background events to be excluded on the ground to a level of at least one part in 10^5. The events downlinked will undergo extensive computer analysis to further reduce the background. For the diffuse isotropic extragalactic gamma-rays, with $E_\gamma > 100$ MeV, we have calculated a signal to noise of about 20 to 1.

The initial cosmic ray spectrum is generated using the NRL code called CREME*. For the analysis presented here, we generated 200 K cosmic ray events.

* CREME code, James A. Adams, Naval Research Laboratory, Washington DC. To be found on EnviroNET, Mail Code 410.1, NASA Goddard Space Flight Center, Greenbelt, MD 20771.

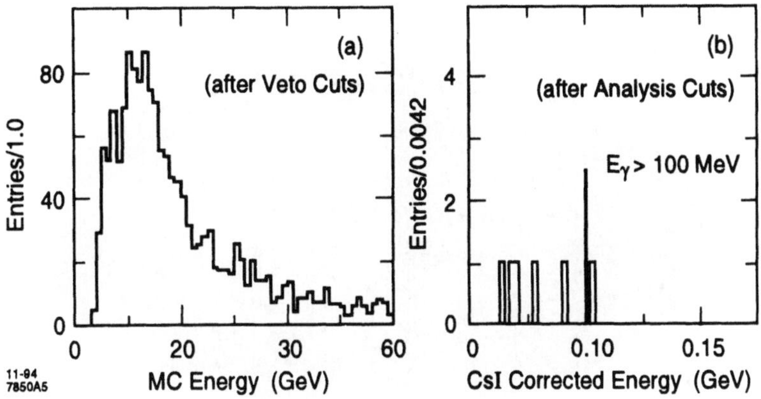

Fig. 6. 200 K cosmic ray events are generated using the CREME code. Part (a) shows the generated cosmic ray proton spectrum after veto cuts. After our additional standard set of simulation cuts, part (b) results. There is one event with a "measured" $E_\gamma > 100$ MeV left in the spectrum.

Figure 6(a) shows the generated cosmic ray proton spectrum after veto cuts. After our additional standard set of simulation cuts, Figure 6(b) results. There is one event with a "measured" $E_\gamma > 100$ MeV left in the spectrum. Using a cosmic ray rate of 0.17 cm^2/sec, integrated over the GLAST acceptance, and an extragalactic diffuse isotropic γ rate of 2×10^{-5} cm^2/sec/sr for $E_\gamma > 100$ MeV, we find a signal to noise of 19.6 to 1. Additional rejection techniques, currently under development, may improve this by about a factor of 10.

Figure 7(a) shows an EGRET all sky map on which is superimposed the GLAST field of view and the EGRET field of view. A dramatic improvement is evident from EGRET to GLAST. Part (b) of the figure shows the single photon projected angle vs energy for GLAST and EGRET. Part (c) of the figure shows the effective area of GLAST, EGRET, and GLAST when using the γ angle measurement from the segmented calorimeter. The flat high energy acceptance of GLAST is achieved by eliminating the veto of good γ events from their charged particle "backsplash". This effect is what limits the high energy acceptance of EGRET due to its monolithic veto "dome". One sees an increase of an additional factor of about 3 in the GLAST effective area by using the segmented calorimeter photon angle measurements. This improvement is important at high γ energy where the rate is low.

Figure 8 shows the MC distributions for photons in the GLAST calorimeter; parts (a) through (d) show 50 to 10,000 MeV. At the highest energy of 300 GeV, the energy resolution is about 25%, which is adequate for these highest energies.

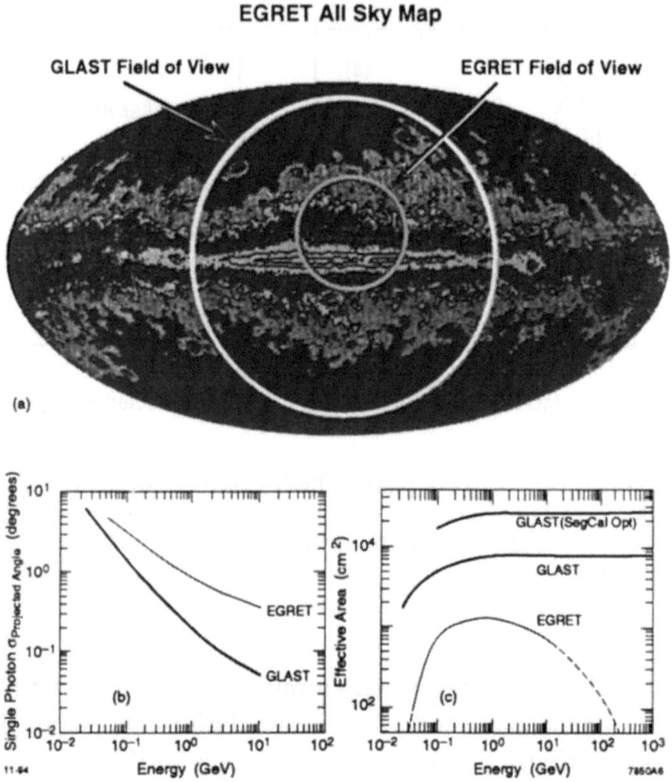

Fig. 7. Part (a) shows an EGRET all sky map on which is superimposed the GLAST field of view and the EGRET field of view. A dramatic improvement is evident from EGRET to GLAST. Part (b) shows the single photon projected angle vs energy for GLAST and EGRET. Part (c) shows the effective areas of GLAST, EGRET, and GLAST when using the γs which interact in the tracker or just in the calorimeter.

Figure 9 even more dramatically shows the power of the GLAST detector. The top part of the figure repeats the EGRET all sky map, with the GLAST (and EGRET) field of view centered on the molecular cloud Rho Ophiuchas. The bottom part of the figure explodes the view of this source using the full detail of the EGRET map. The outer circle shows the achieved EGRET single photon 1 σ position error ellipse at 100 MeV. This ellipse quantifies

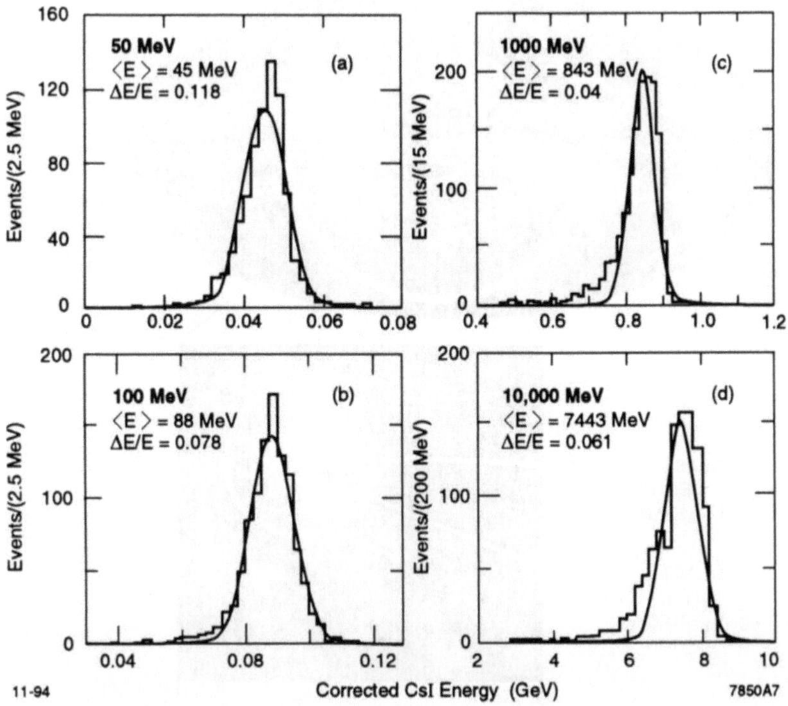

Fig. 8. The MC distributions for photons in the GLAST calorimeter; parts (a) through (d) show 50 to 10,000 MeV.

the relative two source resolution of the detector. The next (white) ellipse in shows the GLAST single photon resolution at the same energy. However, the (black) inner most ellipse gives the GLAST single photon resolution at 1 GeV. At 1 GeV GLAST contains as many photons in this ellipse as does EGRET in its 100 MeV ellipse.

Figure 10 demonstrates this spectacular GLAST resolution (combined with area) in the simulation of results for an all-sky (anti-nadir pointing mode) 1 year GLAST survey ($E_\gamma > 100$ MeV). In the top half of the map, the simulation extrapolates the ln(N) vs ln(S) plot for AGN to lower S as $N(S > S_0) \sim S^{-3/2}$. Included is the galactic diffuse background, plus extragalactic (the known EGRET sources, plus AGN using the above extrapolation for N). The bottom half of the map has the galactic diffuse background, plus extragalactic (the known EGRET sources, plus 100% diffuse background). The Galactic plane is overexposed in the false "grayscale" image. With no special processing, the raw image is shown, the difference

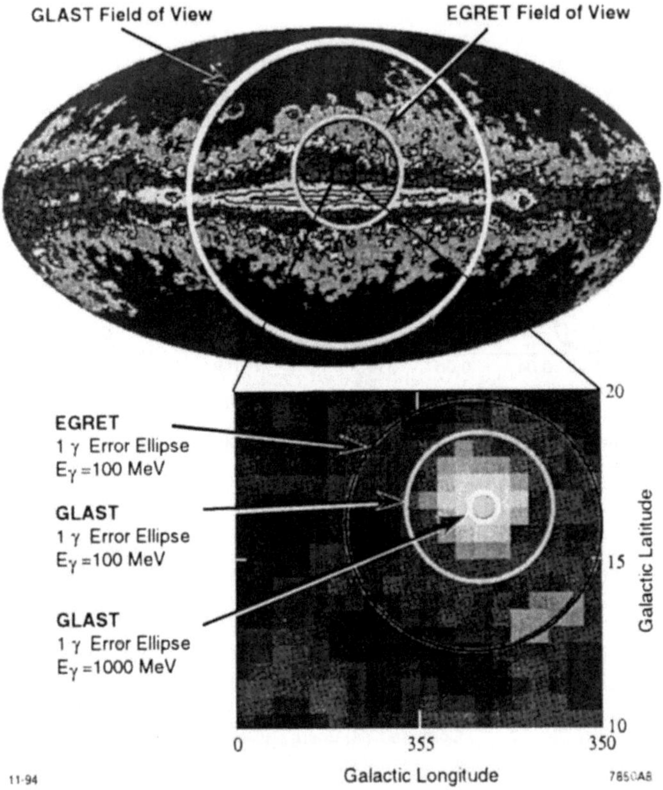

Fig. 9. The top part of the figure shows the EGRET all sky map, with the GLAST and EGRET fields of view centered on the molecular cloud Rho Ophiuchas. The bottom part of the figure explodes the view. The outer circle shows the achieved EGRET single photon 1 σ position error ellipse at 100 MeV. The next (white) ellipse in shows the GLAST single photon resolution at the same energy. However, the (black) innermost ellipse gives the GLAST single photon resolution at 1 GeV. At 1 GeV GLAST contains as many photons in this ellipse as does EGRET in its 100 MeV ellipse.

between top and bottom half is clear (even though the black and white reproduction of the image is poor).

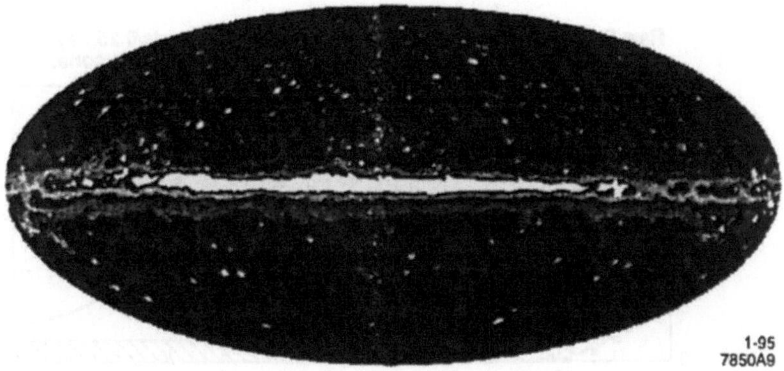

Fig. 10. A galactic coordinate projection of the sky from a GLAST detector simulation. Included in the top half is the galactic diffuse background, plus extragalactic (the known EGRET sources, plus extrapolated sources as discussed in the text). The bottom half of the map has the galactic diffuse background, plus extragalactic (the known EGRET sources, plus diffuse background). The Galactic plane is overexposed in this false "grayscale" image.

3.3. Connections to the ground based observation program

As mentioned previously, the Whipple observatory has cooperated with the CGRO team and others in coordinating space based and ground based observations of high energy gamma-rays (Trevor Weekes, Private Communication). The GLAST collaboration hopes this activity will expand to a worldwide scale when GLAST is in orbit. The major mode of GLAST observations will be in the anti-nadir scanning mode (some pointed observations may also be made). In this mode GLAST will scan most of the sky once per orbit. If a transient of interest should be observed, Earth based observation stations could be contacted very quickly.

Figure 11 demonstrates how this might work. Figure 11(a) shows the GLAST uncorrected point spread function obtained with a $1/E^2$ input photon spectrum. We assume such an energy dependence for Mkn 421 within the GLAST energy range. As part (a) of the figure shows, 68% of the photons are contained within about 2.5° (half angle). Part (b) of the figure shows a simulated flare of Mkn 421, beginning at the start of day 1.5, consistent with past flares. As the figure shows, a 3σ effect for the source flare is seen in less than 0.5 day by GLAST. Part (c) shows the simulated flare spectrum as observed in GLAST. This spectrum includes events from about day 1.5 to day 5. There is useful information in this spectrum to about 10 GeV. If quickly brought on line, the ground based instruments should be able to obtain useful information from about 100 GeV upwards to a few TeV. Note that an extended observation, e.g., 1 year in anti-nadir pointing mode, by

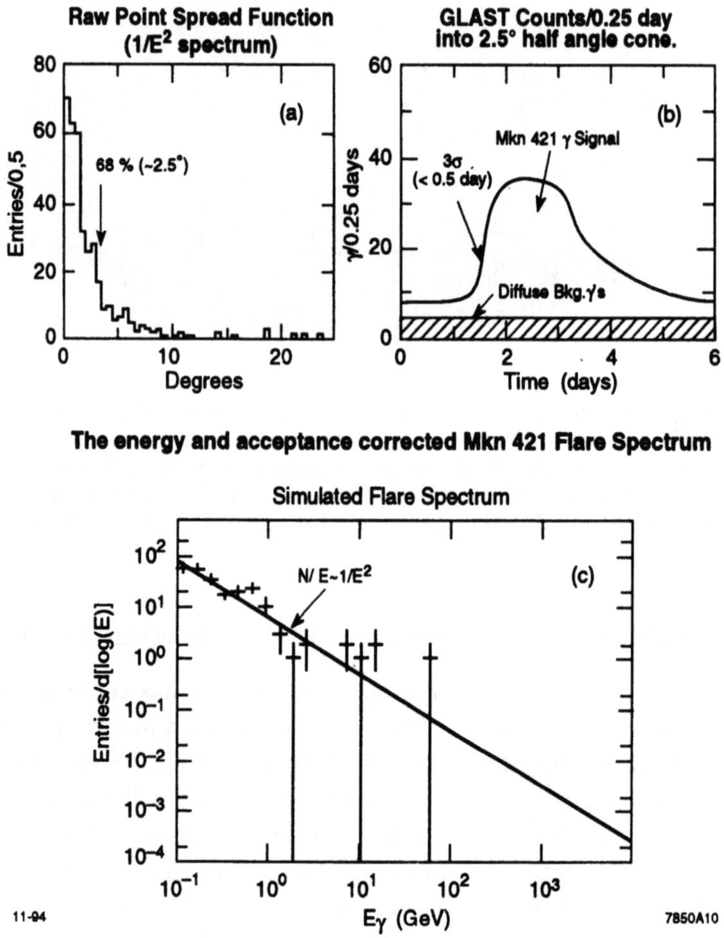

Fig. 11. Part (a) plots the raw point spread function of GLAST for a $1/E^2$ incident photon spectrum. 68% of the photons are contained within about 2.5° (half angle). Part (b) of the figure shows a simulated flare of Mkn 421 consistent with past flares. The ordinate plots GLAST counts/0.25 day into a 2.5° half angle cone. Part (c) shows the simulated flare spectrum as observed in GLAST. This spectrum includes events during the flare, from about day 1.5 to day 5.

GLAST of this source, assuming the source is not flaring during this time, should yield only ~ 6 events > 100 GeV, assuming the same $1/E^2$ spectrum as in part (c) of the figure.

4. Summary and Conclusions

High-energy astronomy in orbit has generated great interest due to recent results from EGRET. The GLAST collaboration is developing the next generation γ-ray telescope, GLAST, that would extend the EGRET discovery reach by about 2 orders of magnitude. This impressive improvement is due to a number of factors:

- Much larger effective area.
- "Full" sky view.
- Factor of 30 in energy reach (10 MeV $< E_\gamma <$ 300 GeV).
- Rapid transient response in anti-nadir scanning mode.
- Much better source resolution.
- Much longer lifetime in space due to no consumables.

GLAST can be built quickly and relatively inexpensively due to its use of modern silicon technology. This technology is in hand as evidenced by its extensive use in large particle physics experiments, the balloon flight of the Wizard collaboration prototype, and the space flight of small detectors by a number of missions.

In addition to its direct contribution to space based γ-ray astronomy, GLAST would significantly contribute to ground based γ-ray astronomy, which works at significantly higher energy, by enabling a fast response to transient phenomenon by the ground based telescopes.

Given all of these factors, we believe that GLAST represents an unusually urgent scientific opportunity that should be realized within this decade.

Acknowledgments

I would like to thank Bill Atwood for extensive discussions and comments.

References

Fichtel, C.E. et al. Ap.J. Suppl., **94**, 551 (1994).
Hurley, K. et al. Nature, **372**, 652 (1994)
Kerrick, A.D. et al. IAU Circular **#5996**, (1994).
Montigny, C von et al. Ap.J., **440**, 525 (1995).
Schneid, E.J. et al. High Energy Solar Phenomena - A New Era Spacecraft Measurements, ed. J.M. Ryan & W.T. Vestrand, AIP Conf. Proc. **#294**, (AIP:New York), p.94 (1994).
Sreekumar, P. et al. Ap.J., **400**, L67 (1992).
Thompson, D.J. et al. Ap. J. Suppl. **86**, 629 (1993)
Thompsen, D.J. et al. Ap.J., **436**, 229 (1994).
Walker, A.H., Ph.D. thesis, Stanford University, Stanford, California (1990).

GAMMA-RAY ASTRONOMY: EXTENSION BEYOND THE GEV DOMAIN BY GROUND-BASED OBSERVATIONS

E. PARÉ
LPNHE-Ecole Polytechnique, 91128 Palaiseau, France

Abstract. Gamma-Ray Astronomy, originated with the OSO-3, SAS-2 and COS-B satellites, has been renewed during the last three years by Compton-GRO's discovery of tens of Galactic and extra-galactic sources up to $\approx$ 10 GeV. Also in the last three years, a ground-based observing technique has emerged for TeV gamma rays: the identification of gamma-induced air showers via their Cerenkov emission. Extrapolating this technique down to the present limit of satellite observations has become a realistic goal, allowing fundamental questions to be tackled which would remain open until the individual spectra were extended.

1. Present situation and Motivations

During the last five years the general situation in High-Energy Gamma-Ray Astronomy changed considerably. As regards space-based Gamma-Ray Astronomy, the C-GRO satellite (launched in 1991) significantly extended the catalogue of γ-ray sources previously available (Fitchel et al.). In particular, more than 30 sources were found to be Active Galactic Nuclei (AGN) whereas only one such source—the quasar 3C273—had been detected in the γ-ray range by a previous satellite which had opened the catalogue to extra-galactic sources: the COS-B satellite (1975-1982). The highest-energy range covered by C-GRO was explored by EGRET and extended up to about 10 GeV, limited above this by the small effective detection area ($\approx 0.15m^2$) and the rapid decrease in γ-ray fluxes with energy.

In the meantime, in 1992, an AGN was observed by the Whipple Observatory group (Punch et al.) in the TeV region: it is a BL-Lacertæ type object, Markarian 421, which was already in the catalogue of sources observed by EGRET. It is significant that the most intense extra-galactic source in this catalogue, the quasar 3C279, has not been observed; it is a very distant object (with a spectral red-shift of z=0.54), whereas Markarian 421 is the nearest source to earth in the catalogue (z=0.03). These observations can be interpreted as an effect of the absorption of TeV γ's, which interact with the diffuse intergalactic infra-red radiation according to: γ + I. R. photon $\rightarrow e^+ + e^-$. By extrapolating the energy spectra of distant sources ($z > 1$) as measured by EGRET while taking into account this absorption, such objects are predicted to be accessible with ACT telescopes if the threshold energy is lowered to below 100 GeV (Stecker et al.). Observations of distant quasars at such energies would confirm the above explanation and provide the first measurement of the inter-galactic infra-red background light. This can lead

Space Science Reviews **75**: 127–136, 1996.

to a new estimation of the era of galaxy and star formation. This is critical to discriminate between Hot or Cold Dark Matter scenarios (MacMinn et al.).

Unfortunately, currently-operating ACT telescopes have thresholds greater than or of the order of 200 GeV. Since no space mission is expected to cover the energy gap between EGRET and ACT data in the near future, lowering ACT thresholds was recently recognised as a major goal in γ-ray Astronomy (Weekes). The 10 GeV to 200 GeV energy region turns out to be critical for many sources: the Crab pulsar signal (periodic signal) is observed in the GeV range but is not seen by ACT telescopes whereas the unpulsed signal (attributed to the Crab Nebula) is detected in both domains. The periodic signal should exhibit a strong cut-off in its spectrum which unfortunately is located in the unexplored region. The spectrum shape in this domain should shed some light on the production mechanism of pulsed γ-rays; this point is also true for other pulsars, and for supernova remnants.

2. The Atmospheric Cerenkov Technique

When viewed through its Cerenkov light, an air shower initiated by a cosmic gamma of a few tens of GeV is a long and straight object, starting in stratosphere—some 25 km above sea level (a.s.l.)—and extending down to about 5 kilometres, with a lateral extension which does not exceed about 20 meters. For an observatory at a typical altitude of 2000 m, the illuminated field on the ground is about 120 m in radius. The angular acceptance of the ACT detectors should be large enough—at least 0.6^0—so as to embrace the entire apparent size of the long rods.

To keep the night-sky light pollution within a tolerable level, the use of fast photon-detectors—such as photo-multiplier tubes (PMT's)—is required so as to take advantage of the quasi-simultaneity (within 2-3 ns) of the photon arrival times. The energy threshold of a detector is determined by the number of the shower photons and by the noise level. The noise can be expressed as the square root of the night-sky light integrated over the time of the coincidence (a few ns). The signal varies roughly as the shower energy. The gamma-ray signal and the noise, respectively, may be characterised by the following numbers (for $350 \leq \lambda \leq 550$ nm):

$$N_{\gamma-signal} = 0.04 - 0.1E_\gamma(GeV)photons/m^2$$

$$N_{sky-light} = 0.4photons/(deg.^2m^2ns)$$

This leads to the following rule-of-thumb relationship between the threshold energy, E_{th}, and the mirror size A_E: $E_{th} \propto A_E^{-1/2}$.

A similar description would hold for a charged primary instead of a gamma. However, the chain of interactions initiated by a hadron differs signif-

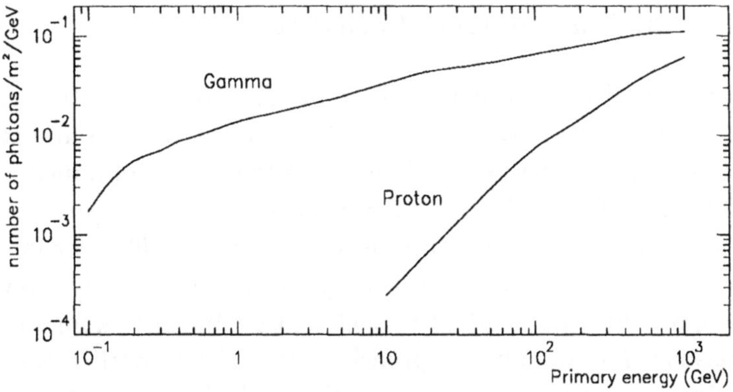

Fig. 1. Cerenkov photon yield divided by incident energy for gamma showers (upper curve) and hadronic showers (lower curve). All Cerenkov photons within 350$\leq \lambda \leq$ 550 nm are summed.

icantly along the shower path through the atmosphere. For these hadron-generated showers, the Cerenkov light is mostly from gamma-showers initiated by secondary π^0 decays. A smaller contribution comes from the charged pions and their resultant muons. Some of those muons may reach the ground and radiate near the detector. The large transverse momenta given to the π's are responsible for a rather chaotic shower shape, as are the penetrating muons. At low energies, at which the shower is fed by just a few pions of moderate momenta, the angular perturbations become even more predominant.

As a result of the $\approx$ 0.1 GeV energy cut-off on the radiating electrons, the photon yield should suffer a smooth cut-off when the primary energy is reduced to a few GeV. This effect, shown in Fig. 1, affects the gammas and the protons quite differently. While for gammas the total number of Cerenkov photons remains nearly proportional to the incident energy down to $\approx$ 20 GeV, for hadrons the cut-off occurs at about 50 GeV. When their energy goes down to the 20 GeV region, their Cerenkov photon yield has essentially vanished. This quite favourable situation, which has been known for some time, is a result of the fast degradation in energy in the hadronic cascade before an electron is generated. Of course, hadrons of higher energies—say above 100 GeV— are still there and producing about as many photons as a 20 GeV gamma. Nevertheless, the spectrum of the CR flux is steep, so this natural cut-off is a blessing.

The diffuse cosmic electrons might become the major contributor to the background. An electron (or positron) shower is just like that of a gamma of the same energy. Good pointing determination can be obtained, as discussed below, which will constitute the only efficient control to this effect.

3. Multi-mirror alternative

To proceed further, the mirror strategy needs to be discussed. Since the A_E versus E_{th} rule of thumb is roughly correct, the total collection area needed for a 20 GeV observation may be extrapolated from the results obtained at higher energies. The most favourable basis for this extrapolation is provided by imaging telescopes. Even so, quite an impressive area of: $2000 - 5000m^2$ is required. Imaging techniques, based on a single mirror or possibly a few mirrors, would be quite difficult to adapt to such a large dimension, for simply mechanical reasons. The sampling approach, that of the THÉMISTOCLE and ASGAT experiments, could be more appropriate. The 7m-diameter mirrors of ASGAT might serve to give a scale. A few hundred such 7m-diameter mirrors would be needed. The sophistication—and cost—of a fine-imaging camera on each mirror could be avoided if the hadron rejection were based on the fine multi-mirror sampling which is available. The chaotic development of the hadronic shower which causes fuzzier images also affects the distribution of the "light-pool" of Cerenkov photons on the ground. For a 20 GeV gamma event, the light-pool distribution is flat and ends at about 150 m from the shower impact point after a ridge in the distribution. This ring is produced by the first generations of electrons in the shower, at about 20 km a.s.l. The ring may be exploited as a gamma signature to reject protons or even electrons although, as it contains not more than 10% of the Cerenkov photons, this is rather demanding in terms of the mirror collection area. A more statistically-efficient signature is based on the homogeneity of the photon distribution over the entire light-pool. Gamma events have a smooth distribution while protons tend to generate patchy patterns. *Most often, the same hadron events survive selection by either the imaging or the light-pool (homogeneity) criteria: for these events a secondary π^0 dominates in the cascade.*

3.1. Observation strategie

To maximise the number of collected shower photons relative to the night-sky noise, the field of view of each individual mirror should be kept small. A 0.6^0 aperture suffices to contain the shower image, provided the image is roughly centred with respect to the optics. This centering can be achieved for all the mirrors by varying the tilt of their optical axes according to their position in the field. Each mirror is oriented toward a central fiducial volume at the common altitude of the shower maxima, at $\approx$ 13 km a.s.l. Of course, the centre of this fiducial volume should vary according to the direction of the cosmic source under investigation, in such a way that its projection along the line of sight falls on the centre of the site. The method combining tilted axes and narrow mirror-acceptances trades an improved signal to noise ratio—and thus a lower energy threshold—against a reduced fiducial area, which still remains as large as $\approx 10^4 m^2$. Furthermore, this method discriminates against the hadrons, for which only 10% of the light is contained within the 0.6^0 acceptance angle compared to 60% for the gammas.

3.2. Angular measurement

Above a few TeV, the wave-front has a conical shape, the axis of which may be determined by timing measurement, as in the THÉMISTOCLE experiment. Such a method cannot be used below a TeV, because the wave-front then reduces to an expanding sphere having its (fixed) centre at the shower maximum. The position of this centre can be determined by the time-samples: the linear term gives its direction and the second-order term its distance (the radius of curvature of the wave front should match the expected altitude, some 13 km a.s.l.). However, the shower axis remains undetermined unless some other information can be used, such as its impact on the ground.

The centre of gravity of the light-pool is an estimator of the axis' impact, at least if the light-pool is fully-contained within the field of mirrors. An alternative method is to fit an ad-hoc distribution—of pre-set size and shape—to the observed population. The occurrence of the sharp edge, with its ring-shaped enhancement, facilitates the fit. But even so, at the lowest energies the method suffers from the effects of the Earth's magnetic field; this tends to make the light-pool edges more diffuse on the East and West sides as the light-pool illumination is influenced by the magnetic deflections of low-energy secondary electrons.

3.3. An ideal set-up

To give a more quantitative evaluation of an ACT multi-mirror array, a rather idealised set-up is now considered (referred below as the "ideal set-up"). The ideal set-up is assumed to sit at a medium altitude of 1.7 km a.s.l.(chosen as that of the Thémis site). The array of telescopes of the ideal set-up will cover a disk 400 m in diameter so as to contain fully the light-pools of γ-showers of $\approx$ 300 m diameter having their impacts within a 100 m diameter central "fiducial" zone.

The size of each mirror and the mirror inter-space in the array define the effective collection area for a shower. Assuming a $38m^2$ area and a mean distance of 14 m (as for the heliostats of the Thémis solar plant), the mirror occupation factor is $A_{mirror}/A_{total} \approx 20\%$. To equip such a 400 m-diameter field, 640 mirrors are needed. The mirror reflectivity is assumed to be limited by the transmissivity of glass to $\lambda \geq 350$ nm. The loss in photon yield is $\approx 25\%$ for the use of more robust back-plated mirrors (or sandwich of glass panes as for the Thémis solar heliostats) instead front-plated ones. It is assumed that the mirrors are isochronous to within 2-3 ns and that their optical aberrations are kept within 0.15^0 (about a fourth of a shower image).

The camera is made of a single PMT subtending 0.6^0 acceptance. The information that may be collected at each mirror includes the PMT output charge integral—which is a measure of the number of collected photons—and the time of these signals. It is here assumed that both data are exploited for the evaluation of the shower energy, the hadron rejection and the direction measurement.

The trigger is assumed to be based on 4-mirror sub-sets, requiring analogue adjustable delays between the elements of a sub-set, but only within a 20 ns total range (this is manageable with existing computer-controlled delay-cable boxes). Per sub-set, the expected average number of detected photons of night-sky light piling up within the 2-3 ns PMT pulse duration is ≈ 7. To be sensitive to a 20 GeV gamma, the trigger at the sub-set level must be validated whenever it sums up to 15 photoelectrons or more. In order to trigger on showers falling in the central fiducial region, only a third of the sub-sets (those around the centre of the field) are taken into account at this level. The final trigger is generated whenever 30 sub-sets respond simultaneously—i.e., within $\leq 10ns$—their respective delays being duly corrected for (as is currently done in the THÉMISTOCLE experiment's electronics). The expected rate of accidental triggers should be negligible as long as the sky-light does not increase by more than 50 %. In addition to the diffuse night-sky light, the brighter stars would distort observations in their vicinity. A 4th magnitude star falling in the field of view of a mirror would double the average night-sky light. The odds for the presence of a star brighter than this in an angular acceptance of a square degree is less then a percent. The tilted-axis strategy partly relieves this difficulty since the telescopes are not looking at exactly the same patch of sky.

3.4. Simulation results

Simulations were run for incident gamma rays (or electrons) and hadrons, following the shower development through the atmosphere and tracing the Cerenkov photons down to the mirrors and up to the cameras. Sky noise was taken into account and timing resolution was assumed to be the same as in the THÉMISTOCLE experiment.

The resolution in the measurement of the shower axis, as shown in Fig. 2, is $\approx 0.1^0$ in the projected angle. The angular resolution is nearly two times better in the parallel direction of the earth's magnetic field than in the perpendicular direction. This is ten times better than EGRET's. The improved angular resolution would permit a better separation of point-like sources in the dense region of the Galactic plane, opening new possibilities for the study of extended sources. From 20 to 200 GeV, the effective collection area remains high, at about $8 \times 10^3 m^2$.

The γ yield expected from the Crab—summing the pulsed and unpulsed contributions—is given in table I, according to the different selection lev-

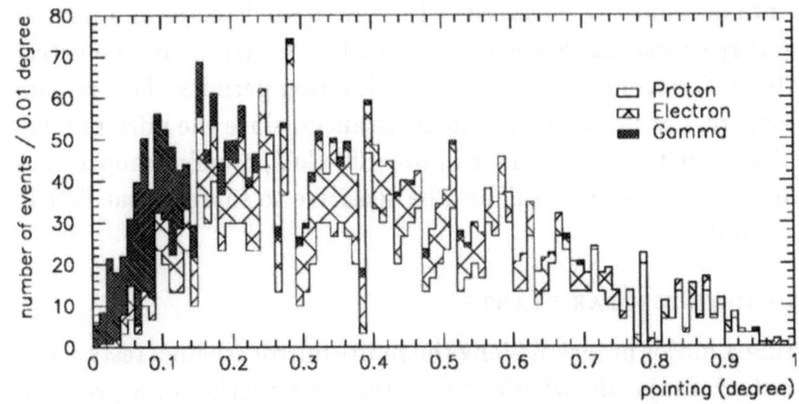

Fig. 2. Distribution of the measured direction of the shower axis for the events selected on shape only: for the gammas (black area), for protons, and for electrons(shaded).

els (trigger and analysis). The expected shower rate due to cosmic primary protons and electrons are also shown. The off-line "shape criteria" selection is based on a chi-squared fit of the number of detected photons per mirror according to a smooth distribution law (constructed as a bi-dimensional Gaussian distribution centred on the impact point together with a Gaussian-shaped ring). A selection is also made on the basis of the timing measurements, by requiring that the wave-front be approximately spherical.

TABLE I

The rates of events for a 400 m diameter field with 650 $38m^2$ mirrors. The energy threshold is expected to be $E_{th} \leq 20GeV$.

	flux $\geq$ 10 GeV $\times 10^8/cm^2s$	Trigger evt/s	Shape sel. evt/s	Ang. sel. evt/s
Gammas from Crab	1	0.53	0.36	0.3
Electrons	$2.5/deg.^2$	1.6	1.0	0.08
Protons	$600/deg.^2$	15.	1.0	0.05
Gamma/(Electrons+Protons)	1/600	1/30	1/6	2/1

The performance of the apparatus as expressed in Table I can also be summarised as follows:

- within an observation time of 5 minutes, the Crab signal should reach a statistical significance of 7 standard deviations.
- within 10 hours of observation, a cosmic source with a flux above 10 GeV of $5 \times 10^{-10} cm^{-2} s^{-1}$ could be seen at the 5 standard-deviation level; this is about one twentieth of the Crab flux.

- taking into account the sensitivity of the EGRET detector and its measured energy spectra, any source observed by EGRET should be detected by the ACT set-up if the source's declination permits. This remains true for sources at large cosmological distances, since the infra-red light absorption, which can be drastic above 100 GeV, remains inoperant at $\approx$ 10 GeV. This telescope will also be sensitive to sources that EGRET only set limits on.

3.5. Using existing solar plants

Before starting a major program along the multi-mirror scheme, tests should be carried out. A major drawback is that the construction of a prototype would be of no value unless it were tailored to run below 50 GeV, the domain where charged cosmic rays are fading out. This would already be quite a major construction program. Following a suggestion by Tümay Tümer (Tümer) to re-use the "Solar-1" solar plant in California, we have investigated the use of the Thémis solar plant in the French Pyrénées, which is the site of the THÉMISTOCLE and ASGAT experiments. These solar plants are extensive fields of large ($\approx 50m^2$ mirror) heliostats, focusing the solar light on a single furnace at the top of a tall tower. Instead of an independent photo-detector at each mirror, there is a unique photo-collecting furnace. As first noted by Tümer, the fact that the focal point is relocated to a common focus does not rule out the independent collection of each heliostat's photons. This is essential to allow the individual adjustment of delays and to avoid an exceedingly large night-sky light collection. Replacing the solar furnace by a secondary optical system—a large Fresnel lens (as proposed by Tümer) or a mirror—would provide a solution. For each heliostat, an individual PMT may be placed at its image position in this secondary optical system which will collect the photons reflected by each individual heliostat. The rather small acceptance of 0.6^0 needed for observation of the shower results in a moderate aperture for the secondary optical system, of the order of 1.5-2.5 m diameter. In practice, this secondary optical system must be divided into several sub-units in order to subdivide the large solid angle of the field as viewed from the top of the tower.

From that point on, everything should be about the same as for the ideal set-up...except that the solar plants exist. The Solar-1 and Thémis solar plants are not identical. The heliostat array is sufficiently large at Solar-1 and somewhat sparse at Thémis. The higher altitude (of 1650 m), better mirrors (with an angular resolution of $\approx 0.15^0$), the free access (with no conflict with a primary user) favour Thémis. The new imaging telescope to be built on this site, the CAT project—with an expected threshold of 200 GeV—will further increase the opportunities for tests. Attempts should be made to validate the scheme for 20 GeV gamma astronomy by the use of either site. Preliminary tests are running on both sites.

As a first step towards reactivating the Thémis heliostats, simulations have been run along the same lines as for the study of the ideal set-up above. If we take into account the collection efficiency of the secondary optics, the mirror effective area per heliostat reduces from nominal $54m^2$ to something more like the $38m^2$ of the 'ideal set-up". Because the diameter of the field is smaller (250 m) angular resolution decreases to 0.15^0. Hadronic rejection is also degraded. An energy threshold of about 25 GeV should be attainable. Even with partial use of the field—using 40 of the 160 heliostats—the signal from the Crab, which could be seen in even a short time of observation, should permit the validation of the method.

TABLE II

Expected energy threshold and yield from the Crab nebula for two phases of exploitation of the Thémis heliostats.

Nb of heliostats	Threshold	Selected γ	γ-Crab/(e+p)	γ-Crab: σ/hr
160 (total)	25 GeV	0.2 / s	1 / 2	12
40	50 GeV	0.08 / s	1 / 2	8

4. Comparison with future satellites and other ACT techniques

On the basis of the great success of EGRET, a new generation of satellite observatories for high energy gamma ray astronomy with improved angular resolution and sensitivity to lower fluxes is under study (Godfrey; Hunter). The characteristics of the future satellite detectors have been based on evaluations given for GLAST. Above 10 GeV, the angular resolution is good enough to completely reject the background from diffuse gammas. The sensitivity is only limited by statistics as the effective area of $1m^2$ is still small. For the weakest sources detectable by Thémis, GLAST will see only 2 events a month above 10 GeV and only 2 events a year above 100 GeV. So quantitative study in this energy region, such as the attenuation of the AGN gamma spectra by the extragalactic infrared light, will be nearly impossible. Nevertheless satellites will be complementary to ground based ACT as they can continuously survey the whole sky whereas ACT telescopes can only look at one source at a time and thus depends on a pre established catalog.

The sensitivity of the "ideal set-up" is limited by the electron background which can only be reduced by improving the angular resolution. Up to now the best performance has been obtained with the imaging technique. To lower the energy threshold to 10 GeV, we need an array of $\approx$100 25m-diameter reflectors, each one equipped with a very high resolution camera (0.1^0 pixel size) to be able to reconstruct the axis of the shower images. The total number of PMT's is $\approx$ 50 000. We have have run simulations with such a set-up taking into account the sky noise but neglecting optical aberrations. The surprising result is that, at 20 GeV, the angular resolution is approximatly the same as the "ideal set-up"(0.1^{o}). This is 5 time worse than a naive statistical extrapolation of the Whipple results. It seems that the earth magnetic bending and the multiple scattering on the very beginning of the shower are responsible for this poor angular resolution. *At low energy, the angular resolution of ACT telescopes seems to be intrinsically limited. Very sophisticated, and very costly, telescopes might not perform much better than solar arrays.*

5. Conclusion

After the Whipple group established that, among the numerous air showers, the few having a common point-source in the cosmos could be pinned down, construction of new experiments and upgrades of existing ones began, aiming at enlarging the domain of investigations. The developments may concern the extension of the energy domain and the improvement in the sensitivity to cosmic sources of lower flux. The efforts are based on variations on the imaging techniques, either by a duplication of the telescope for stereoscopic observation, or by improving the camera performance. Another approach based on the sampling of the Cerenkov light pool by simultaneous use of many heliostats has been validated by the THÉMISTOCLE and ASGAT experiments. This multi-mirror strategy might well be appropriate to the low energy domain for which a huge collection area is needed, as could be tested with the use of solar heliostats. Combined with the CAT imaging telescope, to be built in 1995 on the Thémis site, aiming at a 200 GeV threshold, the fine imaging and the dense sampling techniques could then be used simultaneously on the same events. It is hoped to open investigation in a new range of energies.

References

Fichtel, C.E. et al.: 1994,*ApJS* **94**, 551.

Godfrey, G.L. et al.: 1993, Towards a Major Atmospheric Cerenkov Detetector II, ed. by D. Lamb, Iowa State University, 145.

Hunter, S.D. et al.: 1993, Towards a Major Atmospheric Cerenkov Detetector II, ed. by D. Lamb, Iowa State University, 135.

MacMinn, D. and Primack, J.: 1995 , *ApJ.* submitted.

Punch, M. et al.: 1992, *Nature* **358**, 477.

Stecker, F.W., De Jager, O.C. and Salamon, M.H.: 1992, *ApJ.* **L49**, 390.

Tümer, T. et al.: 1991, 22nd International Cosmic Ray Conference, **2**, 634.

Weekes, T.C.: 1992, Towards a Major Atmospheric Cerenkov Detetector, ed. Frontière, 323.

THE ART AND POWER OF ČERENKOV IMAGING

D.J. FEGAN
University College Dublin, Belfield, Dublin 4, Ireland

Abstract. A detailed account is presented here of the steps involved in processing TeV images of light collected from showers using the Atmospheric Čerenkov technique. Noise treatment, image parameterization and analysis strategies are discussed with emphasis on the power of the imaging technique and it's accomplishments to date. The scientific justification for the development of third generation Čerenkov imaging systems is reviewed and scientific and technical objectives are assessed.

1. Introduction

The development of the Atmospheric Čerenkov imaging technique has been described in detail elsewhere (Fegan, 1992) and the observational status of the field has been summarised in (Weekes, 1992; Fegan, 1994). Major foreseeable realistic developments of the technique before the end of the century would appear to be twofold.

Firstly, it now seems technically feasible to develop and build a facility capable of having an operational threshold energy of 50 GeV or lower, at costs which are very modest in comparison with astrophysical space programs. Secondly, by employing close-pitched arrays of miniature photomultiplier tubes in conjunction with small light collection cones, it would appear possible, in principle, to achieve an angular resolution of better than 0.05°, surpassing the resolution attainable by EGRET at somewhat lower energies.

The scientific justification for the development of third generation Čerenkov detection systems based on efficient imaging cameras operating at very low energy thresholds have been alluded to by many of the contributors to this workshop. The aim of this paper is twofold. Initially the art of Čerenkov imaging will be summarised from the viewpoint of the handling, processing and analysis of images obtained from the Whipple high-resolution imaging camera. Subsequently the power of imaging will be reviewed in terms of what to date is achievable and how this might be improved upon.

2. The Art of Čerenkov Imaging

2.1. The Whipple 10m Reflector and Data Acquisition System

The Whipple Collaboration operates a 10 meter light collector on Mt. Hopkins, Arizona (2320 m a.s.l.). The reflector was originally constructed in 1968 (Weekes, 1972) and numerous modifications have been made to improve the sensitivity and performance of the system (Fegan and et al., 1983; Reynolds, 1989). The reflector in its present configuration has an array of

Space Science Reviews **75**: 137–151, 1996.

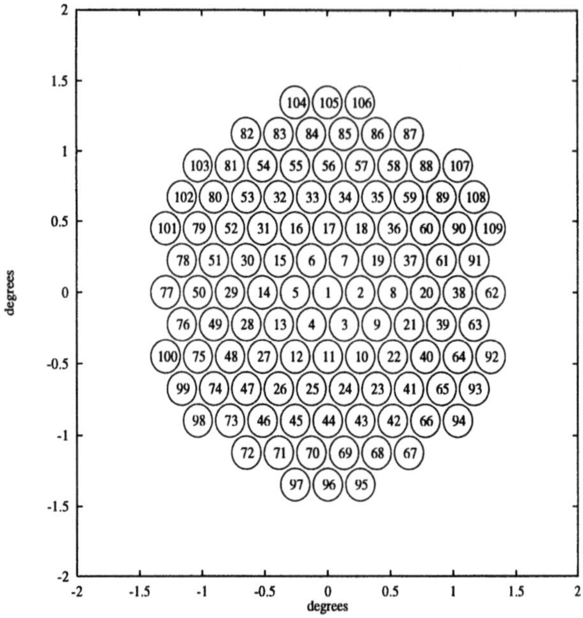

Fig. 1. The 109 tube HRC.

109 photomultiplier tubes (PMT) located at the focal plane of the reflector and is known as the High Resolution Camera (HRC). A full description of the reflector is given by Cawley and et al., 1985.

The HRC camera is illustrated in Figure 1. The array is composed of 109 2.9 cm diameter, fast-response PMTs with a centre to centre spacing of 3.2 cm, each tube viewing a region of the celestial sphere of diameter 0.232°. The angular resolution of the imaging camera is 0.1°

Primary cosmic rays entering the atmosphere initiate showers of secondary particles which propagate down towards the ground based detector. The Čerenkov light from each shower is focussed onto the array of PMTs forming an image. The appearance of this image depends upon a number of factors. The nature and energy of the instigating particle, as well as the point of impact of the light shower on the ground, will determine the initial shape of the image. This image can then become further modified by any defects in the optical quality of the telescope, instrumental noise within the PMTs and subsequent electronics, the presence of bright stellar images in certain tubes, and by spurious signals from charged cosmic rays physically passing through the tubes.

Signals from each PMT are taken, via co-axial cable, to the electronics where they are subsequently amplified. In order for an event to be recorded we define a *hardware* trigger threshold, which demands that at least N out of the 91 inner tubes have a digital count (d.c.) greater than some prede-

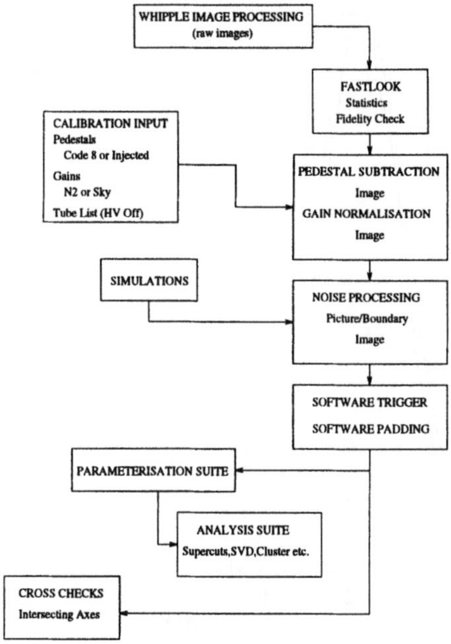

Fig. 2. Data analysis flow diagram.

termined threshold. This is usually set at $N = 2 \geq 40$ d.c. but may vary depending upon the mode of operation.

Each time the camera is triggered by a Čerenkov event, or an external trigger, the event times and Analog-to-Digital (ADC) tube values are read and recorded by an VAXstation computer.

For a 2 out of 91 trigger condition, simulations indicate a γ-ray threshold energy of 0.4 TeV and a collection area of $3 \times 10^4\, m^2$, with a factor of 1.5 uncertainty.

2.2. Off-line Data Analysis

Prior to any meaningful analysis of the raw data images two calibration operations must be performed: the subtraction of the pedestal ADC values and the normalisation of the PMT gains, a process known as flat-fielding. The analysis carried out on the raw images is summarised in the Figure 2.

The pedestal value of an ADC is the finite value which it outputs for zero input stimulus. This is usually set at 20 d.c. so that small negative excursions on the signal line, due to night sky noise fluctuations, will not generate pseudo-negative values in ADC. The pedestal for each PMT is calculated by one of two methods. In the first, the camera is artificially triggered so as to capture ADC values with no genuine input signal. The PMT pedestal and pedestal variance is calculated from simple moments of the pulse-height spectrum (PHS) generated from these injected events. In the second, the pedestals may be calculated using Čerenkov light triggered events, since only

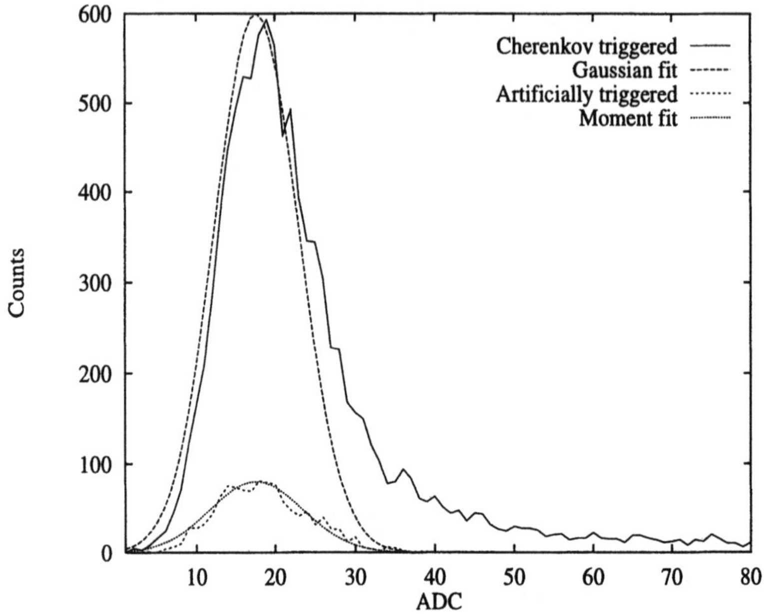

Fig. 3. Calculation of ADC pedestal.

a small number of the PMTs will contain a signal. In this process a Gaussian is fitted to the event PHS and a rough pedestal and pedestal variance is obtained. In a second pass through the data, PMTs having a signal greater than 4 times the pedestal standard deviation after pedestal subtraction, are removed. A second Gaussian is fitted to the resulting cropped pulse-height spectrum and the pedestal estimated.

Figure 3 gives an example of a Čerenkov triggered event PHS and an injected pedestal PHS. A Gaussian corresponding to pedestal and pedestals standard deviation as calculated by both methods is given for comparison.

The determination of relative PMT gains can also be approached by two methods. In the first, files of about a thousand images are recorded twice-nightly using a fast Optitron arc lamp illuminating the focal plane through a diffuser. These Nitrogen pulsar files are used to determine the relative gains by comparing the relative mean signals seen by each PMT. In the second method, gain information is extracted from actual Čerenkov images recorded by the camera. Here we assume that over many thousand events, each PMT is illuminated by Čerenkov light generated by the isotropic background showers and hence will be uniform in average ADC stimulation. A simplified approach to including PMTs having an ADC dominated by Čerenkov light and not the night sky is to use the the larger signal portion of the PHS. In our procedure we use the width of the tail of the integral PHS (Figure 4). For each tube, we define this width to be the number of digital counts from $0.3N_o$ to $0.05N_o$ where N_o is the total number of events used

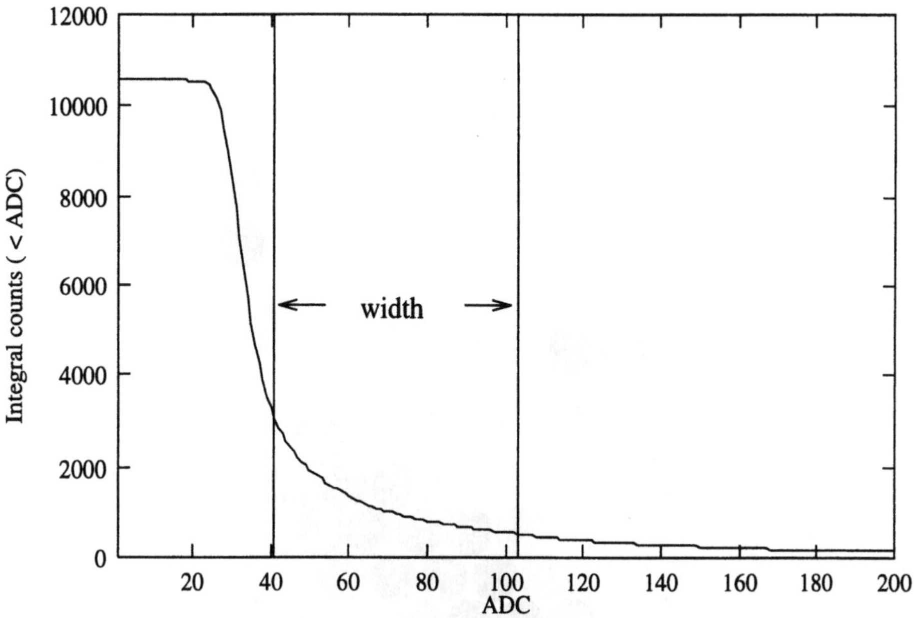

Fig. 4. Calculation of relative phototube gain factors.

in forming the PHS. In general we find that both procedures give similar results in the gain factors calculated and when applied to real data.

Fluctuations in the image usually arise from electronic noise and night-sky background variations. To reject these distortions a PMT is considered to be part of the image if it either has a signal above a certain threshold or is beside such a PMT and has a signal above a lower threshold. These two thresholds are defined as the Picture and Boundary thresholds, respectively. The picture threshold is the multiple of the RMS pedestal deviation which a PMT's signal must exceed to be considered part of the picture, and the boundary threshold the multiple which PMTs adjacent to the picture must exceed to be part of the boundary (Figure 5). The picture and boundary PMTs together make up the image; all others are zeroed. These thresholds were optimised by Punch, 1991 *a posteriori* using the 1988-89 Crab database, consisting of 65 ON/OFF pairs of duration 28 mins each (picture threshold: 4.25, and boundary threshold: 2.25).

After this noise treatment, the Nitrogen gains factors are used to flat-field the camera. The data is then subjected to some low level filtering. A *software* trigger is then applied to each processed image, demanding that at least two of the inner 91 PMTs contain signals greater than 40 digital counts. This eliminates marginal events where noise dominates. Images for which only two of the remaining PMTs contain a signal are excluded as being probably caused by local cosmic-rays physically passing through the phototube array. Images with maxima in the central seven tubes or in the

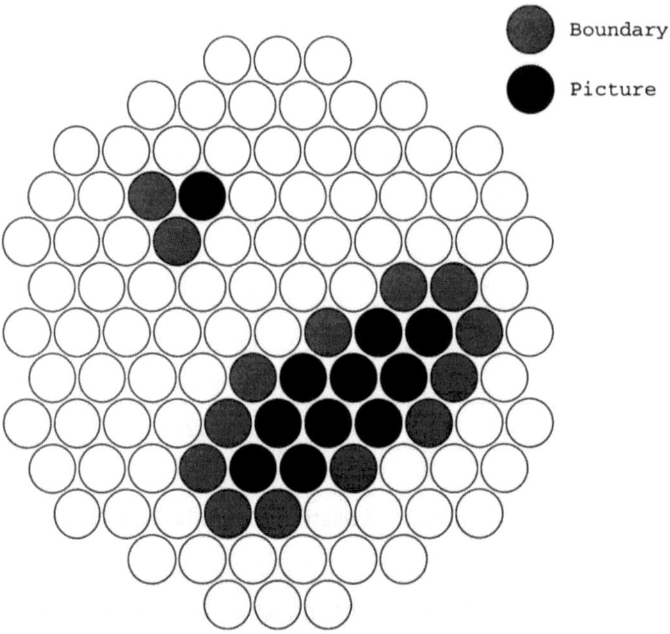

Fig. 5. Illustration of picture and boundary thresholds.

outer ring of PMTs are also excluded, since the orientation of those in the centre is not well defined and those on the periphery are only partially captured by the camera.

Absolute calibration of the PMTs can be made by the use of a muon telescope (Rovero and et al., 1992) working in coincidence with the 10m. For data taken between April and May 1992, an average value of 1.44 ± 0.03 photo-electrons produced at the photo-cathode is obtained for every digital count in the ADCs, implying an energy threshold of 0.5 TeV for the telescope.

2.3. Software Padding

In an ON/OFF data acquisition mode, differences in sky brightness between ON and OFF sky regions can introduce bias. This bias can severely affect the selection of pixels accredited to the boundary region and hence distort the Čerenkov image. For example, if the ON region is brighter than the OFF region then negative bias may arise if fixed threshold image cleaning mechanisms are employed as noise-dominated tubes will sometimes exceed this threshold and be treated as part of the image, increasing its width. Such image broadening reduces the number of events accepted into the γ-ray domain. However, if a scaled threshold is used (as is done in Supercuts (Punch, 1991)) then the noise dominated tubes will have their thresholds correctly adjusted to compensate for the noise levels. This process automatically eliminates the negative bias and no additional processing is required.

Positive bias may arise as follows. Consider tubes which have a combination of small amounts of genuine signal coupled with some noise (e.g. boundary tubes). If the noise level is low the scaled Supercuts threshold is low and most of such tubes will pass the threshold test and be included as part of the image. However, if the noise level is high then the scaled threshold will also be large. Pulse-height spectra for such tubes will tend to be more symmetric with fewer such tubes passing the applied threshold which leads to a narrower image. Such images will be classified as γ-ray images.

A software technique has been developed to correct for the bias introduced by the differing sky brightnesses (Cawley, 1993). This technique, known as *noise padding* works by injecting additional software noise into the events for the *darker* sky region. Let $P_{on}(P_{off}), \sigma_{on}(\sigma_{off})$ be the $ON(OFF)$ pedestal and pedestal deviation values for a particular pixel and $C_{on}(C_{off})$ be the component due to the Čerenkov signal. The total *ON* signal is:

$$ON = P_{on} + \sigma_{on} Gauss(0:1) + C_{on} + \sqrt{C_{on}} Gauss(0:1), \quad (1)$$

where Gauss(0:1) is a random number drawn from a gaussian distribution of 0 mean and unit variance. The noise component due to the night sky light in the *ON* region is:

$$N_{on} = \sigma_{on} Gauss(0:1). \quad (2)$$

Similarly, for the *OFF* region:

$$N_{off} = \sigma_{off} Gauss(0:1). \quad (3)$$

Suppose we are working with pairs where N_{on} is larger than N_{off} then we wish to inject additional noise, N_{add}, in the *OFF* events such that

$$N_{on}^2 = N_{off}^2 + N_{add}^2, \quad (4)$$

or

$$N_{add} = \sqrt{N_{on}^2 - N_{off}^2}. \quad (5)$$

The total *OFF* signal is then

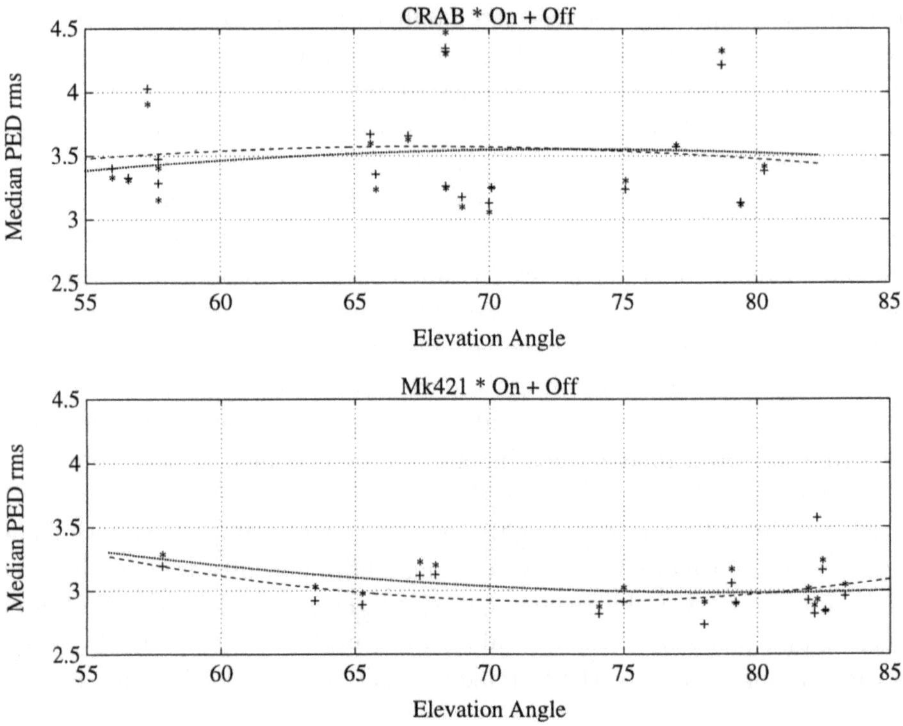

Fig. 6. Comparison of the sky brightness for the ON and OFF regions for both the Crab and Markarian 421 as a function of elevation angle.

$$OFF = P_{off} + \sigma_{off} Gauss(0:1) + N_{add} Gauss(0:1) + C_{off} + \sqrt{C_{off}} Gauss(0:1). \quad (6)$$

In a region where the *OFF* region is brighter than the *ON* region then N_{add} is added to the *ON* pixels. Software padding procedures seem to be very efficient in removing most of the bias induced by the sky brightness without reducing sensitivity.

Examples of sky brightnesses are given in Figure 6 for the *ON* and *OFF* regions of the Crab Nebula and Markarian 421. For both objects the difference in sky brightness in the *ON* and *OFF* region is small and quite constant with zenith angle, hence software padding is not required. However, when the difference between *ON* and *OFF* regions is larger than ≈2% (some X-ray binaries and SNRs), software padding should be applied. Another candidate for padding is the absolute comparison of two objects, such as the Crab and Markarian 421. Figure 6 shows a marked difference in the sky brightness

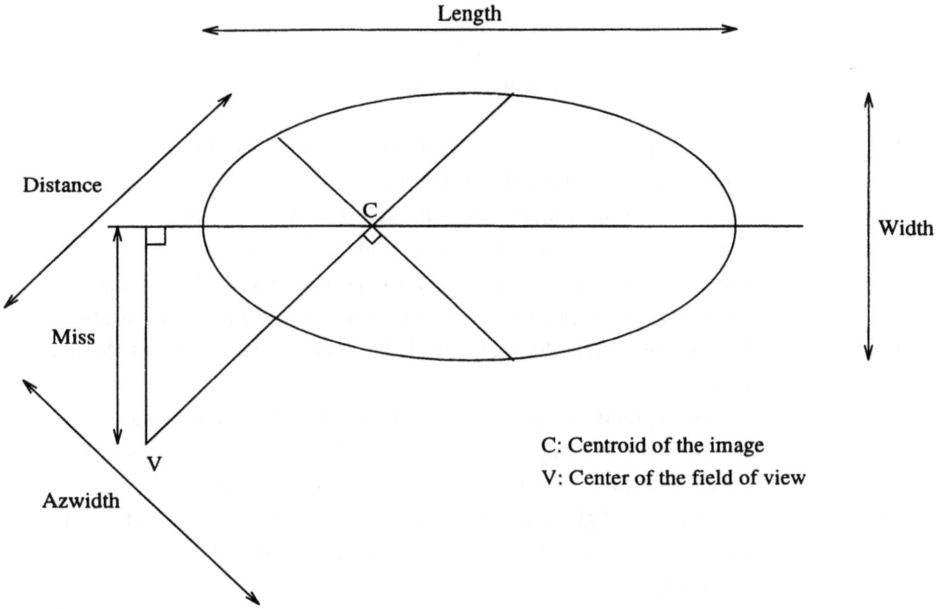

Fig. 7. The definition of the Hillas parameters.

between these sources. In this case Markarian 421 would be padded up to the level of the Crab.

2.4. The Hillas Parameters

For each image, which is essentially elliptical in shape, six initial parameters are calculated to characterise its shape and orientation. Other parameters which are combinations of these initial parameters, such as α and asymmetry, can then be calculated. The Hillas Parameters are defined in Table I and illustrated in Figure 7. Supercuts represents a method of selecting γ-rays on the basis of both shape and orientation of images (Hillas, 1985).

Simulations predict that images from γ-showers tend to point towards the centre of the camera and are narrow and compact in shape, elongating as the impact parameter is increased. Proton images are, on average, broader (due to the emission angles of pions in nucleon collisions spreading the shower), and longer (since the nucleon component of the shower penetrates deeper into the atmosphere) and only occasionally, by accident, orientate themselves towards the centre of the camera, since they are an isotropically distributed within the field of view, whereas γ-rays emanate from a single point.

TABLE I

The Hillas parameters.

Width:	the rms spread of light along the minor axis of the image; a measure of the lateral development of the cascade.
Length:	the rms spread of light along the major axis of the image; a measure of the vertical development of the cascade.
Miss:	the perpendicular distance between the major axis of the image, and the centre of the field of view; a measure of the shower orientation.
Distance:	the distance from the centroid of the image to the centre of the field of view.
Azwidth:	the rms spread of light perpendicular to the line connecting the centroid of the image to the centre of the field of view. This is a measure of both the size and orientation of the image.
Conc:	the degree of light concentration as determined from the ratio of the two largest tube signals to the sum of all signals. (Also referred to as Frac2).
Alpha:	the angle between the major axis of the ellipse and a line joining the centroid of the ellipse to the center of the field of view.
Asymmetry:	a measure of how asymmetric the image is. γ-ray images should have tails which preferentially point away from the source position.

3. The Power of Imaging

3.1. Current Single Telescope Imaging Capabilities

As an indication of the capability of the imaging technique as applied by the Whipple collaboration, results of an analysis of a small representative data set are presented here. The data set consists of 19 ON/OFF file pairs taken on the Crab Nebula during 1993-94. The total ON-source exposure corresponded to 473 minutes of observation taken in perfect weather conditions with a source elevation in excess of 55 degrees. The performance of two independent analysis methodologies, Supercuts and Cluster analysis, are presented for comparative purposes in Table II. These figures for the Crab Nebula observations correspond to a flux of $6.34 \times 10^{-11} cm^{-2} s^{-1}$ at a threshold energy of 0.3 TeV.

The figures presented in Table II are indicative of currently attainable selectivity of the imaging technique in terms of ability to reject hadronic background. The efficiency with which the unwanted background events are rejected,η_{cr}, is in the region 3×10^{-3} to 5×10^{-3} corresponding to about one hadron in every two to three hundred events being mis-identified as γ-rays. Based on simulations the predicted rejection factor for a single imaging telescope would be expected to approach 10^{-3} under ideal conditions. At

TABLE II
Relative performance of Supercuts and Cluster analysis.

	Raw	Shape	Ori	Supercuts	Cluster
ON	192003	6547	12968	1828	1356
OFF	192105	5617	11729	940	622
Excess	-102	930	1269	888	734
σ	-0.16	8.43	7.88	16.88	16.14
η_{cr}	-	-	-	.0049	.0032
σ_{rate}	-	-	-	$\approx 6\sigma hr^{-1}$	$\approx 5.96\sigma hr^{-1}$

the same time, simulations tell us that the efficiency for selections of γ-rays η_g is between 0.5 and 0.6. That this ideal rejection efficiency of 10^{-3} for hadrons has not so far been attained is probably due to a number of factors of which the most important would be our inability to adequately deal with the presence of noise and its influence on the camera images. Other contributory factors would be the inability to (a) collect all the light in the image plane (b) reject with 100% efficiency, charged cosmic ray particles passing physically through the photomultipliers (c) eliminate image arcs and segments of muon rings from charged particles passing local to the telescope (d) deal appropriately with the inherent energy dependence of image parameterization and selection procedures.

3.2. Future Single Telescope Scientific Objectives and Strategies

There are compelling scientific reasons as to why the TeV astrophysics community should attempt to improve the performance of the Čerenkov imaging technique. Perhaps the strongest motivation comes from the EGRET detection of more than forty AGN sources with differential energy spectral indices in the range 1.5 to 2.6, with no strong evidence for spectral breaks at the highest EGRET energies (30 GeV). The discovery of gamma ray emission from Markarian 421 at 400 GeV energies (Punch and et al., 1993) and the subsequent observation of short term variability of this object (Kerrick, 1995) has given a significant impetus to the TeV field.

Bridging the energy gap between EGRETS upper threshold energy (30 GeV) and the lowest energies currently attainable with imaging Cerenkov systems (250 GeV) will be a major objective. Lowering the operational energy threshold will open up a previously inaccessible decade of the electromagnetic spectrum (30 GeV to 300 GeV) with a potentially rich harvest of astrophysical discovery.

The most dramatic way to lower the threshold energy of a Čerenkov detector is by increasing the aperture of the reflector, although in terms of the commonly encountered budgets for TeV astrophysics, a reflector of 20m diameter with associated mount and drive, capable of a tracking accuracy of better than 0.05°, may prove to be beyond the financial capability of any single existing operational group or collaboration. However, while the largest reflector currently in operation has a mirror surface area of $75m^2$, the potential of a $300m^2$ mirror should not be dismissed lightly.

Indeed the argument in favour of such a reflector is compelling when it is realised that such a reflector would facilitate the exploitation of enhanced Cerenkov emission yield from showers of primary energy below 100 GeV. Between 100 GeV and 10 TeV the ratio ρ of Čerenkov light output from gamma-ray showers to proton showers of the same energy is a factor of 2 to 3. At energies below 100 GeV this ratio rises dramatically, since in lower energy cascades the average energy of pions decrease and the pion is more likely to survive to decay to a less efficient Cerenkov producing muon, hence diverting a smaller fraction of primary energy to electrons (Weekes and Turver, 1977). Since nature has presented us with such an opportunity (a factor of ten enhancement at 50 GeV rising to a factor of 100 at 1 GeV) we should attempt to capitalise upon it!

Commensurate with lowering the energy threshold, a secondary objective will be the progression towards finer granulation of the imaging camera through the deployment of increased numbers of smaller phototubes at the prime focus of the mirror. This will facilitate better angular resolution, driven by reduced photon densities as detector threshold energies are forced to lower values. The inherent angular resolution of Čerenkov detectors is better than that available with EGRET at 30 GeV thus offering the possibility of locating sources with sufficient positional accuracy to identify them on the basis of observation at different parts of the electromagnetic spectrum. It should also be appreciated that by employing finer pixellation there is the potential for improved image discrimination based on as yet unexploited image parameters such as asymmetry.

By extending the field of view of Čerenkov imaging detectors the collection area for showers is enlarged giving better characterization of more distant showers. By doubling the typical field of view to a full 6°, the establishment of source spectra should be much more straightforward, vitiating the sampling bias commonly found with existing detectors of which for distant showers suffer moderate to severe image truncation at the edge of the field of view. There are encouraging signs that cluster analysis techniques may become very efficient at detecting higher energy gamma-ray showers falling between 100 and 150m from the detector optic axis while at the same time rejecting higher energy hadronic showers. So, source spectra establishment corresponding to at least two decades of energy above threshold should be quite feasible. Of course it has to be appreciated that the combination of fine pixellation/high angular resolution detectors used in conjunction with wide fields of view, will be a financially costly combination in terms of the multiplicity of photomultiplier tubes and electronic processing channels. A feasible compromise may be to use a layered combination of differing photomultiplier tube sizes, fine resolution at the camera center with coarser grain pixellation at the periphery.

In the foreseeable future, the evolution of the imaging Čerenkov techniques may be expected to progress on the basis of the following aspirations:

Scientific Motivation:

- To exploit the potentially rich source harvest between 30 GeV and 300 GeV - AGN's, pulsars, cataclysmic variables, binaries and serendipitous sources.
- To determine source spectra in the energy range 30 GeV to 10 TeV.
- To study extended sources such as supernova remnants and to resolve possible multiple sources.
- To search for classical burst sources, primordial black holes and cosmic strings.
- To perform all-sky surveys and to measure the diffuse gamma ray background emission.

Technical Objectives:

- To attain improvement in flux sensitivity ($\leq 10^{-12} cm^{-2} s^{-1}$), reducing exposure time and leading to more efficient use of facilities. Improvements are largely determined by enhanced hadron background rejection capability and improved angular resolution.
- To lower the threshold energy of imaging systems $\leq$ 30 GeV.
- To improve the angular resolution to $\leq 0.05°$.
- To increase the field of view of detectors to $5° \leq \Omega \leq 6°$
- To improve the energy resolution of imaging detectors. Stereoscopic viewing may be required to attain a value of 20% or better.

Scientific strategies to enhance the power of imaging:

- Implementation of large aperture (d=20m) reflectors with rigid mountings and pointing/tracking uncertainties $< 0.05°$ accuracy.
- Deployment of pairs of single reflectors working together in stereoscopic viewing mode in order to improve source location accuracy and energy resolution.
- In multimirror reflector systems, reduction of tesselation loss of photons by denser packing.
- Reduction of light loss in the focal plane of the detector by employing light cones to compensate for dead space.

- Employment of cameras with finer pixel resolution in order to better exploit the inherent arc-minute resolution of images.
- Exploitation of better photoelectron conversion technology (Lorenz, 1993).
- Employ fast processing electronics to exploit pulse rise time discrimination.
- Exploit the possibility of using intelligent trigger selection based on Neural network type strategies (Buckley, 1995).
- Examine alternative strategies for software selection of γ-ray events (e.g. Neural network analysis of stereo images).

3.3. Whipple Collaboration - System Upgrades, Current and Future

During 1993-94 modifications pertaining to triggering, optics and focus box improvements have resulted in a reduction of the 10m reflectors threshold energy to about 250 GeV corresponding to an event rate at the zenith of about 20 Hz (Buckley and et al., 1995). Significant improvements in the data acquisition system have made this data rate possible. New observing strategies aimed at increasing the total ON-Source exposure time have now been adopted. Work on upgrading the GRANITE system has also been carried out (Buckley and et al., 1995).

Impending plans involve a further upgrade to the 10m system in terms of increasing the field of view from 3° to 6° while at the same time improving the angular resolution over the central portion of the field. The new focal plane topology envisages the use of 541 tubes. Increased angular resolution has always led to an improvement in the sensitivity of Cerenkov cameras. In 1986 the Whipple collaboration upgraded from 37 (0.5°) to 109 (0.25°) tubes lowering the flux sensitivity by a factor of three (Vacanti and et al., 1991). The CANGAROO collaborators have shown that by employing a 220 element camera (0.13°) the width of the selection parameter α was reduced from 15° to 6° (Kifune and et al., 1995). The CAT project proposes to employ 534 tubes (0.12°) but filling a field of view of 5° (Degrange, 1995).

Other developments envisaged by the Whipple collaboration include (a) faster motors on the 10m mount (b) intelligent trigger real-time systems to select γ-ray like events (c) expanded data-handling facilities to cope with increased rates (d) replacement of 11m optical support structure and mount (e) increased field of view for 11m reflector with the potential for more efficient stereoscopic imaging when run in conjunction with the 10m reflector.

Acknowledgements

The authors acknowledge the contributions of all the members of the Whipple Collaboration in the gathering of the data used for this analysis. In particular, P. Curran, J. McEnery and R.W. Lessard are thanked for help with this manuscript. We acknowledge support from the U.S. Department of Energy, from NASA, from the Smithsonian Scholarly Studies research fund and from Forbairt, the scientific funding agency of Ireland.

References

Buckley, J.H. et al. In *Workshop Towards a Major Cherenkov Detector (Tokyo)*, 1995. In Press.

Buckley, J.H. In *Workshop Towards a Major Cherenkov Detector (Tokyo)*, 1995. In Press.

Cawley, M.F. et al. *Exper. Astron.*, 1:131, 1985.

Cawley, M.F. In *Workshop Towards a Major Cherenkov Detector (Calgary)*, page 176, 1993.

Degrange, B. In *Workshop Towards a Major Cherenkov Detector (Tokyo)*, 1995. In Press.

Fegan, D.J. et al. *N.I.M.*, 211:179, 1983.

Fegan, D.J. *Towards a Major Atmospheric Čerenkov Detector*, page 3. Editions Frontières, 1992.

Fegan, D.J. *High Energy Astrophysics*, chapter 4, page 107. World Scientific, 1994.

Hillas, A.M. In *Proc. 19nd I.C.R.C. (La Jolla)*, volume 3, page 445, 1985.

Kerrick, A.D. *(submitted to Astrophysical Journal Letters)*, 1995.

Kifune, T. et al. In *Workshop Towards a Major Cherenkov Detector (Tokyo)*, 1995. In Press.

Lorenz, E. In *Workshop Towards a Major Cherenkov Detector (Calgary)*, page 182, 1993.

Punch, M. et al. *Nature*, 358:477, 1993.

Punch, M. In *Proc. 22nd I.C.R.C. (Dublin)*, volume 1, page 464, 1991.

Reynolds, P.T. PhD thesis, National University of Ireland, 1989.

Rovero, A.C. et al. In *A.I.P. Conf. Proc. (St. Louis)*, volume 280, page 1176, 1992.

Vacanti, G. et al. *Ap. J.*, 377:467, 1991.

Weekes, T.C. and Turver, K.E. In *Proc. of 12th ESLAB Symposium (Frascati)*, page 279, 1977.

Weekes, T.C. *Ap. J.*, 174:165, 1972.

Weekes, T.C. *Space Science Reviews*, 59:315, 1992.

STEREOSCOPIC MEASUREMENTS OF THE ČERENKOV RADIATION PRODUCED BY TEV GAMMA RAYS

P. M. CHADWICK, J. E. DICKINSON, M. R. DICKINSON, N. A. DIPPER, J. HOLDER, T. J. L. MCCOMB, K. J. ORFORD, S. M. RAYNER, I. D. ROBERTS, M. D. ROBERTS, S. P. TUMMEY and K. E. TURVER
Department of Physics, University of Durham, The Science Laboratories, South Road, Durham DH1 3LE, UK

Abstract.
Current approaches to the identification of VHE gamma rays using stereo images of Čerenkov light flashes are reviewed. The method of analysis developed by the University of Durham group involving a measure of cascade development is described. This method is applied to observations of a burst of VHE gamma rays from AE Aqr and enhancement of the gamma ray signal is demonstrated.

Key words: VHE gamma ray astronomy – gamma ray telescopes

1. Introduction

This paper describes the interpretation of measurements using ground based gamma ray telescopes of atmospheric Čerenkov light produced by 300 GeV cosmic rays. The purpose of the study is to identify those flashes of Čerenkov light which are due to the many charged cosmic rays and those due to the few primary gamma rays, so improving the sensitivity of ground based gamma ray telescopes. The measurements involve sampling the spatial characteristics of the Čerenkov light pool using two separated telescopes; the measured quantities employed in this treatment are the lower moments of the light distribution — the amount of light and the position of the centroid of the light in the field of view of the telescopes.

Useful measurements are therefore possible using telescopes employing medium resolution cameras incorporating 0.5° pixels; the observations do not require the high resolution study of the images with $\leq 0.25°$ pixels which is the basis of the very successful enhancement of telescope sensitivity by the Whipple collaboration (Fegan, 1995) and the proposal for the GRANITE project (Akerlof *et al.*, 1991).

The identification of the nature of the primary of the atmospheric cascade producing the Čerenkov light depends upon estimating a measure of the height of origin of the light; this is possible with varying degrees of dependence on the results from computer simulations of the cascading. The differences between the values for the average depth of maximum and the fluctuations in this depth for nucleon and gamma ray primaries are the basis of the identification of the primary.

Space Science Reviews **75**: 153–167, 1996.

Here we summarize the experimental facilities presently available (or to be completed soon) which will be capable of providing images measured in two separate telescopes. The various approaches to the analysis will be described and the strengths and weaknesses of each method will be discussed. Examples of the successful exploitation of the technique will be presented employing initial data from the University of Durham telescopes at Narrabri (see Chadwick *et al*, 1995).

2. Observatories with Multiple Čerenkov Telescopes

2.1. Introduction

There are five gamma ray observatories which have or shortly will have the capability of recording data on the images detected by separated telescopes. Three of the observatories are in the Northern hemisphere and two are in the Southern hemisphere. The minimum energy studied at these observatories ranges from 200 GeV to $>$ 1 TeV; four of the observatories have telescopes with matched performance whilst the fifth operates an unmatched pair of telescopes. Many of the telescopes have spatial resolution which far exceeds that required for this study, but nevertheless will provide data which may be interpreted using these methods and may allow exploitation of some of the advantages of this technique, *e.g.* applicability to low energy data. One of these facilities consists of two telescopes separated by a small distance (20 m) and so is not well suited for the trigonometric height of maximum (THOM) technique.

2.2. The GRANITE Project at the Whipple Observatory

The GRANITE project, located at the Whipple Observatory on Mount Hopkins, Tucson, Arizona (32° N, 2300 m asl) comprises a pair of matched telecopes, each with a single flux collector of area approximately 60 m^2 (Akerlof *et al.*, 1991).

They are separated by a distance of 140 m in a NS direction; one telescope has been in operation for 25 years and the other is still being developed. Each telescope has a high resolution (91 $\times$ 0.25° pixel size) camera viewing the single flux collector; the trigger requirement for each telescope involves selecting events when, for example, the signal from any 2 of the 91 pixels exceeds a specified level (about 40 photoelectrons). In the initial studies, samples of the same light pool will be identified by time stamps applied to the events from each telescope which trigger independently. The energy threshold of an individual telescope is about 300 GeV (Cawley *et al.*, 1990); the collecting area, event rate and typical energy for the double telescope responses is not yet known. Work is continuing to develop a trigger system from the two separated telescopes with a view to decreasing the threshold energy. No data have yet been reported from the GRANITE project.

2.3. HEGRA Collaboration

The HEGRA experiment is located at La Palma (28.8° N, 2200 m asl) and is in a construction phase. It will be a unique facility with an array of 5 similar high resolution (0.25° pixel size) imaging telescopes with a separation of 80 m between the telescopes (Mirzoyan *et al.*, 1994).

Each telescope comprises a single 8 m^2 flux collector; the trigger for an individual telescope involves the simultaneous response of two or more PMTs in the camera giving an energy threshold in the TeV region. The event rate for simultaneous measurements for a pair of prototype telescopes (pixel size 0.40°) operating in 1994 is about 1 Hz. Additional telescopes will be completed during 1995/96. No analysis of stereoscopic samples has been reported.

2.4. Crimean Astrophysical Observatory

The Crimean Astrophysical Observatory (45° N, 2100 m asl) comprises a pair of similar telescopes separated by 20m on a NS line. The telescopes each consist of $\sim$ 5 m^2 mirrors with a 37-element camera (pixel size 0.4°), triggered when any 2 of the inner 19 channels exceed threshold. The estimated energy threshold is 2 TeV and 2-telescope coincidences are identified by their common gate times (Vladimirsky *et al.*, 1989).

Preliminary results from the telescopes are described in Kornienko *et al.* (1993). The two telescope response rate varies from 55% to 76% of the single telescope triggers. The utility of data from this facility for the type of analysis highlighted in this paper is limited due to the small separation between telescopes.

2.5. CANGAROO Collaboration

The International Astrophysical Observatory at Woomera, South Australia (27° S, near sl) consists of a pair of dissimilar telescopes separated by 100 m on an EW line.

The BIGRAT telescope (University of Adelaide) (Clay *et al.*, 1989) comprises 3×13 m^2 mirrors and is a conventional triple fast coincidence telescope with a multi-PMT camera (pixel size 0.25°) viewing the central mirror and single PMTs viewing the left and right mirrors. The threshold is about 600 GeV. The telescope has been in operation since 1989.

The CANGAROO telescope (Japanese-Australian collaboration) (Hara *et al.*, 1993) is a single mirror (10 m^2) telescope with a very high resolution camera (pixel size 0.125°). The trigger requirement is any 'n of m' camera elements and the event threshold is about 1000 GeV.

The telescopes are operational and each is triggered independently. The rate of response of the two telescopes to the same light flash is 0.5 Hz; few observations have been made with the telescopes recording stereoscopic data and no analysis has yet been reported.

2.6. University of Durham at Narrabri

The Narrabri observatory (31° S, 250 m asl) includes a pair of similar telescopes, each being a conventional triple mirror system with collector area of $3 \times 10\ m^2$ (Brazier *et al.*, 1989, Bowden *et al.*, 1993). The telescopes are located 100 m apart on an EW line. A recorded event in either telescope requires a fast (8 ns) triple coincidence between corresponding pairs of PMTs in clusters of seven 0.7° aperture close packed PMTs viewing a pair of outer $10\ m^2$ mirrors plus a response from any one of nineteen 1″ PMTs (pixel size 0.5°) in the camera viewing the central mirror.

Each telescope triggers independently at a rate of up to 2 Hz with an energy threshold of about 200 GeV. The events to which both telescopes have responded are identified off-line on the basis of a common time derived from a single atomic clock. The proportion of events with a two telescope response is about 45% when the telescopes are pointed at the zenith and this proportion changes little with increase in zenith angle to 60°. The energy of a typical two telescope event recorded near the zenith is about 350 GeV. The telescopes are shown in Figure 1.

3. Interpretation of Stereo Data

3.1. Introduction

Previous proposals for the stereoscopic studies of Čerenkov radiation, *e.g.* the GRANITE project, have involved the measurement of high resolution images with a pair of separated telescopes. The proposed interpretation of the data from such telescopes follows from the success of high resolution imaging with a single telescope (summarised by Fegan, 1995). Pattern recognition (a study of the image shape and orientation) using data from two telescopes was intended to provide an order of magnitude improvement in background rejection from 99% to 99.9% (Meyer *et al.*, 1991). Particular emphasis was given to the interpretation of the intersection of the two orientation vectors.

We will describe here an alternative approach which hinges on the measurement of the longitudinal development of the atmospheric cascade. These measurements are possible using simple low-cost equipment.

3.2. Typical Data from a Pair of Medium Resolution Telescopes

Grey scale representations of the images making up typical events recorded by the medium resolution cameras of the Narrabri pair of telescopes are shown in Figure 2. The response of each PMT has been adjusted to allow for variations in digitizer pedestal and PMT gain. Figure 2(a) shows a event

Fig. 1. The University of Durham stereo pair of telescopes operated at Narrabri, NSW, consisting of the Mark 3A and Mark 5A VHE gamma ray telescopes.

falling more or less equidistant from the two telescopes giving broadly similar images; Figure 2(b) shows an event where the Mk 5A telescope has been close to the core of the cascade and a much brighter image has been recorded than by the more distant Mark 3A telescope.

The lower moments of the distribution of light recorded by each telescope are routinely derived as follows:-

(a) Each PMT response is labelled as (i) an 'image' PMT corresponding to a signal of at least 5σ of noise, (ii) a 'border' PMT corresponding to a signal of at least 2.25σ of noise if it is next to an 'image' tube or (iii) neither of these when the response is set to zero. Categories (i) and (ii) are used to calculate the moments $\overline{x}$, $\overline{y}$, $\overline{x^2}$, $\overline{y^2}$, $\overline{xy}$.

(b) The total amount of light is estimated by the summation of the image and border responses, with allowance for gaps between the tubes.

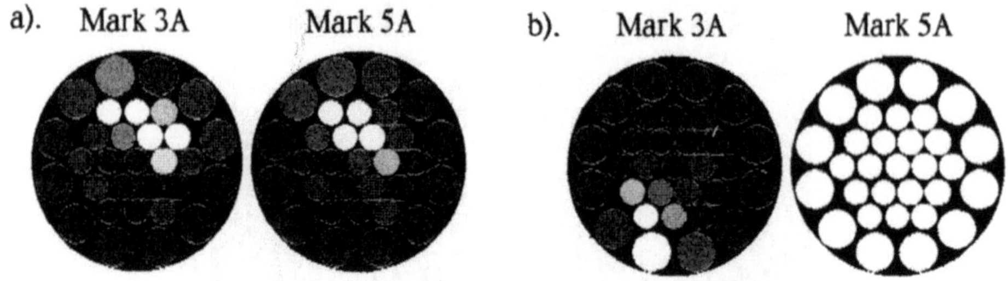

Fig. 2. Typical events recorded by the University of Durham pair of stereo telescopes. (a) shows a Čerenkov event falling more or less equidistant from the two telescopes while (b) shows an event where the Mk 5A telescope has been close to the cascade core and has recorded a bright image.

(c) The centroid of the light (with coordinates (X_{CG}, Y_{CG})) is derived by evaluating the expression

$$X_{CG} = \overline{x}\left(1 + \frac{\exp(dist/1.48)^{3.6} - 1}{dist}\right)$$

where *dist* is the distance of the centroid from the centre of the field of view, with a similar expression for Y_{CG}, to allow for light missing from the image (*i.e.* from outside the field of view).

These lower moments of the distribution are robust; errors of 20% in the calibration of the PMTs result in small changes only. In Figure 3 we show the differences in the (x, y) coordinates of the centroid derived with and without the effects of a 20% error added to the recorded densities.

3.3. Possible Methods of Analysis of Low Resolution Stereo Data

The various approaches to exploiting the differences in longitudinal development of cascades initiated by gamma rays and hadrons depend, in varying degrees, on the input from computer simulations of atmospheric cascades.

In order of decreasing simplicity of approach (and increasing dependence on simulation data) some of the quantities which may be studied using measurements of the lower moments of the light distributions from stereoscopic samples are:-

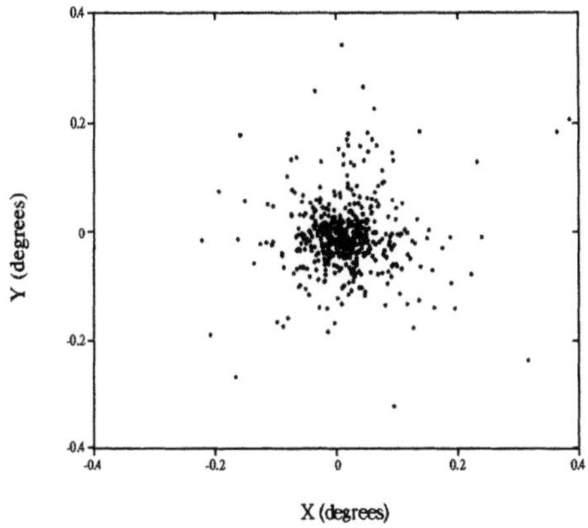

Fig. 3. The differences in the (x, y) coordinates of the centroid of a shower image when derived with and without adding a 20% rms error to the recorded densities.

(1) The height of Čerenkov light maximum, H_c, which may be derived directly from the intersection of lines (or the distance of closest approach of skew lines) drawn through the centroid of the light pattern in each telescope. This procedure makes no assumptions about the cascade development and does not depend on simulation data.

The interpretation of estimates of H_c depends on the fact that variations in the height of maximum of the light are small for gamma ray showers of fixed energy (arising from typically 10% changes in the depth of electron cascade maximum) whereas for background hadrons the longitudinal and transverse fluctuations in the cascade and hence the light are large.

We show in Figure 4 the value of H_c derived from simulated gamma ray events.

A limitation of this method is that no allowance is made for the systematic variation in height of maximum with energy for the primary gamma ray.

(2) The separation, D_{miss}, of the centroid of the light pattern measured in one telescope and the position predicted from measurement of the pattern in the second telescope together with the assumption that the primary was a gamma ray. This quantity makes use of more information than H_c and allowance can be made for the systematic changes in the mean height of maximum of cascades of differing primary energy.

Simulation data for the quantity D_{miss} are shown in Figure 5, again making clear the expected differences between gamma ray and nucleon primaries.

(3) The ratio of the primary energy estimates, R_{ep}, from measurements in the two telescopes. In determining the primary energy of a cascade we

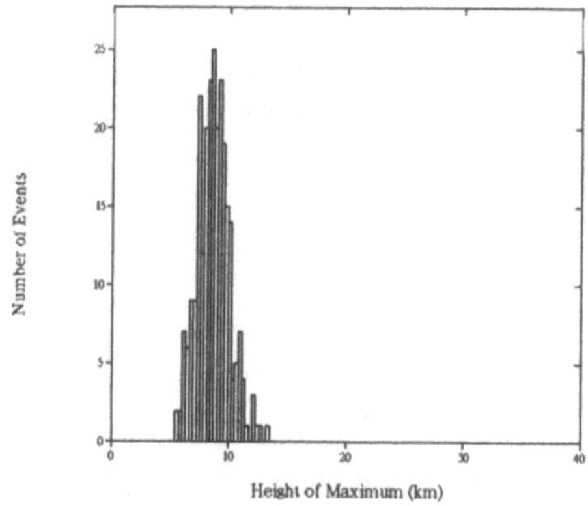

Fig. 4. The distribution of H_c derived from simulated events for gamma ray primaries.

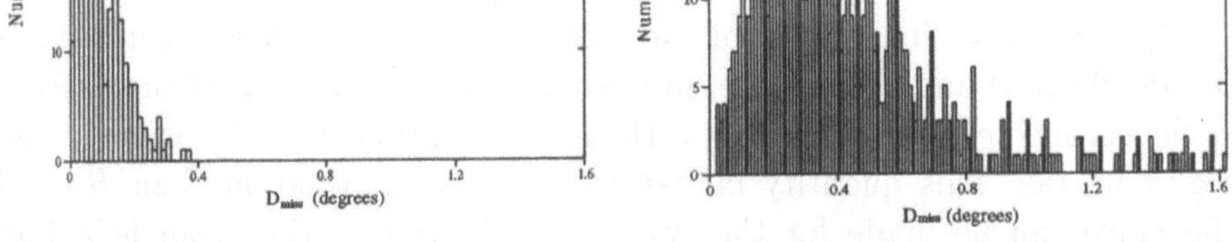

Fig. 5. The distribution of $D_{\rm miss}$ derived from simulated events for gamma ray (a) and hadron initiated events (b).

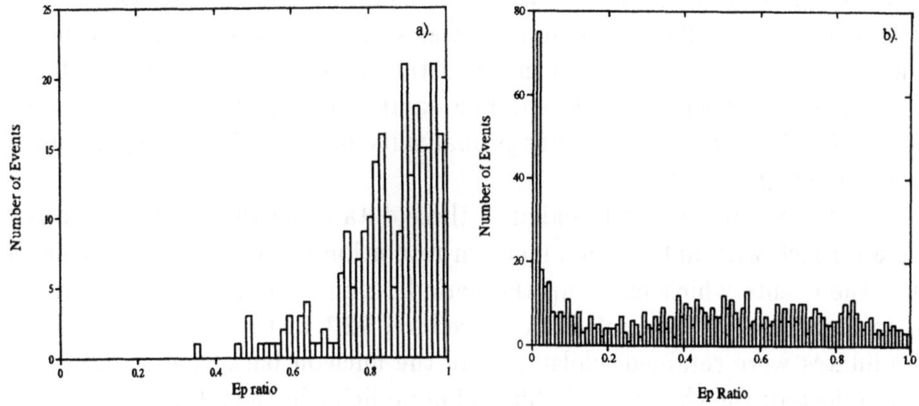

Fig. 6. The distribution of R_{ep} derived from simulated events for gamma ray (a) and hadron initiated events (b).

assume that the light is due to a gamma ray shower which is characterized by:-

(a) small fluctuations in the azimuthal development within a shower,

(b) a good correlation between the impact parameter (the distance of the telescope from the shower core) and the position of the centroid in the field of view, and

(c) a well behaved lateral structure function for the light brightness.

The primary energy of a gamma ray initiated shower should be estimable with reasonable confidence as a result of these features. Two estimates from well separated telescopes should agree and the ratio of the estimates should be near unity.

In the case of hadron cascades, the measured centroid position does not provide a reliable estimate of the impact distance since the arrival direction of the cascade in the atmosphere is unknown. This, together with much larger fluctuations in both the azimuthal and radial features, makes a primary energy estimate imprecise. The ratio is therefore likely to be spread from small to large values in the range 0 – 1.0.

Examples of values of R_{ep} for simulated gamma ray and hadron initiated events are shown in Figure 6.

3.4. The Demonstration Dataset

An episode of periodicity at the second harmonic of the white dwarf rotation period lasting 4200 s was noted during observations of the cataclysmic variable AE Aqr with the Narrabri pair of telescopes on 1993 October 11 (Chadwick *et al.*, 1995). The activity was seen in the total counts from the telescopes. Those events which involved the response of one (but not both) telescopes have been investigated. This sample of 5038 events shows a periodicity (at 16.477 s) with chance probabaility of 1×10^{-3} corresponding to a signal strength of 3.7%.

An attempt was made to enhance these data using the established imaging approach with data from a medium resolution camera. There is evidence that the events which make up the periodic signal show the characteristics of gamma rays. After a selection procedure, 72% of the pulsed gamma ray candidates were retained whilst 77% of the nucleon background candidates were rejected; the chance probablity of periodicity in the interval reduced to 6×10^{-7}. There is thus evidence that AE Aqr was emitting pulsed gamma rays during this interval.

The independent events involving double telescope responses form the basis of a dataset suitable for an investigation of an enhancement using trignometric height of maximum parameters. The interval of periodicity during which stereo samples was available lasted for 4000 s and the dataset comprises 2786 events. Prior to any attempt to enhance the signal, the raw stereo data showed a pulsed signal strength of 4.5% with a chance probability of 3×10^{-3}.

3.5. The Stereoscopic Properties of Gamma ray Initiated Cascades

3.5.1. Height of Čerenkov Light Maximum, H_c

This is the most direct measure of cascade developement and involves no assumptions nor does it depend on simulation results. The height of Čerenkov maximum was evaluated for every 2-telescope event.

In Figure 7 we show the variation of the pulsed signal strength for events selected on the basis of the apparent height of maximum H_c. The expected mean values for gamma rays of energy 300 – 1000 GeV at a zenith angle of 30° are in the range 8.8 – 11.5 km with shower to shower fluctuations of about 10%. Most of the pulsed events are seen to be concentrated in the cells populated by events with height of maximum between 7.5 and 15 km.

3.5.2. D_{miss}

The parameter D_{miss} offers the prospect of improving the enhancement by allowing for the systematic trends in H_c with primary energy.

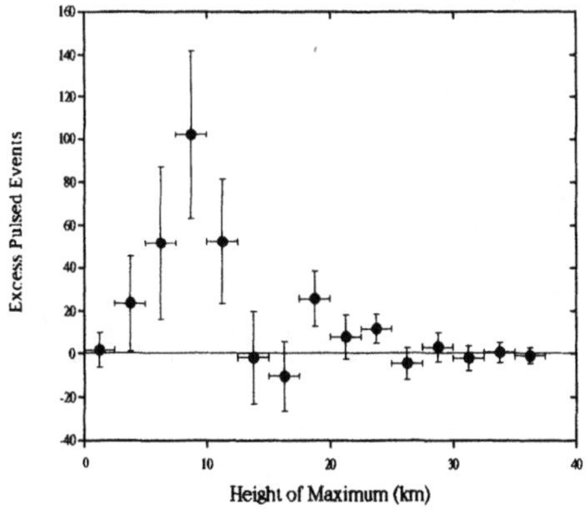

Fig. 7. The variation of the pulsed signal strength at 16.5 s for events selected on the basis of the apparent height of maximum H_c.

The prediction of the position of the centroid of the light in one telescope on the basis of the measurement in the other telescope involves a number of stages:-

(1) Draw a line in space from the telescope through the position of the centroid of its light sample.

(2) Derive the impact parameter (distance to cascade axis along the ground) from the measured position of the centroid. Combine this with the total measured light brightness to estimate the primary energy and hence the most probable height of maximum. All of this is done assuming the primary to be a gamma ray and relies on the results of simulations.

(3) Locate the radiating region in the atmosphere from the intersection of the line drawn in (1) and the height derived in (2).

(4) Re-calculate the primary gamma ray energy from the total optical density and the new value of the shower impact parameter inferred by the line from the source passing through the radiating region.

(5) Iterate steps (2) – (4), if necessary, until convergence.

(6) Predict the position of the centroid of the light pool in the second telescope by drawing a line between the radiating region and the second telescope. Although the calculation of D_{miss} requires an estimate of the primary energy it does not do so in first order.

Each stereo event in the 4000 s data segment was processed as described above and a value of D_{miss} derived. Events were selected according to the

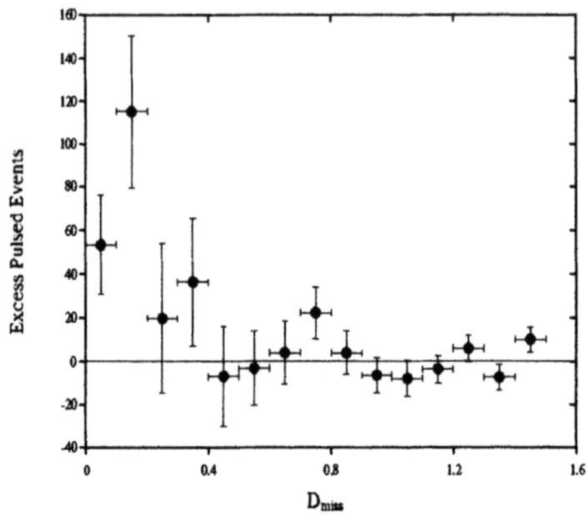

Fig. 8. The variation of the pulsed signal strength at 16.5 s for events selected on the basis of D_{miss}.

value of D_{miss} and the events in each selection were tested for periodicity close to 16.5 s. The results are shown in Figure 8 where the signal is seen to be confined to values of D_{miss} less than 0.20; we expect the signature of gamma rays to be values of D_{miss} close to zero.

3.5.3. Primary Energy Ratio, R_{ep}

The parameter R_{ep} requires the greatest number of assumptions but may have the potential for strong rejection of background hadron events since it uses all of the appropriate information in the first order.

The information in each telescope is used separately to provide, with the assumption that the primary is a gamma ray, two independent estimates of the primary energy by following steps (1) – (5) for calculating D_{miss} above. Every event has a value of the ratio R_{ep} calculated on this basis.

Since gamma rays have very small fluctuations in depth of maximum for fixed primary energy (simulations suggest about 10%) and very small azimuthal asymmetry in a shower (simulations suggest they are close to the Poisson fluctuations in the photon numbers), then they would be expected to cluster near to values of $\log_{10}(R_{ep})$ of 0.0, *i.e.* a value of the ratio R_{ep} of 1.0. We show in Figure 9 the variation in the number of pulsed events in sets selected on the basis of values of $\log_{10}(R_{ep})$. The signal in Figure 9 is confined to a region of $\log_{10}(R_{ep})$ from −0.5 to 0.5 *i.e.* a ratio between 0.33

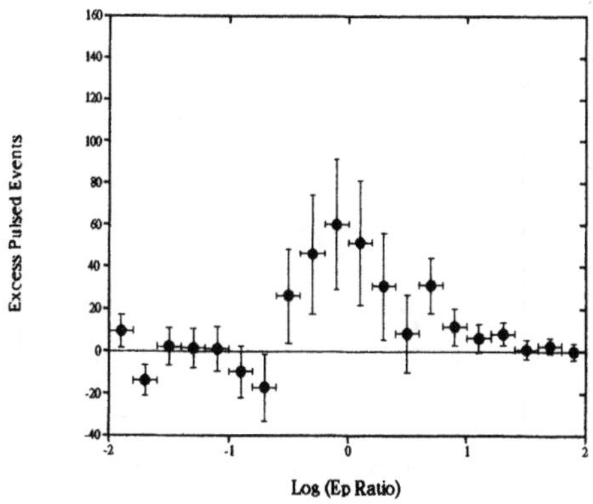

Fig. 9. The variation of the pulsed signal strength (number of gamma rays) at 16.5 s for events selected on the basis of the primary energy ratio R_{ep}.

and 3.0. This relatively wide distribution is in principle reducible if improved telescope pointing, PMT gain calibration and simulations are available.

4. Limitations to the Methods

4.1. Accuracy of Pointing and its Consequences

The main limitation in the data from the AE Aqr burst taken early in the life of the telescopes was the limited accuracy of pointing of the two telescopes.

For example, the histogram of values of D_{miss} for simulated gamma ray primaries shows a sharp peak at a value of about 0.1° with a FWHM $\sim$ 0.1°. The measurement of D_{miss} requires that two telescopes each measure the absolute position in the sky of the centroid of the light flash. The required accuracy for this measurement is consequently much less than 0.1°. The medium resolution telescopes at Narrabri have angular encoders with a resolution of 0.1° each, and an absolute alignment of $< 0.1°$. The resulting error of cross-correlation of the images is about 0.15°. This is not sufficiently accurate for the full exploitation of D_{miss} as a parameter.

The two telescopes' systematic errors in the measurement of D_{miss} have been eliminated by ensuring that the mean values of the orthogonal components of D_{miss} in focal plane (x, y) are zero. This reduces the error to about 0.08°. Further reduction requires measurement of the pointing with higher

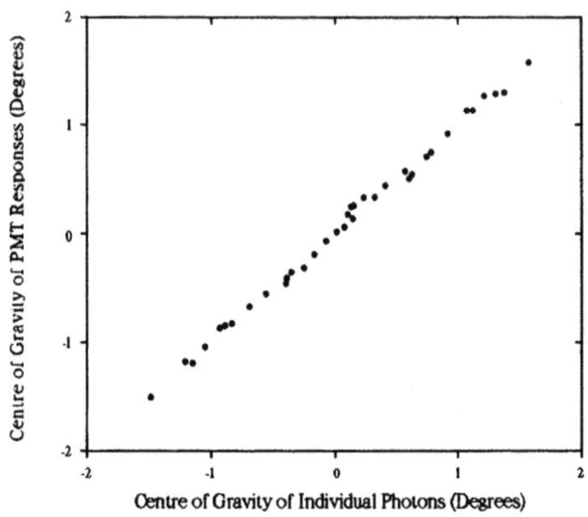

Fig. 10. The x-position of the centriod from the individual photons (X_0) plotted against the x-position of the centroid as calculated from the predicted PMT responses (X_1).

resolution. This has now been provided by using a CCD camera (165×192 pixels, 2.54 mm^2 with an $f/1.4$ 50 mm lens) attached to each telescope. The pixel resolution of about 1 arc min provides sufficient pointing accuracy for the full potential of D_{miss} to be realised.

4.2. PMT Camera Pixel Size

The size of the Čerenkov image is significantly larger than 0.5°. The accuracy required for the determination of the position of the centroid is much smaller than 0.5°. In order to check the effect of measurements using 0.5° pixels on the centroid's accuracy, simulations for gamma-ray primaries providing the exact position of each photon in the images have been subject to the optical aberrations of the mirrors and the image quantization of the PMTs for the Mark 5A focal plane package. The x-position of the centroid calculated on the basis of the position of the individual photons (X_0) is plotted against the x-position of the centroid calculated from the predicted PMT responses (X_1) in Figure 10. The accuracy of the y-position is similar; the effect of the experimental measurement with 0.5° is minimal.

5. Conclusions

Measurement of the lower moments of the light distribution due to low energy primaries using a pair of low-cost telescopes have been shown to provide a worthwhile enhancement in the detection of VHE gamma rays. Numerous experiments are available which may exploit this technique.

Acknowledgements

This work was funded by the U.K. Particle Physics and Astronomy Research Council. We are grateful to the University of Sydney for the lease of the site at Narrabri.

The Mark 3A and 5A telescopes were designed and constructed with the assistance of the technical staff of the Physics department, University of Durham. The efforts of Mr. P. Cottle, Mrs. E.S. Hilton, Mr. W. Hogg, Mr. K. Parkin, Mr. M. Robertshaw and Mr. K. Tindale are acknowledged with gratitude.

References

Akerlof, C.W., *et al.*: 1991, in Cawley, M. *et al.*, eds., *Proc.22nd. Int. Cosmic Ray Conf.*, The Dublin Institute for Advanced Studies: Dublin **2**, 591

Bowden, C.C.G., *et al.*: 1993, in Lamb, R.C., ed., *Towards a Major Atmospheric Cherenkov Detector - II*, 230

Brazier, K.T., *et al.*: 1989, *Exp. Astron.* **1**, 77

Cawley, M.F., *et al.*: 1990, *Exp. Astron.* **1**, 173

Chadwick, P.M., *et al.*: 1995, this conference.

Clay, R.W., *et al.*: 1989, *Proc. Astro. Soc. Aust.* **8**, 41

Fegan, D.J.:, 1995 this conference.

Hara, T, *et al.*: 1993, *Nucl. Inst. Methods Phys. Res.* **A332**, 300

Kornienko, A.P., *et al.*: 1993, *Exp. Astron.* **4**, 77

Meyer, D.I., *et al.*: 1991, in J.Matthews, ed., *High Energy Gamma-Ray Astronomy*, A.I.P.: New York, 253

Mirzoyan, R., *et al.*: 1994, *Preprint MPI-PhE/94-24*, Max-Planck-Institute, Munich.

Vladimirsky, B.M. *et al.*: 1989, in A.A.Stepanian, D.J.Fegan and M.F.Cawley, eds., *Proc. Int. Workshop Very High Energy Gamma Ray Astronomy*, Crimean Astrophysical Observatory: Crimea, 21

WIDE ANGLE AIR CERENKOV DETECTORS

E. LORENZ
Max-Planck-Institut für Physik
Föhringer Ring 6, D-80805 Munich, FRG

Abstract. For a number of important questions in VHE/UHE cosmic ray physics wide angle air Cerenkov detectors offer superior performance compared to scintillator arrays or air Cerenkov telescopes. On hand of the HEGRA air Cerenkov detector AIROBICC the general principle and performance will be discussed. Ideas and prospects for a detector with an energy threshold around 1 to 3 TeV will be presented.

1. Introduction

The field of cosmic ray (CR) physics research is currently rapidly expanding. The research is both driven by new questions in high energy Astrophysics, particle physics (above current accelerator energies) and Cosmology and on the other hand by considerable progress in detector performance. Many questions are linked to the detection of energetic gamma rays as only these can be extrapolated back to their origin that should be in most cases also the origin of the charged CRs. The fraction of VHE/UHE cosmic γ ray flux is extremely low and experimenters have only very modest tools at hand to discriminate them against the hadronic CR. For the search of the origin of cosmic radiation the recent satellite experiments (COS B, EGRET) allow us to make extrapolations to the VHE/UHE range of what sensitivity is at least needed for many studies. Satellite detectors are not suited for observations in the VHE/UHE range because of detection areas well below 1 m^2; therefore physics studies in the VHE/UHE range have to be carried out by ground based detectors. Currently the most successful detectors are air Cerenkov telescopes (ACTs) while the majority of detectors, the wide angle scintillator arrays, have still to report the first high significance observations. Although these detectors are working continuously their main drawback is the high energy threshold close to 50-200 TeV, a modest angular resolution and a very poor gamma/hadron (γ/h) separation power. On the other hand the transmission of the universe is already rather limited around (and above) 50-100 TeV. Interaction of gammas with the photons of the 2.7° K background radiation or with the infrared background radiation, currently of unknown intensity, might prevent observation of UHE γs from any source of cosmological distance with scintillator arrays.

The most promising technique of today in VHE/UHE CR physics is based on the observation of Cerenkov light generated in the atmosphere by extensive air showers. The bulk of the detectors uses large mirror telescopes with light sensors providing for modest image analysis power for γ/h separation.

Space Science Reviews 75: 169–183, 1996.

These ACTs have on the one hand a much lower detection threshold compared to extensive air shower (EAS) scintillator arrays, thus they can see much deeper into the universe, but on the other hand they have only a very limited angular acceptance. Therefore they can only be used for the study of preselected astronomical objects. A possible way out might be the construction of wide angle Cerenkov detectors that sample the Cerenkov light disc at many positions. These detectors can combine the good features of wide angle scintillator arrays like a large angular acceptance with some features of ACTs like a high γ/h separation power. The energy threshold will be lower than that of scintillator arrays but not as low as that of ACTs. This is the consequence of the background from the night sky light being integrated over the larger angular acceptance.

Wide angle detectors are suited to observatorial tasks which either cannot or can only barely be met by ACTs. Some examples are:

i) All sky survey for γ sources.
ii) Search for episodic or sporadic γ ray emission from variable sources, γ ray bursters, evaporating black holes, ect.
iii) Search for diffuse γ radiation, e.g., from the galactic disk, large molecular clouds or interstellar gas 'illuminated' by more energetic charged cosmic radiation.
iv) Search for isotropic γ radiation of cosmological origin, i.e., search for topological defects.
v) Precise measurement of the cosmic ray flux with a few % energy resolution in the energy range where one expects to observe cut-offs due to the influence of the 2.7° K or the infrared cosmic background radiation or due to a significant change in the chemical composition.
vi) Chemical composition above 50 TeV.
vii) Search for collective effects in heavy ion interaction (fire ball events).
viii) Search for indication of 'new' physics above the Z^0 mass scale.
ix) High sensitivity for point source searches in the 10 TeV region because of an excellent angular resolution, high γ/h separation and omission of 'off-source' runs
x) Possible studies of extended sources, for example from nearby SNRs.

There have been some early attempts to view the night sky over a large angular range by means of plain photo multipliers, for example (Bosia, Alimov, Dawson) and further references therein. In all of these experiments only a small number of PMs has been used; thus the potential of the sampling principle has not been fully exploited. In 1989 the first large experiment on wave front sampling with a very restricted angular aceptance of 20-40 mrad, has been built by the Themistocle collaboration (Fontaine) for the search of individual point sources. In 1990 the construction of true wide angle air Cerenkov detectors, based on a large matrix of large diameter, open PMs, has been proposed by (Bott-Bodenhausen, Yakovlev). Only one design, the AIROBICC detector (Bott-Bodenhausen), has been built up to now; therefore many parameters presented here are based on experimental observations or MC predictions linked to this detector.

In the following chapters the detection principle for wide angle air Cerenkov detectors and their performance will be discussed. Partly the performance predictions are based on the experience gained with the AIROBICC detector. Also, a short comparison of advantages and drawbacks compared to wide angle scintillator arrays and ACTs will be given. Finally some prospects for improved wide angle Cerenkov detectors will be presented.

2. The Detector Principle

High energy cosmic particles — hadrons, electrons or γ's — interact with the earth's atmosphere. Fast, charged secondary particles, which are produced in a multistep showering process, generate Cerenkov light which is strongly peaked along the direction of the shower axis. Measurements of the light structure can provide information about the incident particle and energy and about the shower development. Nevertheless, details of the light structure are influenced by the specific interaction properties of the secondary particles (transverse momentum transfer, multiple scattering, momentum above Cerenkov threshold) and the exponential change of the refractive index of the atmosphere. The light pool has a conical shape of approximately 15 mrad cone slope and can extend radially to more than 100 m. The number of photons in a typical UHE shower is many orders of magnitude larger than that of the particles in the shower tail which reaches the ground. The large size of the light pool, the comparatively low fluctuations in arrival time and the small thickness of the wave front (at least within about 100–150 m distance from the core) allow for a much better angular determination of the direction of incidence than that from the particle disc in the shower tail. For light observations it is not necessary that the particles of the shower tail reach ground, thus the threshold can in principle be lower than that of scintillator array detectors.

The basic concept of a wide angle Cerenkov detector is shown in Fig. 1. The large Cerenkov light disk is sampled by a matrix of PMs that view the night sky. The matrix should cover the entire light pool or at least a sizeable fraction of it. In a typical measurement one records both the arrival times of the wave front and the light intensity at many positions. Timing provides information about the direction of incidence while the light intensities provide information about the incident energy and about the longitudinal and radial shower development which offer the possibility to perform a coarse particle identification. In first order such a detector is a fully active calorimeter with imaging capabilities, but, in the language of accelerator physicists, a non-compensating calorimeter, i.e. the integrated

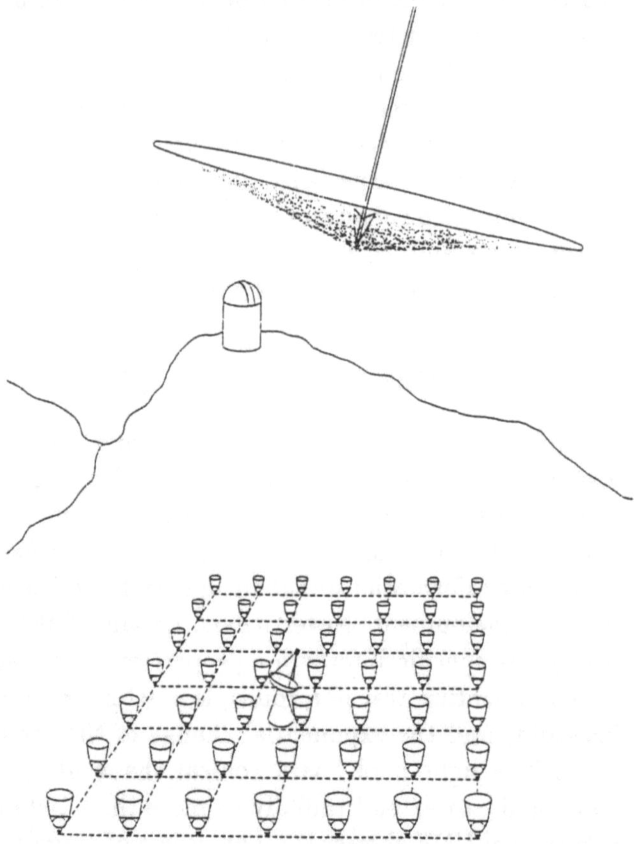

Fig. 1. Basic detection principles of wide angle Cerenkov detectors.

light differs for incident γ's and protons of the same energy. As the detection threshold is set by the presence of the night sky light background (correctly by the fluctuations within the sensitive time) it is necessary to use PMs with a large active area in order to achieve a low detection threshold. In case of AIROBICC we use a 30 m grid spaced 7×7 matrix of 20 cm diameter, hemispherical PMs. The light collection area of each PM is augmented by a 40 cm diameter Winston cone at the expense of restricting the angular acceptance to about 35°, i.e., to about one sterad. The reader interested in details of AIROBICC is referred to (Karle). This detector was built in 1991–92 by the HEGRA collaboration and installed at the HEGRA cosmic ray detector complex on the Canarian Island La Palma (28.8° N, 17.8° W, 2220 m asl). Besides the wide angle Cerenkov matrix the detector complex comprised in 1994 an equal gross area scintillator matrix of 224 stations, 17 large area muon detectors and two ACTs. The combination of many different detector elements at one site allows us, besides the normal task to record many shower parameters at once, to cross check performance data of the first wide angle air Cerenkov detector from simultaneous observations.

In the following chapters the aspect of wide angle Cerenkov detectors as fully active calorimeters is elaborated and then some performance data are discussed.

3. The Combination of the Atmosphere and a Wide Angle Cerenkov Matrix as a Fully Active Calorimeter

Besides an excellent angular resolution another strength of the wide angle Cerenkov detector is its ability to measure in a calorimetric way the energy of the incident particles. For a detector at sea level the atmosphere corresponds to 27 radiation length (rl) and 11 hadronic interaction lengths (X_0). On the one hand the ratio of rl/X_0 is rather unfavorable for γ/h separation because for equal energy the pure electromagnetic and hadronic showers extend to about the same length but, on the other hand, no suppression occurs at the end of the 'calorimeter' for pure electromagnetic showers as in high z calorimeters used in accelerator experiments. The high transmission of the atmosphere in the range of spectral sensitivity of current PMs allows one to collect Cerenkov photons generated over the entire shower development. Thus one can build in principle a fully active calorimeter by observing all these photons with a ground based PM matrix*. The strongly forward peaked Cerenkov light can reach a high density of photons (typically a few 10^4 photons/m^2 close to the core of a 100 TeV shower), therefore only a small fraction of the illuminated area has to be covered by light sensors (in case of AIROBICC the active area coverage is only 0.014%) in order to collect data on the energy flow with high precision. Fully active calorimeters in high energy accelerator physics can have an energy resolution for electromagnetic showers in the range of 1%. At first hand a very good energy resolution in CR physics seems unnecessary but progress in CR physics will soon demand an energy resolution well beyond the capabilities of current scintillator array detectors or ACTs. The precise flux measurements of steeply falling energy spectra or searches for energy dependent cutoffs due to γ interaction with the 2.7° K microwave or infrared background make a high energy resolution of about a few % for electromagnetic showers and, say, below 20% for hadronic showers mandatory.

The Cerenkov radiation is proportional to the ionisation loss in the atmosphere for $\beta = 1$ particles. Therefore a measurement of the total Cerenkov light is in first order a good measurement of the incident energy. There exist subtle effects that can spoil the resolution. The refractive index of the atmosphere changes strongly with altitude. While the proportionality of Cerenkov

* The influence of Rayleigh and Mie-scattering has to be taken into account.

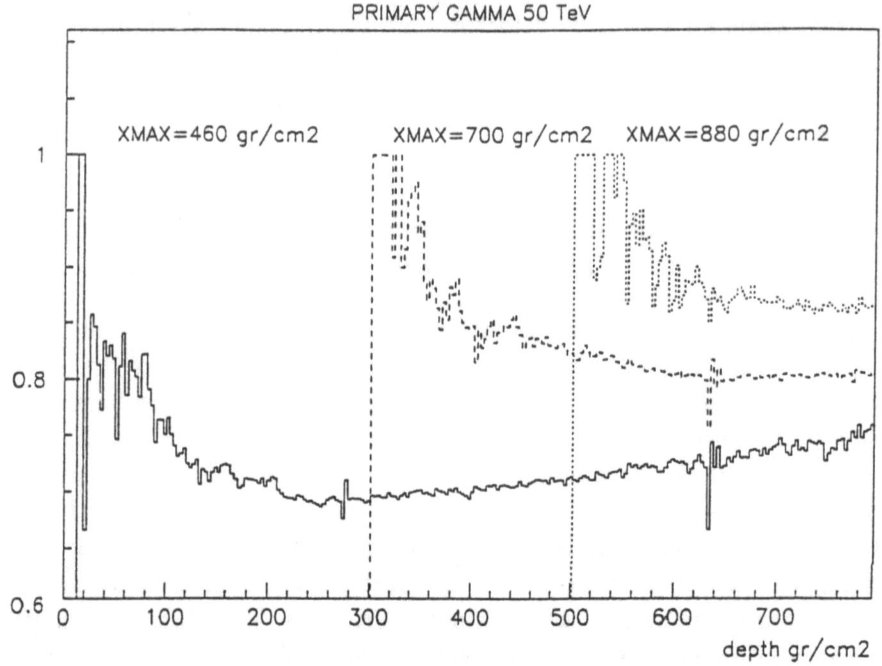

Fig. 2. Fraction of shower particles producing Cerenkov light as a function of altitude for three showers of the same energy but different x_{max}. Courtesy F. Arqueros, HEGRA collaboration.

intensity to ionisation loss is not affected for $\beta = 1$ particles, slower particles will emit less light and eventually no light at all once they are slowed down below the Cerenkov threshold. Nevertheless, the shower particles contain a significant fraction of the total energy. Fortunately, the mean momentum of the shower particles is decreasing with altitude such that the fraction of particles emitting light stays nearly constant along the shower. Fig. 2 shows a Monte Carlo (MC) simulation of this fraction as a function of altitude for three same energies electromagnetic showers developing at different altitudes. Besides the low intensity early phase of the development the fraction of radiating particles is nearly constant but depends clearly on x_{max}, the position of the shower maximum. From the analysis of the radial light intensity one can derive x_{max} and calculate the correct energy. Fig. 3 shows a MC simulation for the correlation of the distance of the shower maximum above a detector at 2200 masl and the so-called light radius (defined as the radius where the light intensity drops to 1/e when fitting a simple exponential to the radial intensity between 30 and 100 m core distance). After an empirical correction of the total light integrated over a disk of 100 m radius it has been found that the energy of electromagnetic showers can be reconstructed with a mean error of 3%. In hadronic showers one is faced with

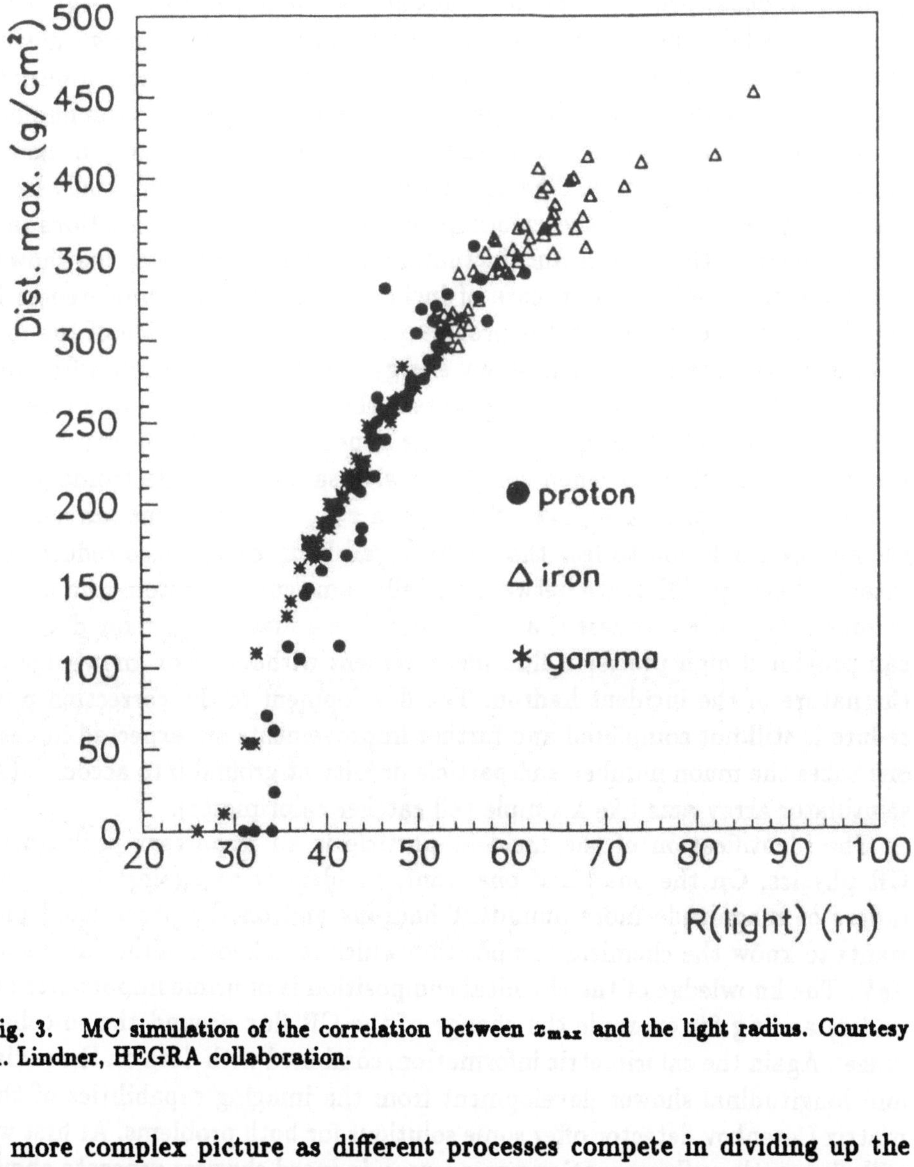

Fig. 3. MC simulation of the correlation between x_{max} and the light radius. Courtesy A. Lindner, HEGRA collaboration.

a more complex picture as different processes compete in dividing up the primary energy. The dominant process is that in each hadronic interaction about two thirds of the energy are transferred to hadronic particles (pions, kaons ...) and one third is transferred to π^0s which after their decay into γs start a high multiplicity electromagnetic cascade. Only a minuscule fraction of this energy is transferred back to hadronic particles through photo production which can therefore be neglected. In hadronic interactions the majority of energy is still transported by energetic hadrons but these particles very often have an energy below their corresponding Cerenkov threshold and do not contribute to the light intensity until they interact again and fuel the electromagnetic component through π^0 production. A sizeable fraction of the charged pions and kaons decay before interacting again. From the prominent decay products the neutrinos are completely lost for energy measurements while muons can sometimes be counted but their energy not be measured. Also some energetic hadrons might reach ground and transport away a fraction of energy. These particles are rare compared to the number of electromagnetic particles, nevertheless they can transport away a significant fraction of the energy. The fluctuations of these effects will finally limit the energy resolution of the calorimeter. On top of these fluctuations one is confronted with a significant fluctuation in the early phase of the shower development, particularly in case of incident protons. For completeness it should be mentioned that other processes like nuclear excitation of the target nuclei and the production of low energy neutrons will absorb a fraction of the incident energy and thus can worsen the energy resolution. In essence only about 50–70% (energy and particle dependent) of the energy is visible in hadronic showers when compared with same energy electromagnetic showers. MC studies have shown that after a x_{max} correction one can narrow the energy resolution to less than 20% for all hadrons and also reduce the measured energy difference between initially same energy protons and heavy primaries (up to Fe) to less than 7%. Therefore a matrix Cerenkov detector can provide a high precision flux measurement without prior knowledge of the nature of the incident hadron. The development of the correction procedure is still not completed and further improvements are expected in case one takes the muon number and particle density at ground into account. (A scintillator array acts like a simple tail catcher calorimeter).

The identification of the incident particle is an important problem in CR physics. On the one hand one wants to identify γs against the many orders of magnitude more abundant hadrons and on the other hand one wants to know the chemical composition which is unknown above about 50 TeV. The knowledge of the chemical composition is of prime importance for understanding for example the change of the CR flux around the so-called 'knee'. Again the calorimetric information, combined with data on the radial and longitudinal shower development from the imaging capabilities of the matrix Cerenkov detector offer some solutions for both problems. At first we will discuss the γ/h separation prospects. γ induced showers generate about twice as much light as hadronic showers. Unfortunately a ground based light measurement alone is useless for discrimination as long as one does not have another independent measurement of another parameter. An independent information might come from the longitudinal development of the showers. Hadronic showers develop slower in the atmosphere. In first order the ratio of the integral light to the shower size N_e provides a tool to separate γs from hadrons. The main drawback of the simple approach is that shower fluctuations both in x_{max} and N_e spoil the separation power. At least the effects due to x_{max} fluctuations can be corrected from the analysis of the radial light intensity. The precise determination of N_e is rather difficult. N_e can be extracted from measurements with a scintillator array, but it should be possible to determine N_e also from the light intensity close to the core, say within 30 m radius. This light is mainly generated by particles in the shower tail up to about 1000 m above the detector. Currently we can only determine N_e with a reasonably small error for showers above 60 TeV because of the coarse sampling of the scintillator array. Using the so-called LES method (Light, Electron number N_e and Slope of the radial light distribution as a measure of x_{max}) we are able to reject $> 97.5\%$ of all charged particles while retaining 50% of the initial γs, i.e., we obtain a quality factor (signal/sqrt(background)) of 5 which is similar to the value of the Whipple telescope for γ showers not being constrained to point sources. Details of the LES method are described in (Karle) and (Arqueros). As one learns more on fluctuation processes one should improve the γ/h separation power considerably.

The measurement of the chemical composition relies more on the imaging power of the detector. In a calorimetric device it is impossible to identify hadrons on the event by event basis without any separate measurement of incident particles before their interaction. The main discrimination methods rely on mean differences which are influencing the shower process and large fluctuations set a limit on the discrimination power. At present it seems that only a coarse composition as a light-heavy or light-medium-heavy composition is determinable. Three observables are promising. a) A measurement of the x_{max} distribution. Showers from heavier particles should have a higher x_{max} than those from light particles due to the increase in cross section for heavy ions. b) A measurement of the fluctuations in x_{max} should also give information about the chemical composition. In the most naive model the fluctuations in a Fe induced shower should be smaller by a factor $\sqrt{56}$ compared to the fluctuations in proton induced showers. c) A measurement of the radial light distribution above 100 m core distance. This light is generated by particles scattered away from the shower axis. In heavy particle interaction the transverse spread of secondary particles is larger than in proton induced interactions due to the lower mean momentum of secondaries. Therefore a measurement of the ratio of the light intensity at short and long distances from the core might be a useful tool. It should be mentioned that a basic weakness of all these methods is their reliance on MC predictions of the showering process.

4. Expected Performance of a Wide Angle Cerenkov Detector

In this paragraph some additional expected performance data will be given. The data are either based on specific MC simulations or results from AIROBICC and can therefore only be considered as a guide line.

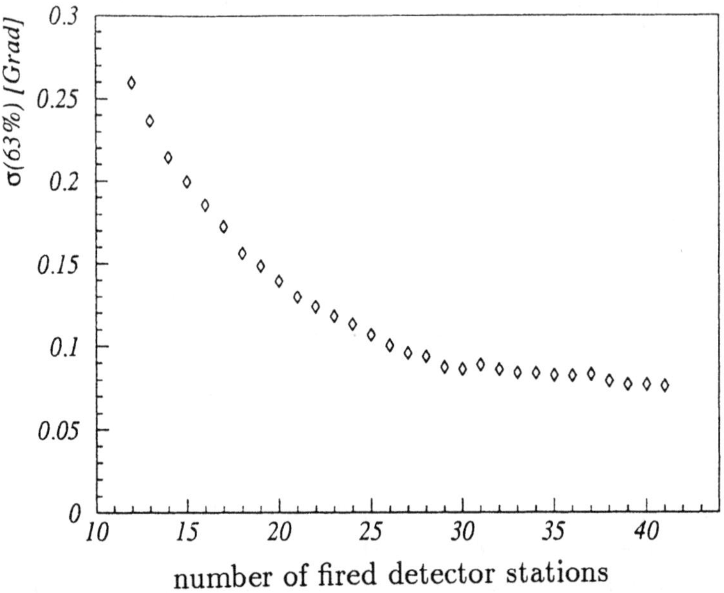

Fig. 4. Angular resolution for AIROBICC as a function of the fired detector stations (Krawcsynski). The energy ranges typically from 26 to about 100 TeV.

A good angular resolution is, besides a high γ/h separation, of prime importance for the search for γ point sources as the strategies are generally based on searches for local excesses above the much more abundant isotropic hadronic background. Cerenkov detectors, based on multiple sampling of the wave front, have a superior angular resolution compared to EAS scintillator arrays because they use the full information from the entire shower development and because the light disc normally has a much larger diameter than the particle disc. The time spread in the light front is much smaller than in the particle disc. In addition, the number of photons per m^2 is at least 3 orders of magnitude larger than the particle density, in turn fluctuations are smaller. As mentioned, the incident shower direction is determined from timing measurements. Provided that one can keep the time resolution for each station below one nsec and detect nearly the entire disk one should be able to achieve an angular resolution well below 0.1° even with a modest sampling as in AIROBICC. Fig. 4 shows the angular resolution of AIROBICC as a function of responding detector stations (Krawczynski). Even for only a partial sampling of the light front (in case of events with their energy close to the detection threshold) one can achieve a high angular resolution while for a full illumination of the array a value below 0.1° has been achieved. We used the so-called Chess-board method, i.e., the comparison of the angles deduced from two interspersed subarrays, for the determination of the angular resolution and found a mean value of σ_θ of 0.17°. It should be noted that at present an important limitation of the AIROBICC angular resolution is given by the error of the exact core position of typically 3–20 mtr.

As the significance of an observation of a given γ point source increases inversely proportional to the angular resolution a wide angle Cerenkov detector with its much shorter up-time but better angular resolution can outperform a scintillator array.

A sub-problem of the angular resolution is the pointing accuracy of the detector. The pointing accuracy must be much better known than the angular resolution in order to make full use of the latter quantity. In scintillator arrays it is now common practice to use the obscuration of CRs by the moon or by the sun for the verification of the pointing. Obviously this is not possible with a Cerenkov detector. In case of AIROBICC we observed a small fraction of events both by the wide angle Cerenkov detector and the ACTs. The latter ones can be calibrated by a drift scan of known stars to a precision of better than 0.02°. From the coincidence observations we were able to confirm that AIROBICC has an alignment error of 0.01°. Other possibilities are based on the generation of a pulsed light spot above the experiment. With a pulsed Nd:YAG laser it is possible to illuminate a high cloud. The back scattered light pulse is detectable by the Cerenkov detector. First tests with a 10 m Joule laser, working in the third harmonic, gave promising results.

The angular acceptance is to first order given by the angular acceptance of the light detector. At larger angles (with respect to the zenith) atmospheric absorption due to Rayleigh and Mie scattering starts to play a major role and will lead to an increase in energy threshold but the effect is much weaker when compared with the attenuation of charged particles in the shower tail.

Another important parameter is the detection area. The detection area is essentially given by the matrix size. The requirement of the shower core position being inside the matrix boundaries results in a well defined area which is independent of energy except for showers close to the threshold. In this case the area can be calculated from the trigger requirement of the minimal size of the shower image or by Monte Carlo methods. In practice one has to be careful to select events with their core position close to the boundary as these calculated positions can be pulled inside the array boundary in case of larger showers with their true core position being outside of the array. For practical reasons one might apply a restricted core cut which must be at least half a grid spacing inside the array boundary. As a consequence one needs a minimal array size of around, say 100×100 m^2 for any useful observations. If one wants to extract shower parameters from the radial light distribution a matrix size of 200×200 m^2 is the bare minimum.

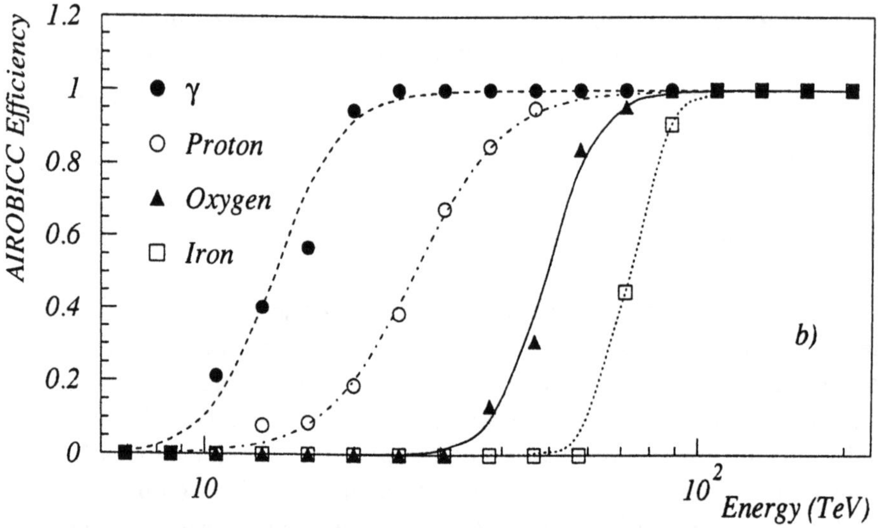

Fig. 5. AIROBICC detection efficiency as a function of energy for different particles (Karle).

As already pointed out the energy threshold of a wide angle Cerenkov detector is set by the night sky light background. Provided one uses large diameter PMs, say greater than 5″, and a restricted angular acceptance, then the threshold can be below that of a scintillator array because particles don't need to reach the ground level. The energy threshold can either be determined purely by MC methods based on atmospheric and detector parameters or can be calculated from the trigger rate and the known cosmic ray flux. Both methods need a coarse knowledge of the chemical composition because the light flux at ground is a function of the primary particle and the height of the shower maximum. Compared to protons, the γ's produce in the UHE range about twice as much light at ground in a disk of about 100 m radius. The factor two in Cerenkov light from γ induced showers is a big bonus for Cerenkov detectors over ground based scintillator arrays as it transfers into a factor two lower energy threshold and, in case of a steep spectrum, in a significant γ enrichment in the recorded sample. Fig. 5 shows the MC calculated detection efficiency of AIROBICC as a function of energy and different incident particles.

A serious limitation of any Cerenkov detector is its restriction of operation during clear and moonless nights. Under best conditions one can achieve a mean up-time of about 15% per year. This disadvantage over permanently running scintillator arrays is normally more than compensated by the lower threshold, higher γ-hadron enrichment, the better angular resolution and the possibility for efficient γ/h separation. Also the much better energy measurement from the calorimetric function and the power to determine x_{max} are a big bonus for Cerenkov detectors.

Compared to ACTs one is penalised by a one to two orders of magnitude higher energy threshold. Nevertheless, as pointed out in the introduction there exist a series of important questions that can barely or cannot be answered by ACTs. Even for point source studies wide angle Cerenkov detectors of the above described detector configuration offer some advantages like a better energy resolution and a clear definition of detection area which makes flux determinations much easier. An additional bonus comes from the omission of additional 'off source' runs and the much better precision in background determination. Based on the experience gained by AIROBICC one can conclude that a wide angle Cerenkov detector can match the best ACTs both in terms of γ/h separation and angular resolution, albeit at higher energies.

5. Prospects for Next Genergation Wide Angle Air Cerenkov Detectors

The current AIROBICC detector has confirmed the usefulness of wide angle Cerenkov detectors. Already from the first ten month period of operation it was possible to determine flux limits of potential γ sources which match or outperform previous long-term measurements of scintillator arrays (Karle). Also it was possible to determine a first direct limit for isotropic γ production in the 60 TeV energy range (Karle).

The experience gained from this detector allows one to make rather safe predictions for more advanced detectors. Clearly, the most important requirements for the future are a lower energy threshold and a better γ/h separation from topological defects. A straightforward procedure to improve the performance and to lower the threshold without any new development would be to built an array of at least 300×300 m^2 extension and to use a finer grid spacing. With 32500 m^2 area coverage AIROBICC is too small to make full use of the large diameter Cerenkov light disc; only for a very restricted sample one can observe the entire central light disc. A fine grid of, say, 10–15 m spacing would allow for a lower energy threshold, a more precise core determination and a better angular resolution for point source searches. Higher precision in the radial light fit will clearly improve the γ/h separation. Another approach would be to use larger diameter PMs. Here the threshold will be in first order proportional to the inverse of the diameters if one chooses to conserve the same angular acceptance. Another improvement would be to bundle 2–4 detector stations and to form local coincidences with overlap times of only a few nsec and with very low discriminator thresholds. On top of the above noted changes one might use PMs with a GaAsP photo cathode originally designed for night vision image intensifiers (Intevac).

Photo cathodes from GaAsP have a 50% peak quantum efficiency (QE) and an extended red sensitivity. Its modest UV sensitivity can be enhanced by coating the glass window with a transparent lacquer loaded with a fast wavelength shifter dye (Eigen). Another approach to lower the threshold is to bundle a few cells of reduced angular acceptance to a unit with the wanted angular acceptance. The reduction of angular acceptance per cell allows for a larger collection area and a reduction of the night sky background.

A design for a very large, active area, light collection system based on a single 20 cm diameter PM, the so-called 'light trap' detector has been described in (Lorenz). This detector should work like a green house. Each detector consists of a flat Plexiglas vessel filled with a liquid which is doped with a fluorescent dye that shifts light from below, say, 450 nm to 550 nm. An dichroic filter is placed on top while the remainder of the vessel is wrapped into a high quality reflector. The filter will be highly transparent to light below 450 nm but will act as a mirror for light above 450 nm; thus the shifted light is trapped. In the centre of the vessel a hemispherical, 20 cm diameter PM is mounted. Again, the light collection area can be enlarged by a Winston cone. Small size test units have given encouraging results.

In summary it seems that a γ threshold of 500 GeV to 1 TeV, i.e., a factor 10–20 reduction in threshold over the current AIROBICC limit is possible. The γ/h separation might be superior to AIROBICC because the 'images' of a γ and a hadron shower become more distinct.

Acknowledgements

I would like to thank my colleagues from the HEGRA collaboration for many interesting ideas and comments and the provision of data from the current AIROBICC analysis.

References

Alimov, T.A. et al.: Proc. 18th ICCR, Bangalore, 11 (1983) 387.

Arqueros, F. et al.: Separation of gamma and hadron initiated air showers with energies between 20 and 500 TeV. 1994, to be published in Nuc. Inst. Meth.

Bosia, G. et al.: *Il Nuovo Cim.* **9B** (1972) 177

Bott-Bodenhausen, M. et al.: AIROBICC – A new angle integrating air Cerenkov counter array for improved gamma/hadron separation in extended air showers. AIP Conf. Proc.: High Energy Gamma Ray Astronomy 220, Ann Arbor (1990) 305 and *Nuc. Inst. Meth.* **A315** (1992) 236

Dawson, B.R. et al., J. Phys. G.: *Nuc. Part. Phys.* **15** (1989) 893

Eigen, G., Lorens, E.: *Nuc. Inst. Meth.* **166** (1979) 165

Fontaine, G.: Aims and Status of the Themistocle Physics Experiment. Laboratoire de Physice Corpusculaire, College de France, F-75231, Paris Cedex, (1989)

Data from Intevac.

Karle, A.: Thesis, Ludwig-Maximilians-Universität, München, 1994, MPI-Preprint MPI-PhE/94-17

Karle, A. et al.: Design and performance of the angle integrating Cerenkov array AIROBICC. 1994, submitted to Astroparticle Physics

Karle, A. et al. : Search for astronomical point sources of γ radiation above 25 TeV energy. to be published in Astroparticle Physics

Karle, A. et al., : Search for isotropic gamma radiation of cosmological origin between 65 and 100 TeV. (1994) MPI Preprint MPI-PhE/94-28, to be published in Phys. Lett. B.

Krawczynski, H: Diploma Thesis. Universität Hamburg, II Institut für Experimentalphysik (1994), HH-HEGRA/94-02

Lorenz, E.: What can we learn from a wide angle air Cerenkov detector like AIROBICC. Proceedings Workshop: Towards a Major Atmospheric Cerenkov Detector III, Tokyo, 1994, Ed. T. Kifune, Universal Academy Press, Tokyo (1994) 333

Yakovlev, V.I: Lebedev Physical Institute Report No 32 (1990), unpublished

LOW THRESHOLD PARTICLE ARRAYS

KONRAD BERNLÖHR
Max-Planck-Institut für Kernphysik
P. O. Box 103980, D-69029 Heidelberg, Germany

Abstract. While atmospheric Cherenkov telescopes have a small field of view and a small duty fraction, arrays of particle detectors on ground have a 1 sr field of view and a 100% duty fraction. On the other hand, particle detector arrays have a much higher energy threshold and an inferior hadron rejection as compared to Cherenkov telescopes. Low threshold particle detector arrays would have potential advantages over Cherenkov telescopes in the search for episodic or unexpected sources of gamma rays in the multi-TeV energy range. Ways to improve the threshold and hadron rejection of arrays are shown, based on existing technology for the timing method (with scintillator or water Cherenkov counters) and the tracking method (with tracking detectors). The performance that could be achieved is shown by examples for both methods. At mountain altitude (about 4000 m or above) an energy threshold close to 1 TeV could be achieved. For any significant reduction of the hadronic background by selecting muon-poor showers a muon detection area of at least 1000 m^2 is required, even for a compact array.

Key words: cosmic rays – gamma rays – air showers

1. Introduction

The use of arrays of particle detectors for observing cosmic rays of PeV energies and above has a long history. Scintillation counters are the traditional type of detectors for these arrays. The arrival directions of the extensive air showers and, thus, the primary cosmic rays are derived from timing measurements between the counters. Reports of an excess of cosmic rays (γ-rays) from the direction of Cygnus X-3, modulated with the 4.8 h orbital period of the binary system (Samorski and Stamm, 1983; Lloyd-Evans et al., 1983), and subsequent reports of sporadic emission from several candidate sources stimulated the build-up of several new arrays of much improved sensitivity and lower energy threshold (typically 30–100 TeV). Despite the improved sensitivity no clear γ-ray source detections have been achieved with these new arrays.

At lower energies, around 1 TeV, the imaging air Cherenkov technique with a significant reduction of the almost isotropic background of protons and nuclei has been applied with success. One source, the Crab nebula, has been observed by several experiments and a few others have been detected by single experiments at quite high significance level (e.g. see Weekes, 1992, and references therein; Baillon et al., 1993; Goret et al., 1993).

In comparison with air Cherenkov telescopes the particle detector arrays have conceptually several important advantages. The particle arrays can be operated at day and night while the air Cherenkov systems (both telescopes and wide-angle counter arrays) are restricted to clear, moonless nights, which

Space Science Reviews **75**: 185–197, 1996.

is about 10–15% of the total time at excellent sites. For Cherenkov systems any candidate source is too close to the sun to be observable at night for about four to six months each year. Arrays observe on- and off-source regions at the same time while telescopes, at least so far, are splitting their valuable observing time between on- and off-source. Arrays can even observe several candidate sources at the same time, due to their large field of view, about 1 sterad. The Cherenkov telescopes have a 100-300 times smaller field of view.

On the other hand, existing particle arrays have higher energy thresholds than Cherenkov telescopes (10–1000 TeV compared to 0.3–3 TeV). The rejection of the hadronic background by particle arrays is much inferior to that achieved by image analysis of imaging air Cherenkov telescopes, at least near threshold energies. The hadron rejection of arrays can improve substantially with increasing energy while that of Cherenkov telescopes hardly improves or even deteriorates. Due to the attenuation of extensive air showers, low threshold particle arrays should be located at rather high mountain altitude. The requirement of large flat areas is, at mountain altitude, another potential disadvantage of particle arrays.

Future particle arrays with thresholds of a few TeV, which could be called low threshold particle arrays in the context of this workshop, would for the reasons mentioned – mainly the inferior hadron rejection – not be built to search for gamma-rays from a few candidate point sources. Atmospheric Cherenkov telescopes are expected to remain unrivalled in this respect. Low threshold particle arrays could be of better use to search for episodic sources with 'bursts' on all time scales (seconds to months), to search for unexpected sources, and to examine source spectra in the multi-TeV range.

Although the planned MILAGRO experiment will be a low threshold particle array, it is not covered by this paper but is addressed in detail by another paper (G. B. Yodh, these proceedings).

2. Existing arrays and requirements for source detections

Before presenting the existing arrays it is instructive to evaluate actual requirements for observing known VHE gamma-ray sources with a particle array. In a simplified way, assuming Gaussian statistics and neglecting the problem of off-source background determination, the signal-to-noise ratio of an array for a point source is proportional to

$$S/N \propto \frac{1}{\sigma}\sqrt{\frac{At}{r_{\mathrm{h}}}},$$

with the angular resolution σ, the effective area A, the measuring time t, and the hadron rejection r_{h}. The hadron rejection depends on the fractions f_γ of gammas and f_{h} of hadrons passing the trigger and experimental shower selection: $r_{\mathrm{h}} = f_\gamma^2/f_{\mathrm{h}}$. In general, σ, A, and r_{h} are functions of energy.

A detailed calculation, assuming a resolution of $\sigma = 0.5°$, a circular source search bin of $1.6\,\sigma$ radius and taking into account the measured fluxes of the cosmic ray background and the Crab nebula results in the following requirements for an array that should be able to observe the Crab at a 5 sigmas significance level in one year with 10000 s observing time each day: Without any hadron rejection an effective area of at least $6 \cdot 10^4$ m^2 above 1 TeV or $7 \cdot 10^5$ m^2 above 10 TeV would be required. In alternative terms the required hadron rejection for an array of 10^4 m^2 effective area would be at least 6 above 1 TeV or at least 80 above 10 TeV.

Existing scintillator arrays fall substantially behind these requirements. Table I shows a selection of scintillator arrays with fairly low threshold energies E_{thresh} and good angular resolution. The quoted thresholds should be used with care because different groups use different threshold definitions. Where available, the thresholds for gamma-rays have been used. Convenient definitions are the mode energy of detected gammas (or protons) or the energy where the array reaches 50% detection efficiency for showers with cores inside the array boundaries. Where possible, the later definition is applied throughout this paper.

TABLE I

Air shower scintillator arrays with $E_{\text{thresh}} \leq 100$ TeV and angular resolution of 1° or better.

Array	Threshold (TeV)	Depth (g cm^{-2})	No. of stations × area (m^2)	Muon det. area (m^2)	Array area (10^4 m^2)
CASA-MIA[1]	70	870	1089 × 1.5	2550	23
CYGNUS-I[2]	50	800	108 × 0.8	110	2.2
EAS-TOP[3]	100	800	29 × 10	—	10
HEGRA[4]	40	800	257 × 1	270	4
JANZOS[5]	100	850	45 × 0.5	—	2
			+ 31 × 1		
SPASE[6]	50	700	24 × 1	—	0.6
Tibet ASγ[7]	10	600	49 × 0.5	—	0.8
			+ 16 × 0.25		
(Tibet II[8]	≤10	600	221 × 0.5	—	3.7)

[1] Borione et al., 1994 – [2] Alexandreas et al., 1991 – [3] Aglietta et al., 1989
[4] Aharonian et al., 1993 – [5] Allen et al., 1993 – [7] Beaman et al., 1993
[7] Amenomori et al., 1993b – [8] Yuda, 1994

Of the arrays shown in tab. I, only the Tibet ASγ array (Amenomori et al., 1993b) can be called a low threshold array in the context of this conference. It has a threshold energy of about 10 TeV and is currently being

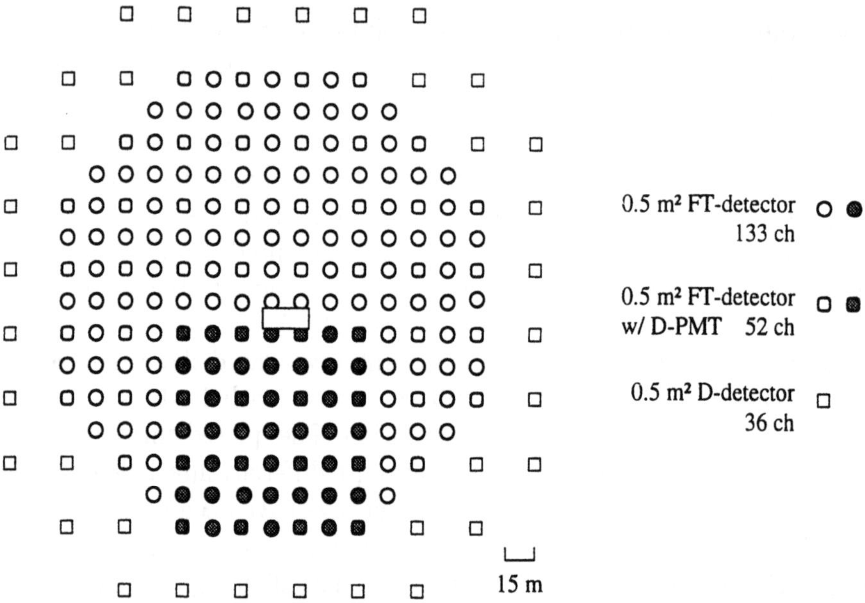

Fig. 1. The new Tibet array (figure kindly provided by T. Yuda). Shaded symbols indicate scintillation counters from the old Tibet ASγ array. The active array area with 185 fast-timing counters will be surrounded by additional 36 density counters.

upgraded into the Tibet II array with a substantially larger area but little reduction in threshold (see fig. 1, Yuda, 1994). The Tibet II array should be operated with a 200 Hz trigger rate when it is complete in 1995. Neither the old nor the new Tibet array have any hadron rejection other than that due to different threshold energies for gammas and cosmic ray nuclei. Arrays with the potential for a substantial reduction of the hadronic background, in particular CASA-MIA (Borione et al., 1994) by selecting muon-poor showers and also HEGRA (Aharonian et al., 1993) by using muons and by other methods, have a much larger threshold energy than the Tibet array.

3. Hadron rejection with particle arrays

The most powerful method for rejecting hadronic showers with particle arrays is to distinguish the showers by their muon content. The average muon content of a gamma shower is expected to be about a factor of 30 less than that of a proton shower of the same energy. The muon content of showers initiated by heavy nuclei is even larger than that of proton showers. At energies of a few TeV just a few tens to a few hundreds of muons are expected to arrive on ground. The detection of a single shower muon should, therefore, be sufficient to reject a hadronic shower.

The detection of the muons from air showers is usually done with particle detectors placed under a thick absorber layer of dirt, lead, or another material in which essentially all particles other than muons, in particular e^+/e^-, should be absorbed. Several practical problems arise when dealing with the detection of just a few muons by low threshold arrays: a) random muons, b) detector noise, c) natural radioactivity, and d) shower electrons misidentified as muons (*punch-through electrons*). Random muons arrive at a rate of 200–300 $m^{-2} s^{-1}$ on mountain altitude, depending on altitude and geomagnetic latitude. Rates due to photomultiplier noise and radioactivity are frequently well above 1 kHz per PM (6 kHz in the case of the CASA-MIA muon detectors, see Borione et al., 1994).

The best cure for these problems (a–d) would be achieved by measuring the direction of the muons which should be almost parallel to the shower axis for genuine shower muons. For scintillation counters or other direction-insensitive detectors one could use coincidences of several layers to resolve the noise problems (b and c), although at a multiple of the cost. A much more cost-effective way to reduce problems a–c is to accept only those muon counters which fired within a short time interval at the shower front. A thicker absorbers would be useful to reduce punch-through electrons (d) but ultimately also absorbs many muons. The usual compromise is an absorber thickness of about 20–25 radiation lengths, corresponding to a cutoff energy for vertical muons of about 1 GeV.

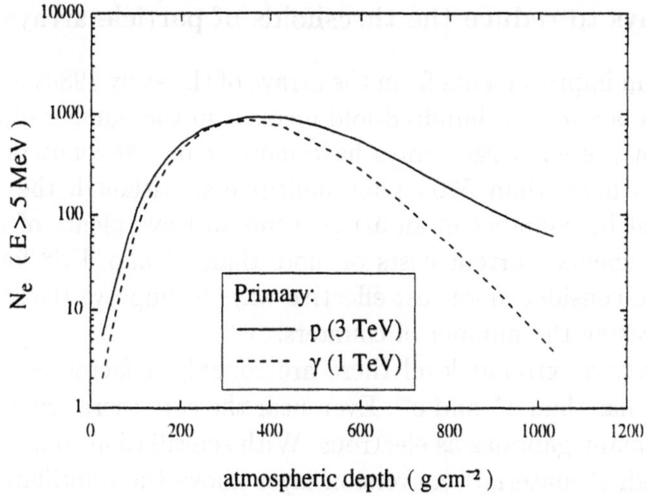

Fig. 2. Average longitudinal development of a 3 TeV proton shower (solid line) and a 1 TeV gamma shower (dashed line) in terms of the number of e^+/e^- with energy above 5 MeV (numerical simulation). Note that, although the 1 TeV gammas have almost the same peak shower size as protons of three times larger energy, the gamma showers are attenuated much faster than the proton showers.

The different longitudinal development of electromagnetic and hadronic showers in the atmosphere (see fig. 2) implies that for arrays situated at high altitudes (above about 4000 m) a substantially lower threshold energy – by more than a factor of two – for gammas than for protons can be achieved. Due to the steep cosmic ray spectrum a reduction of the proton background by about a factor of 3 is achieved. Thresholds for heavier primaries are even higher than for protons. Iron nuclei, for example, would be suppressed by an additional factor of about 6 with respect to protons. At altitudes of 4000-5000 m the composite hadronic background is reduced by a factor of 4–6 with respect to any gamma-ray signal. At an altitude below 2000 m a proton of a few TeV primary energy results, on average, in a larger shower than a gamma of the same energy. A low threshold array at low altitude, as had been proposed for ARGO (Catalano et al., 1992), would thus be enriched with hadronic showers instead of gamma showers.

Other ways to discrimate between hadronic and gamma showers could take advantage of differences in some average shower properties, e.g. the lateral distribution, including fluctuations and asymmetry thereof, or different energy spectra of the secondary electrons. Due to the shower-to-shower fluctuations and the small number of secondaries detected per shower, these other ways are not effective at low threshold. Muon detection and a high-altitude location are the only conceivable ways to reduce the hadronic background with low-threshold particle arrays.

4. Ways to reduce the thresholds of particle arrays

Among the main improvements from the arrays of the early 1980s to those of the early 1990s is a ten- to hundred-fold increase in the number of counters per array. Most present large arrays have more than 100 counters, CASA-MIA even uses more than 5000 photomultipliers. Although the 1% area fraction covered by counters in an array seems to leave plenty of room for further improvements, current costs of more than 10 mio. US$ for a large array lead us to consider more cost-effective ways to improve the sensitivity than just increasing the number of counters.

In air showers at ground level there are roughly a factor of five more secondary gammas than e^+ and e^-. Even near the core there are more than three times as many gammas as electrons. With scintillation counters a lead (or another high Z material) conversion layer above the scintillator causes most energetic gammas to convert to e^+e^- pairs while low-energy electrons are absorbed in the lead. The best compromise between conversion and absorption is a converter thickness of about 1 radiation length. Average signal amplitudes from showers increase by a factor of 1.6. The time resolution and, thus, the angular resolution improves by even a factor of 1.7 because gammas have, on average, a smaller time delay with respect to the shower front than electrons. However, the number of fired stations is little increased and thresholds are only decreased by about 20%. In all arrays of table I this converter technique is already applied.

Water Cherenkov detectors attempt to make use of the signal from converted gammas without loosing that of many electrons. Due to the low light output of the Cherenkov process (at most 100 visible photons per MeV deposited energy, about 1% of the light output from plastic scintillators) and due to the directed Cherenkov emission a very large number of large photomultipliers is needed to detect most of the particles impinging into the water. Without reflectors and multiple scattering an 8-inch PM one meter below the surface would register just about ten photoelectrons from a 10 MeV electron impinging onto a 1 m^2 area ring on the surface. Nevertheless, a considerably lower threshold energy can be achieved by the water Cherenkov technique than is achieved with existing scintillator arrays (G. B. Yodh, these proceedings).

In the case of a large scintillator array one could reduce the threshold by reducing the spacing between stations. A reduction of the spacing by a factor of three, for example from a typical 15 m to a 5 m separation, results in an array where the same number of counters are hit in showers of about three times less energy with, on average, more particles per counter in the smaller scale array. It is estimated that such a reduction in scale would reduce the threshold by a factor of three to, at most, six at 1/9 of the old array area and a 2–3 times worse angular resolution at the reduced threshold. Therefore, the optimum separation is a question of the expected source spectrum. See Gibbs et al. (1988) for the optimization of the CASA array.

As figure 3 illustrates, current state-of-the-art scintillator arrays have thresholds in terms of minimum shower size (N_e) of 5000-10000 and differences in threshold energy are mainly due to different altitudes. A high-altitude location is obviously very important for achieving a low threshold energy. A further increase in altitude beyond that of the Tibet array, however, cannot reduce threshold energies of scintillator arrays much more. Figure 3 also indicates the much lower threshold that could be achieved by an array of tracking detectors, like the proposed Cosmic Ray Tracking (CRT) array (Heintze et al., 1989).

5. Arrays of tracking detectors

Arrays of tracking detectors take advantage of the fact that the arrival directions of secondary particles in air showers are well correlated with the direction of the primary. A good angular resolution can, therefore, be achieved with few detected secondaries. An estimate of the possible shower resolution in each projection is

$$\sigma \approx (3^\circ \text{ to } 4^\circ)/\sqrt{n},$$

where n is the number of detected secondaries. No large spacing between the detectors is needed to achieve that resolution, except that the array should be large enough to locate the shower core. Compact arrays of tracking

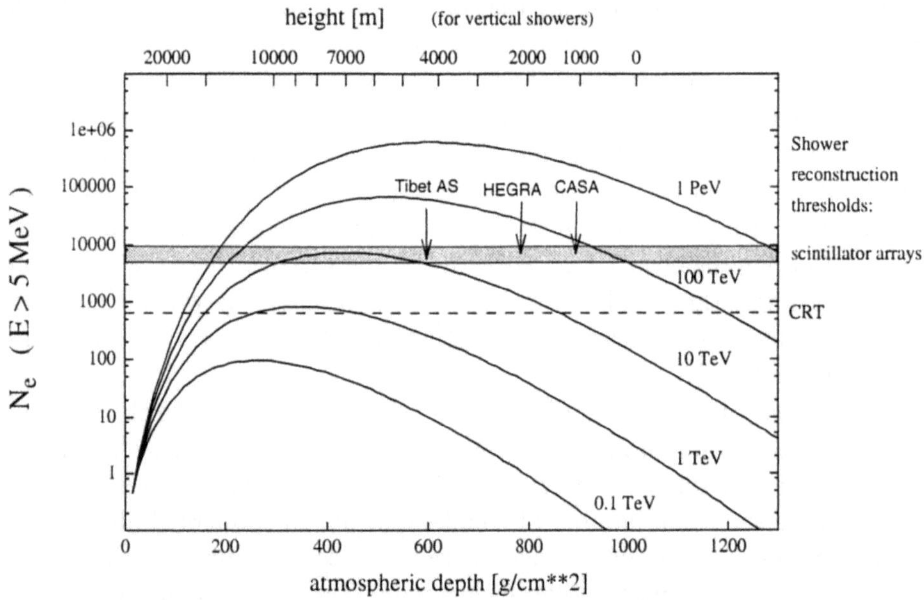

Fig. 3. Average longitudinal development of gamma showers of 0.1 TeV to 1 PeV in terms of the number of e^+/e^- with energy above 5 MeV (analytical approximation). Shower reconstruction thresholds of several scintillator arrays (see table I) in terms of minimum shower size are indicated by the shaded band. Arrays of tracking detectors like CRT could achieve lower thresholds (dashed line).

detectors can be built, if a low threshold is desired. Tracking detectors also have the potential for excellent muon identification. A single detected shower muon can be sufficient to reject a hadronic shower.

The first operational array of tracking detectors is GRAND (Poirier et al., 1988, Poirier et al., 1993), consisting of 64 stations equipped with multiwire proportional chambers (MWPCs). Each station has a stack of three pairs of MWPCs of 1.2 m^2 area above a 5 cm thick steel plate to reconstruct tracks and one pair of MWPCs below to identify muons. Single-track angular resolutions in each projection of 0.26° for muons and 0.35° for electrons are achieved. A subarray of at least 16 stations has been operational for a while and the whole array should be complete by the end of 1994 (Poirier, 1994). In some sense, GRAND is a real low-threshold array for hadronic cosmic rays, due to a single-muon trigger in addition to an air-shower trigger. Event rates of 1500 Hz are anticipated for this array.

For two reasons, at least, GRAND cannot (and is not intended to) be a low threshold array for gamma-ray astronomy. The first reason is its low altitude (220 m) and the second is its rather poor rejection of punch-through electrons. Due to the thin absorber plate and the single pair of MWPCs below it, more than 3% of the shower electrons are misidentified as muons.

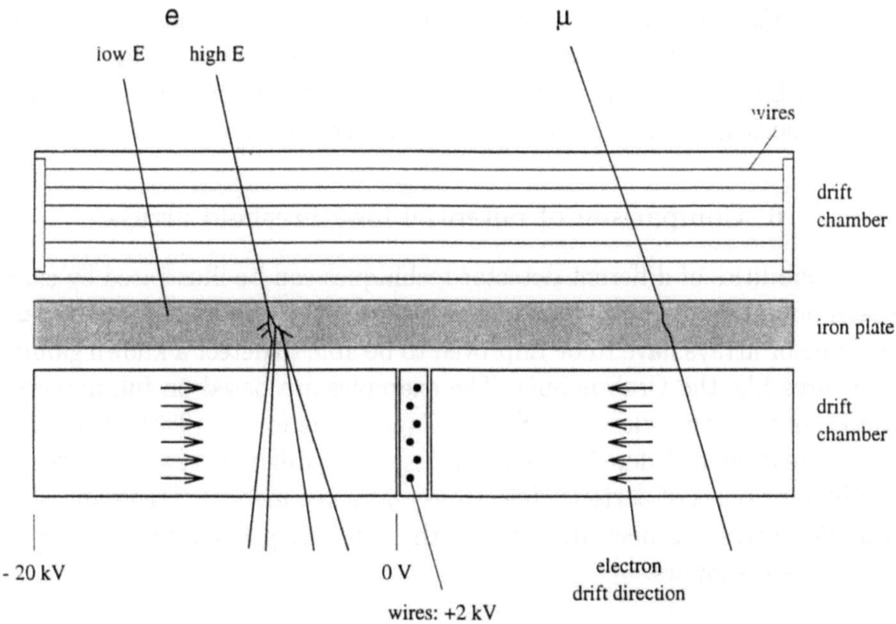

Fig. 4. Schematical view of a CRT detector from the side. Drift chambers and iron plate have a circular cross section of 1.8 m diameter. The wire direction in the upper drift chamber is perpendicular to that in the lower chamber, for easy calibration of the drift velocity. Muons are identified by a matching pair of tracks in each chamber, with little scattering in the iron plate. Electrons are either absorbed in the iron plate or cause an electromagnetic shower extending to the lower chamber.

Another type of tracking detectors, the CRT detectors achieves a much better rejection of punch-through electrons. CRT detectors consist of two circular drift chambers of 1.8 m diameter and a 10 cm thick steel plate inbetween (fig. 4). Each drift chamber has two D-shaped drift cages made of printed-circuit boards (PCBs) and a proportional wire chamber in the middle, with 6 sense wires of more than 1.8 m length supported by an aluminium frame. The drift chambers and the iron plate are in a gas-tight vessel filled with argon-methane. The readout of the wires and segmented-cathode strips is done by a 40 MHz FADC system. Each detector has its own CPU for readout and track reconstruction. For more details and the performance of CRT detectors see Bernlöhr et al. 1993a,b. Because of the thicker absorber plate and a full track reconstruction in the lower chamber, only 0.07% of the shower electrons are misidentified as muons. Therefore, hadronic showers can be identified by single muons.

Ten CRT detectors are in operation at the site of the HEGRA array (Aharonian et al., 1993) on La Palma. By comparison with showers reconstructed from HEGRA array data, a CRT single-muon angular resolution per projection of about 0.4° is determined. Due to the excellent shower muon identification and the good angular resolution the CRT detectors, together with the HEGRA array, seem to be capable to measure the average composition of cosmic rays near the knee of the energy spectrum by the muon angles with respect to the shower axis. The search for gamma-ray sources is beyond the scope of the present small-scale array of ten CRT detectors.

6. Comparison of potential low threshold arrays

The capabilities of different detector techniques can be illustrated by examples of potential arrays. The examples also show by how much present particle detector arrays have to be improved to be able to detect a known gamma-ray source like the Crab nebula. The examples are based on full numerical shower simulations with CORSIKA (Capdevielle et al., 1993), including a simple treatment of detection efficiencies of the different types of detectors for different types of particles (electrons, gammas, muons). Triggering of the example arrays has been included in the simulation but no reconstruction of 'detected' showers.

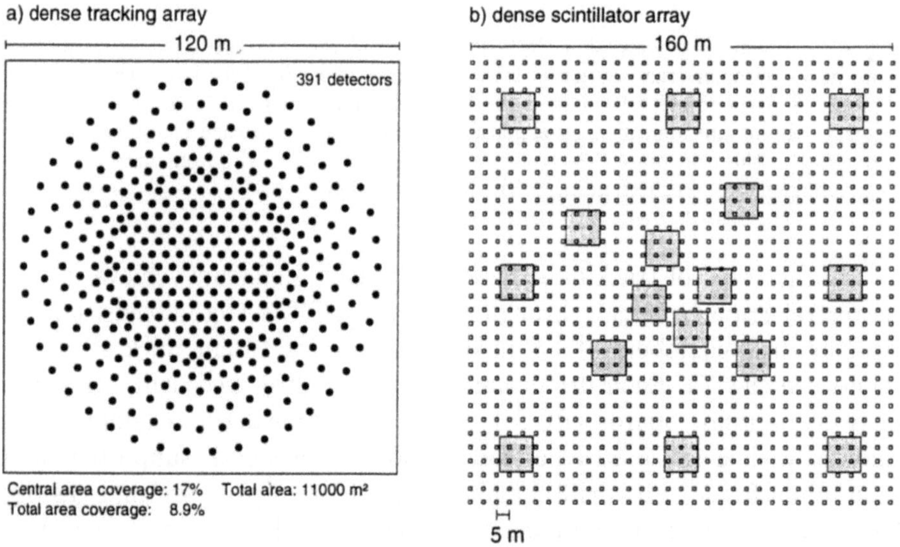

Fig. 5. Potential arrays used in comparison: a) dense CRT array of 391 detectors in a circular area of 11000 m^2, b) dense scintillator array of 1089 surface counters of 1.44 m^2 each and 16 patches of 64 underground muon counters each (shaded squares) of 160 m^2 each.

In order to compare with the existing array with the lowest threshold, the Tibet array, all examples were calculated for the altitude of this array, 4300 m. Arrays of the size and layout of the Tibet array (old and new enlarged version) have been included in the simulations but trigger conditions differ from those used at the real experiment.

Two potential low threshold arrays not actually planned were used for this comparison (see figure 5): a CRT array which has a number of detectors comparable to the original design (Heintze et al., 1989) but is much more compact to achieve a low threshold, and a dense scintillator array. Sizes and numbers of the scintillation counters correspond to the CASA array but separations are only a third of those in CASA. In contrast to CASA, the simulations assume that each station could be fired by a single ionizing particle as, for example, in the Tibet or HEGRA arrays.

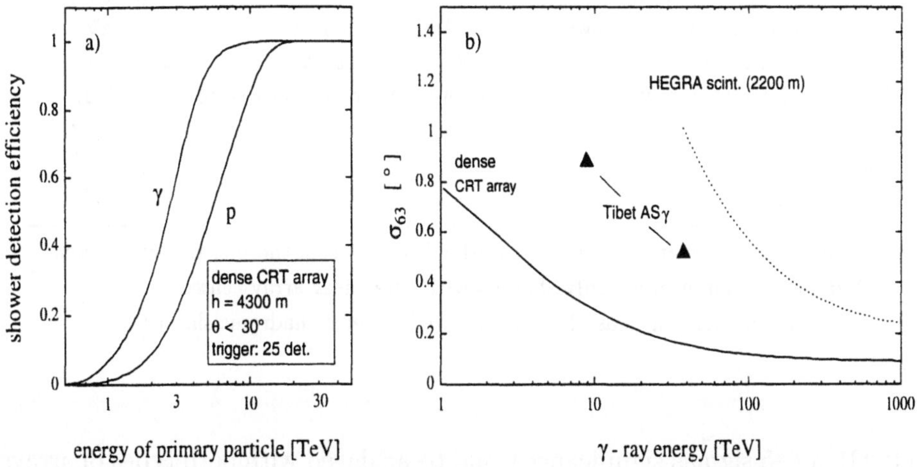

Fig. 6. a) Fraction of detected gamma-ray and proton showers with shower core inside a potential dense CRT array if a 25-of-391 stations trigger is used. b) Estimated resolution of a dense CRT array for γ-ray showers compared with the existing Tibet (Amenomori et al., 1993a) and HEGRA (Aharonian et al., 1993) scintillator arrays.

Figure 6a shows the efficiency of detecting gamma-ray and proton showers with the CRT array, assuming a 25 stations trigger. The efficiency curves of the scintillator array are quite similar and thresholds are comparable to those of the CRT array. In both cases the threshold for gamma showers is almost a factor of two lower than for proton showers. For gammas the 50% efficiency is reached at about 2 TeV for all showers of less than 30° zenith angle and at less than 1.5 TeV for vertical showers. The angular resolution of the potential CRT array has been estimated for the detected showers (fig. 6b). That of the scintillator array is more difficult to estimate without detailed detector simulations but is most probably worse than that of the CRT array. For underground muon detectors made of scintillation counters it would be difficult to get the noise rate of 1024 counters small enough to reject hadronic showers on the basis of one fired counter without loosing most genuine gamma showers as well.

Within the uncertainties of the estimated angular resolutions and hadron rejections, the sensitivities of the assumed dense tracking and scintillator arrays are comparable. Assuming the Crab flux spectrum of Lewis et al.

TABLE II

Potential low threshold arrays assumed at the altitude of the Tibet array (4300 m)

	Tibet II[1]	dense CRT	dense scint.
No. of timing or tracking detectors	185	391	1089
Area of timing/tracking detectors	0.5 m^2	2.5 m^2	1.5 m^2
Total muon detection area	—	980 m^2	2550 m^2
50% efficiency for γ showers	9 TeV	2.5 TeV	2.5 TeV
50% efficiency for proton showers	17 TeV	5.3 TeV	5.0 TeV
Total area of array	36 900 m^2	11 000m^2	25600 m^2
Hadron rejection by muon-poor showers (in addition to that due to thresholds):			
All detected events	—	3 (4)[2]	1.5/4 (2/9)[2,3]
Above 3 TeV	—	5 (7)[2]	2/6 (4/18)[2,3]
Above 10 TeV	—	12 (30)[2]	3/16 (60)[2]

1) Simulated trigger and, therefore, thresholds differ from those of real experiment.

2) Numbers in parenthesis: only showers with cores near array centres.

3) At least 3 muons / at least 1 muon required to reject hadronic shower.

(1993), a 4–5 sigmas significance could be achieved with both types of arrays in about 1–2 years of measurements. Because the improvement in hadron rejection with increasing energy (or shower size) approximately compensates the decreasing flux, the achieved significance level would be quite independent of any energy (or size) cut in the 2–20 TeV range. A possible short-term enhancement as reported for Cygnus X-3 (Dingus et al., 1988) would be detectable within one day. The Tibet array, even after its present enlargement, is not expected to see a significant signal from the Crab nebula, due to a worse resolution and missing hadron rejection by muon-poor showers.

7. Conclusions

Although the thresholds of particle arrays have been lowered by more than an order of magnitude during the last decade, there is still a significant gap between the energy ranges of Cherenkov telescopes and of particle arrays. The existing array with the lowest threshold of about 10 TeV, the Tibet array, is not expected to see a significant signal from the Crab nebula.

With new, denser particle arrays which could be built with tracking detectors or scintillation counters, as demonstrated in the previous section, or with a water Cherenkov system like MILAGRO, thresholds can be reduced close to 1 TeV while having a significant hadron rejection. In order to achieve any hadron rejection at a few TeV, a muon detector area of at least about

1000 m^2 is required, even for a compact (10^4 m^2) array and a very safe shower muon identification as with CRT detectors. If the detector noise level is too large to reject hadronic showers by a single detected muon, much larger muon detection areas are needed. At energies above about 10–20 TeV hadron rejection factors comparable to those of imaging Cherenkov telescopes can be achieved if punch-through electrons can be well suppressed.

Although the purpose of such arrays is not the detection of a known source like the Crab nebula, a detection of the Crab could help to validate the array performance and seems feasible. The main purpose for low threshold arrays would be the search for variable, episodic, or unexpected sources. Spectra of known sources could also be examined at energies where Cherenkov telescopes inevitably run out of statistics. Particle arrays can be valuable all-sky and whole-year monitoring instruments while imaging Cherenkov telescopes, due to the superior hadron rejection, at least below 10 TeV, remain the instruments of choice for looking for gamma rays from a few gamma-ray source candidates. After having established that TeV gamma-ray sources exist, the need for monitoring instruments like particle detector arrays still arises.

References

Aglietta, M. *et al.*: 1989, *Nucl. Instr. Meth.* **A277**, 23
Aharonian, F. *et al.*: 1993, in *Proc. 23rd Intern. Cosmic Ray Conf.*, Vol. 4, p. 291
Alexandreas *et al.*: 1991, *Nucl. Instr. Meth.* **A311**, 350
Allen, W. H. *et al.*: 1993, *Phys. Rev. D* **48**, 466
Amenomori, M. *et al.* (Tibet ASγ Collaboration): 1993a, *Phys. Rev. D* **47**, 2675
Amenomori, M. *et al.* (Tibet ASγ Collaboration): 1993b, in Jones, 1993, p. 219
Baillon, P. *et al.*: 1993, *Astroparticle Physics* **1**, 341
Beaman, J. *et al.*: 1993, *Phys. Rev. D* **48**, 4495
Bernlöhr, K. *et al.*: 1993a, in *Proc. 23rd Intern. Cosmic Ray Conf.*, Vol. 4, p. 199
Bernlöhr, K. *et al.*: 1993b, in *Proc. 23rd Intern. Cosmic Ray Conf.*, Vol. 4, p. 670
Borione, A. *et al.*: 1994, *Nucl. Instr. Meth.* **A346**, 329
Capdevielle, J. N. *et al.*: 1993, in Jones, 1993, p. 545
Catalano, O., Linsley, J., and Scarsi, L.: 1992, *Nucl. Phys. B (Proc. Suppl.)* **28B**, 155
Dingus, B. L. *et al.*: 1988, *Phys. Rev. Lett.* **60**, 1785
Gibbs, K. G. *et al.* (CASA Collaboration): 1988, *Nucl. Instr. Meth.* **A264**, 67
Goret, P. *et al.*: 1993, *Astron. Astroph.* **270**, 401
Heintze, J. *et al.*: 1989, *Nucl. Instr. Meth.* **A277**, 29
Jones, L. (Ed.): 1993, *Very High Energy Cosmic-Ray Interactions, VIIth International Symposium*, Ann Arbor, MI, USA, AIP Conf. Proc. 276
Lewis, D. A. *et al.*: 1993, in *Proc. 23rd Intern. Cosmic Ray Conf.*, Vol. 1, p. 279
Lloyd-Evans, J. *et al.*: 1983, *Nature* **305**, 784
Poirier, J.: 1994, private communication
Poirier, J. *et al.*: 1988, *Nucl. Instr. Meth.* **A264**, 81
Poirier, J. *et al.*: 1993, in Jones, 1993, p. 614
Samorski, M. and Stamm, W.: 1983, *Astroph. J.* **268**, 117
Weekes, T.: 1992, *Space Sci. Rev.* **59**, 315
Yuda, T.: 1994, private communication

WATER CHERENKOV DETECTORS:MILAGRO

GAURANG B. YODH
Department of Physics, University of California Irvine
Irvine, CA, USA,92717

Abstract.
This paper discusses the properties of using the water Cherenkov technique to detect air showers in the few hundred GeV to 100 TeV energy range. The responses of a 6 m^2 2 m deep water Cherenkov counter and that of a 6 m^2 10cm thick scintillator-lead sandwich counter to air shower electrons and photons is described. The advantages of water Cherenkov detector is outlined. Its application to do VHE gamma ray astronomy is discussed with particular reference to the MILAGRO telescope currently under construction. Milagro, a water-Cherenkov detector to do gamma ray astronomy above 100 Gev, uses an existing pool 60m x 80m by 8m, located in the Jemez mountains near Los Alamos, NM. The threshold of the MILAGRO detector is comparable to atmospheric Cherenkov detectors, however it has several advantages over these optical detectors. MILAGRO can operate 24 hours a day in all weather conditions and it has an open aperture which allows it to view the entire northern sky every day. These capabilities allow for a systematic all-sky survey to be done for the first time at these energies. MILAGRO will measure the Crab spectrum with high significance over a wide energy range, it will detect and measure the spectra from AGN's such as MRK 421 and it will search for short duration bursts from GRBs and possibly evaporating PBHs.

Key words: very high energy gamma ray astronomy,water Cerenkov telescopes

1. Introduction

The Compton Gamma Ray Observatory(CGRO) has observed gamma ray emission from various sources from a few Mev to 10s of GeV, including that from the Crab, from Geminga, from Gamma Ray Bursters(GRB) and from Active Galactic Nuclei(AGNs) (Fichtel 94a, Fichtel 94b). Two general features of gamma ray emission can be discerned from their data: (1) Many of the observed spectra are non-thermal and hard, with differential spectral indices near -2 and (2) the intensity is variable over periods from hours to days. This suggests existence of particle acceleration mechanisms that are efficient and emission regions which are small. Some of the energy spectra observed by the EGRET instrument extend up to 10 GeV The sensitivity range of EGRET does not allow one to determine whether these spectra continue to higher energies. Recently, TeV emission from the AGN Markarian 421 was observed by the Whipple experiment(Punch et al.1993, Kerrick et al 1995). It is of great interest to find out if the emission from other CGRO sources extends to higher energies (Aharonian 1994a, Aharonian 1994b, Kirk et al., 1995). If it does, then one can put constraints on models of production of gamma rays and understand better the nature of particle acceleration in these sources. For AGN sources, if emission extends into the TeV energy range there is a possibility of determining the extra-

Space Science Reviews **75**: 199–212, 1996.

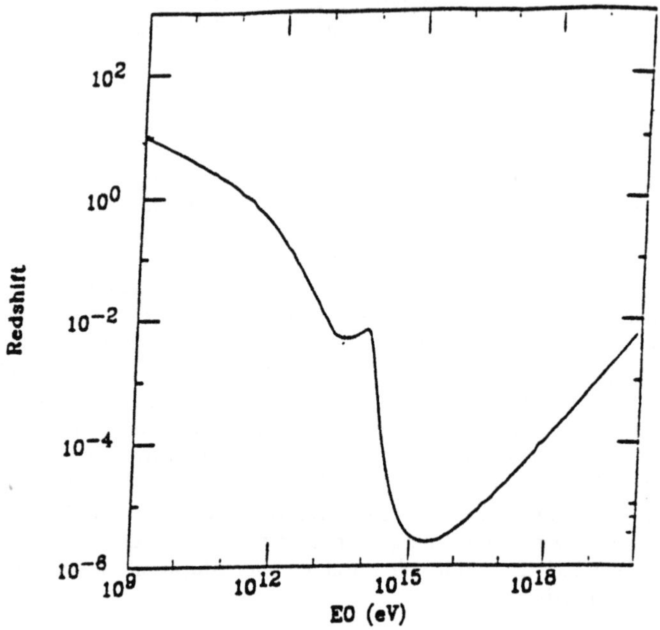

Fig. 1. Distance at which the intergalactic infra-red and universal CMBR radiation fields yields and optical depth of unity versus gamma ray energy

galactic infra red background radiation which can lead to energy dependent attenuation of the gamma ray flux in the few TeV energy region. In figure 1 is shown the distance at which the absorption by infra-red background radiation (Stecker et al.,1977, Stecker et al.,1992) and cosmic microwave background radiation,CMBR,(Gould and Schreder 1967) would reduce the gamma ray flux by e-fold versus gamma ray energy. Observation of gamma ray sources in the TeV and PeV energy ranges is, also, important for understanding the origin of cosmic rays and the nature of the accelerators of cosmic rays. Search for bursts in the higher energy range could establish whether GRBs are galactic or cosmological. Another interesting problem is to search for evaporating Primordial Black Holes(PBHs)(Halzen et al. 1991). If gamma rays are produced via nuclear collisions high energy neutrinos should be produced also which should be detected by the current large neutrino telescopes being built, such as AMANDA and DUMAND. In this paper, I shall review features of the water Cherenkov technique which make it especially suited to do gamma ray astronomy from 100 GeV to 100 TeV and outline the design and capabilities of a new water Cherenkov detector, MILAGRO, currently being built at 8700 feet in New Mexico.

2. Water Cherenkov technique

An ideal detector to cover the energy range beyond that accessible to EGRET type experiments (1) should span the energy range from 10 GeV to 100 TeV, (2) should have sufficient collection area and efficiency to have sensitivity to extrapolations of EGRET fluxes, (3) should have continuous all weather operation, (4) should have all sky coverage, (5) should have an angular resolution better than $\sim 0.5°$ and (6) should have reasonable energy resolution.

Ground based air shower detectors currently operating to search for gamma rays from astrophysical sources use relatively small scintillation detectors spread out over a large area to sample the air shower particles produced by particles incident on the atmosphere(Cronin et al., 1993, Alexandreas et al., 1992). They sample less than 1 percent of the shower particles and have a relatively high energy threshold of greater than about 10 TeV. The Water Cherenkov technique (Lawrence et al.,1991, Kellerman et al.,1970, Alexandreas et al., 1991, Yodh 1992) offers the possibility of significantly increasing the fraction of shower particles detected, thereby lowering the energy threshold of an array that uses water Cherenkov detectors below that using scintillation counters, even if the counters are of the scintillator-lead sandwich type which increase the efficiency for detecting photons.

The water Cherenkov technique takes advantage of the following properties of air shower particles: (1) the photon density is >> electron density at all radial distances, (2) there are many more photons with energies below 100 MeV than electrons and (3) the average electron or photon energy at large distances is in the tens of MeV range. In particular, the special features of water Cherenkov detectors are: (1) they have sensitivity to all charged particles above the Cherenkov threshold of β=0.75, (2) 1 or more meters of water depth will convert the photons into electronic particles which will Cherenk, (3) with a reasonable number of PMTs a water detector is sensitive over its entire area, (4) with enough PMTs hit by a shower one can achieve good angular resolution, and (5) large area water detectors are cheaper ($ 600 per m^2) and can be operated reliably. In addition, such detectors should be placed as close to shower maximum as possible to obtain the lowest possible energy threshold. Convenient access to power and other utilities is desirable. The land surrounding the detector should be reasonably terrained to be able to place air shower and air Cherenkov detectors to complement the water detector. The latter is important to cross calibrate the atmospheric Cherenkov technique with the air shower technique. These considerations make it possible to design a water Cherenkov telescope which approaches the ideal detector described in the first paragraph.

The superior properties of a water Cherenkov detector is illustrated by comparing the response to air shower particles of (1) an 8ftx8ft and 2m deep water tank with diffuse reflective walls with that of (2) an 8ftx8ft by

10cm thick scintillator counter sandwich with one radiation length of lead to enhance the efficiency for incident shower photons. In these prototypes, both counters are viewed by a single 8 inch PMT. For the scintillator a response similar to that of the scintillation detectors of CYGNUS is assumed(23 pes for a thru going muon).

Air showers were generated on an integral spectrum of $E^{-1.7}$ starting at 500 GeV and the detectors were placed at an atmospheric depth of 750 gm/cm^2. Each particle (e, γ or muon) was thrown over the area of the counters. The response of the counter was calculated in terms of (1) number of pes per particle and as (2) efficiency of detection. This was studied as a function of particle energy and distance from the core of the shower. The results are summarized in figures 2 thru 5.

The mean number of particles(for all showers generated) as a function of particle energy summed over all distances is shown in figure 2(a) and the mean number of particles as a function of core distance summed over all energies is shown in figure 2(b). One observes two features: (1) that photons outnumber electrons by a large factor and (2) their energy distributions are very similar.

The efficiency, per particle, of detecting an electron as function of energy and core distance is displayed in figures 3(a) and 3(b). Similar curves for photons are shown in figures 4(a) and 4(b). Figure 3(a) shows that scintillator is better for low energy electrons, while figure 4(a) shows that water is better for photons of all energies. In figure 3(b) one sees that due to showering of higher energy electrons at small core distances water is better than scintillator and this advantage of water does not disappear until one gets to large distances where the electron energy is low. Figures 4, however, show that water is better for photons at all energies and all distances.

Finally, in figure 5 the efficiency for producing 1 or more photo-electrons, either due to photons or electrons in a shower, as a function of the core distance is shown. Water is superior to scintillator by a factor of two over all distances. Water Cherenkov detectors have already been used in the CYGNUS experiment. The CYGNUS experiment had five pools installed into the array, four on corners of an approximate square 80 m on the side and one in the middle. These 44 m^2 detectors, each with 7 PMTs have been operational for the last three years. The angular resolution as determined from the pool signals for CYGNUS triggers is found to be 0.4 degrees as compared to 0.67 degrees for the scintillation counter array. This is primarily due to the detection of more particles and thereby determining more precisely the arrival time of the air shower.

On the basis of such studies the MILAGRO detector was designed which uses the water Cherenkov technique to make a VHE-UHE gamma ray telescope. This is described next.

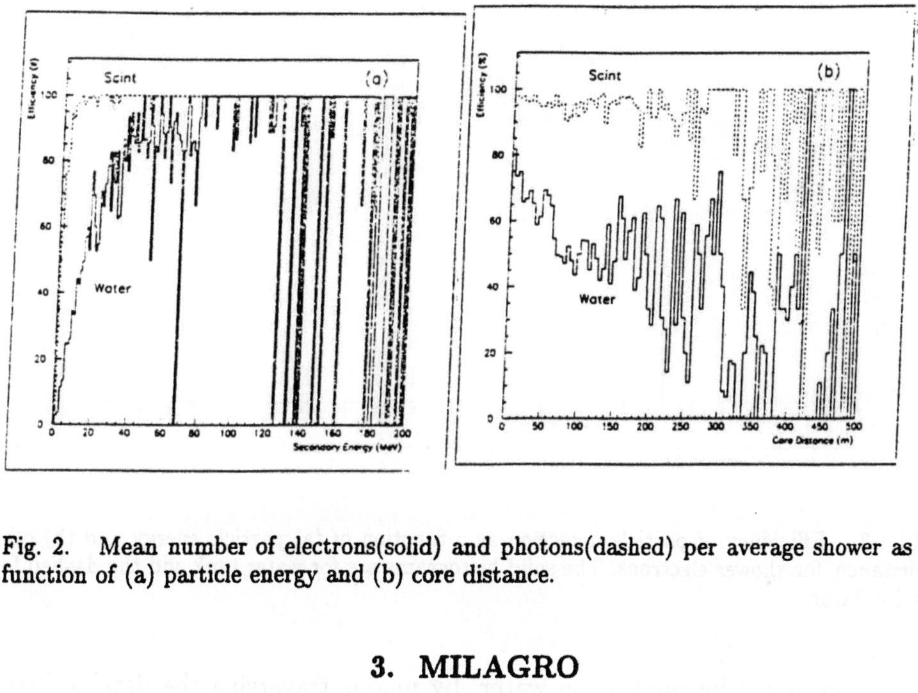

Fig. 2. Mean number of electrons(solid) and photons(dashed) per average shower as a function of (a) particle energy and (b) core distance.

3. MILAGRO

Multi Institution Los Alamos,Gamma-Ray Observatory

The MILAGRO instrument has been designed to explore the energy range above 100 GeV extending into the UHE range. The MILAGRO telescope will have the capability to observe over 1 sr of the the sky continuously and be sensitive to a wide range of time scales.(Williams et al., 1991)

3.1. Characteristics of MILAGRO telescope:

3.1.1. Layout

The telescope is based on the technique of using water Cherenkov detector of large area, fully sensitive to all components (except neutrinos) of the air shower. Located at high altitude it can detect air showers with energy down to 100 GeV with good efficiency. The detector measures energy flow and samples a large number of particles in the shower front to minimize timing fluctuations. The detector is to use an existing large covered pond, 60 m by 80 m and 8 m deep, located at Fenton Hill at an altitude of 8700 ft near Los Alamos. It will be instrumented with three layers of fast 8 inch phototubes. The Cherenkov light emitted by electrons and positrons, by

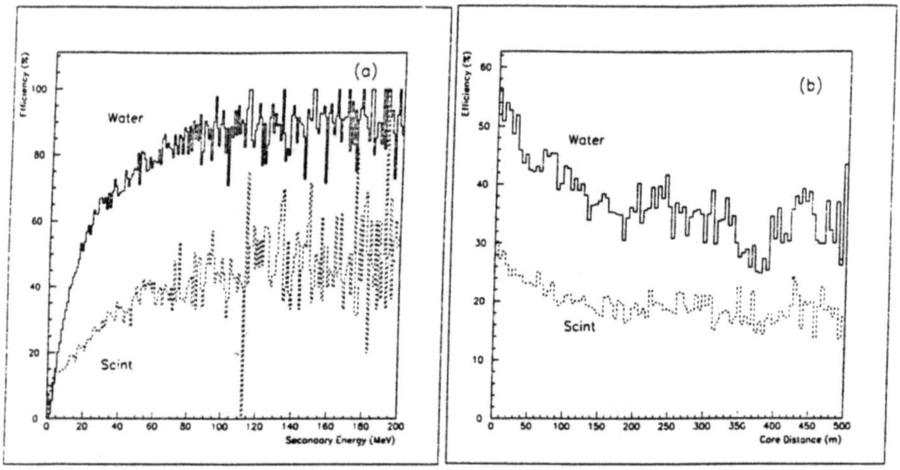

Fig. 3. Efficiency of particle detection as a function of (a) particle energy and (b) core distance, for shower electrons. The solid histograms are for water tank and the dashed for scintillator.

pairs produced by photons in water, by muons traversing the detector and by hadrons making cascades will be detected by the phototubes.

A schematic of the cross section of the pool showing the placement of PMTs in the three layers is shown in figure 6. The top layer of tubes is sensitive primarily to the electro-magnetic component of the air shower hitting the pool. The middle layer will detect light from the electromagnetic, the hadronic and the muonic components of the shower. The bottom layer is optically isolated from the top two layers and is sensitive mainly to the muons in the shower hitting the pond.

3.1.2. Energy sensitivity and Trigger Efficiency

The energy sensitivity of the MILAGRO detector is best illustrated by a plot of the effective area as a function of primary energy for three different trigger requirements for two different species of primary particles: protons and gammas. This is shown in figure 7. Air showers, generated on the cosmic ray energy spectrum from 0 to 45 degrees in zenith, were thrown over an area of 500 m by 500 m around the detector and the number of PMTs hit were determined. With a trigger requirement of 15 tubes in the shower layer the trigger rate is expected to be about 1 KHz. These curves also show a very

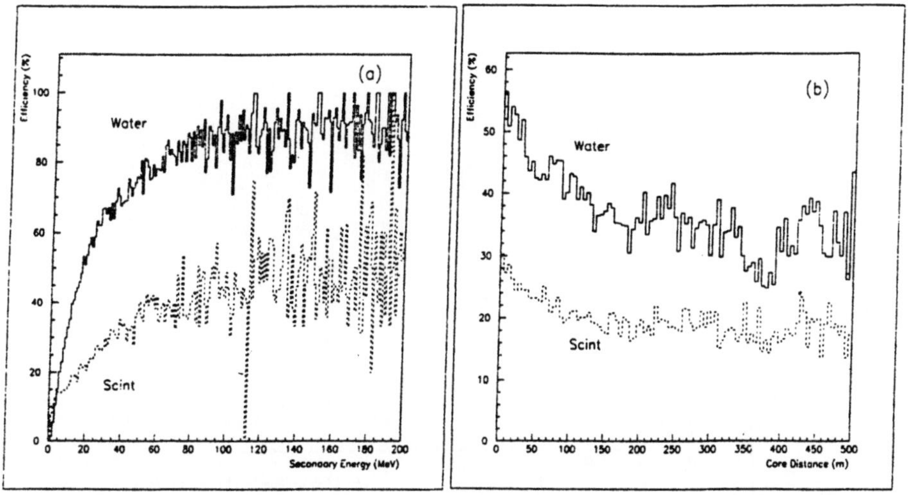

Fig. 4. Efficiency of particle detection as a function of (a) particle energy and (b) core distance for photons. The solid histograms are for water tank and the dashed for scintillator.

important feature of MILAGRO at lower energies: **The trigger efficiency for low energy protons is at least an order of magnitude smaller than that for gamma rays at energies below 300 GeV.**

3.1.3. Angular resolution

The arrival direction is determined by timing. Because of the large number of particles detected per shower, the angular resolution will be $\leq 0.4°$ at 10 TeV, a considerable improvement over present values. The simulation of angular resolution was based upon the time spread measurements made with the five pools in CYGNUS experiment. Angular resolution as a function of PMT threshold is shown in figure 8. For large showers (large number of PMT hits) the angular resolution improves to 0.25°.

3.1.4. Muon and Hadron detection:

A second layer of phototubes, looking upwards, will be sensitive to not only tails of electromagnetic showers but also to hadrons of high energy near the cores of air showers produced by cosmic ray nuclear primaries. A third layer of PMTs, near the bottom of the reservoir, will count muons in the shower. These tubes are shielded from downward coming Cherenkov light by opaque

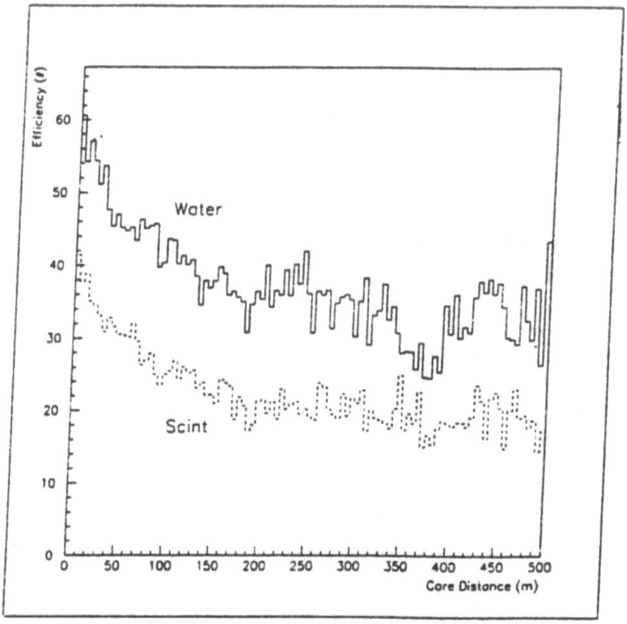

Fig. 5. Efficiency of producing one or more photo-electrons by shower particles as a function of core distance for water tank(solid) and scintillator(dashed).

barrier and each tube will have its own 3mx3mx1m diffusing muon cell. The muon layer will provide excellent rejection of hadronic showers at energies above few TeV.

3.2. Sensitivity and Physics Capabilities

3.2.1. Point Source Sensitivity:

The ability of MILAGRO to detect a point source depends on both the intensity and the spectrum of the source. The significance of a signal from a source is given by

$$\frac{S}{\sqrt(B)} = \frac{0.72 \int A_\gamma(E) I(E) dE}{(\Omega T \int A_p(E) 0.1 (E_{TeV})^{-2.67} dE)^{0.5}} \tag{1}$$

where the measured proton flux is explicitly inserted (5.1×10^{-2} m^{-2} s^{-1} sr^{-1}) in the denominator. In the above equation $A_\gamma(E)$ and $A_p(E)$ are the effective areas of the detectors for gamma rays and protons as a function of primary energy, Ω is the solid angle of the source bin, and I(E) is the source spectrum. The factor 0.72 is the fraction of source events that fall within the optimal angular bin and T is the observation time on the source. Background rejection of cosmic ray showers by muon detection can by included in the effective areas.

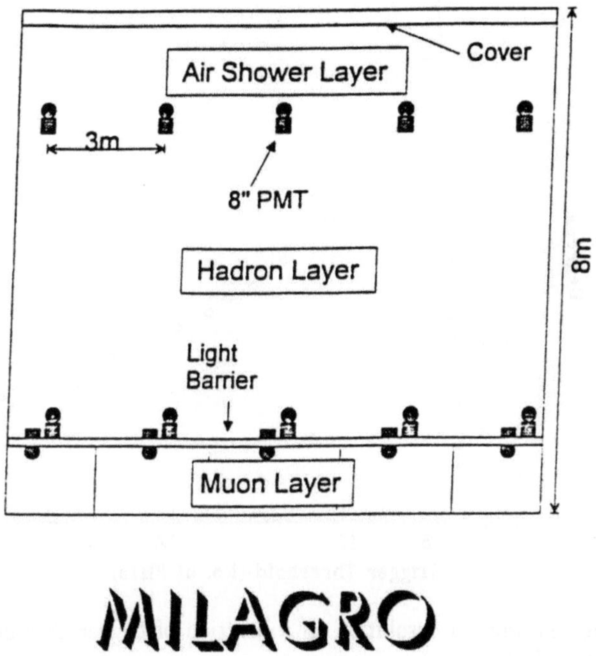

Fig. 6. Schematic cross section of the MILAGRO detector showing the approximate deployment of PMTs for shower, hadron and muon layers

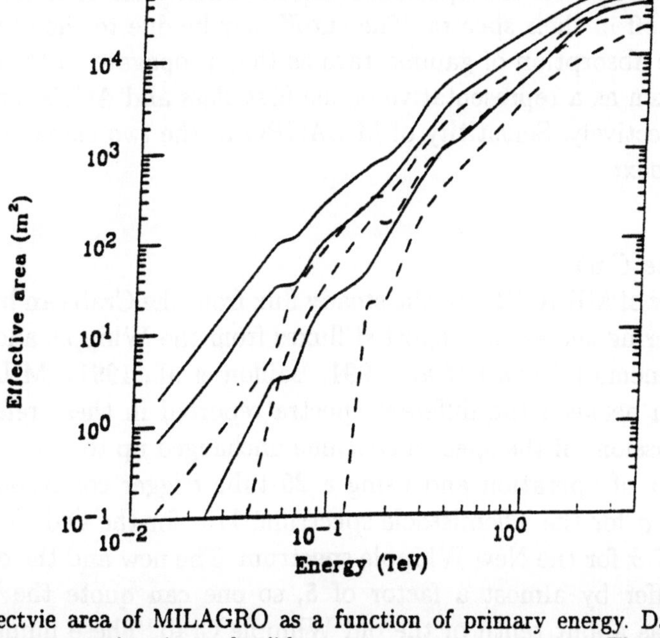

Fig. 7. Effectvie area of MILAGRO as a function of primary energy. Dashed curves correspond to proton primaries and solid curves to gamma primaries. Three sets are for three different conditions. Lowest for greater or equal to 50 PMT, next for 25 PMT and topmost for 15 PMTs, respectively.

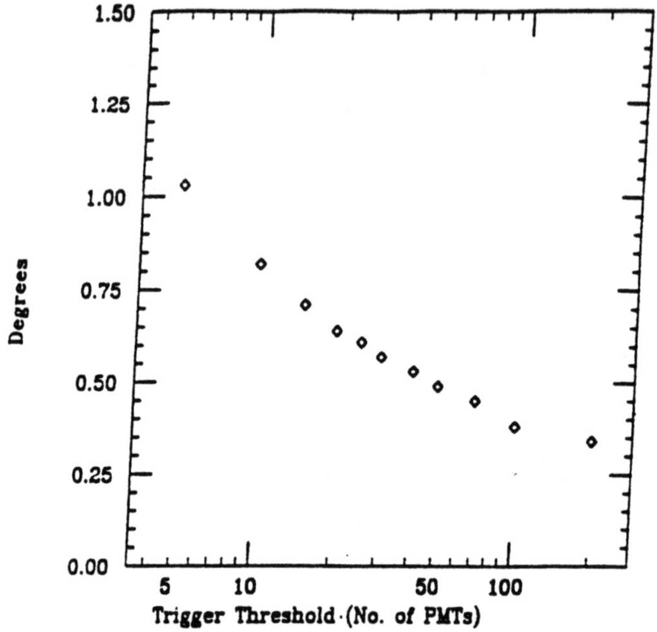

Fig. 8. Expected angular resolution as a function of trigger threshold for illuminated photomultiplier tubes.

There are two classes of constant sources of interest. The first is a source with a simple power law spectrum and the other class of sources are those with a cutoff in their spectra. The cutoff may be due to the source itself or due to the absorption of gamma rays as they propagate to the earth. Crab can be taken as a representative of the first class and AGNs for the second class, respectively. Sensititity of MILAGRO to the two classes of sources is discussed next.

3.2.2. The Crab

The ability of MILAGRO to the steady flux from the Crab can be illustrated by considering the earlier reported fluxes from the Whipple and Themistocle experiments (Vacanti et al., 1991, Baillon et al., 1991). MILAGRO can distinguish between the different spectra reported in these references. For steady emission , if the spectra continue unchanged up to several TeV, then in one year of operation and using a 25 tube trigger condition we should observe 25 σ for the Themistocle spectrum, 11 σ for the Old Whipple spectrum and 7 σ for the New Whipple spectrum. The new and the old Whipple spectra differ by almost a factor of 5, so one can quote the MILAGRO sensitvity as about tenth of the old Whipple Crab. These numbers do not include any rejection of hadronic showers based upon muons or other techniques. If we see the signal without muon rejection and if the signal disap-

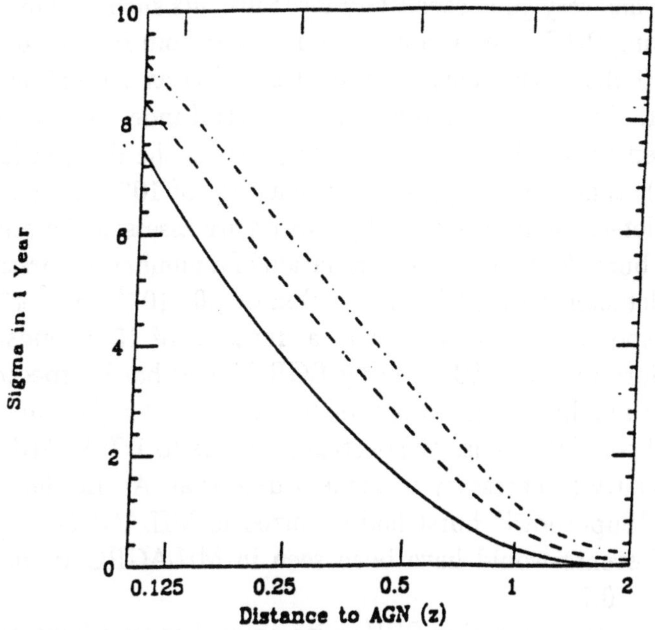

Fig. 9. Number of standard deviations that MILAGRO would observe in 1 year from an AGN with a source flux of $3.6\times10^{-11}/E$ cm^{-2} s^{-1} TeV^{-1}, versus distance to the AGN. The dot-dash curve is for a 15 PMT trigger, the dashed curve for a 25 tube trigger and the solid curve for a 50 tube trigger.

pears when we cut on events with muons then it would indicate onset of some significant new physics.

3.2.3. Active Galactic Nuclei

From Figure 1 it is clear that low energy threshold is needed to detect distance sources. In figure 9 is shown the expected statistical significance of a signal from an AGN in MILAGRO after 1 year of observation versus distance to the AGN. The three curves correspond to the three different trigger requirements of Figure 7. The dot-dash curve is for a 15 PMT trigger, the dashed curve for a 25 PMT trigger and the solid curve for a 50 PMT trigger. The source flux used was 3.6×10^{-11}/E cm^{-2}, s^{-1},TeV^{-1} where E is in TeV. This roughly five times the flux of Mrk 421 and more typical of the observed AGNs by EGRET. One of the current models for the extra-galactic IR field was used in generating this figure. The effect of a lowering the trigger threshold is dramatic, MILAGRO with 15 tube trigger can see nearly twice as far as with a 50 PMT trigger.

3.2.4. Sensitivity to Gamma Ray Transients

MILAGRO has very good sensitivity for transients such as those from GRBs or evaporating PBHs over a range of time scales from 100 ms to 100 seconds.

Figure 10 shows the sensitivity of MILAGRO as a function of duration of the burst. This assumes a differential spectral index of -2.5(i.e. the spectrum is given by $A_0(E(TeV))^{-2.5}$), The y-axis is the differential burst flux (A_0 above), that yields a pre-trial probability of 10^{-8}. This guarantees a significant detection after any trial penalties are assessed (because of poorly determined burst location, the examination of a number of bursts, etc.). The first GRB detected by EGRET had a flux of 1.0×10^{-8} cm^{-2} s^{-1} TeV^{-1}, a differential spectral index of 2.5 and a duration of 15 seconds(Schneid et al.,1992). More recent GRBs seen by EGRET had harder spectra (spectral index $\sim$ 2.0) making them 2-3 orders of magnitude brighter at 1 TeV than this burst. If the GRB emission spectrum extends to 1 TeV, MILAGRO will have the sensitivity necessary to make a detection. As another example, if the famous "Superbowl" burst had occurred in MILAGRO's field of view, hundreds of events would have been seen in MILAGRO if the GRB was located at $z < 0.1$.

For the rate of evaporating PBHs with about 1 second burst time, MILAGRO will have about a 1000 times lower sensitivity than that achieved in the CYGNUS experiment(Alexandreas et al 1993), which is currently the best limit. The 90 % confidence limit will be about several 100 evaporating PBHs per cubic parsec per year.

3.3. Concluding remarks:

The water Cherenkov telescope, MILAGRO, represents a major step forward in ground based gamma-ray astrophysics. MILAGRO will have an energy threshold near 100 GeV, just beyond the range of current space based detectors. The detector is under construction and we expect the telescope to become operational by end of 1997. With the operation of the MIlAGRO instrument we will enter a new era in the study of high energy gamma sources such as AGNs, GRBs and other point sources, opening up a new energy range and providing continuous coverage of the overhead sky.

4. Acknowledgements

This work was supported in part by the National Science Foundation, Physics division. I want to thank Heinz Volk and Felix Aharonian for their excellent organization of the workshop. I would like to thank my colleague, Anthony Shoup, for the study of relative sensitivity of water Cherenkov detectors and scintillation counter detectors to air showers. This presentation is on behalf of the MILAGRO collaboration: H.S. Ahulwalia,[9] S. Barwick,[2] D. Bauer,[6] S.D. Biller,[2] D. Caldwell,[6] M. Cavalli-Sforza,[4] C.Y. Chang,[1] M.L.Chen,[1]

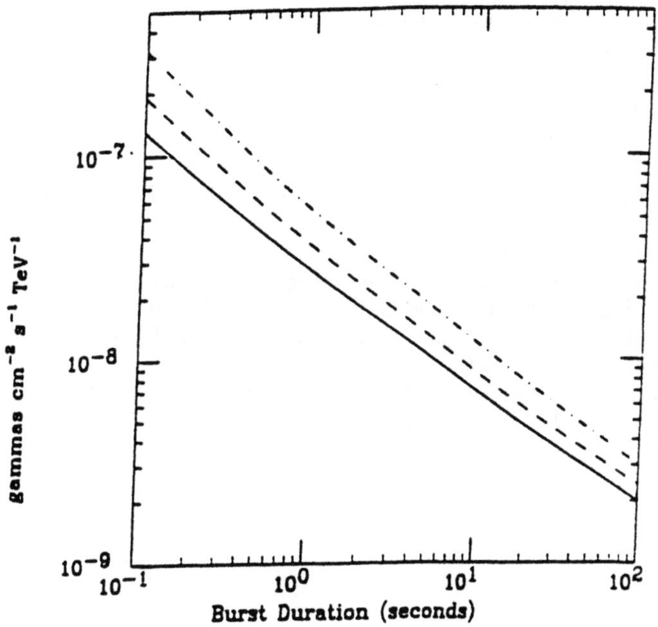

Fig. 10. Sensitivity of MILAGRO to bursts as a function of burst duration

P. Chumney,[2] D.G. Coyne,[4] B.L. Dingus,[11] C.L. Dion,[1] D.E. Dorfan,[4] R.W. Ellsworth,[5] S.J. Freedman,[10] B.K. Fujikawa,[10] T.J. Haines,[3] C.M. Hoffman,[3] L.A. Kelley,[4] S. Klein,[4] A. Lu,[6] A.I. Mincer,[8] D.E. Nagle,[3] P. Nemethy,[8] T.J. O'Neill,[7] V.D. Sandberg,[3] G. Sanders,[3] S. Schaller,[3] D.M. Schmidt,[3] A. Shoup,[2] C. Sinnis,[3] O.T. Tumer,[7] D.H. White,[3] D.A. Williams,[4] T. Yang,[4] S. Yellin,[6] G.B. Yodh,[2] and A. Zych[7]

[1] *The University of Maryland, College Park,* [2] *The University of California, Irvine,* [3] *Los Alamos National Laboratory, Los Alamos, New Mexico,* [4] *University of California, Santa Cruz,* [5] *George Mason University, Fairfax, Virginia,* [6] *University of California, Santa Barbara,* [7] *University of California, Riverside,* [8] *New York University, New York,* [9] *University of New Mexico, Albuquerque,* [10] *University of California, Berkeley,* [11] *Goddard Space Flight Center, Greenbelt, Maryland*

References

Aharonian F. A. et al., 1994a, *Ap. J.*, L5, 423

Aharonian F. A. et al., 1994b, *Preprint*

Alexandreas D. E., et al., 1991, *Proc. of the 22nd Int. Cosmic Ray Cong.(Dublin)*, **2**, 680.

Alexandreas D. E., et al., 1992,*Nucl. Inst. and Methods*, **A311**, 350.

Alexandreas D. E., et al., 1993,*Phys. Rev. Lett.*, **71**, 2524.

Baillon P., et al., 1991 *Proceedings of 22nd Int. Cosmic Ray Conf.(Dublin)*, vol 1, p 220; and P. Baiilon et al., 1993, *Proc. of the 23rd Int. Cosmic Ray Conf. (Calgary)*, vol 1., p 267.

Cronin J. W., et al., 1993,*Ann. Rev. of Nuclear and Particle Science*, **43**, 883
Fichtel C. E.,1994a, *Ap. J. Suppl.*, **90**, 917
Fichtel, C.E., et al., 1994b, *Ap. J. Suppl.*, **94**, 551
Gould, R. J. and Schreder, G., 1967,*Phys. Rev.*, **155**, 1404 and 1408.
Halzen F., et al.,1991, *Nature*, **353**, 807.
Kellerman E. W., 1970, *J. of Phys A, Gen. Physics*, **3**, 284.
Kerrick A. D., et al., 1995, *Ap. J.*, in press
Kirk J. W., et al., 1995,*Astron. and Astrophys.*, to appear
Lawrence M. A., Reid R. J. O. and Watson A. A., 1991, *J. Phys. G, Nucl. Part. Phys.*, **17**, 733.
Punch M., et al., 1992,*Nature*, **358**, 784.
Schneid, E. J., et al., 1992, *Astron. and Astrophysics*, **255**, L13.
Stecker, F. W., Puget J. L. and Fazio G. G., 1977, *Ap. J. Lett.*, **214**, L51.
Stecker, F. W., Puget J. L. and Fazio G. G., 1992, *Ap. J. Lett.*, **390**, L49.
Vacanti G., et al, 1991,*Ap.J.* **377**, 467.
Williams D. A. et al., 1991, *Proc. of 22nd Int. Cosmic Ray Conf. (Dublin)*, **2**, 684.
Yodh G. B., et al., 1992,*Nucl. Phys. B(Proc. Suppl.)*, **28B**, 123

HIGH ENERGY NEUTRINO TELESCOPES

L. K. RESVANIS
Physics Laboratory, University of Athens
104 Solonos St., Athens 10680, Greece

Abstract. A review of the potential of Neutrino Telescopes to contribute to Astrophysics and Elementary Particle Physics is presented. The status of the four neutrino telescopes presently under construction is reviewed. An inexpensive modular design of how to reach an 1 km^2 telescope with an energy threshold of a few GeV is proposed.

1. Introduction

The cosmic ray spectrum shows clearly that there exist cosmic rays with energies of at least up to 3 $\times 10^{20}$ eV. These highest energy particles must be extragalactic because they are too energetic to be trapped in our galaxy by the galactic magnetic field. So neutrinos and gamma rays at high energies must exist because they are the ultimate decay products of interacting cosmic rays in the Cosmos.

Astronomy means "pointing back" to the source, i.e. the origin of production. From the charged cosmic rays only those of the highest energies (probably protons) travel rectilinearly i.e., they are not bent significantly by the galactic magnetic field. Therefore only very few of the cosmic ray protons are useful in "pointing back" to their origin. Neutrons are not very useful either, because only those with energies bigger than 10^{18} eV live long enough to cross our galaxy. Unfortunately there are very few of those too. So, they can not be helpful if their origin is outside our galaxy. In any case, the physics is understood and one can calculate that protons (or heavier nuclei) with energies above 4 $\times 10^{19}$ eV interact significantly with the primordial 2.7 K microwave background. Their energy is rapidly degraded and eventually these events of the highest energies (originally) get buried in the background of lower energies. So, charged hadrons cannot provide us with information for distances further away than 20-30 Mpc.

EGRET observations of gamma rays with energies up to the tens of GeVs and the recent observation of Mk421 by the WHIPPLE, Mount Hopkins Atmospheric Cherenkov Telescope, around 1 TeV have stimulated calculations on the interaction of very high energy gamma rays with the ambient intergalactic photons [Stecker 92]. The calculations can be interpreted to show that the mean free path of gamma rays with energies above some hundreds of GeV is around 100 Mpc due to scattering with the intergalactic ambient infrared and ultraviolet starlight. This conclusion combined with the well known calculations that show that gamma ray attenuation due to scattering with the 2.7 K background is very serious for gamma rays of hun-

Space Science Reviews **75**: 213–234, 1996.

dreds of TeV energies, leaves the neutrinos as the only promising particles for TeV or higher energies Astronomy. The neutrinos not only go through interstellar space without suffering any attenuation but they also escape their progenetor's acceleration and target sites without suffering any absorption. Therefore the only way to measure the emission/production spectrum at the source is to measure the spectrum of the neutrinos arriving on Earth.

The inescapable conclusion is that with neutrinos one can see "further" and "deeper" in the Cosmos than with any other known to date particle. The question is "what is the neutrino flux"? Or conversely, how big a detector do we need in order to start the field of "Neutrino Astronomy"? The only reasonable way to answer this question is to use the experience gained over the last 30 years or so by the construction and operation of the first and second generation neutrino telescopes and scale up by a few orders of magnitude in well defined steps. Each step should have the potential to contribute significantly to Astrophysics or Particle Physics: Relic neutrinos from the Big-Bang fill the Universe but no one has yet proposed a practical way to detect them. Low energy neutrinos of keV to $\approx$ 1 MeV are emitted continuously from the interior of stars like our Sun. At this moment there are at least half a dozen solar neutrino telescopes in operation, but there is no proposal to detect neutrinos produced in the interior of other stars. Slightly higher energy neutrinos ($\approx$ 14 MeV) are produced during the explosions of supernovae and one can argue successfully that the birth of Neutrino Astronomy happened with the detection of supernova 1987A by KAMIOKANDE and IMB, although no one has yet proposed a practical way to detect supernovae neutrinos further away than the immediate neighbourhood of our galaxy. Neutrinos in the higher energies are produced from decays of particles produced from cosmic rays interacting and the subsequent cascades in the Earth's atmosphere. The shape of their spectrum is caused by the competing processes of meson production and decay in a target medium (the atmosphere) of continuously changing density. Neutrinos with energies in the range of one to a few hundred GeV for example may also come from the annihilation of WIMPs (Weakly Interacting Massive Particles) in the Sun or the central core of the Earth. Neutrinos from point sources are decay products of particles produced in hadronic interactions, in potential cosmic accelerators such as neutron stars, black holes and young supernova remnants. Our own galaxy is full of such candidates. Further the Cosmos is full with Active Galactic Nuclei (AGN) and they are very promising sources of Ultra High Energy neutrinos. Detection of neutrinos from point sources would unambiguously establish the existence of high energy hadronic interactions e.g.

$$
\begin{array}{llll}
p + target \rightarrow & \pi^0 \quad , & \pi^{\pm} & + \cdots \\
 & \downarrow & \downarrow & \\
 & \gamma\gamma & \mu^{\pm}\nu_\mu & \\
 & & \downarrow & \\
 & & e^{\pm}\nu_\mu\nu_e &
\end{array}
$$

The observation on Earth of γ's from π^0 decay can only be a rare event, because the target must be thick enough in order to produce π^0's and at the same time thin enough (≈ 5 - $100\ g/cm^2$) so as not to absorb the π^0 decay γ's and then these γ's should not interact with interstellar matter or the ambient starlight ultraviolet and infrared photons or the primordial 2.7 K microwave photons during their very long trip to the Earth. On the other hand high energy photons (up to a few TeV) can be produced not only in hadronic processes via π^0 decay but also by purely electromagnetic processes e.g. inverse Compton scattering or synchrotron radiation of energetic electrons and then they must escape the general environment of the target before they get absorbed.

2. Physics Scope

The physics aim of neutrino telescopes covers one or more of the following topics (in order of descending neutrino energy):

1. Neutrino Astronomy (galactic and extragalactic) and the search for cosmic accelerators [Berezinski 90, Stenger 92].
2. Particle Physics beyond the Standard Model (a few examples follow):
   a) Search for dark matter particles. Their annihilation or decay will eventually give neutrinos e.g. neutralinos trapped in the Sun or the Earth [Bottino 92].
   b) Ultra High Energy neutrinos have energies beyond 10^7 GeV in the Laboratory. No terrestrial accelerator can produce these energies. If the neutrino telescope is large enough the limitation of low flux can be, in part, overcome and this might be the only way for High Energy Physics to reach these Ultra High Energies. In any case we can start right away, the experimental problems of neutrino telescopes are well understood.
   Multiple W/Z production [Ringwald 93].
   Search for possible substructure of the elementary particles i.e. compositeness of quarks and leptons [Kovesi-Domokos, Domokos 92, 93].
3. Neutrino oscillations (and thereby contributing to the question of "dark matter" in the Universe) using neutrinos produced in the atmosphere. The large range of the available oscillation length (15 km to 13000 km) gives an extremely good sensitivity of $\Delta m^2 \geq$

$10^{-4}eV^2$ [Suzuki 88, 93], [Resvanis 92, 93], [Katsanevas 92], [Kourkoumelis 93].

4. Long baseline neutrino oscillations using one of the high energy physics accelerators [Resvanis 92, 93], [Katsanevas 92].
5. Proton decay (in the sense that the ultimate background to proton decay experiment is atmospheric neutrino interactions and therefore a low threshold neutrino telescope can also be a proton decay detector).
6. Supernovae detection.
7. Magnetic monopoles.
8. The unexpected. A new observational window will open up with these neutrino telescopes. No one has ever viewed sites in the Universe shielded by more than a few hundreds grams of matter. One should keep in mind that every time a new brand of astronomy opened up, a new class of phenomena was discovered.

The experimental requirements for the study of the above physics topics look incompatible at first glance e.g. low vs high energy threshold, good angular accuracy vs coarse, high fractional coverage with sensitive photocathode area vs maximising the detector's sensitive area, etc. After a couple of decades of operating small prototypes and R+D the High Energy Physics community seems to have the right tools to build a detector to study all the above. The approach must be stepwise and modular. The detector of today's measurements is the prototype for tomorrow's experiment. The background of today's measurement is the signal of tomorrow's experiment.

3. Production of Astrophysical High Energy Neutrinos

There exist some excellent review articles of the subject and the reader is referred to Berezinski 1990, T. Gaisser 1986 and K.S.Thorne 1977, 1991. The point one should stress here, is that according to Stenger's calculations [Stenger 92] the optimal condition for high energy neutrino production in a cosmic beam dump (e.g., in a X- ray binary system) is a column density $z \geq 100\,\mathrm{gcm}^{-2}$, target density $< 10^{-8}\,\mathrm{gcm}^{-3}$ and a primary proton spectral index $a \leq 2$. Then, a beam of 1 TeV neutrinos would be produced at a rate of 10% of the protons that strike the source. Under these conditions the neutrino flux emitted is at least 3 times bigger than the corresponding gamma ray flux produced by the same hadronic interactions. The photon interaction cross section is many orders of magnitude larger than the neutrino cross section, therefore high energy gamma rays can be destroyed much easier than neutrinos either near the source of their production by interacting with the matter or during their vast flight distance to Earth by interacting with the 2.7 K background or the intergalactic infrared and ultraviolet starlight

[Stecker 92]. The inescapable conclusion is that for very high energies we should expect more copious neutrino sources than gamma ray sources.

Another class of sources which has become rather popular in the last few years is neutrinos produced in the vicinity of Active Galactic Nuclei. In this mechanism, neutrinos originate from the decay of mesons which in turn are the decay products of photoproduced Δ resonances.

$$\begin{array}{ccccccc} p\gamma \rightarrow \Delta^+ \rightarrow & \pi^+ n & \text{or} & \Delta^+ & \rightarrow & \pi^0 p \\ & \downarrow & & & & \downarrow \\ & \mu\nu_\mu & & & & \gamma\gamma \\ & \downarrow & & & & \\ & e\nu_e\nu_\mu & & & & \end{array}$$

Recent calculations show that although signals from individual AGNs will not be detectable with the neutrino detectors presently under construction, the sum of all AGNs [Hawaii 92] should be detectable. Finally, we should mention that high energy neutrinos may also originate from heavy (hundreds of GeV) dark matter particles e.g., WIMPs which are trapped in the Sun or the Earth and which eventually annihilate with their antiparticles producing neutrinos amongst other particles.

4. Choosing the Neutrino Energy

In order to detect neutrinos from the previously mentioned sources it is advantageous to focus on high energy neutrino detection. The reason is simple. The mere word astronomy implies that one ought to be able to point back to the source with high angular accuracy. In neutrino detection we cannot achieve the angular accuracy of traditional astronomy. Neutrinos are detected by observing mainly the muon which is produced from the charged current neutrino interactions with matter, in the vicinity of the detector. In high energy neutrino astronomy usually, one is NOT required to detect the vertex of the neutrino interaction, this way one maximizes the available detection volume. The physics of the interaction is such that, the angle between the detected muon and the parent neutrino direction [DUMAND 93] for 63% of the charged current events is $\approx 1.5^o/\sqrt{E(TeV)}$. So, if a source is observed with a dozen events or so, one can greatly improve the directional determination by determining the centroid of the distribution and reach accuracies of a fraction of a degree. The pointing improves also with increasing energy not only because the muon goes further forward but also because the statistics improve due to the increase of the cross section ($\sim E_\nu$ until 10 TeV and then $\sim \log E_\nu$). Another parameter which improves with energy is the range of the muon ($\sim E_\mu$ until 1 TeV and then $\sim \log E_\mu$) and thus the effective detection volume of the detector is increased proportionately. Further, the signal to noise improves with energy for the following reason. The inherent background comes from atmospheric neutrinos, which are the result of cosmic ray interactions in the atmosphere. The spectral index γ of the flux of atmospheric neutrinos (Stanev 92) follows that of the cosmic ray spectrum and is ≈ 2.7 up to 100 GeV or so, but then at higher energies it becomes ≈ 3.7, while neutrinos produced extraterrestrially follow the hard core ($\gamma \approx 2.0$ - 2.2) cosmic ray spectrum. So, for neutrino energies larger than 100 GeV the signal to noise improves with energy.

For all the above reasons it is advantageous to optimise the telescope for detection of high energy neutrinos. One should keep in mind though, that the design of neutrino telescopes in the water with phototubes directed towards the Earth (in order to minimize background from downcoming cosmic rays that reach the detector) can be disadvantageous when one detects very high energy neutrinos (e.g., from AGNs) because then the Earth is no longer transparent to neutrinos (e.g., the mean free path for a 500 TeV neutrinos is about one Earth diameter). Therefore the upward coming neutrinos of energy larger than a few hundred TeV will be severely attenuated.

5. Backgrounds in Neutrino Telescopes

The first large neutrino telescopes, were designed for proton decay experiments e.g., IMB, KAMIOKANDE, but yesterday's background is today's signal. The neutrino interactions in or around the big proton-decay experiments were considered to be a source of background. In the quest for neutrino astronomy, part of it becomes the signal. But, on the Earth's surface the background due to the downcoming cosmic ray muon is overwhelming, the signal (for calculational purposes now we consider the atmospheric neutrinos as the signal) to noise (called up to down ratio) is of the order of 10^{-12}. Therefore, for shielding purposes, neutrino telescopes are located in deep mines inside mountains or in deep water. For instance, 4000 mwe of shield reduces the up to down ratio to about 10^{-4}. So, it is essentially impossible in the energy regime below 10 TeV to do neutrino astronomy by looking for downcoming muons, because neutrino induced muons are indistinguishable from downcoming cosmic ray muons. This is the reason that for those muons which originate outside the detector (and therefore no vertex determination is possible) only muons coming up from the lower hemisphere are useful. Then the only remaining background is due to the omnipresent atmospheric neutrinos because the earth is transparent to most of them. In general, well shielded detectors can "look" up at about 20° *above* the horizon, while shallow detectors (1000 mwe) can look up only 20° *below* the horizon. The kinematics of neutrino production and the multiple scattering is such that the best overall angular resolution is $\approx 1^o$. So in order to find point sources of neutrinos one would at best divide the sky into 1° square pixels and plot the detected muon direction in a Declination versus Right Ascension plot.

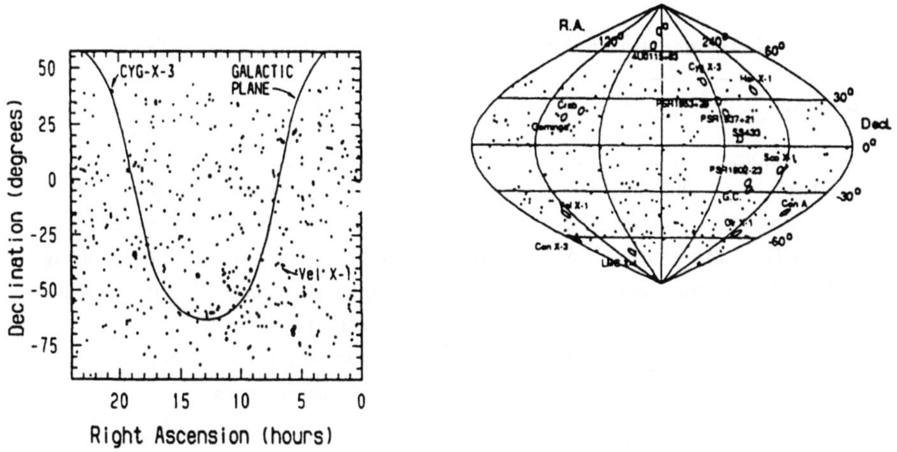

Fig. 1. A sky plot of the upward muon data for (a) the IMB experiment and (b) the KAMIOKANDE experiment.

A source would manifest itself as standing above the background caused by the atmospheric neutrino interactions, which constitute a flat background (with a slight zenith dependence visible over tens of degrees).

6. IMB - KAMIOKANDE - MACRO Results

There exists an excellent two year old review of High Energy Neutrino Astronomy [Learned 92]. So I will mainly focus on the progress in the field during the last two years.

Following the procedure outlined above, IMB and KAMIOKANDE have plotted their data [Giacomelli 92], figure 1, with granularity which is commensurate with their angular resolution. Clearly, there is no event clustering of statistical significance. KAMIOKANDE quotes limits from point sources which range from 2×10^{-14} $cm^{-2}s^{-1}$ for Vel X-1 to 8.5×10^{-14} for Her X-1. In the '93 Calgary conference, the MACRO collaboration (Di Credico, 1993) reported on 230 days of data taking and calculated limits at the 95% c.l. of steady muon emission of 5.8×10^{-13} $cm^{-2}s^{-1}$ for Cyg X-3 and Her X-1. For the Crab they find 6.0×10^{-13} $cm^{-2}s^{-1}$. Although they improved these upper limits by a factor of 2 by taking advantage of the temporal characteristics of these sources which are known from photon astronomy, it is obvious that the field needs larger detectors.

TABLE I
Large underground detectors operating in the 80's and 90's

Detector	Location	Area(m^2)	Technique	Primary Purpose
IMB (o)	Ohio, USA	400	Water Cerenkov	Proton decay
Kamioka	Japan	120	Water Cerenkov	Proton decay
Soudan II	Minnesota,USA	100	drift tubes + Fe	Proton decay
MACRO	Gran Sasso,Italy	800	liquid scintillator + streamer tubes	monopoles
LVD	Gran Sasso, Italy	500	liquid scintillator + streamer tubes	supernovae ν

(o) It has been terminated

7. How Big a Detector?

Let us change line of approach now. Consider the recently reported gamma ray observations (WHIPPLE) and use the detected gamma ray flux to calculate the required detector sensitive area.

In the TeV range of energies the WHIPPLE Air Cherenkov γ-ray experiment has detected the following fluxes:
Crab Nebula: 2.5×10^{-11} gammas cm^{-2} s^{-1}
Markarian 421: 7.5×10^{-12} gammas cm^{-2} s^{-1}

Most likely the gamma ray flux from the Crab is electromagnetic in origin but no one doubts the hadronic origin (i.e. π^0 decay products) of the gamma rays from Markarian 421, which is a well known AGN. Therefore one can safely assume the pessimistic case of equal rate of production of gamma and neutrinos at the site of Markarian 421 and keep in mind that it is much easier to destroy gamma rays than neutrinos in their $\approx$ 100 Mpc journey from Markarian 421 to Earth. One should mention here that Markarian 421 is the closest blazar on the GRO catalogue of AGNs that emit GeV gamma rays.

Stenger [Stenger 92] has calculated that a deep underwater detector with a geometrical area of 20,000 m^2 (e.g. DUMAND II or the first phase of NESTOR) will be able to detect, in one year's observation, a source with a neutrino flux greater than $1.2 \approx 10^{-10}$ neutrinos cm^{-2} s^{-1} for neutrinos with energy greater than 1 TeV. (This does not mean that the detector has a threshold of 1 TeV, it is just a handy reference point).

Comparing this with the gamma ray flux from Mk421 we draw the conclusion that we need a detector that should quickly reach an active area of 200,000 m^2. Conventional particle detectors e.g., scintillators, drift tubes, resistive plate chambers etc. cost about 10,000 to 20,000 Swiss francs per square meter. This means that a 10^5 m^2 detector will cost $\approx$ 2000 million Swiss francs which is twice the CERN budget for one year. The conclusion is that the cost must be reduced to 100 - 200 swiss francs per square meter. This is where the water Cherenkov technique comes in.

8. Large Volume Water Cherenkov Detectors

The index of refraction of sea water is 1.35, this gives a Cherenkov angle of 43°. In other words, relativistic particles will "illuminate" the sea water with a 43 degree cone, emitting about 250 Cherenkov photons/cm (between 3000 - 5000 Å) a range where most photomultipliers have a reasonable sensitivity. Then, depending on the water transmissivity (ranging from 20 m in clear lakes to 50 - 60 m in clean deep ocean waters), one can instrument large areas of water fairly cheaply, at a cost of about 100 Swiss francs per square meter. The long range of muons in water helps tremendously, a 1 TeV muon has a range of about 2.5 km in water! A variance to this technique is using the Antarctic ice as a Cherenkov radiator. The high energy physics community has accumulated considerable experience with the water Cherenkov technique mainly from the three big proton decay experiments IMB, HPW and KAMIOKANDE and the various test-deployments down to 4,000 m as done with the DUMAND I and the NESTOR tests, lake BAIKAL at 1,000 m and AMANDA in the ice. The deep water large Cherenkov neutrino experiments will also deploy hydrophones in the hope of determining very high energy neutrino interactions acoustically, the threshold of this technique is in the PeV range. The acoustic pulse is produced by the thermal expansion of the water when heat is deposited by the interaction. The energy transfer is very inefficient, about 10^{-9} [Learned 92], but the advantage is that the attenuation length, in the frequency range of 10 - 20 kHz, is of the order of a few kilometers. So, one could instrument literally, many cubic kilometers at low cost. DUMAND II is deploying hydrophones, while in the acoustic version of NESTOR, called SADCO, preliminary tests show that the usual 14°C ambient temperature of the bottom of the Mediterranean is more advantageous for acoustic detection (lower threshold) than the 4°C ocean bottom temperature. In any case it is far too early to make any comments on the acoustic detection technique. In order to demonstrate some of the advantages of the water Cherenkov technique we show in figure 2 the reconstruction efficiency as a function of the distance from the central axis of the NESTOR detector for vertical muons of energy 1 TeV, 10 TeV and 100 TeV. One should note that a reconstruction efficiency of 80% is achieved with vertical muons that cross 50 to 100 m outside the detector.

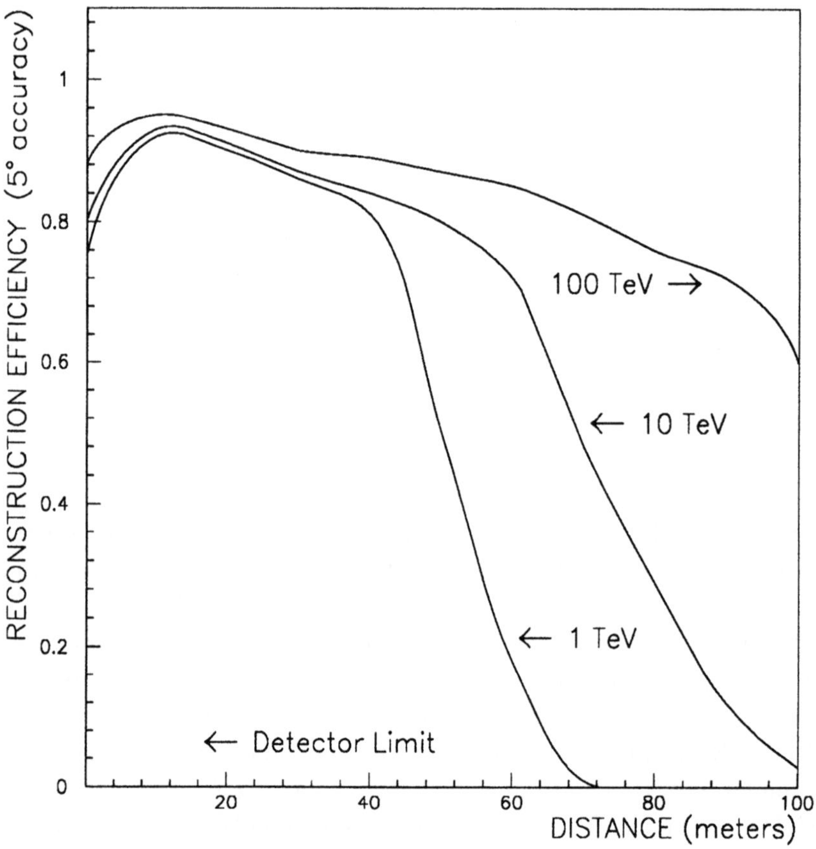

Fig. 2. Reconstruction efficiency versus distance from the axis of the tower (vertical incidence of muons) for one NESTOR tower (5° accuracy).

Even if a 10^5 m^2 detector finds neutrino sources, the physics of this field will not be harvested until square kilometer detectors come into operation. For this reason the quest for neutrino astronomy has three benchmarks.

- The reliable operation over a few years of a detector of 10,000 to 20,000 m^2 (i.e. a detector ten times to twenty bigger than the largest neutrino telescope in operation today).
- The quick growth of this detector to cover a sensitive area of 100,000 m^2.
- The subsequent expansion of the detector in order to instrument an 1 km^2 in a timely and affordable fashion.

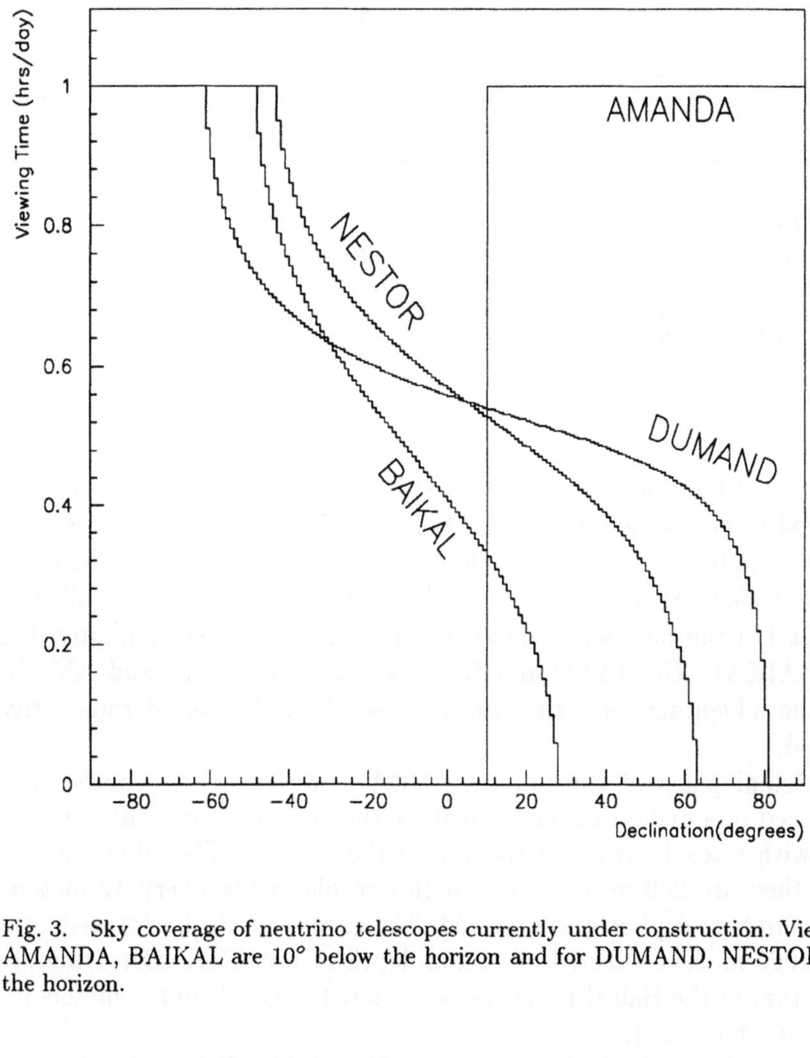

Fig. 3. Sky coverage of neutrino telescopes currently under construction. View limits for AMANDA, BAIKAL are 10° below the horizon and for DUMAND, NESTOR 10° above the horizon.

9. Description of Current Projects

The sky coverage by each of the four neutrino telescopes under construction, table II, is shown in figure 3. Note that DUMAND and NESTOR are 177° degrees apart in longitude, so their combination will monitor the Cosmos continuously.

10. AMANDA

The AMANDA project aims at building a neutrino telescope in the ice of the South Pole at a depth of 1000 m. Although the transparency of the ice had not been measured in situ, it was expected to be around 20 m. In this

TABLE II

High energy neutrino telescopes currently under construction

Detector	Location	Date- E.Y.O.(o)	Area (10^{3} m^2	Depth (mwe)	Technique	Threshol (GeV)
Baikal NT-200	Siberia	'96	3	1000	WC	10
DUMAND II	Hawaii	'96	20	4760	WC	20
NESTOR	Greece	'97	20	3700	WC	Few
AMANDA	Antarctic	??	8	???	WC in ice	Few

(o) Expected Year of Operation

experiment, 1.2 km long holes were to be drilled in the ice, the phototubes to be placed in and some hours later they would be frozen in. The cost and the limited supply of fuel in the South Pole is such that only 8 inch diameter holes were drilled. As a result of this, the AMANDA phototubes will only have about 1/4 the photocathode area of the 15 inch diameter phototubes used by BAIKAL, DUMAND and NESTOR. On the other hand AMANDA has the advantage that the Antarctic sea is fairly free of radioactive background.

The original plans of the full AMANDA array (figure 4) consist of 9 strings placed in a cylindrical configuration, on the surface of a 30 m radius cylinder, with a tenth string on the axis of the cylinder. The 10 strings are identical, they are 200 m long each, with one phototube every 12 meters, the total number of phototubes is 200. The array was to be located at a depth of 1000 m. The effective area and depth of the AMANDA array are similar to that of the Baikal telescope (i.e. a few thousand square meters for neutrinos of a few TeV).

During the winter (Australian summer) of 93-94 the collaboration deployed four strings to a depth of 1000 m, the deployment of a fifth string at the end of January 94 was not completed due to the breakdown of the drill and the approaching bad weather [Morse 94]. They have studied [Halzen 94] the ice transmissivity using their laser calibration system. Contrary to claims [Halzen 93] the ice does not seem to be bubble free, at least not at depths of 1000 m. Unfortunately the scattering length is extremely short (it varies from 11 cm at a depth of 800 m to 20 cm at a depth of 1000 m). This means that after traversing a few meters the direction of the Cherenkov light is totally randomised and therefore one cannot reconstruct the direction of the muon. However by doing an elaborate study of the calibration data they fit the absorption length to be 59 meters at 515 nm. In a shallow

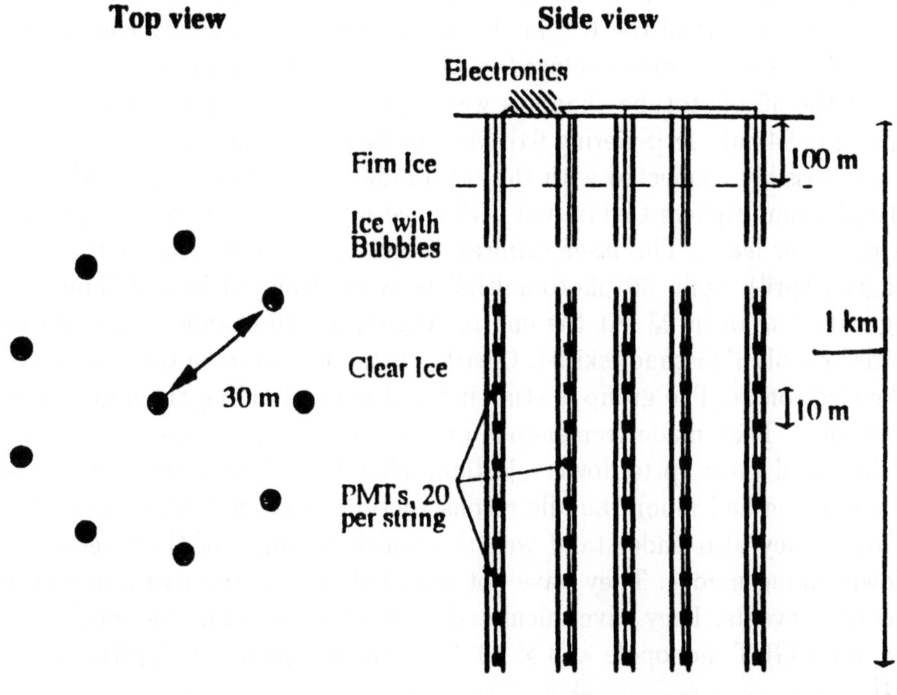

Fig. 4. The AMANDA array as originally planned.

detector this is a mixed blessing because it shows that the effective area is much larger than what was originally anticipated. Given the shallow depth of 800 - 1000 m, the downcoming high flux of cosmic ray muons cannot be rejected because the very short scattering length does not permit any reconstruction of direction. Since the four deployed strings are frozen in (i.e. cannot be retrieved) the collaboration plans to use them together with their future array. Since one is concerned that the short scattering length might be caused not only by air bubbles alone but by dust, volcanic ashes or fluorescence, a program of measuring the optical properties of ice bellow 1000 m seems prudent.

The collaboration does not plan to do any drilling in the South-Pole during the 94-95 Australian summer [Halzen 94].

11. Lake BAIKAL

The lake Baikal experiment will be composed of 192 large (15 inch) phototubes (figure 5). The deployment is made in the spring when the lake is frozen, so a solid platform is available. In the spring of 93 part of the detector was deployed, consisting of 36 phototubes, the darkened phototubes in figure 5. The water transmission length is $\approx$ 20 m and it varies with the seasons. Further, there seems to be an optical background in the water similar in magnitude of the K^{40} in the ocean. The data is transferred to shore via cable. The detector deployed in March 93 operated for almost one year, 30 of the 36 phototube channels were operating by August 93 and by the spring of 94 only 18 [Spiering 94]. Most of them had some intermittent problems, possibly connected with the substandard manufacture of a switch. All the photomultipliers themselves performed well and there were no problems with water leaks. The laser calibration system worked well. In the spring of 94 (April) again 36 photomultipliers were deployed in a different configuration than in 93. At the end of August 94, 30 channels were working satisfactorily [Wischnewski 94. Clearly they must improve the reliability of the electronics. The group is studying and understanding the detector performance. They made tremendous progress with their reconstruction algorithms and their up to down rejection. They have demonstrated that their single muon and muon- bundle reconstruction agrees well with their Monte Carlo. They also understand well the zenith response of their detector to downcoming muons. They have not reported yet the reconstruction of any neutrino events. They have calculated preliminary limits on Monopole fluxes i.e. for a GUT monopole $< 3 \times 10^{-16}$ cm^2/s.sr [Spiering 94], [Wischnewski 94].

The group deserves the whole hearted congratulations of the international scientific community because they managed to deploy and operate the world's first underwater neutrino telescope under the very difficult conditions that prevail in Russia at the present. They plan to deploy another 70 phototubes in the spring of 95 and to complete the detector in the spring of 96.

12. DUMAND II

The DUMAND experiment [DUMAND 93] is the progenitor of today's neutrino telescopes. They have been studying the problem for more than one dozen years now. They have offered a number of solutions and have developed and made available the technology required to operate and deploy neutrino telescopes in the water. The full DUMAND II array (figure 6) will be deployed at a depth of 4800 m, near Hawaii, and it will consist of an octagon of nine strings (eight on the perimeter and one in the center). A total of 216 large (15 inch) phototubes will be deployed facing downwards. There will be 24 phototubes per string in 10 meter intervals. The radius of the octagon will be 53 m, matching the 45 m long light transmission length that they have measured in the blue part of the spectrum. The total enclosed instrumented mass is about 2.7 Mtons. The sensitive area for a muon of one TeV energy will be 20,000 m^2 with an angular resolution of better than one degree (standard deviation).

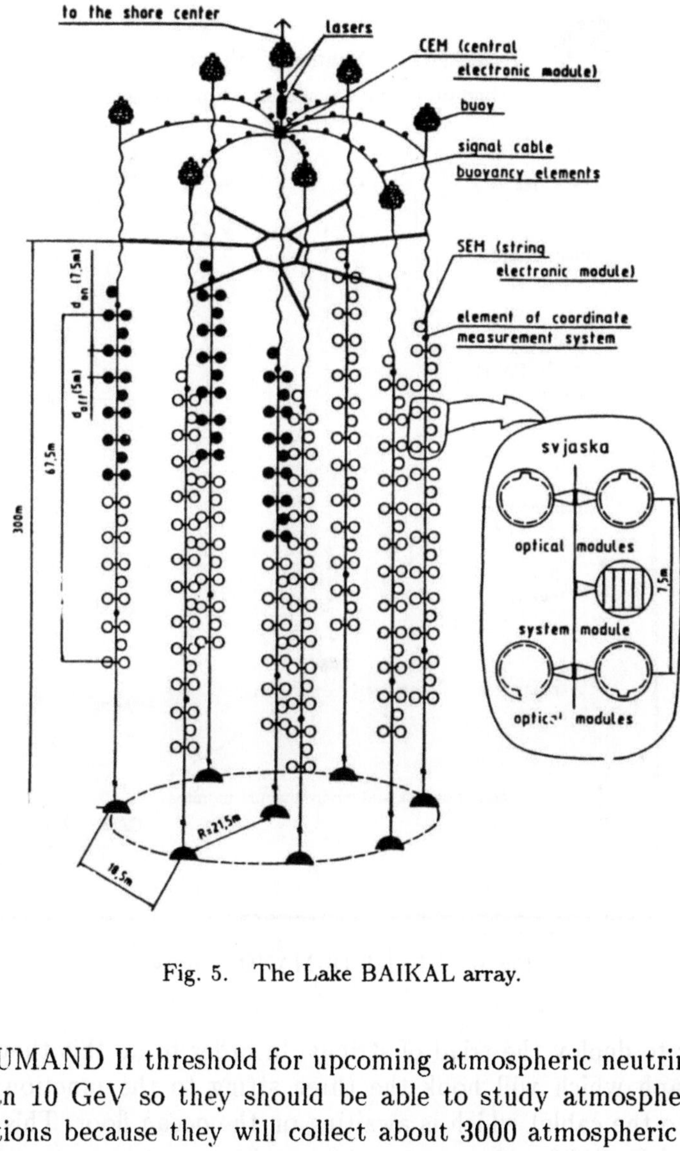

Fig. 5. The Lake BAIKAL array.

The DUMAND II threshold for upcoming atmospheric neutrinos will be larger than 10 GeV so they should be able to study atmospheric neutrino oscillations because they will collect about 3000 atmospheric upcoming neutrino events per year in the full nine string array.

The DUMAND collaboration deployed successfully in December 93 its 36 km long electrooptical cable from the detector to the shore, together with the so called junction box (the plug in box for distribution of electrical power and receipt of data via optical fibers). One string of 22 photomultipliers was also attached to the junction box as well as the environmental and control instruments. Data were taken for a while and in fact one double muon event was reconstructed (cosmic ray). Due to a faulty connector water entered inside the housing which controled the string electronics and the power supplies shorted out. The string was recovered in January 94 and was repaired. The other two strings are ready now. The collaboration is

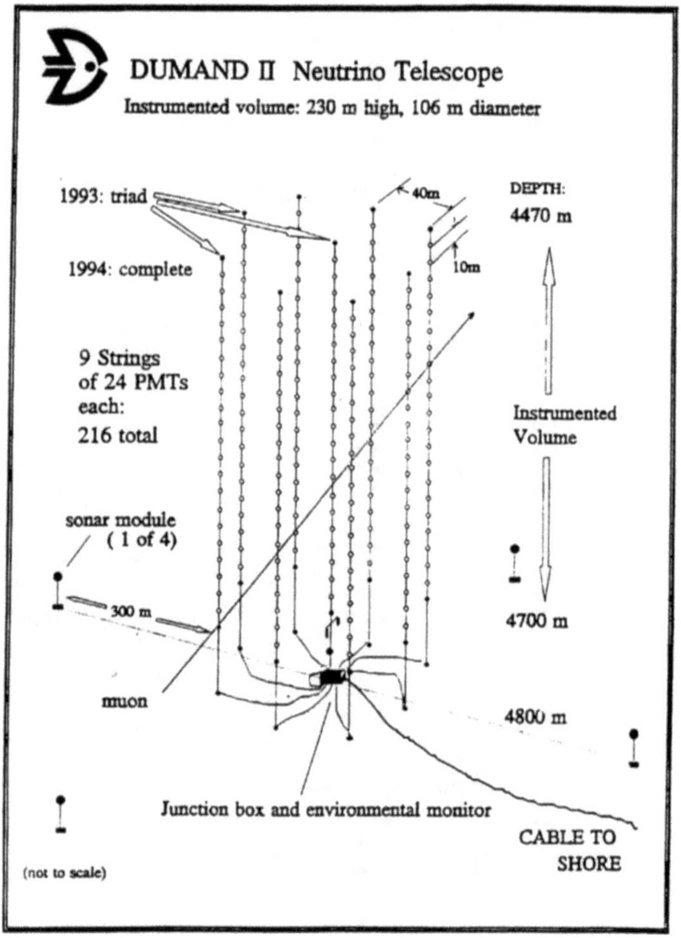

Fig. 6. The full DUMAND II array.

ready now to deploy the triad of strings. In order to do this they must use a bathyscaph which will hook the three string to the junction box (and therefore to the cable) which is awaiting on the ocean floor. This operation is scheduled for 1996.

13. NESTOR

The NESTOR collaboration has located near the S.W. of Greece a 8 km by 9 km horizontal plateau at a depth of 3800 m [Resvanis 92]. The depth is constant to within ± 50 m over the entire plateau. The plateau is at a mere distance of 7.5 miles from the shore. Extensive studies of the water transmissivity were made employing I) spectrophotometric analysis of a very large number of samples, II) deployment of photometers in situ (i.e. down to 4000 m) and III) large acceptance photometric measurements in situ.

These measurements show that the water transmission length in the blue part of the spectrum is about 55 m. [Khanaev 92], [Anassontzis 94]. The underwater currents have been measured over five years and they have been found to be minimal [Demidova 92] i.e. a few centimeters per sec. Last, the sedimentology analysis is completed [Trimonis 92]. The NESTOR design differs from that of DUMAND and AMANDA in that it deploys half the phototubes looking upwards, like Baikal, thus having a 4π sensitivity and at the same time will be shielded by 3500 mwe. They employ a total of 168 large 15 inch phototubes like DUMAND but they are clustered closer so that they have a much lower energy threshold.

They have been making tests in situ, since 1991, of a half scale model of their basic detector element. This is a horizontal rigid hexagon made out of titanium, of 16 m radius. At each one of the corners and at its center there is a pair of two 15 inch phototubes (one looking up and the other one down). By stacking 12 of these hexagons in the vertical, with a distance between hexagons of 20 m, they create a tower (figure 7). Figure 8 shows that the effective area of a single NESTOR tower is equivalent to the full DUMAND II. The whole tower will be deployed in a single operation, currently planned for the spring of 97, its sensitive area for TeV muons will be ≈ 20000 m^2 (figure 8). The collaboration hopes to find quickly additional funds to build another six towers in order to deploy them in a hexagonal fashion around the first tower and at a distance of 100 - 150 meters from it (figure 9). This array would have a sensitive area larger than 10^5 m, it would provide an overall angular resolution better than 1°, it would have an enclosed mass of > 20 Megatons and a total cost of about 20 million dollars. Within each one of the 7 towers the energy threshold is a few GeV, i.e. a low threshold active target of 1.5 Megaton mass.

With one NESTOR tower they claim to be able to do almost the same astrophysics of DUMAND II plus two experiments unique to NESTOR: One is atmospheric neutrino oscillations by doing a full zenith fit of the atmospheric neutrinos in an internally normalisable way, independent of cosmic ray flux calculations. They claim to have the sensitivity with only one tower to make a definitive statement to confirm or not the KAMIOKANDE neutrino oscillations claim [Resvanis 92, 93], [Katsanevas 92], [Kourkoumelis 93]. The other is a long baseline neutrino oscillations experiment from CERN in conjunction, perhaps, with another experiment at Gran Sasso which on a two dimensional projection falls on a straight line drawn between CERN and NESTOR [Resvanis 92, 93], [Katsanevas 92, 93].

14. An Affordable 1 km^2 Detector

Assuming operational success of the 7 tower hexagon of hexagonal NESTOR towers the next step is obvious. Employ another six towers in a hexagonal

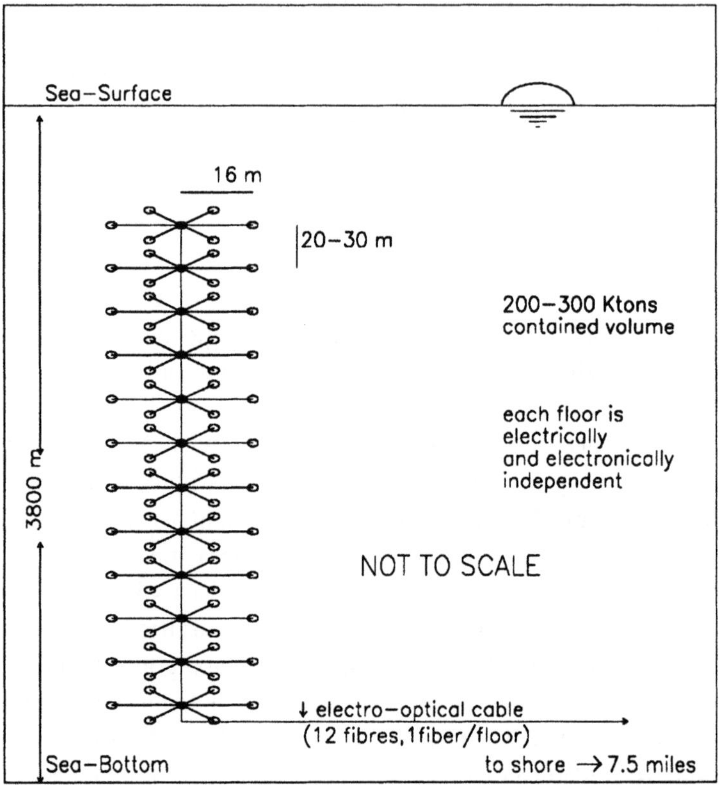

Fig. 7. One NESTOR tower.

fashion 150 meters around the first hexagon and fill in the gaps between them in the perimeter with six strings. Each string has 24 phototubes (15 inch each) clustered in pairs and spaced with the same spacing between floors of the towers. This addition of 6 towers and six strings will cost less than 25 million dollars and the array will have a sensitive area of 430,000 m^2. For the following step it suffices to use strings only with 24 phototubes each. Employing 18 strings ≈ 150 meters outside the last perimeter will only cost 5 million dollars and the total sensitive area will be 850,000 m^2 for a total cost of 50 million dollars. The enclosed mass will be about 300 Megatons, of which 4 or 5 will be contained within each tower with local energy threshold of a few GeV, figure10.

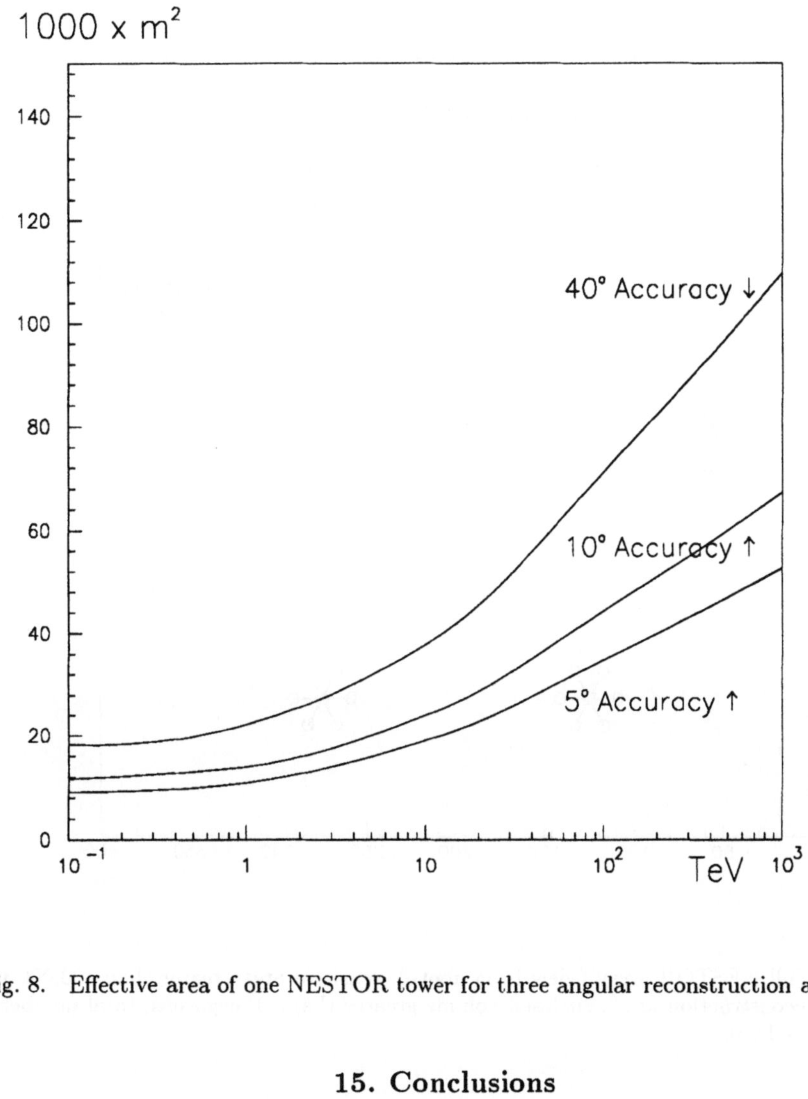

Fig. 8. Effective area of one NESTOR tower for three angular reconstruction accuracies.

15. Conclusions

In the next couple of years four large ($\approx 10^4$ m^2) neutrino telescopes should be operational and we should learn a lot. In order to study the physics of copious sources of astrophysical neutrinos one would most likely need larger telescopes. It would make sense to pool resources and expertise in order to built jointly an 1 km^2 detector.

Acknowledgements

The author wishes to thank the organisers of this great workshop, especially Heinz Volk and Felix Aharonian. The author thanks A.Bottino, S.Kovesi-Domokos, G.Domokos, J.Learned and V.Stenger for many useful discussions.

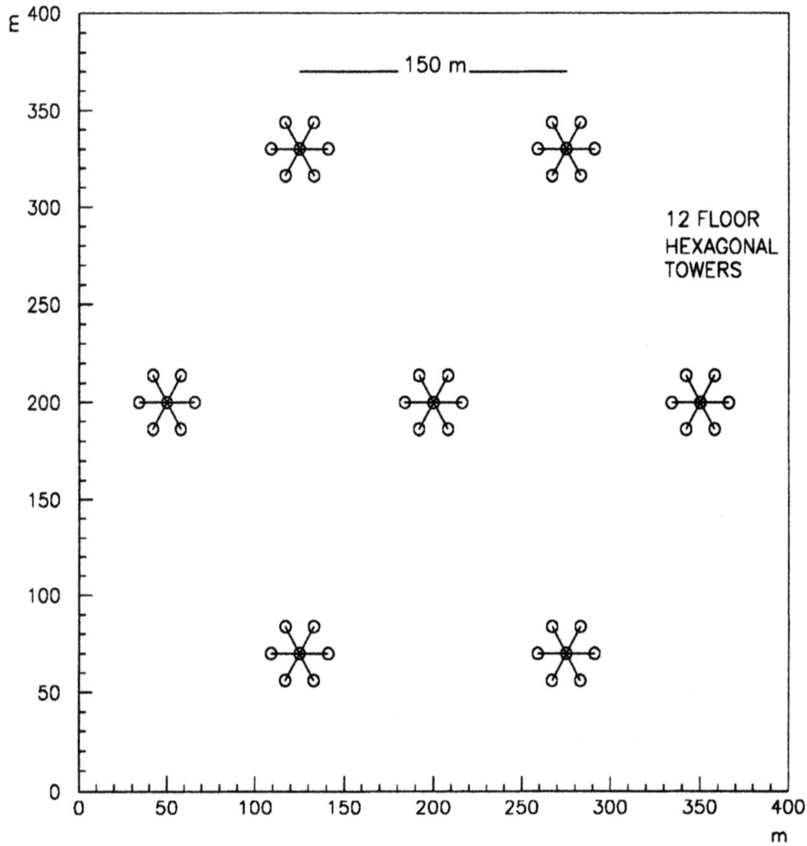

Fig. 9. Full NESTOR array (view from above), sensitive area greater than 100,000 m^2, angular reconstruction $\approx 1^\circ$, enclosed volume greater than 20 megatons, total number of phototubes 1176.

Many thanks are also due to his colleagues of the NESTOR collaboration for myriads of useful discussions.

References

E.Anassontzis, 1994, *Nuclear Instruments and Methods* **A349**, 242.

Berezinski, 1990, *Astrophysics of Cosmic Rays*, North Holland.

A.Bottino, 1992, *Fourth International Workshop on NEUTRINO TELESCOPES*, M.Baldo Ceolin, editor, Venice.

Demidova, 1992, *2nd NESTOR International Workshop*, L.K.Resvanis, editor.

Di Credico, 1993, *contributed paper to the Calgary 1993 cosmic ray conference by the MACRO collaboration.*

DUMAND 89, 1993, *Proposal the DUMAND collaboration.*

T.Gaisser, F.Stecker, A.Harding and J.Barnard, 1986, *Ap. J*, **309**, 674.

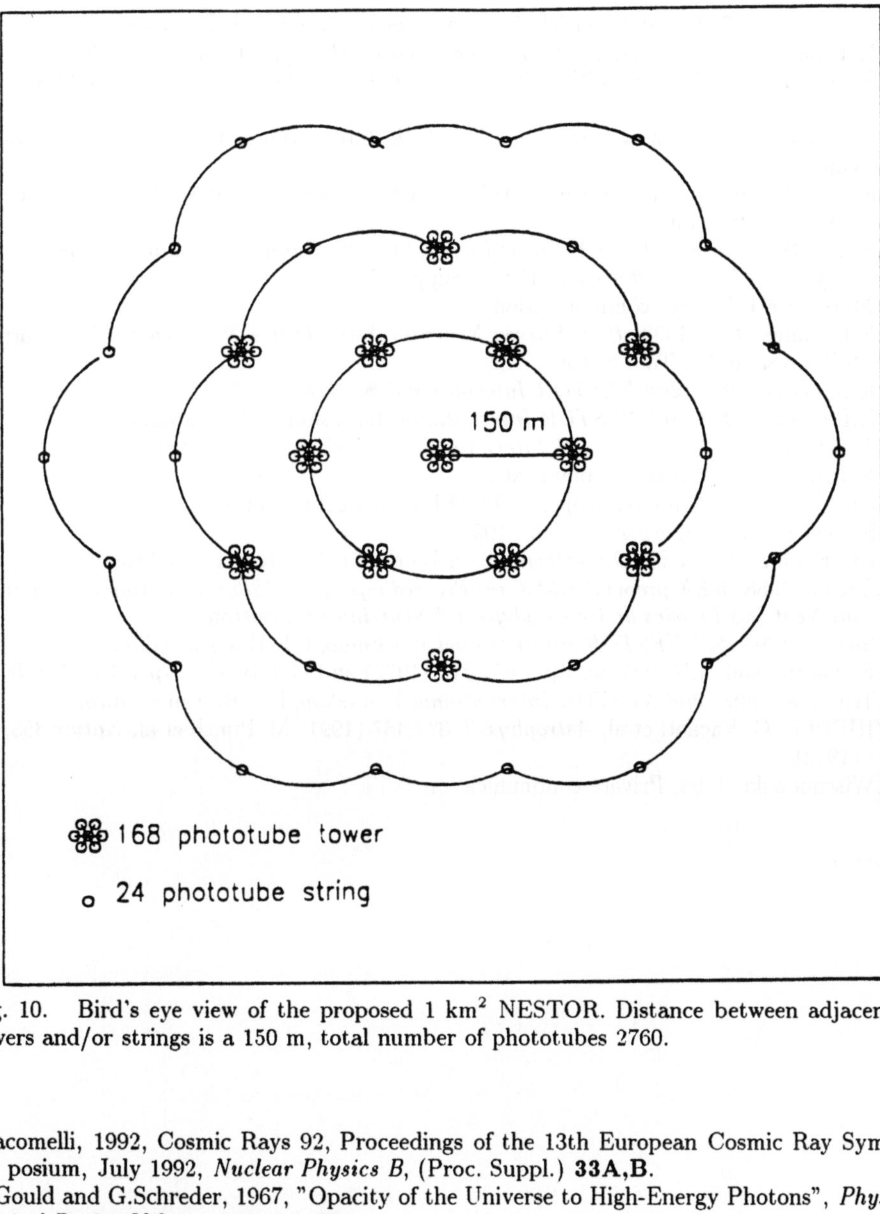

Fig. 10. Bird's eye view of the proposed 1 km^2 NESTOR. Distance between adjacent towers and/or strings is a 150 m, total number of phototubes 2760.

Giacomelli, 1992, Cosmic Rays 92, Proceedings of the 13th European Cosmic Ray Symposium, July 1992, *Nuclear Physics B*, (Proc. Suppl.) **33A,B**.

R.Gould and G.Schreder, 1967, "Opacity of the Universe to High-Energy Photons", *Physical Review* Volume **155**, p.1408.

F.Halzen, 1994, Private communication.

F.Halzen, 1993, "Particles and the Universe", *Proceedings of the 17th Johns Hopkins Workshop*, Budapest 1993, p.206, World Scientific.

Hawaii, 1992, High Energy Neutrino Astrophysics, V.J.Stenger, J.G.Learned, S.Pakvasa and X.Tata editor.

IMB, D.Casper et al, 1991, *Phys. Rev. Lett.* **66**, 2561.

O.De Jager, F.Stecker, M.Salamon, 1994, "Estimate of the intergalactic infrared radiation field from γ-ray observations of the galaxy Mrk421", *Nature*, vol **369**, p294.

KAMIOKA, K.S.Hirata et al, 1992, *Phys. Lett.*, **B280**, 146.

KAMIOKA 94, 1994, "Atmospheric ν_μ/ν_e Ratio in the Multi GeV Energy Range", Fukuda et.al., ICRA Report 321-94-16 (June 94).

Khanaev, 1992, *2nd NESTOR International Workshop*, L.K.Resvanis, editor.

S.Katsanevas, 1992, *2nd NESTOR International Workshop*, L.K.Resvanis, editor

S.Katsanevas, 1993, *3rd NESTOR International Workshop*, L.K.Resvanis, editor.

C.Kourkoumelis, 1993,*3rd NESTOR International Workshop*, L.K.Resvanis, editor.

S.Kovesi-Domokos, G.Domokos, 1992, *2nd NESTOR International Workshop*, L.K.Resvanis, editor.

S.Kovesi-Domokos, G.Domokos, 1993, *3rd NESTOR International Workshop*, L.K.Resvanis, editor.

Learned, 1992, Cosmic Rays 92, *Proceedings of the 13th European Cosmic Ray Symposium*, July 1992, *Nuclear Physics B*, (Proc. Suppl.) 33A,B.

R.Morse, 1994, Private communication.

L.K.Resvanis et al, 1992, *High Energy Neutrino Astrophysics*, V.J.Stenger, J.G.Learned, S.Pakvasa and X.Tata editor.

L.K.Resvanis, 1992, *2nd NESTOR International Workshop*, L.K.Resvanis editor.

L.K.Resvanis, 1993, *3rd NESTOR International Workshop*, L.K.Resvanis editor.

A.Ringwald, 1993, *3rd NESTOR International Workshop*, L.K.Resvanis editor.

C.Spiering, 1994, Private communication

Stanev, 1992, *Neutrino Telescopes 1992*, M.Baldo- Ceolin, editor.

F.Stecker et al, 1992, *Ap.J* 390, L49, 1992.

Stenger, 1992, *2nd NESTOR International Workshop*, L.K.Resvanis, editor.

A.Suzuki, 1988, *KEK preprint 88-38, in: Proceedings of the 13th International Conference on Neutrino Physics and Astrophysics, "Neutrino 88"*, Boston.

A.Suzuki, 1993, *3rd NESTOR International Workshop*, L.K.Resvanis editor.

K.S. Thorne and A.N.Zytkow, *Ap.J* 832,212 (1977) and G.T.Biehle, *Ap.J.*380,167(1991).

E.Trimonis, 1992, *2nd NESTOR International Workshop*, L.K.Resvanis, editor.

WHIPPLE: G. Vacanti et al, *Astrophys. J.* 377,467 (1991) M. Punch et al, *Nature* 358,477 (1992).

R.Wischnewski, 1994, Private communication

PULSARS AS GAMMA RAY SOURCES: NEBULAR SHOCKS AND MAGNETOSPHERIC GAPS

JONATHAN ARONS
Astronomy Department, University of California at Berkeley

ABSTRACT I summarize the results of recent research on the structure and particle acceleration properties of relativistic shock waves in which the magnetic field is transverse to the flow direction in the upstream medium, and whose composition is primarily electrons and positrons with an admixture of heavy ions. Shocks which contain heavy ions that are a minority constituent by number but which carry most of the energy density in the upstream medium put $\sim$ 20% of the flow energy into a nonthermal population of pairs downstream, whose distribution in energy space is $N(E) \propto E^{-2}$, where $N(E)dE$ is the number of particles with energy between E and $E+dE$. Synchrotron maser activity in the shock front, stimulated by the quasi-coherent gyration of the whole particle population as the plasma flowing into the shock reflects from the magnetic field in the shock front, provides the mechanism of thermalization and non-thermal particle acceleration. The maximum energy achievable by the pairs is $\gamma_{\pm} m_{\pm} c^2 = m_i c^2 \gamma_1 / Z_i$, where γ_1 is the Lorentz factor of the upstream flow and Z_i is the atomic number of the ions. The shock's spatial structure contains a series of "overshoots" in the magnetic field, regions where the gyrating heavy ions compress the magnetic field to levels in excess of the eventual downstream value. These overshoots provide a new interpretation of the structure of the inner regions of the Crab Nebula, in particular of the "wisps", surface brightness enhancements near the pulsar. The wisps appear brighter because the small Larmor radius pairs are compressed and radiate more efficiently in the regions of more intense magnetic field. This interpretation suggests that the structure of the shock terminating the pulsar's wind in the Crab Nebula is spatially resolved, and allows one to measure $\gamma_1 \sim 4 \times 10^6$, the upstream magnetic field B_1 to be $\sim 3 \times 10^{-5}$ Gauss, as well as to show that the total ion flow is $\sim 3 \times 10^{34}$ elementary charges/sec, in good agreement with the total current flow predicted by the early Goldreich and Julian (1969) model. The total pair outflow is shown to be about 5×10^{37} pairs per second, in good agreement with the particle flux required to explain the nebular X—ray source.

The energetics of particle acceleration within the magnetospheres of rotation powered pulsars and the consequences for pulsed gamma ray emission are also briefly discussed. The gamma ray luminosity above 100 MeV is shown to scale in proportion to $\dot{E}_R^{1/2}$, as is in accord with some of the simplest ideas about "polar cap" models. Models based on acceleration in the outer magnetosphere are also briefly discussed.

Space Science Reviews **75**: 235–255, 1996.

1. INTRODUCTION

Pulsars appear as high energy gamma ray sources in two distinct forms. The surrounding nebulae (often known as "plerions" [Weiler 1978] for their filled image in radio, optical and X-ray maps) formed by the dissipation of the unseen carrier of the rotational energy from a pulsar reprocess much of the energy into nonthermal synchrotron radiation, with photon energies as high as a few GeV. In addition, inverse Compton scattering of the synchrotron photons can create observable photons with energy above 1 GeV (for a summary of the observations, see deJager and Hardin 1992, Gallant and Kirk 1995). These nebulae, most famously typified by the Crab Nebula, allow us to indirectly investigate the physics of the main carrier of the pulsar's energy. Pulsed gamma rays, now observed by the Compton Observatory from half a dozen rotation powered pulsars (Ulmer 1994), provide a diagnostic of high energy particle acceleration under the direct rotational control of the pulsar, presumably from somewhere within a radius of order the light cylinder distance from the neutron star.

Besides the interest in the physics of these objects in their own right, the understand ing of rotation powered pulsars and their nebulae has implications for the knowledge of the whole zoo of nonthermal radiation sources powered by central compact objects Other examples include the excitation of extragalactic radio sources by jets, and some aspects of the emission from active galactic nuclei themselves.

2. PULSAR WINDS, SHOCK STRUCTURE AND PARTICLE ACCELERATION

The Crab Nebula and its pulsar are the best studied pulsar-plerion system. It is typical of a ''diffuse'' plerion (a category introduced by Arons and Tavani 1993), is which synchrotron radiation and adiabatic expansion dominate the energy losses from relativistic electrons and positrons, and these losses are slower than the rate of the basic particle acceleration process. Relativistic, magnetohydrodynamic (MHD) models of the coupling between pulsar and nebula are the most successful at the present time. Building on earlier ideas (Piddington 1957; Rees and Gunn 1974; Kundt and Krotscheck 1980 Kennel and Coroniti (1984a,b) showed that such a wind from the pulsar, terminated by standing, transverse magnetosonic shock wave, could give a good account of the nebula dynamics and of the nebular spectrum of high energy photons (near infrared, optical X-rays and gamma rays), which are an instantaneous probe of the pulsar's outflow There are two prominent results of their model. 1) The post shock flow fits into the slowly expanding nebula only if the upstream flow is almost entirely dominated by kinetic energy, with

$$\sigma \equiv \frac{B_1^2}{4\pi m N_1 \gamma_1 c^2} \ll 1. \tag{1}$$

Here B_1 is the upstream magnetic field, N_1 is the upstream density, and γ_1 is the Lorentz factor of the flow. 2) If the upstream flow is entirely composed of pairs (an assumption), and if the post-shock energy distributions are power laws with $N(E) \propto E^{-s}$ (another assumption), then the synchrotron emission from the downstream flow can be fit to the observed X- and γ–ray spectra of the Nebula, taken as an unresolved object, if $s \approx 2.3$. Once this fit is made, the optical and near infrared spectra of the nebula (again taken as an unresolved object) are *predicted* (more or less) correctly by the model.

2.1. Wind Parameters and Geometry

Such winds are characterized by several parameters. The first is the total outflow rate in electron-positron pairs, $\dot{N}_{\pm}$. The MHD model assumes $\dot{N}_{\pm} \gg 8\pi^2\mu/ecP^2 \sim 2 \times 10^{34}$ s^{-1}, where μ is the magnetic moment and P is the rotation period, and the numerical value is that appropriate to the Crab pulsar. The spindown energy carried by the relativistic wind is partitioned into several components, as is revealed by the expression for the total rotational energy lost from the star

$$\dot{E}_R = \int_{\Omega_w} d\Omega \left\{ \frac{\mathbf{E} \times \mathbf{B}}{4\pi} + c\beta_w\gamma \left[(n_+ + n_-)m_\pm c^2 + n_i m_i c^2 \right] \right\}, \tag{2}$$

where Ω_w is the solid angle subtended by the wind, $\mathbf{E}$ and $\mathbf{B}$ are the electric and magnetic fields in the flow, n_+, n_- and n_i are respectively the lab (= nebular) frame densities of positrons, electrons and ions, respectively, $m_\pm$ and m_i are the rest masses of the pairs and heavy ions respectively, and $c\beta_w\gamma_w$ is the spacelike part of the wind our velocity. Quite a while ago, Aschenbach and Brinkmann (1975) showed from the spatial distribution of the X-ray emission in the Nebula that the energy outflow from the Crab Pulsar is largely confined to the plane of the pulsar's rotational equator, a result confirmed by Einstein (Brinkmann, Aschenbach and Langmeier 1985) and ROSAT (Trümper 1993) observations. A cartoon representation of the structure is shown in Figure 1.

Subsequently, other authors have found that $\sigma \ll 1$ is required by more elaborate dynamical and spectral considerations. Emmering and Chevalier (1987) incorporated the time dependence introduced by nebular expansion into the dynamical model, and sound $\sigma \approx 1.5 \times 10^{-3}$ is required in order to match the observed rate of expansion, within the context of a spherical model of the nebula. By contrast, Kennel and Coroniti's steady flow model yields $\sigma \approx 3 \times 10^{-3}$. Begelman and Li (1992) introduced the first non-spherical MHD model for the nonrelativistic expansion of the nebula, and found $\sigma \approx 3 \times 10^{-3}$ gives the best approximation to the ellipsoidal shape of the nebula. Their model neglects the interaction between the expanding "bubble" of relativistic particles and fields and the embedded filaments of non-relativistic gas from which the optical emission lines arise. On the spectral front, de Jager and Harding (1992) showed that if one adopts the magnetic field profile of Kennel and Coroniti's (1984a) $\sigma \approx 3 \times 10^{-3}$

model, one can construct a successful phenomenological model of the hard photon emission from the nebula, at photon energies from MeV through GeV to TeV, based on synchrotron emission at photon energies up to a GeV and inverse Compton emission from the same population of relativistic electrons at photon energies above a GeV.

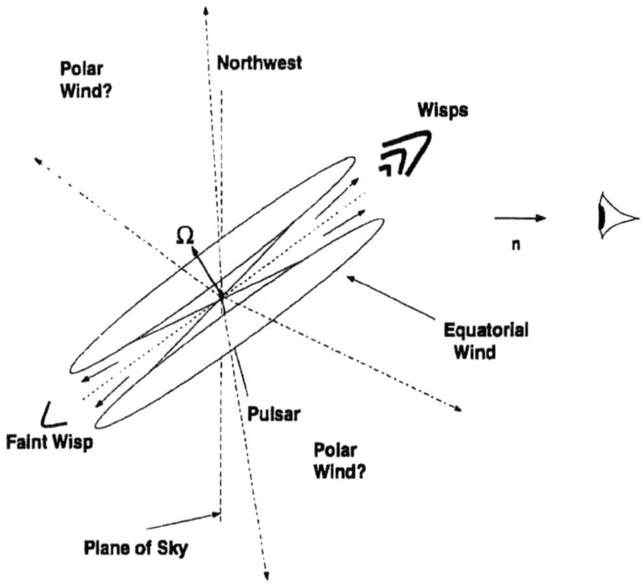

Figure 1: Cartoon representation of the termination region of the relativistic wind lying in the rotational equator of the Crab pulsar. The wind's magnetic field is toroidal with respect to the pulsar's rotation axis; therefore, it is transverse to the outflow direction The wisps are identified with the regions of magnetic compression within the shock structure, where the volume synchrotron emissivity is enhanced. The faint wisp has lower surface brightness because of the relative Doppler boost of the intensity between flow approaching and receding from the observer. The wisps appear as arcs rather than rings because the radiating pairs' momenta are not isotropically distributed with respect to the magnetic field direction, but instead are predominantly transverse to B. Therefore emission is observable only from the arcs of the toroidal field which are close to being transverse to the line of sight.

This theory is sufficiently successful as a macroscopic explanation of the phenomenology to provide a setting for investigation of the microscopic physics. Several questions are implied by the macroscopic flow model. Does a relativistic, magnetosonic shock in a pair plasma actually create the assumed power law distributions downstream? If so, what is the mechanism of acceleration in such shocks? Are there other observable consequences of this mechanism, other than the desired particle acceleration, which allow one to test the shock hypothesis directly? We emphasize that one cannot appeal to the ever popular ideas of diffusive shock acceleration proposed for particle acceleration by interplanetary and nonrelativistic interstellar shocks (Bell 1978; Quenby and Lieu 1989) even as extended to relativistic shocks with magnetic field parallel to the flow *e.g.*, Ellison *et al* 1990), since the termination shocks of pulsar winds must be fairly close to having the magnetic field perpendicular to the flow upstream when the plerion surrounds the pulsar. Indeed, for the Crab, a laminar Archimedean spiral in the outflow from the pulsar would have Θ_{Bn}, the angle between the flow direction and the magnetic field in the shock frame where we observe the system, differing from 90° only by one part in 10^9! In addition, in relativistic shocks, the particles can stream ahead of the shock and participate in the back and forth motion across the front as required in diffusive shock acceleration only when $\Theta'_{Bn} \ll 1/\gamma_1$, where Θ'_{Bn} is the angle between the flow direction and the magnetic field in the proper frame of the upstream fluid (Begelman and Kirk 1990, Gallant *et al* 1992). I will argue that $\gamma_1 \sim 10^6$ in the Crab Nebula. Then the magnetic field is still almost exactly transverse even in the upstream fluid frame. Therefore, some other mechanism of shock acceleration must apply.

2.2. Shock Structure and Particle Acceleration Properties

The details of the structure and particle acceleration properties of transverse, relativistic magnetosonic shocks, as revealed by theory and computational particle-in-cell (PIC) simulations, have been described elsewhere (Alsop and Arons 1988, Langdon, Arons and Max 1988, Hoshino and Arons 1991, Gallant, Hoshino, Langdon, Arons and Max 1992, Hoshino, Arons, Gallant and Langdon 1992). Here, I briefly summarize the results.

Pulsar outflows must have extremely high Mach number. The Alfven Mach number in the flow is $M_A = \beta_1\gamma_1/\sqrt{\sigma} \cong 10^7$! Under these circumstances, magnetic reflection of particles from the shock front plays an essential role in the shock dynamics. Figure 2 shows a cartoon of the particle phase space, superposed on a plot of the electromagnetic field amplitudes, for a plane parallel shock wave. For definiteness, and because it is the most observationally relevant case, I focus on a flow composed of electron-positron pairs and heavy ions. The heavy ions are presumed to be a minority by number, but carry most of the upstream flow energy and momentum.

The flow momentum in the reflected ions gets temporarily stored in compressed magnetic field and pair plasma. The pairs have already been compressed in a thin shock with similar reflected particle structure, which forms at $x = 0$ in Figure 2. The simulations reveal the details of this compression, which shows up as a series of magnetic "overshoots", in which the magnetic field becomes larger than its final downstream value.

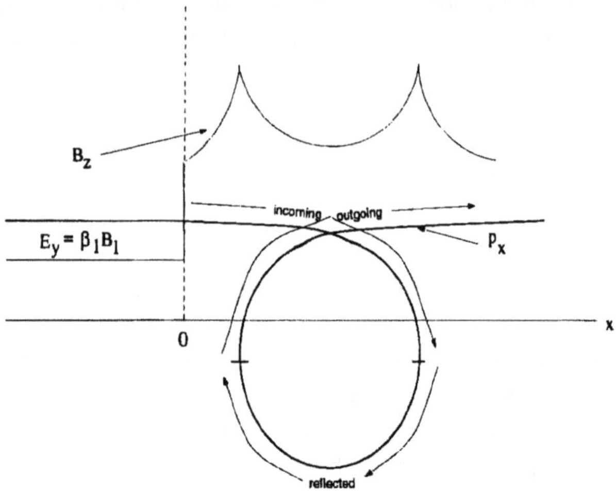

Figure 2: Sketch of geometry for the plane-parallel case. The plot shows a qualitative sketch of the magnetic field B_z and the ion momentum p_x, while the transverse electric field stays constant at the fixed value $E_y = \beta_1 B_1$ throughout the steady state structure Here β_1 is the upstream velocity measured in units of c, and B_1 is the constant upstream magnetic field. The shock in the pairs occurs at x = 0, where the ions are set into Larmor gyration by the increased magnetic field. In the region where the ions are reflected the ion "fluid" has a three valued flow velocity, comprising incoming, reflected and outgoing particles. The magnetic field enhancements occur at the turning points of the ions' orbit, where the transverse current maximizes.

The particle reflection process is "coherent", in the sense that all the particles gyrate around the magnetic field lines with small momentum dispersion. This coherent gyration is unstable. The ion ring undergoes gyrophase bunching and collectively radiates cyclotron and synchrotron waves at the harmonics of the relativistic ion Larmor frequency $\omega_{ci} = ZeB/m_i\gamma_1 c$. Within the context of the theory completed to date, these synchrotron modes are magnetosonic waves propagating strictly across the magnetic field in the relativistically hot electron-positron plasma. At the leading edge of the ion reflection region, the pairs have a relativistic Maxwellian distribution in energy space with a temperature $T_\pm \approx m_\pm c^2\gamma_1/k$, where γ_1 is the Lorentz factor of the upstream flow. Within the frequency range $\omega_{ci} \ll \omega \ll \omega_{c\pm} \equiv eB/(kT_\pm/c)$, small amplitude magnetosonic waves in the pairs have the simple dispersion relation

$$\omega^2 = k_\perp^2 c^2 \frac{c_s^2 + v_A^2}{c^2 + v_A^2}, \tag{3}$$

where c_s is the relativistic speed of sound and $v_A \equiv B/(8\pi m_{eff} N_\pm)^{1/2}$ is the format Alfven speen in the pair plasma. Here $N_\pm$ is the space density of electrons and positrons and $m_{eff} = m_\pm + kT_\pm/c^2$. One can readily show that even if one includes slight obliquity of the magnetic field to the **k** vectors, cyclotron absorption by the pairs is much larger than Cerenkov absorption.

The pairs preferentially absorb these magnetosonic waves at their relativistic cyclotron frequency $\omega_{c\pm} = eB/m_\pm\gamma_1 c$, corresponding to harmonic number $n \sim m_i/Zm_\pm \gg 1$. The energy gains cause the relativistic masses of the pairs to increase, allowing them to resonantly absorb lower frequency waves until finally resonance with the fundamental ion cyclotron wave terminates the process. Thus the maximum energy in the pair spectrum corresponds to a Lorentz factor $\gamma_{\pm,max} = m_i/Zm_\pm\gamma_1$, which is the same as saying that the maximum energy achieved by an electron or a positron equals the upstream flow energy per particle of the much more massive ions. The energy spectrum of the magnetosonic waves is proportional to k^{-1}, and it is straightforward to show that the cyclotron absorption

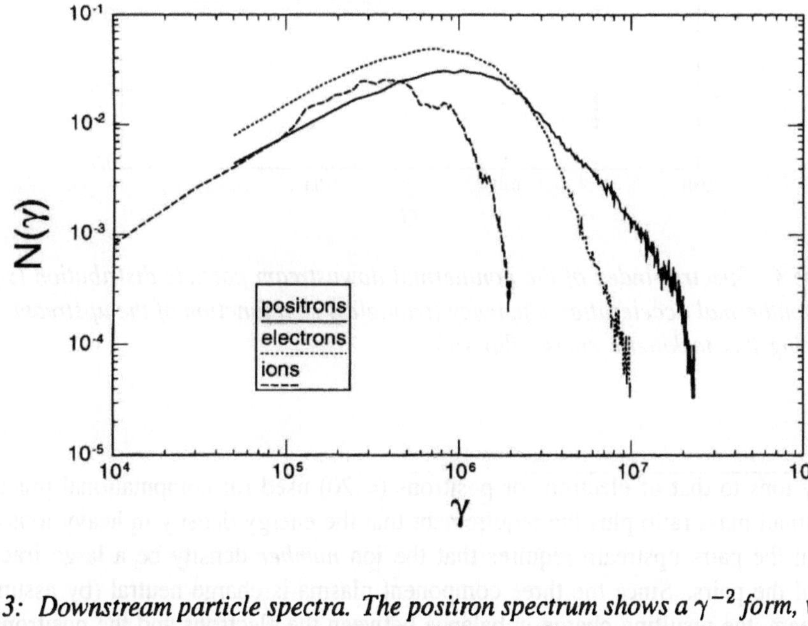

Figure 3: Downstream particle spectra. The positron spectrum shows a γ^{-2} form, while the small mass ratio used in the simulations ($m_i/Zm_\pm = 20$) causes the electron slope to e steeper. Further simulations with larger mass ratios show that the electron spectrum approaches that of the positrons, as is expected theoretically.

of this wave spectrum leads to a downstream energy distribution of the pairs $N_\pm(\gamma) \propto \gamma^{-2}$, a result confirmed by the simulations and illustrated[1] in Figure 3. Here $N_\pm(\gamma)d\gamma$ is the number of electrons or positrons with energy between $m_\pm c^2\gamma$, $m_\pm c^2(\gamma + d\gamma)$.

2. Figure 3 suggests that this mechanism preferentially accelerates positrons over electrons. This preference is an artifact of the small ratio of the rest mass of the

Furthermore, this acceleration process is efficient. Figure 4 shows the fraction of the upstream flow energy converted into the nonthermal part of the downstream distribution It has long been assumed (e.g., Kennel and Coroniti 1984a,b) that relativistic shock can produce particle spectra with these characteristics as a result of applying MHD models to the Crab Nebula. The results summarized here are the first to uncover the physics of how this acceleration happens.

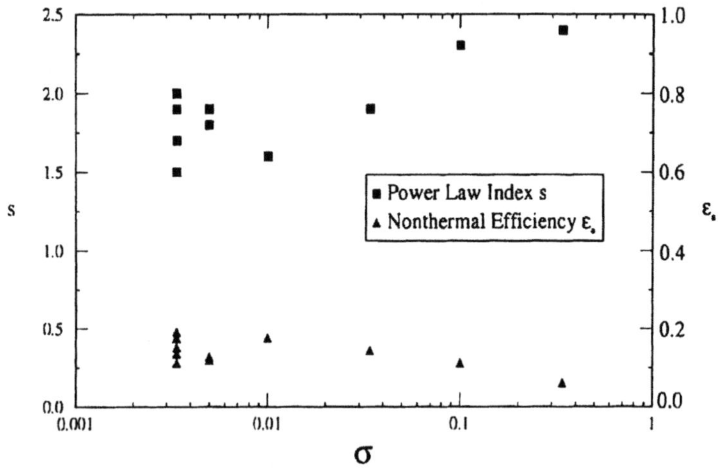

Figure 4: Spectral index of the nonthermal downstream particle distribution (squares) and nonthermal acceleration efficiency (triangles) as a function of the upstream ratio of Poynting flux to kinetic energy flux (σ).

---

heavy ions to that of electrons or positrons (= 20) used for computational practicality The small mass ratio plus the requirement that the energy density in heavy ions exceed that in the pairs upstream requires that the ion *number* density be a large fraction of that of the pairs. Since the three component plasma is charge neutral (by assumption) upstream, the resulting charge imbalance between the electrons and the positrons give the magnetosonic waves a strong left-handed ellipticity, with consequent preferential cyclotron resonant absorption by the positrons. With realistic mass ratios, the wave are almost exactly linearly polarized, which yields the equal acceleration rate for the electrons and positrons $\dot{\gamma}/\gamma = 0.03\omega_{ci}(\delta B/B)^2$ in quasi-linear theory (Arons and Gallant, in preparation). Here ω_{ci} is the relativistic cyclotron frequency of the gyrating ion ring in the shock front, and δB is the amplitude of the fluctuating magnetic field formed by the spectrum of magnetosonic waves. In the simulations, where $\delta B/B \gg 10$ the observed acceleration rate was $\dot{\gamma}/\gamma \approx \omega_{ci}$ (Hoshino and Arons 1991, Hoshino *et al.* 1992.)

2.3. THE WISPS IN THE CRAB NEBULA AS RESOLVED SHOCK STRUCTURE

One can get the flavor of astrophysical models based on these results from the following simple considerations. Suppose the magnetosphere of the Crab pulsar emits dense flow of pairs, with $\dot{N}_\pm \sim 10^{38}\dot{N}_{\pm,38}$ coming out in the broad sector around the rotational equator where the pulsar must emit the wind required to feed the torus of X- and gamma-ray emission in the Nebula (Aschenbach and Brinkmann 1975; Pelling *et al* 1987). A number of magnetospheric models suggest such a flow is possible. In addition, suppose the magnetospheric electrodynamics causes an outflow of heavy ions to be pulled up from the stellar surface and accelerated to high energy in the outflow around the magnetic equator (Arons 1983a). The ion flux is constrained by the electrodynamics to be close to the Goldreich-Julian flux, $\dot{N}_{GJ} = 2\Omega^2\mu/Zec \approx 2\times 10^{34}/Z$ for this pulsar, where Z is the charge on the ions. The resulting ratio of the mass densities in the wind is

$$\rho_i \equiv \frac{m_i N_{i1}}{m_\pm(N_{1+} + N_{1-})} = 918\frac{A}{Z}\frac{\Omega^2\mu}{ec\dot{N}_\pm}, \qquad (4)$$

where A is the atomic number of the ions, μ is the pulsar's magnetic moment and Ω is its angular velocity. Solving for Z yields $Z \approx 3(A/56)(1/\dot{N}_{\pm,38})(\rho_i/2)$ in the case of the Crab. More refined analysis (Gallant and Arons 1994) suggests the charge on the ions as they are extracted from the stellar surface is comparable to the ions' charge as they enter the shock structure in the Nebula, that they are fairly close to being fully stripped, and that helium is the most likely candidate for the ion composition.

These ions might be accelerated through a fair fraction of the total magnetospheric potential $\Phi_{open} \approx \Omega^2\mu/c^2 \sim 4\times 10^{16} \approx (4\dot{E}_R/c)^{1/2}$ Volts, corresponding to $\gamma_1 \sim 4\times 10^7(Z/A)(\Phi/\Phi_{open})$. If the shock puts $\sim 20\%$ of the outflow energy into the accelerated pairs, as our simulations suggest, and if these pairs radiate essentially all of this in X-rays and gamma-rays, as is indicated by the macroscopic flow models, then one would conclude, from the hard photon luminosity of the nebula, that the ions experience an acceleration potential $\Phi/\Phi_{mag} \sim 0.2/I_{45}$, where I_{45} is the moment of inertia of the neutron star in units of 10^{45} cgs.

We can go much further. From momentum conservation, one expects the shock to occur roughly $10^{17.5}$ cm from the pulsar (Rees and Gunn 1974). Optical observations, schematically represented in the outflow model in Figure 1, reveal the presence of the curious "wisps" in the surface brightness in this region, a series of enhancements in the surface brightness elongated in the direction perpendicular to the flow lying to the Northwest of the pulsar; there is also a fainter wisp to the Southeast (Scargle 1969, van den Bergh and Pritchet 1989). These features are shown in Figures 5 and 6.

If the pulsar accelerates the wind to an energy/particle comparable to that estimated above, the ion Larmor radius in the magnetic field typical of the outflow would be $r_{Li} \simeq 0.15(\gamma_1/10^{6.7})(10^{-4.5}\ \text{Gauss}/B_1)$ parsecs, a length corresponding to several seconds of arc at the distance of the Crab Nebula. Within an overshoot, the Larmor radius is somewhat smaller. The bolometric synchrotron emissivity in such magnetic overshoots is readily found to be proportional to B^4, assuming the downstream pitch angle scattering has not largely isotropized the pairs, as turns out to yield a better mode of the wisps' surface brightness (Gallant and Arons 1994); the continuum emissivity in a fixed narrow band is proportional to B^2. The essential new astrophysical idea is to identify the enhanced surface brightness expected in these compressions with the observed wisps.

In order to test this idea, Gallant and Arons (1994) developed a simplified theoretical model which captures the essential physics revealed in the simulations, but in the outflow sector geometry shown in Figure 1. We threw away the detailed kinetic properties which lead to particle heating and nonthermal particle acceleration, but retained the effects of the pressure in the pair plasma by assuming adiabatic fluid flow of the pairs, which are well frozen to the magnetic field. We retained the mechanical structure induced by the reflection of the heavy ions from the shock front by modeling the ions as a stream of charged particles with finite Larmor radius, all with same energy/particle at each point. One dimensional steady flow in the geometry of a spherical sector was assumed. The surface brightness of the model was calculated assuming the spectral emissivity of a local relativistic Maxwellian, since the plane parallel simulations show the pair spectrum becomes fully nonthermal only after the pairs traverse at least two of the magnetic overshoots.

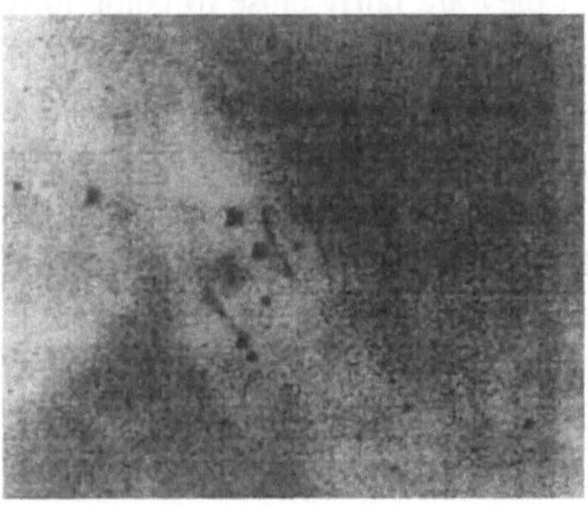

Figure 5: I band image of the inner $1' \times 1'$ of the Crab Nebula in October 1988, from van den Bergh and Pritchet (1989). The contour plot derived from this image in Figure 6 identifies the features.

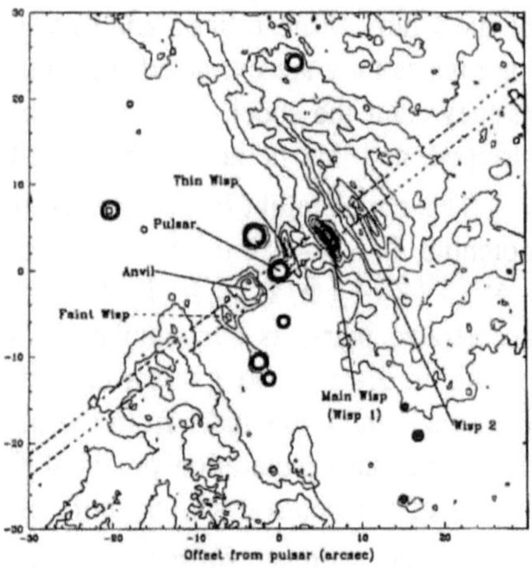

Figure 6: Contour plot of the central $1' \times 1'$ of the Nebula, from the I band image of van den Bergh and Pritchet (1989). The contours are spaced in intervals of 25 μJy/arcsec^2, starting at 250 μJy/arcsec^2. For clarity, contours representing surface brightness levels below the starting value were omitted, since these represent foreground and background Nebular emission. The dashed lines show the 2″ wide slice from which the intensity profile shown in Figure 8 was derived.

The free parameters of the model (injection rate of pairs, tip angle of the outflowing "disk" of plasma and field with respect to the plane of the sky, ratio of the energy density of heavy ions to the energy density of pairs in the upstream flow, upstream magnetic field and upstream flow Lorentz factor) are overdetermined by the obervations of the bright wisps to the pulsar's Northwest *and* the fainter counterwisp to the Southeast. A typical model is shown in Figure 7, and the "best" fit is in Figure 8.

From this identification, one can extract a host of conclusions. The details of these are presented elsewhere (Gallant and Arons 1994). 1) The pulsar injects $\sim 10^{37.6}$ pairs/s and $\sim 3 \times 10^{34}/Z$ ions/s with charge $Z \sim 2$ into the wind and with $\gamma_1 \sim 4 \times 10^6$. This rate of pair injection is in good accord with the number of particles per second required to feed the approximately toroidal X–ray source at $\sim 30''$ radius from the pulsar. 2) The series of downstream, large amplitude compressions of the plasma are spaced roughly one ion Larmor radius apart. This spacing is similar to that seen between the first few wisps in van den Bergh and Pritchet's (1989) image if $\gamma_1 \sim 4 \times 10^6$, the same value as leads to a proper representation of wisp 1's thickness. 3) The EUV and X-ray spectrum of wisp 1 should be close to that of Maxwellian synchrotron emission; however, at the lower particle energies which give rise to optical and infrared emission, nonthermal acceleration can steepen the spectrum even at radii interior to wisp 1, since only partial deflection of the ions is required to emit high harmonic magnetosonic waves. The spectrum for a uniform medium is well known (Jones and Hardee 1979). With $B \sim 10^{-4}$ Gauss and pair temperature $\sim 4 \times 10^6 m_{e^\pm} c^2$, the characteristic decline of the spectrum from a relativistic Maxwellian should appear above $\sim$ 25 eV. 4) Successive wisps further from the ion overshoot should show a progressively more nonthermal spectrum, as the ions transfer their energy to the pairs. 5) The flow velocity of the pairs throughout this region is between 0.3c and 0.5c, which provides a Doppler asymmetry to the luminosity just about right to explain the brightness ratio seen between wisp 1 and the one faint wisp observable to the pulsar's southeast. Possible partial anisotropy of the pairs with respect to the local magnetic field enhances this conclusion. 6) The shocks observed in our simulations are unsteady. Translated into the shock frame, which is where we observe the Crab Nebula, the unsteady shock velocity becomes either anharmonic oscillation of the position of the first ion overshoot (= wisp 1), or disappearance and reformation of wisp 1, on a time scale of several months, a phenomenon consistent with time variability of the wisps reported long ago by Scargle (1969). Such variability should be uncorrelated with pulsar glitches and other rotational anomalies. In addition, such unsteady behavior may imply unsteady nebular emission at energies exceeding 1 MeV, as has been occasionally reported. 7) The time variable overshoots can transfer further energy from the gyrating ions to the pairs in the downstream medium, over and above that found in the spatially 1D simulations. Because of the finite rate of pitch angle scattering present downstream, magnetic pumping can increase the amount of energy transferred to the pairs, and can also act to transform the Maxwellian low energy pair spectra into non-thermal form, thus allowing the same shock model to explain both the high energy photons and the radio synchrotron emission from plerions (Arons, in preparation). 8) The maximum

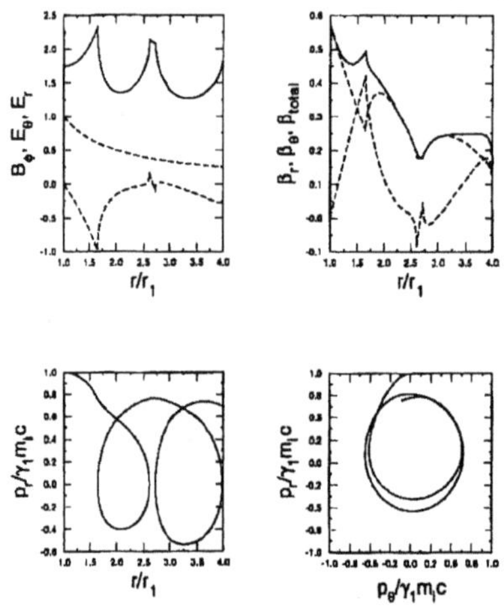

Figure 7: Upper left panel: Magnetic field (top curve), transverse electric field (E_θ - middle curve) and radial electric field (bottom curve) of a model with ion Larmor radius twice the radial distance of the leading edge of the shock wave from the pulsar, energy density in the ion flow upstream twice that contained in the pairs, energy density in the upstream plasma (ions plus pairs) 25 times that in the fields, and upstream Lorentz factor $\gamma_1 = 10^6$ as a function of r/r_1, where r_1 is the radius where the leading edge shock in the pairs occurs. The fields are measured in units of the upstream magnetic field B_1. Upper right panel: radial (middle curve), transverse (bottom curve) and total (top curve) flow velocities of the pairs, measured in units of c. Note the large component of transverse velocity in the first magnetic overshoot. Lower left panel: Radial momentum of the ions, measured in units of $\gamma_1 m_i c$. Lower right panel: Radial ion momentum as a function of transverse ion momentum.

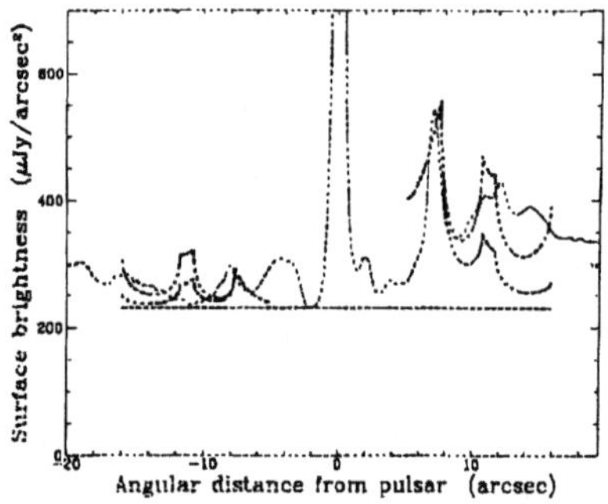

Figure 8: Predicted intensity profile for the "best-fit" parameters to the main wisp and counter-wisp, versus the observed brightness profile. Dashed line with central spike: I band surface brightness from van den Bergh and Pritchet's (1989) data, from the 2″ wide strip shown in Figure 6. Horizontal Dashed line: Intensity level due to the optical nebula surrounding the wisp region. Upper dashed curves: mode predictions with the width of the pitch angle distribution assumed constant with radius, $\Delta\alpha(r) = 8°$. Lower dashed curve: model predictions with pitch angle width increasing with radius at a rate such that the finite Northeast-Southwest extent of the wisps to the Northwest of the pulsar is explained as being due to pitch angle anisotropy in the pairs The asymmetry in brightness between the Northwest (postive angle with respect to the pulsar, flow approaching the observer) and the Southeast (negative angle with respect to the pulsar, flow receding from the observer) is due to Doppler boosting and faintening.

energy reached by pairs in the theoretical model of the shock structure is equal to the energy per particle of the ions in the upstream flow, $E_{max\pm} = \gamma_1 m_i c^2 \approx 2 \times 10^{15}$ eV when the ions are fully stripped Helium (drawn electrically from the top of the neutron star's atmosphere). This upper cutoff of the accelerated pair distribution yields synchrotron photons whose spectrum cuts off exponentially at energies above 100 MeV, as found in de Jager and Harding's (1992) phenomenological model. The details of the formation of the gamma ray spectrum are somewhat different, however, since the magnetic structure both includes the strong overshoots introduced by the reflected ions, and also has an average profile differing from that used by de Jager and Harding because of the anisotropic character of the pairs' pressure. Finally, 9) the pulsar's injection rate of ions inferred from the fit of the model to the observed surface brightness, $\dot{N}_i \approx 3 \times 10^{34}/Z$/sec, is almost exactly equal to the number of ions per second required from the pulsar as the return current needed to prevent the star from charging up, which provides the first empirical (if model dependent) evidence that such return currents exist.

Therefore, we suggested that the structure of a relativistic shock in a relativistic wind composed of pairs plus heavy ions has its spatial structure revealed to us in the detailed morphology of the wisp region in the Crab Nebula. That this may be so is a consequence of the very high rigidity of the outflow from the pulsar, a result in accord with the enormous voltages developed in the magnetosphere combined with the negligible radiation drag on the flow near this photon poor compact object.

2.4. TIME VARIABLE HARD GAMMA RAY EMISSION FROM THE CRAB NEBULA?

Points 6) and 8) deserve a bit of amplification, in the context of this meeting. Absorption of the magnetosonic waves causes the electron-positron pairs to accelerate to energies comparable to $\gamma_1 m_i c^2 = \eta Z e \Phi_{open}$, with $\eta \sim 0.3-1$. Here $\Phi_{open} \approx 4 \times 10^{16}$ Volts is the electric potential drop across the open field lines of the pulsar. The time to accelerate pairs to this energy is $t_a \approx \omega_{ci}^{-1} = m_i c \gamma_1 / Z e B_2 \sim 3$ months. The thickness of this acceleration region, located at $r_s \sim 10^{17}$ cm, is about $\Delta r_a \approx v_2 t_a \approx 10^{17}$ cm also. The synchrotron spectrum emitted from these particles will have the form of a power law up to a critical energy $(\hbar e B_2 / m_\pm c)(m_i / Z m_\pm)^2 \gamma_1^2 \approx 150$ MeV; above this energy, the synchrotron spectrum declines quasi-exponentially, in good agreement with phenomenological models (de Jager and Harding 1992). For radii exceeding $r_s + \Delta r_a \sim 10^{17.5}$ cm, the pair acceleration is largely complete, and the particles cool through synchrotron losses, as in the Kennel and Coroniti (1984b) model. The highest energy pairs lose one half their energy in the synchrotron loss time $t_s = (6\pi m_\pm c / \sigma_T B_2^2 \gamma_1)(m_\pm Z / m_i) \approx 6$ months. Thus the radial extent of the region emitting $\varepsilon > 100$ MeV synchroton gamma rays is comparable to that of the acceleration region, $r_s + \Delta r_{synch} \sim 3 \times 10^{17}$ cm, corresponding to an angular size of $10''$ seconds of arc at the 2 kpc distance of the Crab Nebula. This is far to small to be resolved by existing gamma ray telescopes, but it is the *same* dimension as the wisp region itself.

The whole particle acceleration process relies on the cyclotron instability of the gyrating ions to create the magnetosonic waves whose absorption energizes the pairs The long wavelength end of this wave spectrum extends to the scale of the ion Larmor radius itself, causing the whole wisp structure to fluctuate on the the ion Larmor time $\omega_{ci}^{-1} =\sim 3$ months $\sim t_s$ with magnetic ampitude $\delta B > B$ (Hoshino *et al.* 1992 Li and Arons 1995). Such fluctuations, on time scales comparable to the acceleration and synchrotron loss time, should cause fluctuations in the 100 MeV to 1 GeV gamma ray emissivity of the Crab Nebula. Variations in the wisp positions and brightness on this time scale have been known for a long time (Lampland 1921, Scargle 1969) and were recently discovered in the infrared (Graham 1995). A search for such gamma ray variability, along with correlated time variability in the wisps, would shed substantial further illumination on the energization physics of the Crab Nebula.

By contrast, the TeV emission from the Crab (*e.g.*, Vacanti *et al.* 1991) should show little if any of the variability seen in the optical and likely in the 100 MeV gamma rays. Inverse Compton scattering of the Nebular optical and IR is the most likely origin of these very high energy gamma rays. These photons occupy a much larger volume than the wisp region, as do the particles which do the scattering (de Jager and Harding 1992, Gallant and Kirk 1995). Therefore, the semi-coherent variability of the large scale fields in the shock acceleration zone should be lost in the averaging over a larger region that contributes to the inverse Compton emission.

3. ENERGETICS OF (POPULAR) MODELS FOR PULSED GAMMA RAYS

The Compton Gamma Ray Observatory, coupled to discoveries in other wave bands, has expanded our knowledge of pulsed gamma ray emission from rotation powered pulsars three fold, from two clearly detected objects to six. For a recent review of the observations, see Ulmer (1994), for example. For my purposes, the EGRET data can be summarized by the efficiency diagram shown in Figure 9, not previously published. $\varepsilon_\gamma \equiv L_\gamma / I\Omega\dot{\Omega}$ defines the efficiency. The gamma ray luminosity estimates assume distances from dispersion measures or other means appropriate to the star, and a beaming solid angle of one steradian. To do better obviously requires a better beaming model (as well as better distances), topics to which I return as I discuss the significance of Figure 9.

In the West, two kinds of models have been popular in trying to understand such emission, involving either relatively low altitude emission from the accelerated particles and outbound pair plasma formed low over the magnetic poles, or at and exterior to the surface in the outer magnetosphere where $\mathbf{\Omega} \cdot \mathbf{B} = 0$, in a region close to the boundary between the open and the closed field lines. These models are frequently called "polar cap" and "outer gap", respectively.

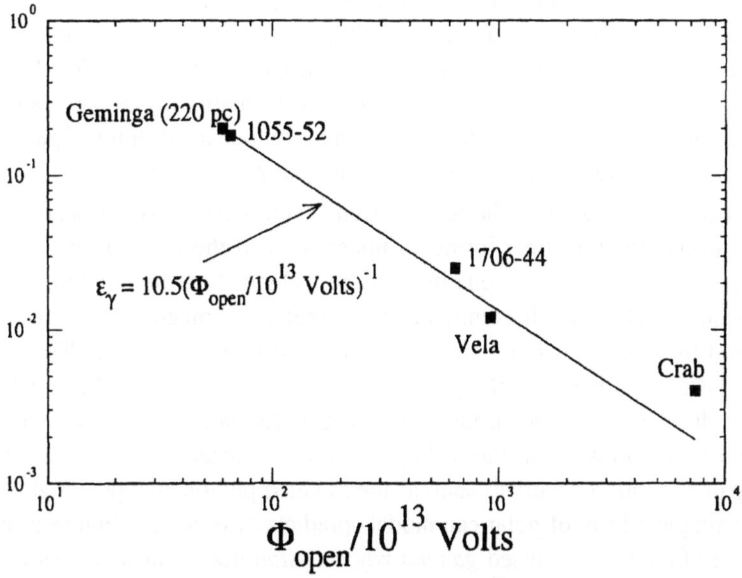

Figure 9: ε > 100 MeV gamma ray efficiency as a function of open field line voltage $\propto (\dot{P}/P^3)^{1/2}$.

3.1. Polar Cap Models

There is a moderate amount of evidence that the surface magnetic fields of rotation powered pulsars are dipolar (Arons 1993). The evidence does not exclude the possibility of the formal center of the dipole field being substantially displaced from the center of the star, a particular kind of magnetic anomaly which has been advocated on theoretical grounds (Ruderman 1991, Chen and Ruderman 1993). Rotation of the star creates an electric field which in a vacuum causes the extraction of a relativistic particle beam from the surface (electrons if $\mathbf{\Omega} \cdot \mathbf{B}_{\mathbf{pole}} > 0$, ions if $\mathbf{\Omega} \cdot \mathbf{B}_{\mathbf{pole}} < 0$). On the assumption that this extraction corresponds to the electric current which allows the fields to carry away the star's angular momentum, one can show that under a wide range of circumstances, the beam leaves a residuum of the vacuum electric field (a "starvation" field) suffcient to accelerate the beam charges to ultrarelativistic energies (Michel 1974, Fawley, Arons and Scharlemann 1977, Scharlemann, Arons and Fawley 1978, Muslimov and Tsygan 1990, 1992, Beskin 1990). Under somewhat more restrictive circumstances, the beam particles *copiously* radiate gamma rays (either through curvature emission or through inverse Compton scattering of the thermal EUV and X-rays from the surface) with sufficient energy to form electron-positron pairs, either through one photon pair creation in the strong magnetic field (Sturrock 1971) or through photon-photon absorption in the thermal radiation field. These pairs alter the charge density sufficiently to short out the strong accelerating electric field (Ruderman and Sutherland 1975) at and above a well defined "pair formation front" (Arons and Scharlemann 1979), located at a height H_{PFF} above the stellar surface. This height is a strong function of the pulsar's voltage (as well as rapidly increasing with increasing distance from the magnetic axis), ranging from a few hundred meters for a high voltage pulsar like the Crab to a few kilometers for a more typical star (Arons 1983b). Above the pair formation front, the beam coasts creating more gamma rays whose subequent cascades radiate a spectrum of gamma rays observable by an observer able to look down onto the polar cap region.

The total power invested in the acceleration of the beam between stellar surface and the pair formation front then forms an upper limit to the gamma ray luminosity $L_\gamma \leq L_{accel}$ or $\varepsilon_\gamma \leq \varepsilon_{accel}$; equality can be achieved in high voltage pulsars where the beam acceleration becomes radiation reaction limited. Since the current density is fixed by electrodynamics to be closely equal to B/P, one readily finds that $L_{accel} = (c/4)\Phi^2_{open}(\Delta\Phi(\mathrm{PFF})/\Phi_{open}) = \dot{E}_R(\Delta\Phi(\mathrm{PFF})/\Phi_{open})$, where $\Delta\Phi(\mathrm{PFF})$ is the voltage drop between the surface and the pair formation front. So long as the pair formation front forms at all, this voltage is almost constant, independent of Φ_{open} itself, since pair formation occurs as soon as the radiated photon energy is high enough Therfore, the simplest form of polar cap models predicts $\varepsilon_a \propto \Phi^{-1}_{open}$. Figure 9 suggests that indeed the efficiency of pulsed *gamma ray emission* does follow this simple rule which corresponds to $L_a \propto \dot{E}_R^{1/2} \propto (\dot{P}/P^3)^{1/2}$. Approximate equality $L_\gamma \approx L_a$ between the particle acceleration luminosity and the gamma ray luminosity is typical of these relatively high voltage pulsars, in which field aligned acceleration of an electron beam is limited by radiation reaction. However, Table 1 shows that when one uses a detailed model, the scaling doesn't always work as well.

Name	P (sec)	Voltage (V)	ε_a	ε_γ
Crab	0.033	8×10^{16}	0.001	0.004
Vela	0.089	9×10^{15}	0.008	0.012
1706	0.102	6.5×10^{15}	0.010	0.025
1055	0.197	6.5×10^{14}	0.050	0.18
Geminga	0.237	6×10^{14}	0.056	0.20

Table 1: Comparison of beam acceleration efficiency to gamma ray emission efficiency.

Here, I have plugged in the particle acceleration luminosity calculated by Arons (1983b), with the clear result that this particular low altitude model is not energetic enough to explain the "observations". The discrepancy for Vela contributed to my conclusion some time ago (Arons 1983b) that this version of a polar cap model by itself doesn't contain enough acceleration energy to explain the observations of the pulsed emission from the Crab and Vela pulsars. More modern versions, incorporating the dragging of inertial frames effect on the accelerating electric field uncovered by Muslimov and Tsygan (1990, 1992) and by Beskin (1990), may improve the match up.

Beaming is the main problem with a low altitude gamma ray emission model at present. All of the pulsars shown in Table 1 have most of their energy in widely separated pulse components (phase separation $\Delta\phi > 100°$). One way out of this apparent problem has been pointed out by Daugherty and Harding (1994). The beaming discrepancy would be alleviated if the beaming solid angle were much less than the two steradians assumed in constructing Figure 9 and Table 1, a solid angle suggested by the naive interpretation of the wide separation of the pulse components. Daugherty and Harding observe that if the gamma ray pulsars are almost aligned rotators, one can still get widely separated pulses with only a small fraction of the sky irradiated. If this interpretation is true, the number of gamma ray pulsars is much smaller than the number of unidentified EGRET sources.

As Daugherty and Harding themselves point out, their suggestion does violence to the kinematical interpretations of other aspects of these pulsars' emission. For example, the sweep of the polarizartion in the Crab pulsar's optical pulse strongly suggests an angle between the rotation and magnetic axes approaching 90°, while the similarity of the optical and gamma ray pulse profiles suggests the emission regions to be the same, which undercuts the aligned rotator interpretation of the gamma rays. Daugherty and Harding attempt to undercut the argument for spatially coincident emission regions by pointing out that the main radio pulses do not show a polarization sweep like that in the optical, even though the pulse shapes are the same. However, this is not a strong argument against coincident emission regions, since as was pointed out some time ago by Barnard (1986), the dense pair plasma causes the radio polarization to reflect the structure of the magnetic field well beyond the light cylinder, where it is almost entirely toroidal, while the optical polarization does reflect the magnetic structure of the emission zone. Therefore, while the energetics of an improved polar cap model may turn out to be consistent with the observed gamma ray emission, the detailed kinematics of all the emission mildly argues against low altitudes being the emission site, suggesting that the accelerator behind the gamma rays might be somewhere in the outer magnetosphere, a view consistent with the likely sites of the powerful ion accelerator which energizes the wind.

3.2. Outer Gap Models

The most popular idea for an outer magnetospheric model of gamma ray emission has been the suggestion by Cheng and Ruderman (1977), that vacuum like electric fields developing near the surface in the magnetosphere where $\mathbf{\Omega} \cdot \mathbf{B} = 0$ (Holloway 1973) might be "unstable" to the formation of pair creation discharges parallel to the magnetic field. The pair creation could limit the extent of the vacuum field zone both along and across (Cheng, Ho and Ruderman 1986) the magnetic field. This idea automatically produces a beaming pattern in which gamma rays illuminate a large fraction of the sky. The initial theories, confined to a distinct variety of model for each of the two gamma ray pulsars then known, said little about the beaming fraction but yielded gamma ray spectra with some modicum of resemblance to the observations. Recently Chiang and Romani (1994) have impressively developed the beaming aspects of the model showing that if one uses a model magnetic field with a plausible toroidal component (Smith 1986), essentially all the beaming patterns of the known gamma ray pulsars can be understood, including the phasing of the gamma ray to the lower frequency pulses so long as the emission comes from a relatively thin boundary layer closely following the separation between the open and closed field lines.

The complications of the beaming calculations are beyond the scope of my discuss sion here. However, the energetics of the "outer gap" version of this boundary layer deserves comment. A calcuation essentially the same as I used above for the polar caps yields a maximum "gap" luminosity $L_{accel} = \dot{E}_R(\Delta\Phi(\text{gap})/\Phi_{open})(\Delta\Omega_{gap}/4\pi)$ where $\Delta\Omega_{gap}$ is the solid angle into which the open field lines occupied by the gap point. To explain sharp pulses, one needs $\Delta\Omega_{gap} \ll 4\pi$, but then it is quite difficult to understand gamma ray emission efficiencies of 50% or more. Furthermore, these are symmetrical accelerators, in the sense that just as much particle acceleration energy flows back to the stellar surface as goes off into the gamma rays emitted toward the outside world. The high energy particles and photons that collide with the star cause the outer edge of the polar caps to glow in X-rays, suggesting that a one-to-one correlation of gamma ray luminosity (which must be inferred through the beaming model) and the soft X-ray emission from the polar cap must exist. It is too early to tell whether this is a problem or a virtue of this kind of model, but such correlations must exist if this theory is correct.

So called "slot gaps" (Arons 1981b, 1983b) offer a variant on the outer magneto sphere acceleration idea, in which the acceleration is more unidirectional. These are based on the observation that pairs do not completely short out the beams accelerated from the surface, but leave an annular rim of accelerating beam around the dense pair plasma filling most of the polar column. Such unidirectional acceleration also can create gamma rays, although the pure curvature spectrum emitted from a beam accelerating from a cold star does not look much like the spectra of the observed objects (Arons 1981a). A Compton scattering version of the model may be more promising, which has rather different predictions for the relation between soft X-ray emission from the stellar surface and the gamma rays. Finally, a variety of more exotic ideas, based on possible dissipation layers near the light cylinder (*e.g.*, Bogovalov and Kotov 1992, Mestel and Shibata 1994) have hardly been explored at all as possible models of the data.

Whatever the final answer may be to the nature of the gamma ray emission from the gamma ray pulsars, gamma ray astronomy has finally stimulated real advances in the investigation of the magnetospheres of rotation powered pulsars.

ACKNOWLEDGMENTS

I acknowledge support by NSF grant AST-9115093 and NASA grant NAGW-2413, both to the University of California at Berkeley. Part of the work was performed under the auspices of the U.S. Department of Energy at the Lawrence Livermore National Laboratory under contract W-7405-Eng-48.

REFERENCES

Alsop, D., and Arons, J. 1988, *Phys. Fluids*, **31**, 839.

Arons, J., and Scharlemann, E.T. 1979, *ApJ*,**231**, 854

Arons, J. 1981a, in *Plasma Astrophysics*, T.D. Guyenne and G. Levy, eds. (Paris: ESA SP-161), 273

Arons, J. 1981b, in *IAU Symp. No. 95, 'Pulsars'*, W. Sieber and R. Wielebinski, eds. (Dordrecht: Reidel), 69

Arons, J. 1983a, in *Proc. Workshop on Electron-Positron Pairs in Astrophysics*, ed. M.L. Burns, A.K. Harding, and R. Ramaty (New York:American Institute of Physics), p. 163.

Arons, J. 1983b, *ApJ*, **266**, 215

Arons, J. 1993, *ApJ*, **408**, 160

Arons, J., and Tavani, M. 1993, *ApJ*, **403**, 249

Aschenbach, B., and Brinkmann, W. 1975, *A&A*, 41, 147.

Barnard, J.J. 1986, *ApJ*, **303**, 280

Begelman, M.C., and Kirk, J.G. 1990, *ApJ*, **353**, 66

Begelman, M.C., and Li, Z.-Y. 1992, *ApJ*, **397**, 187

Bell, A.R. 1978, *MNRAS*, **182**, 147

Beskin, V.S. 1990, *Pis'ma Ast. Zh.*, **16**, 665 (*Soviet Astr. — Letters*, **16**, 330)

Bogovalov, S.V., and Kotov, Yu.D. 1992, *MNRAS*, **257**, 537

Brinkmann, W., Aschenbach, B., and Langmeier, A. 1985, *Nature*, **313**, 662

Chen, K.-Y., and Ruderman, M.A. 1993, *ApJ*, **408**, 179

Cheng, A.F., and Ruderman, M.A. 1977, *ApJ*, **216**, 865

Cheng, K.S., Ho, C., and Ruderman, M.A. 1986, *ApJ*, **300**, 500; *ibid.* , 522

Chiang, J., and Romani, R.W. 1994, *ApJ*, **436**, 754

Daugherty, J.K., and Harding, A.K. 1994, *ApJ*, **429**, 325

de Jager, O.C., and Harding. A.K. 1992, *ApJ*, **396**, 161.

Ellison, D.C., Jones, F.C., and Reynolds, S.P. 1990, *ApJ*, **360**, 702.

Emmering, R.T., and Chevalier, R.A. 1987, *ApJ*, **321**, 334

Gallant, Y.A., Hoshino, M., Langdon, A.B., Arons, J., and Max, C.E. 1992, *ApJ*, **391** 73

Fawley, W.M., Arons, J., and Scharlemann, E.T. 1977, *ApJ*, **217**, 227

Gallant, Y.A. and Arons, J. 1994, *ApJ*, **435**, 230

Gallant, Y.A., and Kirk, J. 1995, submitted to *Astr. and Ap.*

Goldreich, P., and Julian, W.H. 1969, *ApJ*, **157**, 869

Graham, J. 1995, submitted to *ApJ*

Holloway, N.J. 1973, *Nature (Phys. Sci.)*, **246**, 6

Hoshino, M., and Arons, J. 1991, *Phys. Fluids B*, **3**, 818.

Hoshino, M., Arons, J., Gallant, Y.A., and Langdon, A.B. 1992, *ApJ*, **390**, 454.

Jones, T.W., and Hardee, P.E. 1979, *ApJ*, **228**, 268.

Kennel, C.F., and Coroniti, F.V. 1984a, *ApJ*, **283**, 694.

Kennel, C.F., and Coroniti, F.V. 1984b, *ApJ*, **283**, 710.

Kundt, W., and Krotscheck, E. 1980, *A&A*, **83**, 1

Lampland, C.O. 1921, *Pub. A.S.P.*, **33**, 79

Langdon, A.B., Arons, J., and Max, C.E. 1988, *Phys.Rev.Lett.*, **61**, 779

Li, Z.-Y., and Arons, J. 1995, submitted to *ApJ*

Mestel, L., and Shibata, S. 1994, *MNRAS*, **271**, 621

Michel, F.C. 1974, *ApJ*, **192**, 713

Muslimov, A.G., and Tsygan, A.I. 1990, *Astr. Zh.*, **67**, 263 (*Soviet Astr. — AJ*, **34** 133; 1992, *MNRAS*, **255**, 61)

Pelling, R.M., Paciesas, W.S., Peterson, L.E., Makashima, K., Oda, M., Ogawara, Y. and Miyamoto, S. 1987, *ApJ*, **319**, 416.

Piddington, J.H. 1957, *Aust. J. Phys.*, **10**, 530.

Quenby, J.R., and Lieu, R. 1989, *Nature*, **342**, 654.

Rees, M.J., and Gunn, J.E. 1974, *MNRAS*, **167**, 1

Ruderman, M.A., and Sutherland, P.G. 1975, *ApJ*, **196**, 51

Ruderman, M.A. 1991, *ApJ*, **366**, 261

Scargle, J.D. 1969, *ApJ*, **156**, 401

Scharlemann, E.T., Arons, J., and Fawley, W.M. 1978, *ApJ*, **222**, 297

Smith, F.G. 1986, *MNRAS*, **219**, 729

Sturrock, P.A. 1971, *ApJ*, **164**, 529

Trümper, J. 1993, in *Texas/Pascos '92: Relativistic Astrophysics and Cosmology*, C.W Akerlof and M.A Srednicki, eds. (New York: New York Academy of Sciences), 260

van den Bergh, S., and Pritchet, C.J. 1989, *ApJ*, **343**, L69.

Ulmer, M. 1994, *ApJ (Supp.)*, **156**, 401

Vacanti, G., Cawley, M.F., Colombo, E., Fegan, D.J., Hilas, A.M., Kwok, P.W., Lang, M.J., Lamb, R.C., Lewis, D.A., Macomb, D.J., O'Flaherty, K.S., Reynolds, P.T., and Weekes, T.C. 1991, *ApJ*, **377**, 467

Veiler, K.W. 1978, *Mem. Soc. Astr. Ital.*, **49**, 545

INVERSE-COMPTON GAMMA RAYS FROM PLERIONS

ALICE K. HARDING
NASA/Goddard Space Flight Center
Greenbelt, MD 20771

Abstract. Young pulsars surrounded by supernova remnants can power synchrotron nebulae through the injection of relativistic particles. Inverse Compton scattering by the high-energy electrons and positrons can produce TeV gamma-ray emission strong enough to be detectable by ground-based telescopes. The Crab nebula is the archetypical example of a gamma-ray plerion and was the first detected TeV source. The observed spectrum is consistent with predictions of synchrotron-self Compton models. This paper will review such models for the Crab and other plerions. Inverse-Compton scattering on other soft photon sources, particularly the 2.7K microwave background, may also be detectable in older remnants.

1. Introduction

The source that has undoubtedly played the major role in the coming of age of TeV gamma-ray astronomy is the Crab nebula. The prototypical example of the class of sources known as plerions, or filled supernova remnants, the Crab was the first predicted (Gould 1965) source of TeV emission. It was also the first TeV source to be detected, at first with marginal significance (Fazio et al. 1972) and later confirmed at high significance with the advent of imaging Cherenkov telescopes (Weekes et al. 1989) . There are now two additional sources detected with significance of 10σ or greater and one of these, PSR1706-44 (Kifune et al. 1995), may be a plerion as well. Clearly this type of object will be important to study in future searches for sources of TeV emission.

Plerions such as the Crab nebula are thought to be powered by the rotational energy loss of a young pulsar. They typically emit a broad-band spectrum of synchrotron emission from radio to X-ray wavelengths which is probably due to relativistic electrons and positrons supplied by the central pulsar. Plerions form when the relativistic wind from a pulsar is confined by a more slowly expanding ($v_{\rm exp} \ll c$) shell of the surrounding supernova remnant. The spin-down energy of the pulsar may then be dissipated in a shock which accelerates the particles and randomizes their pitch angles. The emission at TeV energies from the Crab nebula can be explained as inverse Compton scattering of the synchrotron radiation by the relativistic particles in the plerion. Calculated synchrotron-self Compton spectra provide a very good fit to the observed γ-ray spectrum of the Crab (DeJager & Harding 1992).

There are about half a dozen other plerions associated with pulsars, all older than the Crab. But predictions of synchrotron-self Compton (SSC) fluxes from these sources indicate that they would not be detectable at TeV

Space Science Reviews **75**: 257–268, 1996.

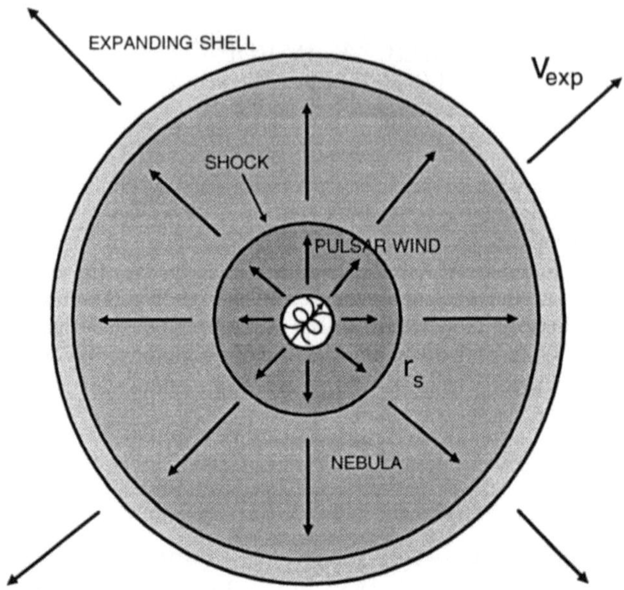

Fig. 1. Schematic illustration of the pulsar wind model of a plerion

energies. However, inverse-Compton scattering by the synchrotron emitting electrons on other sources of soft photons, including infrared radiation from dust and the 2.7K microwave background, may produce detectable fluxes (DuPlessis et al. 1994). As the equipartition magnetic fields decrease with age in plerions, the importance of the SSC flux will diminish relative to the inverse Compton flux from external soft photon sources. We will review the SSC model for the Crab nebula based on the pulsar wind model of plerions, as well as predicted TeV fluxes from inverse Compton scattering in other plerions.

2. Pulsar Wind Model

Most of the rotational energy loss of a pulsar appears in the form of a relativistic wind of electron-positron pairs. If the pulsar is surrounded by a supernova remnant, which expands at velocity $v_{\rm exp} \ll c$, then a reverse shock will form in the pulsar wind (see Figure 1). The location of this shock may be estimated (Rees & Gunn 1974) by balancing the wind ram pressure with the total nebular pressure (the accumulated energy input from the pulsar over its lifetime) to give

$$r_{\rm s} = v_{\rm exp}\,\tau \sqrt{\frac{v_{\rm exp}}{3c}} \tag{1}$$

where $r_{\rm s}$ is the shock radius (i.e. distance of the shock from the pulsar) and τ is the age of the pulsar. The shock radius in the Crab is $r_{\rm s} \approx 2\times 10^{17}$ cm or

8", which matches the size of an underluminous region observed surrounding the pulsar, where the particles upstream of the shock have zero pitch angle. The region downstream of the shock, between r_s and outer radius r_N, is the synchrotron nebula where the particles have been heated and accelerated by the shock.

The average magnetic field in the nebula is reasonably well estimated by the equipartition field, B_{eq}, found by assuming that the energy density of the magnetic field is half of the total pressure (equal to the pulsar wind ram pressure):

$$\frac{B_{eq}^2}{8\pi} = \frac{1}{2}\frac{L}{4\pi r_s^2 c}. \tag{2}$$

Here, L is the spin-down luminosity of the pulsar. This gives a value,

$$B_{eq} = 3 \times 10^{-4}\,\mathrm{G}\, B_{12}\, P^{-2}\, \tau_{yr}^{-1}\, v_8^{-3/2} \tag{3}$$

where $B_{12} \equiv B_o/10^{12}$ G is the pulsar surface field strength, P and τ_{yr} are the pulsar period and age in years, and $v_8 \equiv v_{exp}/10^8\,\mathrm{cm\,s^{-1}}$. For the Crab nebula, with $v_{exp} = 1500\,\mathrm{km\,s^{-1}}$, $B_{eq} \simeq 4 \times 10^{-4}$ G, in good agreement with $\langle B \rangle \approx 3 \times 10^{-4}$ G determined from the cooling break in the synchrotron spectrum between radio and optical (Marsden et al. 1984).

Building on the basic ideas of Rees & Gunn (1974), Kennel & Coroniti (1984a) developed an MHD wind model of the Crab nebula where the MHD flow solutions are determined by the shock jump conditions and by the nebular pressure and expansion velocity. They found that the wind is strongly particle dominated: the ratio σ of the magnetic field to particle energy density in the wind, upstream of the shock, is determined from their best fit solution to be $\sigma \approx .003$. They also derived the magnetic field distribution in the nebula as a function of radius, which depends primarily on the ratio σ. The field is a minimum at the shock and increases roughly as $B(r) \propto r$ until equipartition is reached, after which it decreases as $B(r) \propto 1/r$. The magnetic field at the shock is much smaller than equipartition value for small σ but for large values of $\sigma > 1$, the field approaches equipartition everywhere and $B(r) \propto 1/r$. As will be discussed in the following section, the magnetic field is extremely important in computing the inverse-Compton γ-ray spectrum from plerions, to the extent that the observed TeV flux from the Crab can significantly constrain the value of σ in the MHD nebular flow.

3. Inverse Compton Gamma Rays

The first step in computing the inverse Compton γ-ray spectrum from a plerion is to derive the spectrum of relativistic particles. This has been done for the Crab nebula by two different approaches: 1) obtain the electron spectrum by inverting the observed synchrotron spectrum, assuming a magnetic field distribution and 2) compute the electron spectrum through a kinetic equation that includes an assumed injection, radiation and adiabatic losses, as well magnetic field spatial distribution, matching to the observed synchrotron spectrum. Neither method provides a unique solution, and several assumptions are required in both cases.

The first approach was followed by Gould (1965), Grindlay & Hoffman (1971), Rieke & Weekes (1969) and DeJager & Harding (1992, DH). If we have good measurements of the synchrotron spectrum $F_{\rm syn}(\nu)$ over a broad energy range, which we do for the Crab (but not the other plerions), the electron spectrum $N_e(\gamma, r)$ can be obtained through inversion of the expression,

$$4\pi d^2 F_{\rm syn}(\nu, r) = \int P(\nu, r)\, N_e(\gamma, r)\, d\gamma, \tag{4}$$

where d is the distance to the source. There are two assumptions that are required to solve for $N_e(\gamma, r)$. The synchrotron spectral emissivity $P(\nu, r)$ must assume a nebular magnetic field distribution as a function of radius r, and the geometry of the nebular emission must be modeled in $F_{\rm syn}(\nu, r)$. Gould (1965) and Rieke & Weekes (1969) assumed a constant field strength throughout the nebula and a nebular size independent of frequency. Grindlay & Hoffman (1971) and DH introduced a radial dependence in the magnetic field and a frequency dependent size of the nebula, resulting in a radially dependent electron spectrum. DH assumed the field distribution from the MHD flow solutions of Kennel and Coroniti (1984a).

The second approach was first used by Stepanian (1980, 1990), who neglected adiabatic losses. More recently, Gallant & Kirk (1995) modified Kennel & Coroniti's (1984b) electron distribution derived from their MHD flow solution. Their computed SSC spectrum of the Crab agrees well with that of DH.

With an electron spectrum obtained by either method, the inverse Compton spectrum is calculated by

$$F(\epsilon) = \epsilon \int d\nu \int dr \int d\gamma\, U(\nu, r)\, N_e(\gamma, r)\, \frac{d^2 N_{\gamma,\epsilon}}{dt d\epsilon}, \tag{5}$$

where $U(\nu, r)$ is the density of soft photons and $d^2N_{\gamma,\epsilon}/dtd\epsilon$ is the production rate of scattered photons with energy ϵ. For each plerion, $U(\nu, r)$ will include several different soft photon sources: (1) synchrotron radiation (2) 2.7K microwave background radiation (3) infrared radiation from dust and (4) galactic infrared background radiation.

3.1. Synchrotron-Self Compton

Relativistic electrons scattering their own synchrotron radiation will give a synchrotron-self Compton (SSC) component of the inverse Compton γ-ray spectrum. Table 1 lists several plerions associated with pulsars whose ages and angular sizes give an estimate of their equipartition magnetic fields (from Eqn [3]). The Crab nebula is the only plerion with a presently observable SSC flux. LMC0540-69 has a larger value of B_{eq} but is 25 times more distant. Vela X, with a field 15 times smaller, has a predicted SSC flux about a tenth that of the Crab. However, older remnants with lower magnetic fields can have significant fluxes of inverse Compton scattering of external photon sources.

TABLE I
Plerions with Associated Pulsars

SNR	PSR	Age $\log(\tau_{yr})$	B_{eq} (Gauss)	U_B $(\mathrm{erg\,cm^{-3}})$	Dust U_{IR} $(\mathrm{erg\,cm^{-3}})$	ref.
CRAB	0531+21	3.1	3×10^{-4}	3.6×10^{-9}	5×10^{-13}	1
VELA X	0833-45	4.06	2×10^{-5}	1.6×10^{-11}	1×10^{-12}	2,3
MSH15-52	1509-58	3.19	1.8×10^{-5}	1.3×10^{-11}	5×10^{-14}	4,5,6
CTB80	1951+32	5	7×10^{-6}	1.6×10^{-13}	-	7,8
LMC0540-69	0540-69	3.22	8×10^{-4}	2.6×10^{-8}	-	9

1. Strom & Greidanus (1992) 2. Weiler & Panagia (1980) 3. Harnden et al. (1985) 4. Seward et al. (1983) 5. Seward et al. (1984) 6. Caswell et al. (1981) 7. Fruchter et al. (1988) 8. Safi-Harb et al. (1994) 9. Clark et al. (1982)

3.2. External Soft Photon Sources

Plerions that are more extended than the Crab, either because of higher initial expansion velocities as in MSH15-52 or because they are older and have had more time to expand, have lower magnetic fields. The energy density of external soft photon sources can become comparable to the magnetic energy density and can be a dominant source of inverse Compton (IC) emission (DeJager 1994, DuPlessis et al. 1994, Aharonian & Atoyan 1995).

The energy density in the 2.7K microwave background is $U_{2.7\mathrm{K}} = 4 \times 10^{-13}\,\mathrm{erg\,cm^{-3}}$, which actually exceeds the energy density of the equipartition field in some plerions (see Table 1). The ratio of the energy loss of a relativistic electron scattering 2.7K background photons relative to synchrotron losses is,

$$\frac{\dot{\gamma}_{\mathrm{IC}}(U_{2.7K})}{\dot{\gamma}_{\mathrm{syn}}} \sim 0.1 \left(\frac{B}{10^{-5}\,\mathrm{G}}\right)^{-2} \tag{6}$$

provided that $\gamma < 10^8$; otherwise higher-order corrections to the Thomson cross section or the full Klein-Nishina cross section must be used. In the

Thomson limit, electrons of energy γ will scatter microwave background photons to energy,

$$\epsilon \sim \gamma^2 (2.7K) = 3.8 \times 10^{12}\,\mathrm{eV}\, E_{\mathrm{keV}} \left(\frac{B}{10^{-5}\,\mathrm{G}}\right)^{-2}, \tag{7}$$

where E is the incident photon energy and the final expression relates the electron energy to the synchrotron characteristic energy. Therefore, the inverse-Compton flux at TeV energies will originate from electrons that radiate X-ray synchrotron photons. Thus, one can obtain an estimate of the inverse Compton flux of a plerion due to scattering of microwave background radiation from an observed X-ray synchrotron energy spectrum of the form, $F_{\mathrm{syn}} = S_{1\mathrm{keV}}\, E_{\mathrm{keV}}^{-\alpha_x}\,\mathrm{keV\,cm^{-2}\,s^{-1}\,keV^{-1}}$. The spectrum of the relativistic electrons can easily be obtained from the X-ray spectrum assuming a constant value of B and that $\dot\gamma_{\mathrm{IC}} < \dot\gamma_{\mathrm{syn}}$ (e.g. Lang 1974). The inverse-Compton scattered photon spectrum in the Thomson limit (Blumenthal & Gould 1970) is then (DeJager 1994, DuPlessis et al. 1995)

$$F_{\mathrm{IC}}(\epsilon) = 6.9 \times 10^{-17}\,(2.7K)^{\alpha_x} \left[\frac{F(p)}{(1\times 10^{-11})^{\alpha_x}\,\alpha(p)}\right] B^{-(\alpha_x+1)}\, S_{1\mathrm{keV}}\, \epsilon^{-\alpha_x} \quad (\mathrm{TeV\,cm^{-2}\,s^{-1}\,TeV^{-1}}) \tag{8}$$

The functions $\alpha(p)$ and $F(p)$ of the electron spectral index $p = 2\alpha_x + 1$ come from the inversion of the synchrotron spectrum and may be found in Lang (1974).

Several sources of infrared photons may also contribute to the IC flux of plerions. Some supernova remnants have swept up a significant amount of dust grains from the interstellar medium (Dwek & Arendt 1992). Collisions heat this dust to temperatures between 20K and 100K and the infrared emission from the dust can act as target photons for IC scattering by the relativistic electrons. The infrared energy density in plerions with associated IR sources are shown in Table 1. In some cases, U_{IR} may exceed $U_{2.7\mathrm{K}}$.

The galaxy as a whole is a source of diffuse infrared and optical emission. In the Solar neighborhood, this radiation field has an energy density of $U_{\mathrm{gal}} \simeq 10^{-13}\,\mathrm{erg\,cm^{-3}}$, so will contribute less than the microwave background radiation. However, near the galactic center the energy density of the diffuse IR emission increases to $U_{\mathrm{gal}} \simeq 10^{-11}\,\mathrm{erg\,cm^{-3}}$, which may exceed U_{B} in many cases.

As may be seen from Eqn (8), the IC flux from a thermal soft photon source will depend on plerion magnetic field as $F_{\mathrm{IC}} \propto B^{-(p+1)/2}$. Because F_{IC} has a weaker dependence on field strength than the SSC flux component, the IC flux from external photon sources will dominate for older plerions with weaker B_{eq}. When the external soft photon density is larger than U_{B}, the energy losses from inverse Compton scattering exceed that from synchrotron radiation. In this case, the IC losses should be taken into account when deriving the electron spectrum.

4. SSC Model of the Crab Nebula

The SSC spectrum of the Crab nebula can be modeled quite accurately because the synchrotron spectrum has been defined over a wide frequency range from radio to high energy γ-rays. There are three breaks in the observed synchrotron spectrum at $\approx 1.8 \times 10^{13}$ Hz, $\approx 9 \times 10^{15}$ Hz and $\approx 3.6 \times 10^{19}$ Hz. There is also a dust component that contribute about 50% of the total IC flux at 1 TeV. Figure 2 shows the observed γ-ray spectrum of the Crab nebula with model fits including both a synchrotron component and the computed IC spectrum. The contribution from scattering on the 2.7K microwave background is included and is important above $\sim$ 1 TeV (see also, Aharonian & Atoyan 1995). The synchrotron cutoff at 60 MeV was determined theoretically by equating the synchrotron loss time at the pulsar wind shock with a minimum acceleration timescale, and including a Doppler boosting factor due to the wind particles moving at $v \sim c/\sqrt{3}$ and 30^0 to our line of sight. The EGRET measurement of the spectrum in the energy range 70 MeV to 10 GeV defines the transition from the synchrotron to the IC components of the emission with a high degree of significance (De Jager et al. 1995). The spectrum in this range can be fit to two combined power laws having indices $-4.3^{+1.2}_{-0.6}$ and -1.6 ± 0.36.

The SSC model spectra depend on the parameter σ from the MHD flow model of Kennel & Coroniti (1984a), and thus on the magnetic field distribution in the nebula. The measurements in the TeV range give a good lower limit on $\sigma \gtrsim .003$, but the model flux becomes relatively insensitive to higher values of σ. This is because the IC emission at TeV energies and below originates from the entire nebula and thus measures the average magnetic field strength, which is only weakly dependent on σ. A more accurate determination of σ would be provided by measurements above 10 TeV, which begin to probe the inner parts of the nebula nearer the shock, where the field strength depends more strongly on σ. However, measurements at TeV energies with $1'$ resolution could differentiate between IC models with $\sigma \sim 1$, where $B(r) \propto 1/r$ in which case the nebula would be resolved, and $\sigma \ll 1$, where the nebula would have a much smaller angular size (DH).

5. Inverse Compton Emission from PSR1706-44?

A TeV source with flux $1 \times 10^{-11}\,\mathrm{ph\,cm^{-2}\,s^{-1}}$ has been detected at the location of the gamma-ray pulsar PSR1706-44 by the CANGAROO telescope (Kifune et al. 1995) with a significance of $\sim 10\sigma$. Since the emission does not appear to be pulsed at the period of the pulsar, is it possible that it

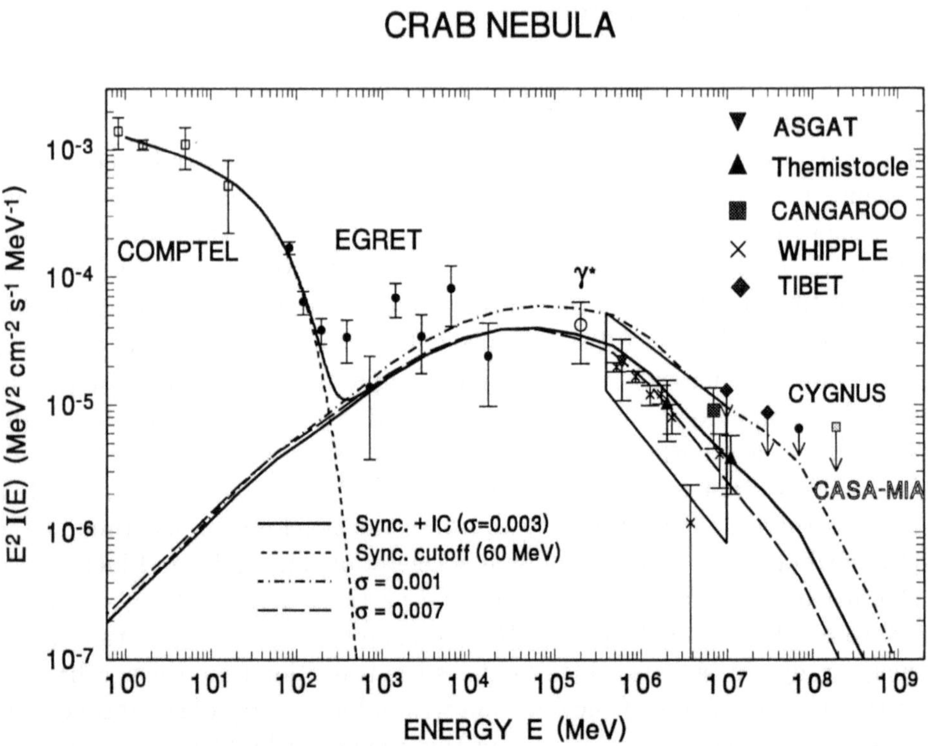

Fig. 2. Gamma-ray spectrum of the Crab nebula from De Jager et al. 1995. Data is from COMPTEL: Much 1994, EGRET: DeJager et al. 1995, WHIPPLE: Lewis et al. 1993, γ^*: Akerlof et al. 1989, TIBET: Amenomori et al. 1993, CYGNUS: Alexandreas et al. 1993, CASA-MIA: Borione et al. 1993, ASGAT Goret et al. 1993, Themis: Djatanni-Atai 1994, CANGAROO: Tanimori et al. 1994

could be IC radiation from a surrounding plerion? There are several recent measurements of this source at other wavelengths that suggest that this could be the case. Becker et al. (1994) reported the detection of an unpulsed X-ray source with ROSAT at the position of the pulsar and their power law spectral fit gives $F_{\rm x}(E) = 2 \times 10^{-3}\, E_{\rm keV}^{-1.4}\,{\rm keV\,cm^{-2}\,s^{-1}\,keV^{-1}}$. In addition, a compact radio nebula surrounding PSR1706-44 has been detected at 20 cm by Frail et al. (1994). If we interpret the X-ray source as synchrotron emission from a plerion, we can estimate the IC flux from scattering of the relativistic electrons on the 2.7K microwave background. From Eqn (8),

$$F_{\rm IC}(\epsilon) = 0.94 \times 10^{-11} \left[\frac{5 \times 10^{-6}\,{\rm G}}{B}\right]^{2.4} \epsilon^{-1.4}\,{\rm TeV\,cm^{-2}\,s^{-1}\,TeV^{-1}}. \qquad (9)$$

Since the expansion velocity of the compact radio nebula is not known, we cannot compute the equipartition field of the plerion, and so B is left as

TABLE II

TeV Flux from Scattering of 2.7K Microwave Background

SNR	B_{eq} (Gauss)	S_{1keV} ($cm^{-2}\,s^{-1}$)	α_x	$F_{2.7K}$(1TeV) ($ph\,cm^{-2}\,s^{-1}$)
CRAB	3×10^{-4}	5.53	1.15	1.7×10^{-12}
VELA X	2×10^{-5}	0.133	1.3	1.6×10^{-11}
MSH15-52	1.8×10^{-5}	0.079	1.15	7.6×10^{-11}
CTB80	7×10^{-6}	3.45×10^{-3}	1.1	2.8×10^{-12}
LMC0540-69	8×10^{-4}	0.019	0.8	1.1×10^{-15}
PSR1706-44[a]	5×10^{-6}	0.002	1.4	1×10^{-11}
SGR1806-20	4×10^{-6}	6.25×10^{-3}	1.2	2.4×10^{-11}

[a] B_{eq} needed to give observed flux.
References for S_{1keV} and α_x are given in Table I or in the text.

a free parameter in the estimate of F_{IC}. From this estimate, however, it seems possible that IC scattering can account for the observed TeV flux if $B \lesssim 5 \times 10^{-6}$ G. An equipartition field of this value is consistent with an expansion velocity $v_{exp} > 1100\,km\,s^{-1}$ from Eqn (3). However, since $\dot{\gamma}_{IC} \gtrsim \dot{\gamma}_{syn}$ in this source, one should also include IC losses in deriving the electron spectrum, which may lower the above estimate of F_{IC}. Since, from Eqns (9) and (8), the index of the TeV spectrum is predicted to be the same as that of the observed X-ray spectrum (assuming IC losses are not important), a measurement of the spectrum of PSR1706-44 in the TeV range will be a critical test of the inverse Compton emission model.

6. MSH15-52 and Other Plerions

Several plerions other than the Crab have well enough measured synchrotron spectra over at least part of the waveband from radio to X-rays to compute their inverse Compton γ-ray spectrum. As discussed in Section (3.2), the IC component from scattering on 2.7K microwave background radiation alone requires a measurement of only the X-ray spectrum around 1 keV. Table II lists some of the plerions having measured X-ray energy spectra, $F_X = S_{1keV}\,E_{keV}^{-\alpha_x}\,keV\,cm^{-2}\,s^{-1}\,keV^{-1}$, and estimates of the IC flux from scattering of the synchrotron emitting electrons on the microwave background using Eqn (8). This flux is a lower limit on the IC flux from each plerion, because there can be other sources of soft photons in addition to the microwave

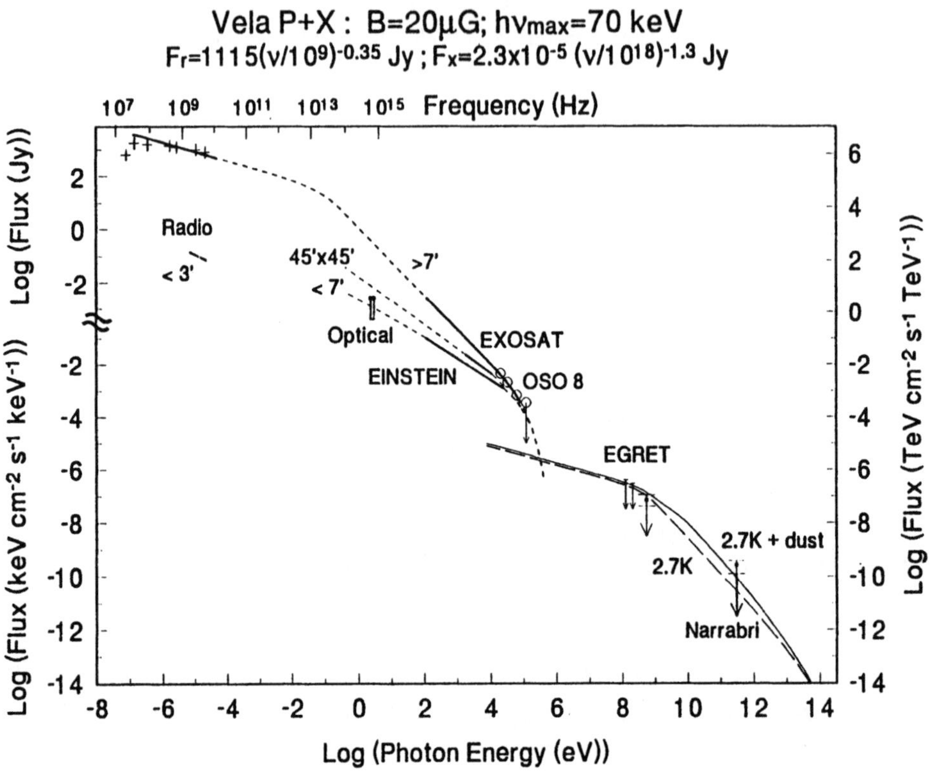

Fig. 3. Broad-band spectrum of the VELA X nebula (De Jager & Harding 1995). The dashed and solid lines above 1 MeV are the IC spectrum calculated from the observed synchrotron spectrum, interpolated through the optical band.

background. However, in several sources having low magnetic fields such as MSH15-52, $F_{2.7K}(1\text{TeV})$ alone gives a detectable predicted flux.

MSH15-52 is associated with the pulsar 1509-58, young as determined by its spin-down age but much more extended than the Crab, possibly due to a larger value of v_{exp}. Consequently, the equipartition field as derived from Eqn (3) is only 1.8×10^{-5} G. Du Plessis et al. (1995) have computed the IC spectrum of MSH15-52 and find that the microwave background component exceeds the dust component (the SSC component is negligible due to the low field). The synchrotron spectrum is measured only in the radio and X-ray regions, so they made some assumptions to interpolate the spectrum in the optical. They find that the IC flux at TeV energies is detectable if the magnetic field is near or below the equipartition field.

VELA X also has B_{eq} much less than the Crab and therefore an SSC component too low to be detectable. However, in this plerion a surrounding IRAS dust shell could be the dominant and detectable component of the IC spectrum at TeV energies. Figure 3 shows the measured spectrum of the

VELA X nebula with the calculated IC flux (from $10^3 - 10^{14}$ eV), assuming an interpolation between the radio and X-ray synchrotron spectra (DeJager & Harding 1995). The 2.7K background component is not detectable in this source. Detection of TeV emission from VELA X will enable us to study the scattering of dust in plerions and possibly put more accurate constraints of the distance to the source, i.e. its location relative to the dust shell.

The source SGR1806-20 is a compact X-ray source recently detected by ASCA (Murakami 1993) at the position of an X-ray and γ-ray burst source. Although there is no pulsar associated with this source, the steady X-ray luminosity, age and size give an estimate of $B_{eq} = 4 \times 10^{-6}$ G, which yields a detectable TeV flux. The distance to the source is not well determined, but if it is near the galactic center, U_{gal} will be high enough to increase the IC flux significantly beyond than in Table II.

7. Summary

Observations at TeV energies has opened an important window for the study of plerions. The well-known sources such as the Crab have for decades raised intriguing questions concerning neutron star evolution and energetic particle production by pulsars. Since they efficiently trap and radiate nearly the entire spin-down luminosity of the compact source, we obtain information on the fraction of the energy input appearing as particles and that appearing as magnetic field (i.e. the composition of the pulsar wind), the field distribution in the nebula and how closely equipartition is achieved.

The predicted IC emission from known plerions indicate that several in addition to the Crab may be detectable at TeV energies by existing atmospheric Cherenkov telescopes. In plerions whose γ-ray emission is dominated by IC scattering on external soft photon sources, we may study the effects of this interaction on the spectrum and energy losses of the relativistic electrons. For one of these, MSH15-52, the dominant emission component may be due to scattering on the microwave background, a soft photon source that is well determined. In this case, a measured flux and spectrum will yield information on the relativistic electron spectrum that is independent of assumptions about the soft photon density. Plerions with very low magnetic fields, where IC scattering on external photon sources dominates both the emission and the particle energy loss, will form a new class of source, Compton nebulae.

8. Acknowledgements

I thank M. G. Baring and O. C. De Jager for valuable discussions and input.

References

Aharonian, F. A. & Atoyan, A. M. 1995, A & A, in press.
Akerlof, C. W. et al. 1989, Proc. GRO Workshop (GSFC), 4-49.
Alexandreas, D. E. et al. 1993, Proc. 23rd Int. Cosmic Ray Conf. (Calgary), 1, 353.
Amenomori, M. et al. 1993, Proc. 23rd Int. Cosmic Ray Conf. (Calgary), 1, 342.
Becker, W., Brazier, K. T. S. & Trumper, J. 1994, A & A, submitted.
Blumenthal, G. R. & Gould, R. J. 1970, Rev. Mod. Phys., 42, 237.
Borione, A. et al. 1993, Proc. 23rd Int. Cosmic Ray Conf. (Calgary), 1, 286.
Caswell, J. L., Milne, D. K. & Wellington, K. J. 1981, MNRAS, 195, 89.
Clark, D. H. et al. 1982, ApJ, 255, 440.
De Jager, O. C. 1994, Currents in High Energy Astrophysics (ERICE), ed. M. M. Shapiro, R. Silberberg & J. P. Wefel (Dordrecht) in press.
De Jager, O. C. & Harding, A. K. 1992, ApJ, 396, 161.
De Jager, O. C. & Harding, A. K. 1995, in preparation.
De Jager, O. C. et al. 1995, ApJ, submitted.
Djatanni-Atai, A. 1994, Ph.D. Thesis, Univ. P. et M. Curie (Paris).
DuPlessis, I. et al. 1995, ApJ, in press.
Dwek, E. & Arendt, R. G. 1992, Ann. Rev. Astr. Astrophys., 30, 11.
Fazio, G. G. et al. 1972, ApJ, 175, L117.
Frail, D. A., Goss, W. M. & Whiteoak, J. B. Z. 1994, ApJ, submitted.
Fruchter. A. S. et al. 1988, Nature, 331, 53.
Gallant, Y. & Kirk, J. G. 1995, this proceedings.
Goret, et al. 1993, A & A, 270, 401.
Gould, R.J. 1965, Phys. Rev. Lett., 15, 577.
Grindlay, J.E. and Hoffman, J.A. 1971, Astrophys. Lett, 8, 209.
Harnden, F. R. et al. 1985, ApJ, 299, 828.
Kennel, C.F. and Coroniti, F.V. 1984a, ApJ, 283, 694.
Kennel, C.F. and Coroniti, F.V. 1984b, ApJ, 283, 710.
Kifune, T. et al. 1995, ApJ, 438, L91.
Lang, K.R. 1974, Astrophysical Formulae, Springer-Verlag (New York, Heidelberg, Berlin), p. 35.
Lewis, D. A. et al. 1993, Proc. 23rd Int. Cosmic Ray Conf. (Calgary), 1, 279.
Marsden, P.L., Gillett, F.C., Jennings, R.E., Emerson, J.P., de Jong, T., and Olnon, F.M. 1984, Astron. Astrophys., 278, L29.
Much, R. P. 1994, PhD Thesis, Technischen Universitat Munchen.
Murakami, T. et al. 1993, Nature, 368, 127.
Rees, M.J. and Gunn, J.E. 1974, M.N.R.A.S., 167, 1.
Rieke, G.H. and Weekes, T.C. 1969, ApJ, 155, 429.
Safi-Harb, A., Ogelman, H. & Finley, J. P. 1994, ApJ, submitted.
Stepanian, A.A. 1980, Izvestiya Krymskoi Astrofizicheskoi Observatorii, 62, 79.
Stepanian, A.A. 1990, in Proc. Vulcano Workshop 1990: Frontier Objects in Astrophysics and Particle Physics, Italian Physical Society, Bologna - Italy, p.377.
Seward, F. D. et al. 1983, ApJ, 267, 698.
Seward, F. D. et al. 1984, ApJ, 281, 650.
Strom, R. S. & Greidanus, H. 1992, Nature, 358, 654.
Tanimori, T. et al. 1994, ApJ, 429, L61.
Weekes, T.C. et al. 1989, ApJ, 342, 379.
Weiler, K. W. & Panagia, N. 1980, A & A, 90, 269.

SHELL TYPE SUPERNOVA REMNANTS

L O'C DRURY
Dublin Institute for Advanced Studies
School of Cosmic Physics
5 Merrion Square, Dublin 2
Ireland <ld@cp.dias.ie>

Abstract. The possibility of observing gamma ray emission from supernova remnants is discussed. It is shown that this could be possible in the 100 MeV band accessible to satellite instruments, but that confusion with the Galactic background is a major problem. At TeV energies and with modern imaging atmospheric cherenkov telescopes the situation should be much better and at least some of the nearby remnants may be detectable. Positive detections in both bands would provide a decisive test of current theoretical ideas on particle acceleration in supernova remnants and the origin of the Galactic cosmic rays.

1. Prolegomena

One of the main motivating factors in the early days of gamma-ray astronomy was the hope of directly detecting cosmic ray sources in our Galaxy (Pinkau 1970). However this appears to have been rather lost sight of in the recent development of gamma-ray astronomy. One influence was undoubtedly the ambiguities encountered in the interpretation of the COS-B data which concentrated attention more on problems associated with the global distributions of gas and cosmic rays in the Galaxy. Another was (and is) the excitement generated by extragalactic sources. However the problem of cosmic ray origin is still far from solved despite nearly three generations of work; it would be a major triumph for gamma-ray astronomy if it could contribute towards its resolution.

The purpose of this talk is to review some recent work on the currently most popular model for the origin of the bulk of the Galactic cosmic rays, namely that they are produced by diffusive shock acceleration in the shock waves associated with supernova remnants, and the resulting implications for gamma-ray production in supernova remnants. The detailed arguments have been published recently (Drury, Aharonian and Völk, 1994; see also Aharonian, Drury, Völk 1994) and essentially identical conclusions have been presented by Naito and Takahara (1994). In fact the basic ideas are not particularly original. Very similar calculations of the gamma-ray emission from supernova remnants, but without comparison with instrumental sensitivities, have been presented by Dorfi (1990, 1991).

2. Introduction

Why might one expect supernova remnants (SNRs) to be associated with cosmic ray production? The main reason is a simple argument about the

Space Science Reviews **75**: 269–277, 1996.

energy required to sustain the galactic cosmic ray (GCR) population against loss by escape, nuclear interactions and ionization energy loss. This is easily estimated to be of order 10^{41} erg s^{-1} with an uncertainty of perhaps a factor two or three. The mechanical energy input to the Galaxy from each supernova (SN) explosion is canonically taken to be 10^{51} erg so that with a rate of about one every 30 years the total mechanical power input from SNe is of order 10^{42} erg s^{-1}. Thus the Galactic SNe have enough power to drive the GCR acceleration mechanism if there exists a mechanism for channeling about 10% of the mechanical energy into relativistic particles. It is in fact very hard to find any other plausible energy source to drive the acceleration.

However it is important to note that the SNe cannot directly produce the GCR either in or immediately after the explosion because of the "adiabatic loss" problem. Were a significant amount of the explosion energy converted quickly to relativistic particles in a small volume surrounding the explosion site, one would have a bubble containing a very high pressure of relativistic particles. In attempting to stream away from the explosion site these particles would excite plasma instabilities which couple them to the surrounding medium so that, in expanding, the cosmic ray bubble has to do pressure work. The result is that the particles suffer adiabatic losses and their energy decreases in inverse proportion to the radius of the bubble. Clearly if the initial trapping radius is less than 10% of the radius at which the bubble comes into pressure equilibrium with the interstellar medium, an overall conversion efficiency of 10% is impossible. This is one reason why, in discussions of cosmic ray origin, interest has shifted from supernovae to the resultant supernova remnants.

The high velocity ejecta produced in the supernova explosion (typically a few solar masses moving at a few percent of the speed of light) interact with the ambient medium to produce a system of strong shocks. The shocks in turn can pick up a few particles from the plasma flowing into the shock fronts and accelerate them by the diffusive shock acceleration process to high energies. This is the currently most popular theory for the origin of the GCR. However it is important to emphasise that, at least with conventional parameters for the interstellar medium, the process can only accelerate protons to energies of about 10^{14} eV. The cosmic ray spectrum continues however to much higher energies, so that there must, on this picture, be a second source which takes over (and takes over smoothly!) just before the so-called "knee" at around 10^{15} eV. Also, although the SNR shock acceleration model can explain the power and spectrum of the GCR at lower energies, there are odd features of the chemical composition (the so-called FIP biases) which do not seem to fit in although the overall composition is reasonably well explained.

Thus, although a pretty theoretical picture, shock acceleration in SNRs certainly does not explain everything about the GCR. In addition, there is no strong direct evidence for the acceleration of nuclei (as distinct from electrons) in SNRs. However if SNRs do contain significant acclerated proton populations (and other nuclei) these will inevitably interact with the swept-up interstellar matter in the remnants to produce π^0 gamma-rays. Thus one critical test of the SNR shock acceleration model for the origin of cosmic rays is to calculate the associated gamma-ray production and see whether this should be observable.

3. Evolution of SNRs

It is conventional to describe the evolution of a SNR in terms of three consecutive phases. In the first phase the ejecta expand more or less ballistically into the surrounding medium. Ahead of the ejecta a strong shock is driven into the ambient medium. The high pressure behind this "forward" shock drives a second "reverse" shock into the ejecta, but although travelling into the ejecta the velocity of this shock at early times is much less than the velocity with which the ejecta themselves are expanding; a simple momentum balance argument shows that the velocity ratio will be the square root of the density ratio between the ejecta and the ambient medium. Thus this system of forward and reverse shocks is carried outwards by the general expansion of the remnant. This ballistic or "free expansion" phase ends when the total amount of material swept up by the forward shock has a mass comparable to that of the ejecta, or equivalently the mean density of the ejecta drops to that of the surrounding medium. As this condition is approached the reverse shock accelerates and implodes inwards through the ejecta heating them.

The remnant evolution then continues through a phase where the total mass in the remnant is much greater than that ejected in the explosion, but the external pressure is very much less than the internal pressure. The evolution is then very close to the self-similar solution for a strong point explosion in a cold gas obtained analytically by Sedov (and numerically by von Neumann and Taylor). This phase is usually called the Sedov phase. It lasts until either the external pressure becomes significant and the shock weakens, or radiative cooling becomes important inside the remnant. In the final dying phase the shock weakens and the remnant breaks up and merges into the general interstellar medium.

This simple conventional picture is probably quite a good description of the evolution of the remnant formed by the explosion of a type Ia supernova, though to represent the deflagration of a low-mass star. The picture is certainly more complicated in the case of type Ib and II SNe. Here the progenitor is thought to be a high mass star which has undergone considerable mass-loss prior to the explosion. The ejecta then do not expand into a relatively uniform unstructured ISM but into the wind bubble blown by the progenitor. In addition the explosion may lead to the formation of a pulsar which then acts as an internal energy source for the remnant. In this talk I will ignore these complications and concentrate on simple shell-type SNRs.

It is instructive to look at the mechanical power processed in the SNR shocks as a function of the age of the remnant. For a spherical shock of radius R with material of density ρ flowing into the shock at velocity V the mechanical power is $\Phi = 2\pi\rho R^2 V^3$. In the first ballistic phase the forward shock velocity is almost constant so that, apart from any fluctuations resulting from density variations in the surrounding medium, the forward shock power increases as t^2 where t is the remnant age; a crude estimate gives

$$\Phi_{\rm F} \approx \frac{3E_{\rm SN}}{t_{\rm SW}}\left(\frac{t}{t_{\rm SW}}\right)^2$$

where $E_{\rm SN}$ is the mechanical energy of the SN explosion and $t_{\rm SW}$ is the "sweep-up" time when the first phase ends. Thus an alternative interpretation could be that the first ballistic phase ends when an amount of energy equal to the explosion energy has been processed through the forward shock. Assuming pressure balance the reverse shock power $\Phi_{\rm R}$ is related to that of the forward shock by

$$\Phi_{\rm R} \approx \left(\frac{R_{\rm R}}{R_{\rm F}}\right)^2 \left(\frac{\rho_0}{\rho_{\rm ej}}\right)^{1/2}$$

where $R_{\rm R}$ is the radius of the reverse shock, $R_{\rm F}$ that of the forward shock, ρ_0 the density of the ambient medium into which the forward shock is propagating and $\rho_{\rm ej}$ that of the ejecta into which the reverse shock is propagating. Thus, with the possible exception of the very end of the ballistic phase, the power of the reverse shock is less than that of the forward shock. Simple estimates suggest that the total amount of energy processed through the reverse shock during the ballistic phase is probably no more than 50% of $E_{\rm SN}$.

During the subsequent Sedov phase there is only the one forward shock, and its power decreases as

$$\Phi \propto \frac{E_{\rm SN}}{t}$$

(this follows from the self-similarity of the solution). The numerical Sedov solution shows that the constant of proportionality is quite close to one so that we can take $\Phi = E_{\rm SN}/t$. Thus the total amount of energy processed through the shock in the Sedov phase is of order $\ln(t/t_{\rm SW})E_{\rm SN}$, or for typical remnants which enter the Sedov phase after a few hundred years and last until several 10^4 y about $5E_{\rm SN}$.

In summary therefore we expect, in units of the explosion energy, that the forward shock processes about $E_{\rm SN}$ during the ballistic phase and about $5E_{\rm SN}$ during the subsequent Sedov phase while the reverse shock processes perhaps $0.5E_{SN}$ during the ballistic phase. Both the forward and reverse shock are at their peak power towards the end of the ballistic phase. Note that there is no violation of energy conservation even though the total amount of energy processed through the shocks is about 6.5 times the explosion energy; the energy is simply being "recycled" several times through the shock system.

4. Particle acceleration in SNRs

The key points to notice from the above discussion are that there is very little energy available for shock driven acceleration of energetic particles until near the end of the first ballistic phase and that the peak power occurs roughly at the transition from the first phase to the Sedov like phase. Models of diffusive shock acceleration in SNRs can be calculated in considerable detail (see eg Kang and Jones 1991, Markiewicz et al 1990, Berezhko et al 1994 and references therein). Although all the detailed models still contain substantial *ad hoc* parametrizations of parts of the physics the main qualitative features of the solutions are easily understood and appear to be relatively robust predictions. The important quantity, from the point of gamma-ray production, is the amount of energy transferred to accelerated particles in the remnant. If we define this to be a fraction ϑ of the total explosion energy E_{SN}, then the detailed model calculations show that in general ϑ is small during the ballistic phase, then rises rapidly at the start of the Sedov phase to a value of order 0.2–0.7 and remains roughly constant during the remainder of the Sedov phase (the ongoing acceleration just balancing the adiabatic losses). The main uncertainty relates to the so-called "injection" of particles into the acceleration process. If this is set to a very low value ϑ remains small, but then this is not a viable model for cosmic ray production because the Galactic cosmic ray power requires $\vartheta \geq 0.1$ when the remnants die. It is possible to produce rather contrived models where the injection and subsequent acceleration is delayed until towards the end of the Sedov phase, but only by assuming that the injection is strongly suppressed at high Mach number shocks. Thus the general picture, of $\vartheta \approx 0.3$ during the Sedov phase and $\vartheta \approx 0$ in the ballistic phase appears to be a rather robust and model independent prediction of the idea that the GCR are accelerated by shock acceleration in SNRs.

As well as the total energy transferred, the other important aspect of the model calculations is that the acclerated particle energy spectra are predicted to be hard. The differential energy spectra are not exact power-laws, but approximate to $E^{-2.1}$ between proton energies of about 1 GeV and 10^{14} eV (or perhaps even harder as in the calculations of Berezhko et al 1994). This should be compared with a spectral exponent of about 2.6 for the interstellar GCR population in the solar neighbourhood. In fact propagation models for the GCR independently require the production spectra in the sources to be harder to fit the observed energy dependence of the relative abundance of secondary spallation nuclei in the GCR.

Finally, the accelerated particle spectra do not extend to arbitrarily high energies. There is an upper cut-off to the spectrum determined by the finite size and age of the remnant. Assuming that the diffusion coefficient of the accelerated particles in the neighbourhood of the shock is driven down to a value close to the Bohm limit by strong wave excitation, the upper cut-off energy rises linearly during the ballistic phase, and then remains more or less constant during the subsequent Sedov phase at a value of order a few times 10^{14} eV (for protons).

5. Gamma-ray production in SNRs

If SNRs contain relativistic protons (and other nuclei; but the protons are the energetically dominant component) there will inevitably be nuclear interactions between the accelerated particles and the nuclei of the ejecta and the swept-up interstellar matter. These interactions, at energies above about 1 GeV, produce pions and the neutral pions decay to produce gamma rays. In addition, there will of course be bremsstrahlung and inverse Compton radiation both from the directly accelerated electrons and from secondary electrons produced by the charged pion decays. In principle this gamma ray production associated with the electron component can be estimated from the radio synchrotron emission, and indeed the combination of gamma ray and radio data may be a powerful way of constraining the magnetic fields in SNRs (cf Cowsik and Sarkar). However we are interested here in evidence for proton acceleration and will thus concentrate on the π^0 channel.

We define an emissivity factor $q_\gamma(E)$ by writing the production rate of gamma rays above some threshold energy E per unit volume as

$$Q_\gamma = q_\gamma(E) n E_C$$

where n is the number density of the target atoms and E_C is the cosmic ray energy density. Clearly the exact value of q_γ depends on the assumed energy spectrum of the cosmic rays. For power-law energy spectra the emissivity factors can be calculated using the method of spectrum weighted moments (eg Gaisser 1990). It turns out that for production above 100 MeV and for power law spectra with exponents in the range expected in SNRs

$$q_\gamma \approx 0.5 \times 10^{-13} \,\mathrm{photons\, s^{-1} erg^{-1} cm^3 (H\text{-}atom)^{-1}}$$

with a dependence on spectral index of no more than about 20%. However this value is about a factor of three lower than that conventionally used for the ISM, the difference being entirely due to the assumed low-energy form of the interstellar GCR spectrum. At higher energies the spectrum of the π^0 gammas mirrors that of the parent particles and there is thus a strong dependence on the spectral slope.

The total gamma ray luminosity of a SNR is simply $\int q_\gamma n E_C d^3r$ which, because the cosmic ray energy density is rather smoothly distributed, can be approximated as

$$L_\gamma \approx q_\gamma \vartheta E_{\rm SN} n$$

where n is the mean number density of the remnant (which in the Sedov phase will be almost identical to the mean density of the medium into which the remnant is propagating). The target matter is concentrated in a shell immediately behind the outer shock so that with sufficient angular resolution a SNR in gamma rays should appear as a limb brightened shell.

Inserting numerical values, for a SNR at a distance d away the gamma ray flux expected at the Earth is

$$F(\geq 100\,{\rm MeV}) \approx 4.4 \times 10^{-7} \vartheta \left(\frac{E_{\rm SN}}{10^{51}{\rm erg}}\right) \left(\frac{d}{1\,{\rm kpc}}\right)^{-2} \left(\frac{n}{1\,{\rm cm}^{-3}}\right) {\rm cm}^{-2}\,{\rm s}^{-1}.$$

For the same parameters and assuming a differential energy spectrum approximately proportional to $E^{-2.1}$ inside the remnant the high energy gamma ray flux expected in the TeV region is

$$F(\geq E) \approx 9\times 10^{-11} \vartheta \left(\frac{E}{1\,{\rm TeV}}\right)^{-1.1} \left(\frac{E_{\rm SN}}{10^{51}{\rm erg}}\right) \left(\frac{d}{1\,{\rm kpc}}\right)^{-2} \left(\frac{n}{1\,{\rm cm}^{-3}}\right) {\rm cm}^{-2}\,{\rm s}^{-1}.$$

6. Observability of SNRs in gamma-rays

The key question of course is whether these predicted flux levels are likely to be observable. In fig 1 the expected integral flux from a "typical" near-by SNR is compared with the various backgrounds and instrumental sensitivities.

We see that in the 100 MeV band accessible to satellite instruments such as the EGRET on the Compton-GRO SNRs are potentially observable, but that the general Galactic background is a major problem. In this connection it is interesting to note that Sturner and Dermer (1995) have recently made a statistical argument identifying some of the unidentified Galactic plane point sources with SNRs. For one of their more probable associations, that of the shell-type remnant IC 443 with GRO J0617+22 the observed flux (Fichtel et al, 1994) is of order $3 \times 10^{-7}\,{\rm cm}^{-2}{\rm s}^{-1}$ in reasonable agreement with our estimates if, as appears probable, this remnant is expanding into a relatively dense region. There have also been interesting, but not entirely convincing, claims for cosmic ray enhancements in the Loop-1 region, though to be an old nearby SNR (Bhat et al, 1985; Lebrun and Paul, 1985; Wolfendale and Zhang, 1994).

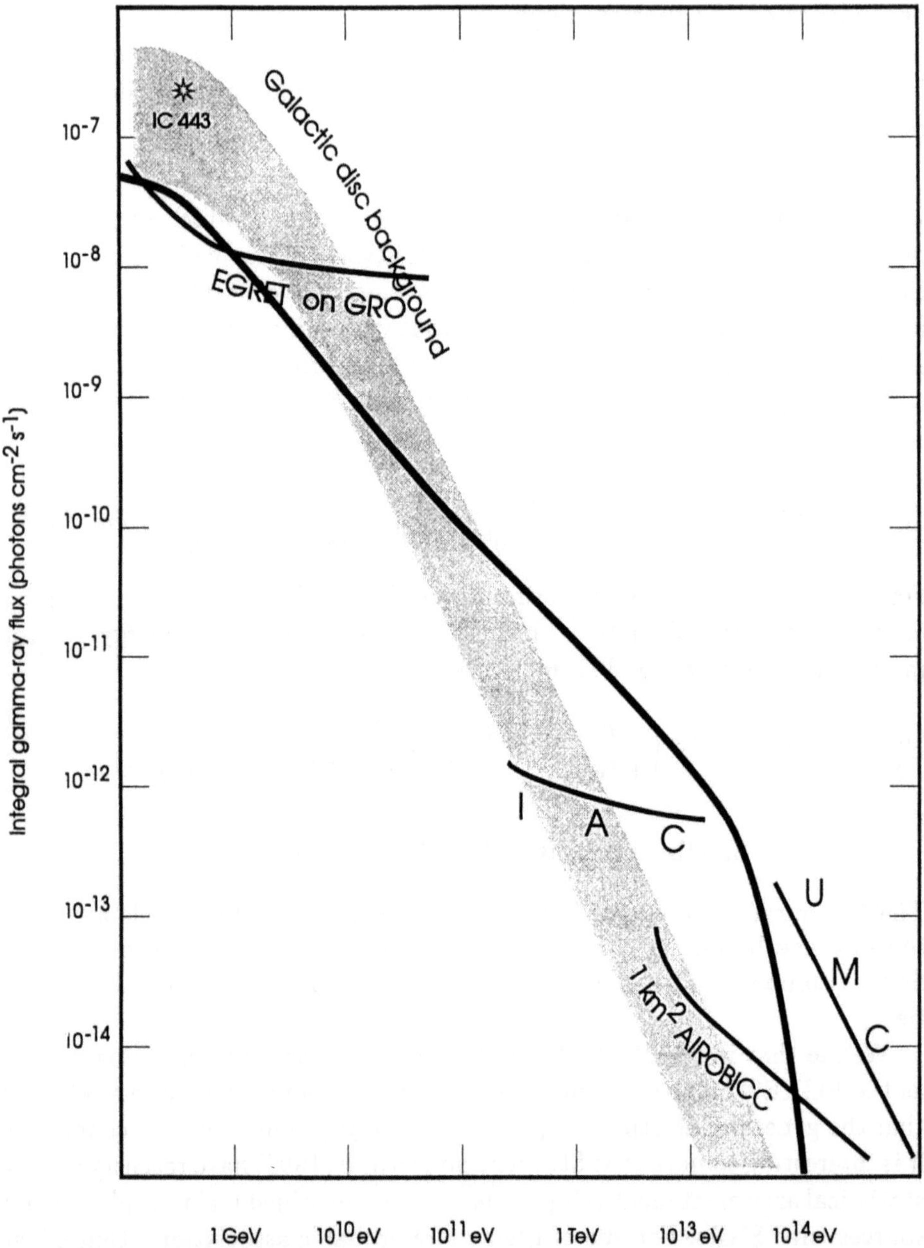

Fig. 1. The expected integral gamma ray flux from a SNR at 1 kpc with the canonical energy of 10^{51} ergs expanding into a medium of number density $1\,\mathrm{cm}^{-3}$ with conversion efficiency $\vartheta = 0.1$ and spectral exponent of 2.1. Also indicated is the approximate level of the general Galactic disc emission within an acceptance angle of 1°, the sensitivity level of imaging atmospheric Cherenkov telescopes, of the UMC air shower array of the $1\,\mathrm{km}^2$ AIROBICC detector array and the EGRET instrument on the Compton-GRO and the possible detection of IC 443.

If we turn to the TeV band accessible to modern imaging atmospheric Cherenkov (IAC) telescopes we find that the situation looks very interesting. As has been emphasised recently by Weekes (1991), for hard "equal energy per decade" spectra ($N(E) \propto E^{-2}$) the sensitivity of the IAC technique is comparable to that of the best satellite instruments at 100 MeV. However, because the general GCR spectrum, and thus the resulting general Galactic disc emission, is substantially softer ($\propto E^{-2.6}$) the contrast between sources and background should improve by a factor of about 30–100. In fact the main background would appear to be that due to the high energy cosmic ray electrons which induce atmospheric cascades indistinguishable from gamma induced ones. However this is an unstructured isotropic background (unlike the Galactic plane emission) and thus can be subtracted if adequate statistics are obtained.

It should be noted that although in these estimates we have treated SNRs as point sources, they have typically an angular size of 1°. In principle the IAC technique can resolve down to 0.1° with an array of telescopes, at which point the limb brightned structure should be easily visible; as well as confirming the identification, this should improve the signal to noise ratio substantially. If definite measurements at TeV energies and GeV energies can be obtained, we will have a very good handle on the energy spectrum which will allow a crucial test of acceleration theory, especially as the photon spectra are expected to start to cut-off at around 10 TeV corresponding to a proton cut-off at about 10^{14} eV.

7. Conclusions

Gamma ray observations of SNRs in the TeV region may in the next few years give us a convincing confirmation of current theoretical ideas on particle acceleration and the origin of Galactic cosmic rays.

References

Aharonian F, Drury L O'C and Völk H J: 1994, *AA* **285**, 645
Berezhko E G, Yelshin V K and Ksenofontov L T: 1994, *Astroparticle Phys* **2**, 215
Bhat et al : 1985, *Nature* **314**, 511
Cowsik R and Sarkar S: 1980, *MNRAS* **191**, 855
Dorfi E A: 1990, *AA* **234**, 419
Dorfi E A: 1991, *AA* **251**, 597
Drury L O'C, Aharonian F and Völk H J: 1994, *AA* **287**, 959
Fichtel et al: 1994, *ApJS* **94**, 551
Lebrun F and Paul J: 1985, *Proc 19th Int Cosmic Ray Conference (la Jolla)* **1**, 309
Naito T and Takahara F: 1994, *J Phys G* ,
Pinkau K: 1970, *Phys Rev Let* **25**, 603
Sturner S J and Dermer C D: 1995, *AA* **293**, 17
Weekes T C: 1991, *Proc 22nd Int Cosmic Ray Conference (Dublin)* **5**, 59
Wolfendale A W and Zhang L: 1994, *J Phys G* ,

THE NONTHERMAL ENERGY CONTENT AND GAMMA RAY EMISSION OF STARBURST GALAXIES AND CLUSTERS OF GALAXIES

H. J. VÖLK, F.A. AHARONIAN AND D. BREITSCHWERDT
Max-Planck-Institut für Kernphysik Postfach 103980, D-69029 Heidelberg Germany
vlk@vampi.mpi-hd.mpg.de

Abstract. The nonthermal particle production in contemporary starburst galaxies and in galaxy clusters is estimated from the Supernova rate, the iron content, and an evaluation of the dynamical processes which characterize these objects. The primary energy derives from SN explosions of massive stars. The nonthermal energy is transformed by various secondary processes, like acceleration of particles by Supernova Remnants as well as diffusion and/or convection in galactic winds. If convection dominates, the energy spectrum of nonthermal particles will remain hard. At greater distances from the galaxy almost the entire enthalpy of thermal gas and Cosmic Rays will be converted into wind kinetic energy, implying a fatal adiabatic energy loss for the nonthermal component. If this wind is strong enough then it will end in a strong termination shock, producing a new generation of nonthermal particles which are subsequently released without significant adiabatic losses into the external medium. In clusters of galaxies this should only be the case for early type galaxies, in agreement with observations. Clusters should also accumulate their nonthermal component over their entire history and energize it by gravitational contraction. The pion decay γ-ray fluxes of nearby contemporary starburst galaxies is quite small. However rich clusters should be extended sources of very high energy γ-rays, detectable by the next generation of systems of air Cherenkov telescopes. Such observations will provide an independent empirical method to investigate these objects and their cosmological history.

1. Introduction

That high energy γ-ray astronomy is not confined to Galactic sources has been impressively confirmed by the detection (von Montigny *et al.*, 1995) of about 40 active galactic nuclei (AGNs) with the EGRET instrument of the Compton Gamma Ray Observatory. Subsequently the close-by (d=100 Mpc, z=0.03) AGN Mrk 421 has also been detected at TeV energies by the Whipple Observatory (Punch *et al.*, 1992). As a rule, the EGRET AGNs are highly time variable, with jets directed towards the observer, and indeed a TeV outburst of Mrk 421 has also been seen on a time scale of days (Kerrick, 1995). So far Mrk 421 is the only TeV source of this kind detected. This is not too surprising, since the proximity of that object minimizes the problem of intergalactic γ-ray extinction by pair production on background photon fields. Other, *suggested* types of extragalactic TeV γ-ray sources may be rather quasi-steady. This should be true for the giant pair halos that probably surround extragalactic "point" sources of very high energy ($E \geq$ 100 TeV) primary gamma-radiation, presumably due to luminous distant AGNs (Aharonian *et al.*, 1994). Such halos have angular sizes of the order of

Space Science Reviews **75**: 279–297, 1996.

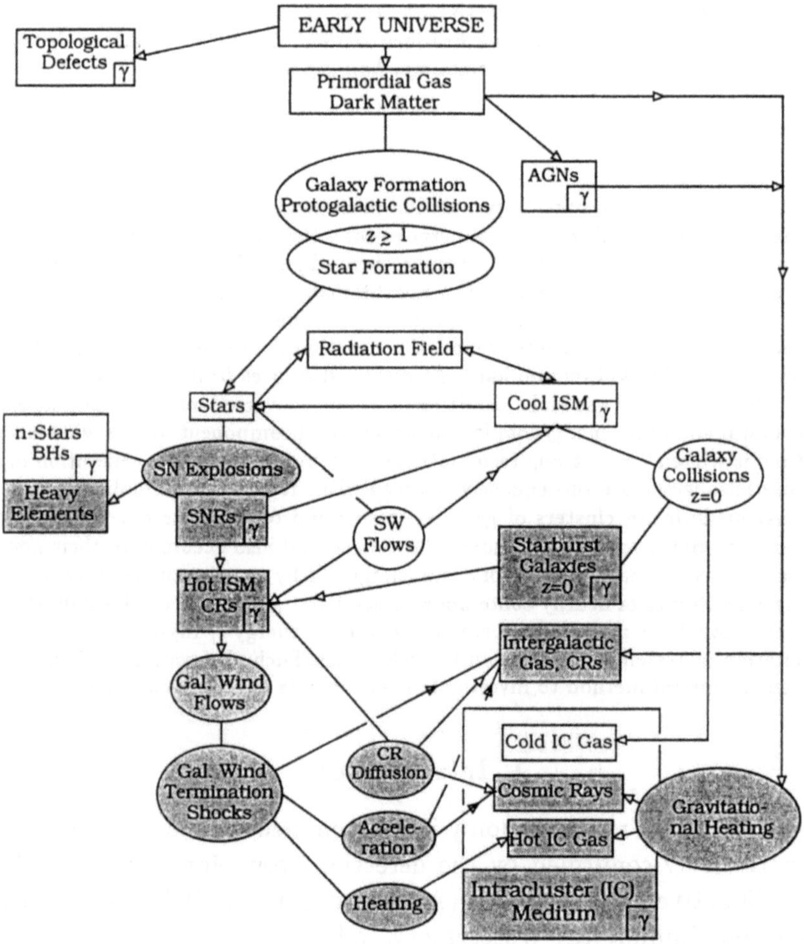

Fig. 1. Known and expected extragalactic nonthermal sources of gamma rays (indicated by γ) and connecting physical processes (encircled by ellipses) in a simplified schematic form. The emphasis is on starburst galaxies and galaxy clusters, and the main topics of the paper are shaded. Apart from topological defects as possible relics from the big bang, and active galactic nuclei (AGNs), star formation in galaxies is the underlying process. In contemporary late type galaxies (at redshift $z \simeq 0$) matter mainly cycles between stars and the cool interstellar medium (ISM) through Supernova (SN) explosions and stellar winds (SWs). However in starbursts, due to contemporary or protogalactic (at $z > 1$) collisions, a very high pressure hot ISM is produced in SN remnants (SNRs), with comparable energy release in Cosmic Rays (CRs). Especially in clusters, galactic winds then play a central role in removing mass and in heating the external medium. They may be the strongest sources of extragalactic nonthermal particles over the history of the Universe.

a degree. Topological defects, as hypothetical relics from the big bang, might on the other hand provide a diffuse background of TeV photons (Aharonian *et al.*, 1992, Xi *et al.*, 1993).

In this paper we shall discuss two further types of potential quasi-steady, extragalactic TeV sources. These are ***clusters of galaxies*** and, as a corollary, ***contemporary nearby starburst galaxies***. Clusters of galaxies as the largest gravitationally bound objects in the Universe would therefore also be the largest confined γ-ray sources, on the scale of a few degrees. Individual starburst galaxies, even if they are very nearby, like the starburst nucleus of M 82, would only have a source extension of the order of an arcmin. Both sources are characterized by very strong star formation activity, either in the distant past or at present, with a proportionally strong Cosmic Ray (CR) nucleon production and thus presumably also a high γ-ray luminosity. They act as signposts for the extragalactic sky in X-rays (clusters) and in the infrared (starburst galaxies). As a consequence they have become objects of cosmological interest in the last decade.

Fig. 1 is an attempt to describe the general context for these sources. The associated physical processes will be discussed in the following sections: Sec. 2 concerns the formation of early type galaxies and the likely role of their winds. In Sec. 3 the star formation history, iron enrichment, and Supernova (SN) rate are evaluated for the evolution of the Perseus cluster which we use as the prototype of a rich cluster. Sec. 4 deals with the production and energetics of nonthermal particle components (CRs) and their confinement. In Sec. 5 we discuss galactic winds from various types of galaxies which emphasize the role of early type galaxies in clusters. Sec. 6 estimates the pion decay γ-ray fluxes, and Sec. 7 contains the discussion and conclusions.

2. Formation of Early Type Galaxies and the Role of their Winds

Galaxies and their initial stellar populations form at redshifts $z > 1$ by gravitational collapse of primordial density fluctuations. In regions where the density of (proto)galaxies is high, i.e. in incipient clusters of galaxies, there is a particularly large probability for such objects to interact and to merge. This is often assumed to be the explanation for the large density of elliptical (E) and lenticular (S0) galaxies, especially in the cores of clusters. These early type galaxies nowadays outnumber spirals by at least a ratio of 10:1 in rich clusters ($z \simeq 0$), (e.g. Dressler, 1993). Often, as in the case of the nearby ($d \simeq 20\,\mathrm{Mpc}$) Virgo cluster, the visual impression is that many of the presently observed cluster spirals are in fact freshly falling in rather than being initial constituent cluster members (B. Binggeli, private communication). In the field between clusters, on the other hand, the ratio between early and late type galaxies is essentially the opposite. Such observations suggest that at least the massive isolated protogalaxies primarily (or even exclusively) form spirals, in agreement with recent numerical simulations (Steinmetz, 1993) and the assumption that massive early type galaxies are *interaction products*. We shall adopt this point of view here since it implies the minimum number of ad hoc assumptions. Then it is to be expected that the initial violent formation of the massive early type galaxies in clusters (Tinsley, 1980) is accompanied by an extended burst of star formation: the massive stars, exceeding 5 to 8 $M_\odot$, explode within less than 100 Million years in the form of core collapse supernovae, so-called Type II SNe, releasing heavy elements, especially iron, heating the ambient gas to X-ray temperatures, and producing CRs in their gaseous remnants (SNRs). In contrast the less massive stars evolve on much longer time scales and partly still exist today. The associated Type Ia SNe should therefore occur on a comparatively quasisteady basis in early type galaxies and they are indeed observed at the present epoch.

To which extent the initial starburst in clusters (or protoclusters) has consumed most of the available gas to form stars is not very well known. Our view is that most of the primordial gas will have been *prevented* to go into galaxies in the first place by the enormous and violent energy release in the form of stellar winds (SWs) and SNRs from the starbursts due to the first generation of mergers which have generated massive galactic winds. As a smaller additional effect, galactic winds and interactions with other galaxies may have removed a good part of the interstellar medium (ISM) from the already existing galaxies. This naturally explains the known inefficiency of galaxy and star formation in rich clusters, where only about 20 percent of the luminous matter is contained in the galaxies, the rest being in the intracluster (IC) gas (Blumenthal *et al.*, 1984; David *et al.*, 1990 a). Such a high ratio of IC mass to galaxy mass has also been deduced in detail by ROSAT observations (Böhringer, 1994) for a number of nearby rich clusters.

We will come back to the gas dynamics of galactic winds quantitatively in Sec. 5. Here we shall describe roughly their role for the propagation and generation of CRs: The galactic SNRs enrich the ISM with nonthermal particles (e.g. Axford, 1981). These CRs, together with the hot shocked thermal gas, will drive a galactic wind, mainly depending on the SN rate and the IC pressure. In any case the CRs will leave the galaxy either diffusively and without energy loss if there is no wind, or convectively together with the wind suffering *severe energy losses* due to adiabatic expansion in the process. The latter situation may possibly be studied by γ-ray observations of nearby contemporary ($z \simeq 0$) starburst galaxies which derive their activity from recent interactions with other galaxies. The adiabatically cooled gas in galactic winds will be reheated and CRs accelerated in the *galactic termination shock* through which the wind has to adjust to the finite overall pressure in the environment (Jokipii and Morfill, 1985; 1987). In clusters the IC pressure is so large that winds from spiral galaxies should not develop, except in a starburst phase (Breitschwerdt *et al.*, 1991, 1993; Breitschwerdt, 1994; Sec. 5). This favours the early formation phase of E/S0 galaxies and their winds as being responsible for the iron and CR enrichment of the IC medium. The gravitational contraction and adiabatic heating of clusters presumably overlaps with that epoch, but it continues until today, as morphological studies of clusters of various types (Briel *et al.*, 1992, Schwarz *et al.*, 1992) show. Such a prolonged timescale of cluster formation is in agreement with currently popular cold dark matter models (e.g. White *et al.*, 1987) and their extensions. It must lead to a sizeable *adiabatic* increase in the nonthermal energy content of the IC gas.

The role of elliptical galaxies, their winds, and the evolution of the thermal IC medium has been modeled extensively by David *et al.* (1990 b, 1991a,b). We shall not go into that in detail, given the greater uncertainties of nonthermal processes. An independent discussion of the importance of galactic winds for particle acceleration in galaxy clusters is contained in an interesting recent paper by Pietsch *et al.* (1994), focussing mainly on recent ROSAT results. As far as the nonthermal CR component is concerned these authors emphasize particles of the highest energies, as they are observed in air showers. We concentrate on the overall energetics and the γ-ray emission in the TeV range and below.

Clusters are probably closed systems of thermal gas with several hundred members in the case of rich clusters. We shall deal with such large systems here and shall argue that also the nonthermal component is *diffusively confined* by them since it was formed. Because CR nucleons suffer but little energy losses in the low density, high pressure IC gas, rich clusters should accumulate nucleonic CRs and we "see" the entire dose of CR production preserved over cosmological times. In other words, *clusters have preserved their own history* also in their main nonthermal content. Since CRs should penetrate hot and cold IC gas alike, we expect also the resulting γ-rays to act, at least in principle, as probes for the total amount of baryonic matter in the IC medium, regardless of its thermodynamic state.

3. Star Formation History, Iron Enrichment, and SN Rate in Clusters of Galaxies

Nearby galaxy clusters are typically at distances $d \sim 100\,\mathrm{Mpc}$ with a radial extent R of the hot IC gas of a few Mpc, i.e. angular diameters of a few degrees. The Perseus cluster, which we shall take as our example for numerical estimates, has been observed by ROSAT to have a radius of the X-ray gas $R_{\mathrm{cl}} \simeq= 3\,\mathrm{Mpc}$, with an estimated total hot gas mass $M_{\mathrm{cl}} = (3\text{–}4)\cdot 10^{14} M_{\odot}$ at a distance of 97 Mpc (adopting a Hubble constant of $H_0 = 50\,\mathrm{km\,s^{-1}Mpc^{-1}}$ throughout). Taking $M_{\mathrm{cl}} = 4\cdot 10^{14}\,M_{\odot}$ as the more likely value, this leads to a mean gas density $n_{\mathrm{cl}} = 1.4\cdot 10^{-4}\,\mathrm{cm^{-3}}$, corresponding to about one tenth of the gas density in the halo of the ISM of our Galaxy. In the central regions of the cluster this density should reach values that are at least one order of magnitude larger, but this is not our concern here. Assuming a temperature $T_{\rm cl} = 7\,{\rm keV}$ of the hot gas (Fabian *et al.*, 1981), which is uniform across the cluster, we obtain a mean $(nT)_{\rm cl} = 1.1 \cdot 10^4\,{\rm K\,cm^{-3}}$, about a factor of 2 higher than the mean value for the ISM of our Galaxy. This corresponds to a total thermal gas pressure of $p_{\rm th}^{\rm cl} = 1.9$ eV cm^{-3} and a total internal gas energy $E_{\rm g}^{\rm cl} = 1.2 \cdot 10^{64}$ erg, disregarding subsequent radiative cooling in the dense central regions. Not taking into account CRs and magnetic fields, the necessary dynamical mass exceeds the luminous gas mass by a factor ranging between 3 and 10 (Böhringer, 1994).

A quantity of major interest for the following is the *magnetic field* in the IC medium. If we assumed a magnetic field strength B_{cl} in equipartition with the thermal gas pressure, the above value for $p_{\rm th}$ would suggest $B_{\rm cl} \approx 10^{-5}$ G. Such a large field strength is probably not maintainable by any dissipative dynamo process in the IC medium (M. Rees, private communication). On the other hand, various lines of recent evidence suggest fields of the order of 1 μG in most clusters (Kronberg, 1994), and we shall adopt 1 μG as a representative value; it still allows a dominant (thermal plus nonthermal) gas pressure.

For an estimate of the CR intensity in clusters, also the *iron mass* is an essential quantity. Arnaud *et al.* (1992) have used EXOSAT (Edge, 1989) and GINGA (Hatsukade, 1989) spectroscopic data to infer a mean iron abundance of $\epsilon_{\rm cl} = 0.35 \pm 0.02\,(1\sigma)$, relative to the solar value $[{\rm Fe}]_\odot = 4 \cdot 10^{-5}$. It varies very little from cluster to cluster and also depends only weakly on the gas temperature. This implies that the iron mass essentially traces the gas mass in clusters. Moreover, these authors found that the total gas mass correlates statistically quite well with the sum of the visual luminosities of elliptical and lenticular galaxies in clusters (Fig. 2), but not with the corresponding luminosities of the cluster spirals. Although using a very small sample, they found in addition a proportionality between iron mass and the E plus S0 luminosities (Fig. 2) and concluded from both these results that only elliptical and lenticular galaxies have contributed to the iron enrichment of the IC gas.

E and S0 galaxies are presumably characterized by violent initial starbursts, generating a strong galactic wind that can mix at least part of the iron generated by SN explosions with the IC gas. With a knowledge of the iron production per SN and the iron retention factor in the galaxies we can then infer the total number of SN events, provided no iron has been lost from the cluster, e.g. by a cluster wind. Assuming a given CR luminosity per SNR this determines the total amount of CRs produced in the cluster volume.

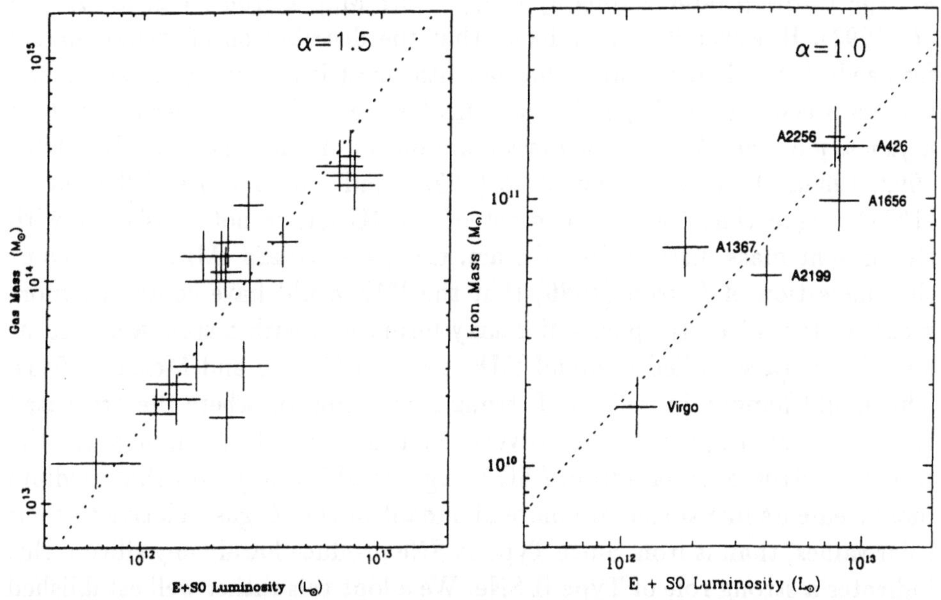

Fig. 2. (left): Correlation of the hot IC gas mass with the sum of the visual luminosities of the E and S0 galaxies for a sample of clusters. (right): Correlation of total IC iron mass with the sum of visual luminosities of the E and S0 galaxies for six clusters, suggesting a near proportionality. The slopes of the dotted lines are denoted by α (from Arnaud *et al.*, 1992).

A simulation of the star formation history for a large elliptical galaxy, assumed to be a protogalactic merger event, is shown in Fig. 3 (Steinmetz, 1993). The major increase in the star formation rate occurs at redshift $z \simeq 3$ and lasts from $z = 3$ to $z = 2$, over a Gigayear period with a maximum rate of about 200 $M_\odot$ yr^{-1}. In the simulations this number is typically an order of magnitude larger than that for late type galaxies. However these simulations disregard any mass loss in the form of winds and assume all available gas to be ultimately incorporated into stars. We argue that violent winds due to the high star formation rate will in reality *quench* the merging process. The process should therefore stop earlier and have a much reduced integral mass consumption relative to the originally available mass. Nevertheless, the order of magnitude of the maximum star formation rate may be the same. Such a situation is to be contrasted with a present star formation rate of the Galaxy of the order of 1 $M_\odot$ yr^{-1}. The generation of hot gas is probably a quite nonlinearly increasing function of the star formation rate. This may be the reason why early type galaxies are so much more effective in shaping their cluster environment.

The relation of the iron production to the SN rate is rather complex. Whereas Type II SNe may typically produce a $\delta M_{\rm Fe} \leq 0.1\,M_\odot$ per event, a Type Ia SN produces roughly $\delta M_{\rm Fe} = 0.7\ {\rm M}_\odot$ of iron (e.g. Renzini *et al.*, 1993). However it is very likely that the distribution of stellar masses for newly formed stars (IMF) during a starburst is favouring massive stars, i.e. the precursors of Type II SNe. This has been shown to occur due to a *deficiency* of the IMF in low-mass stars during a starburst (e.g. Bernlöhr, 1992, 1993a,b). Also Arnaud *et al.* (1992), and most recently Elbaz *et al.* (1994), argue that the iron content of the IC gas is not consistent with the present mass distribution of stars, using a standard IMF. They favour the suggestion of Larson (1986) that the IMF could have contained more massive stars at early epochs of galaxy formation, with a standard IMF at later times (a so-called bimodal IMF, see also Güsten and Mezger, 1983). This might happen in regions of strong star formation where the preheated clouds do not fragment down to very small masses. Most importantly, the recent spectroscopic data from ASCA (e.g. Ohashi, 1994) show that medium mass elements like silicon are more abundant in the IC gas, relative to their solar values, than is iron. Since Type Ia SNe produce hardly any silicon, this indicates a strong role of Type II SNe. We adopt this rather well established point here and prefer $\delta M_{\rm Fe} = 0.1\,{\rm M}_\odot$, i.e. making the assumption that essentially all SNe that have initially occurred in early type galaxies are of Type II. Renzini *et al.* (1993) however favour an average of 0.2 $\rm M_\odot$ per SN and for completeness we shall make numerical estimates for both values 0.1 and 0.2. The total number of SNe over the cluster history is then given by

$$N_{\rm SN} = \int_{T_{\rm H}}^{0} dt \nu_{\rm SN}(t) = \frac{M_{\rm cl}\,[{\rm Fe}]_\odot\,\epsilon_{\rm cl}}{\delta M_{\rm Fe}}, \tag{1}$$

where $\nu_{\rm SN}$ denotes the SN rate. For the Perseus cluster this gives

$$N_{\rm SN}^{\rm Per} = 3.1 \cdot 10^{12} \left\{ \left(\frac{\epsilon_{\rm cl}}{0.35}\right) \left(\frac{M_{\rm cl}}{4 \cdot 10^{14} {\rm M}_\odot}\right) \left(\frac{\delta M_{\rm Fe}}{0.1\,{\rm M}_\odot}\right)^{-1} \right\}, \tag{2}$$

which is about 12.5 times more than if all the 500 cluster galaxies produced SNe over a Hubble time $T_{\rm H} = 1.5 \cdot 10^{10}$ yrs with the present Galactic rate of 1/(30 yrs). If the amount of iron retained in the galaxies is equal to that in the IC gas (e.g. Arnaud *et al.*, 1992), and the initial starburst lasted only 10^9 yrs, then the SN rate per galaxy during the early burst is $12.6\,{\rm yr}^{-1}$. For $\delta M_{\rm Fe} = 0.2\,{\rm M}_\odot$ these numbers are decreased accordingly (and given in brackets below). Assuming that 50 $\rm M_\odot$ are needed to ultimately produce one Type II SN of an average mass of, say 10 $\rm M_\odot$, then the mean mass consumption per galaxy for star formation dM_*/dt corresponds to 630 (315) $\rm M_\odot\,yr^{-1}$, over 10^9 yrs, in rough agreement with Fig. 3*. Roughly, the given

* Assuming 5 times more mass per galaxy to be present initially, as suggested by the present ratio of gas to galaxy mass, and a crudely linear increase of dM_*/dt with available

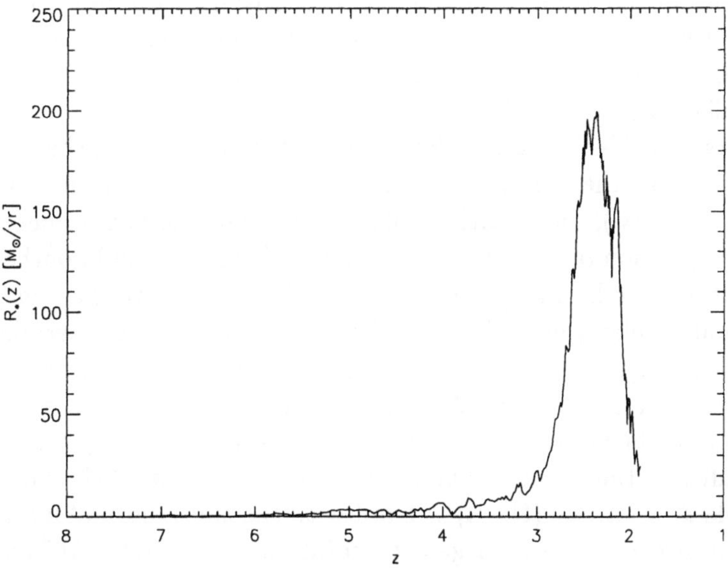

Fig. 3. **Star formation rate $R_*(z) = dM/dt$ as a function of redshift z for a merging event of two (proto)spirals of mass $M = 8 \cdot 10^{11} M_\odot$ each (from Steinmetz, 1993)**

SN rates exceed that of our present Galaxy by more than 2 orders of magnitude, and those of contemporary starburst galaxies like M 82 by more than 1 order of magnitude.

4. Nonthermal Particle Components

We shall assume CRs to be generated either at SNRs inside galaxies or at galactic termination shocks, neglecting all other minor sources. For the time being we shall also confine the discussion to nucleonic CRs which we assume to dominate the energetics like they do in our Galaxy. Pure diffusive escape of particles produced inside implies no adiabatic energy losses. Since all other energy losses are negligible for relativistic nuclei, essentially all the CR energy will end up in the IC medium or beyond, with an energy spectrum $f(E) \propto E^{-2.1}$, that corresponds to the hard SNR source spectrum (e.g. Berezhko *et al.*, 1994); the present CR energy inside the galaxies, compared to that in the IC medium, is of the order of one percent, if a galactic escape

density, we could have an average $\nu_{SN} = 63\,(31.5)\,\mathrm{yr}^{-1}$ over an initial starburst period of 200 Million years, keeping N_{SN} fixed. Systems of this kind might indeed exist (e.g. Elbaz *et al.*, 1992). Individual galaxies could certainly be somewhat more or less active, depending mainly on their masses. The most active objects might then even reach the magic rate of $\nu_{SN} \sim 1\,\mathrm{day}^{-1}$!

time $\leq 10^8$ yrs is assumed. On average 10 to 30 percent of the entire hydrodynamical SN energy must go into the CR component in order to fulfill the CR energetics of our Galaxy, numbers also consistent with observations of the starburst galaxy M 82 (Völk *et al.*, 1989). We shall use a 30 percent efficiency for the numerical estimates as a likely value from particle injection theory (Malkov and Völk, 1994).

Let us first consider the *energetics*. If the galaxy has a wind, then, except for energies where the particles cannot be confined any more in a galaxy (probably $\geq 10^{17}$ eV), the particles will undergo a fatal adiabatic energy loss due to the expansion of the gas in the wind: the thermal (and nonthermal) gas is essentially cold, having spent all its enthalpy to drive the wind. Gas reheating and acceleration of *freshly injected* CRs occurs at the termination shock which is a quasisteady shock of high Mach number and large radius, of the order of the size of the galaxy or even much larger. It can accelerate CRs even more efficiently than SNR shocks which have a very limited life time. In addition, the mean magnetic field vector in the wind should have a typical angle $\phi = \arctan(\Omega r/u) \simeq 1$ rad with the shock normal for a wind standoff distance $r = 50\,\mathrm{kpc}$, a galactic rotation time $2\pi/\Omega = 10^8$ yrs, and a wind speed $u = 500\,\mathrm{km/s}$ (cf. Zirakashvili *et al.*, 1993). This means that the termination shock is a *quasi-parallel* rather than a perpendicular shock, and that the standard theory of diffusive shock acceleration applies.

The total amount of CR energy produced in the galactic SNRs is

$$E_{\mathrm{CR}}^{\mathrm{SN}} = 9.4 \cdot 10^{62}\mathrm{erg} \left\{ \left(\frac{\mathrm{N_{SN}}}{3.1 \cdot 10^{12}} \right) \left(\frac{\Theta_{\mathrm{CR}}^{\mathrm{SN}}}{0.3} \right) \left(\frac{\mathrm{E_{SN}}}{10^{51}\mathrm{erg}} \right) \right\}, \tag{3}$$

$\Theta_{\mathrm{CR}}^{\mathrm{SN}}$ being the CR production efficiency per SN. Without radiative cooling the rest goes into thermal gas energy with efficiency $\Theta_{\mathrm{g}}^{\mathrm{SN}} = 1 - \Theta_{\mathrm{CR}}^{\mathrm{SN}}$

$$E_{\mathrm{g}}^{\mathrm{SN}} = 2.2 \cdot 10^{63}\,\mathrm{erg} \left\{ \left(\frac{N_{\mathrm{SN}}}{3.1 \cdot 10^{12}} \right) \left(\frac{\Theta_{\mathrm{g}}^{\mathrm{SN}}}{0.7} \right) \left(\frac{E_{\mathrm{SN}}}{10^{51}\,\mathrm{erg}} \right) \right\} \tag{4}$$

which is still only 18 percent of $E_{\mathrm{g}}^{\mathrm{cl}}$. Clearly this is an upper limit.

In addition the hot gas (and the CRs) will drive a galactic wind. Also in such a wind radiative cooling will remove some of this energy. However at the large gas temperatures and wind speeds such wind cooling is negligible if the example of M 82 is a guide (see Sec. 5). Thus we will neglect radiative cooling in the wind. The efficiency Θ_{gal} of conversion of SN energy into wind energy is not very well determined. For M 82 we estimate it as 0.2 but we will also consider the ideal case of $\Theta_{\mathrm{gal}} = 1$, i.e. adiabatic conversion of SN energy into wind energy. Taking for a moment $\Theta_{\mathrm{gal}} = 0.2$, and assuming for simplicity that the CR production efficiency $\Theta_{\mathrm{CR}}^{\mathrm{gw}}$ at the wind termination shock has the same value 0.3 as for a galactic SNR, we obtain a value for the gas internal energy transferred to the IC medium $E_{\mathrm{g}}^{\mathrm{gw}} = 4.4 \cdot 10^{62}$ erg.

The corresponding energy in CRs is $E_{\rm CR}^{\rm gw} = 1.9 \cdot 10^{62}$ erg. $E_{\rm g}^{\rm gw}$ is by a factor 27 smaller than the observed IC internal gas energy. In other words, sizeable adiabatic heating through subsequent gravitational contraction must have occurred. We assume that adiabatic heating has occurred after the main event of CR production and obtain

$$\frac{E_{\rm CR}^{\rm cl}}{E_{\rm CR}^{\rm gw}} = \left(\frac{E_{\rm g}^{\rm cl}}{E_{\rm g}^{\rm gw}}\right)^{\frac{\gamma_c - 1}{\gamma_g - 1}}, \tag{5}$$

where $E_{\rm CR}^{\rm cl}$ is the total present day CR internal energy, and $\gamma_{\rm g} = 5/3$ and $\gamma_{\rm c}$ denote the adiabatic indices of thermal gas and CRs, respectively. Taking $\gamma_{\rm c} = 4/3$ corresponds to an ultrarelativistic gas and to the smallest possible exponent on the *rhs* of eq. (5) due to the softness of the CR equation of state. Thus the CR enhancement is the *smallest possible* one. It corresponds to a contraction of the cluster gas by a linear scale factor of 6.1. In terms of the assumed parameters of the system, we obtain the enormous value of

$$E_{\rm CR}^{\rm cl} = 1.2 \cdot 10^{63} {\rm erg} \left\{ \left(\frac{\rm T_{cl}}{7\,{\rm keV}}\right)^{1/2} \left(\frac{\epsilon_{\rm cl}}{0.35}\right)^{1/2} \left(\frac{\rm M_{cl}}{4 \cdot 10^{14}\,{\rm M_\odot}}\right) \right.$$
$$\left. \left(\frac{\delta M_{\rm Fe}}{0.1\,{\rm M_\odot}}\right)^{-1/2} \left(\frac{\Theta_{\rm gal}}{0.2}\right)^{1/2} \left(\frac{\Theta_{\rm CR}^{\rm gw}}{0.3}\right) \left(\frac{1 - \Theta_{\rm CR}^{\rm gw}}{0.7}\right)^{-1/2} \left(\frac{E_{\rm SN}}{10^{51}\,{\rm erg}}\right)^{1/2} \right\}. \tag{6}$$

The CR energy density $E_{\rm c} \simeq 0.25\,{\rm eV/cm^3}$, calculated in this way is about equal to that in our Galaxy and corresponds to 10 percent of the gas thermal energy density, as estimated above.

This calculation is meaningful only if we assume the IC medium to be a closed system also for the CRs. As a next step we therefore must consider the question of *CR confinement*.

In order to find out whether CRs are indeed confined to the cluster, we estimate their diffusion properties. (It is relatively easy to see that a *cluster wind* might develop despite a de facto confinement of the gas. However it cannot have expanded very far during a Hubble time. Thus it cannot have transported much mass over distances of the order of several cluster radii and cannot have spread confined CRs to any significant extent either). The *resonant diffusion* of energetic particles in magnetic field fluctuations with a power spectrum $P(k)$ in wave number k is given by

$$\kappa_{(p)} = \frac{1}{3} w r_{\rm g}(p) \frac{B^2}{\int_k^\infty P(k')dk'}; \quad k \simeq \frac{1}{r_{\rm g}(p)}, \tag{7}$$

where $B, w \simeq c$, and $r_{\rm g}$ denote the mean field strength, the particle speed and the gyro radius, respectively. Field fluctuations can be present due to a variety of causes. In particular the virial motion of the galaxies in the gravitational potential of the cluster should completely stir up the magnetic field in the IC gas on scales of the intergalactic distance $d_0 = (4\pi/(3N_{\rm gal}))^{1/3}R_{\rm cl}$, where $N_{\rm gal}$ is the number of galaxies in the cluster. Assuming the field on this scale to have a fluctuation level of order unity, we have $P(k_0)k_0 = B^2$, where $k_0 = d_0^{-1}$. For the Perseus cluster with $N_{\rm gal} = 500$ this corresponds to a basic scale of $d_0 = 0.7$ Mpc. Assuming $P(k) = P(k_0)\cdot(k/k_0)^{-n}$, we choose $n = 3/2$, corresponding to the situation in the Galaxy, but also consider the case of a Kolmogorov spectrum, $n = 5/3$. For $n = 3/2$, κ is proportional to $r_{\rm g}^{1/2}$, leading to

$$\kappa \simeq 1.2\cdot 10^{28}\left(\frac{w}{c}\right)\left(\frac{p}{mc}\right)^{1/2} \mathrm{cm^2 s^{-1}} \left(\frac{Z\,B}{10^{-6}\,\mathrm{G}}\right)^{-1/2}, \tag{8}$$

where p denotes particle momentum, and Z its nuclear charge. For $n = 5/3$ we obtain instead

$$\kappa = 1.4\cdot 10^{30}\left(\frac{w}{c}\right)\left(\frac{p}{mc}\right)^{1/3} \mathrm{cm^2 s^{-1}} \left(\frac{Z\,B}{10^{-6}\,\mathrm{G}}\right)^{-1/3}. \tag{9}$$

These values are not unlike those assumed for our Galaxy. The resulting diffusive escape time from the cluster, $T_{\rm esc} \sim R^2/\kappa$, is then $2\cdot 10^{14}\,(p/mc)^{-1/2}$ yrs for $n = 3/2$, and $2\cdot 10^{12}(p/mc)^{-1/3}$ yrs for $n = 5/3$.

$T_{\rm esc} \simeq T_{\rm H}$ for a field of 10^{-6} G, and for protons with an energy of $2\cdot 10^{17}$ eV and $2\cdot 10^{15}$ eV for $n = 3/2$ and $5/3$, respectively. Lower energy particles remain confined in the IC medium. We conclude that all particles of interest here will accumulate and remain confined in the cluster until the present epoch, allowing us to "see" *cosmological* CRs in clusters (see also Pietsch *et al.*, 1994). Energy losses of ultrarelativistic CR nucleons in the gas are negligible: the time between nuclear collisions corresponds to about $30\,T_{\rm H}$. TeV electrons, on the contrary, will lose energy very quickly through inverse Compton collisions and synchrotron radiation on time scales of less than 10^6 yrs. Clearly, electrons are *contemporary.*

5. Galactic Winds in Clusters

Supersonic galactic winds can only occur if the total gas plus CR pressure in the intergalactic medium is not too high to prevent a subsonic/supersonic transition. Otherwise the galactic halo will be essentially static. The IC pressure of rich clusters like the Perseus cluster is indeed very high, compared for example to the local environment of our Galaxy. Therefore it is an open question whether all cluster galaxies will be able to sustain a stationary wind. This is particularly true for *spirals* like our present Galaxy. Except in a starburst phase they cannot maintain a wind by the action of the thermal gas alone because of the strong radiative cooling of the ISM. This has been calculated for instance by Habe and Ikeuchi (1980), who found that thermally driven winds for our Galaxy could only exist for gas temperatures exceeding $4 \cdot 10^6$ K even for vanishing extragalactic pressure. On the other hand it was argued by Breitschwerdt *et al.* (1991) that the Galactic CR component can indeed provide the extended pressure force in the halo for a wind to exist against the probable local intergalactic pressure of about $5\,\mathrm{K\,cm^{-3}}$. Fig. 4 shows a typical solution of this kind with a subsonic-supersonic flow transition at a critical point located at 34 kpc from the Galactic disk for a flow tube starting at a Galactocentric radius of 10 kpc. The integrated Galactic *mass loss rate* of about half a solar mass per year is not at all negligible. It can be quite significant over a Hubble time. However, under contemporary conditions, the *ram* pressure of the Galactic wind is so small that the balance of wind ram pressure against the extragalactic pressure pushes the termination shock towards the critical point already for $(nT) \sim 100\,\mathrm{K\,cm^{-3}}$. For a star formation rate at early epochs of a spiral galaxy like ours which may exceed the present rate by an order of magnitude (e.g. Steinmetz, 1993), the *ram* pressure of the wind may, in the simplest approximation, rise proportionally . Also the critical point will move inwards, so that the maximum permissible value of presently $(nT) \sim 100$ may be raised by one to two orders of magnitude. For a spiral galaxy in a cluster this results in a maximum IC pressure (nT) of a few $10^3\,\mathrm{K\,cm^{-3}}$ that still allows a wind. Such a number is marginal, even for the spatially averaged cluster pressure in a system like Perseus at the present time. It may be also marginal for the more central regions of the cluster at early epochs, when the overall IC pressure was lower.

The situation is very different for *early type galaxies.* As we have seen, their initial starburst strength exceeds the present star formation rate of our Galaxy by more than a factor of a hundred. For such high star formation rates the hot gas production may rise quite nonlinearly with the SN rate. An example may be the starburst galaxy M 82. Fig. 5 shows a hydrodynamic calculation on the same physical basis as Fig. 4 (Breitschwerdt, 1994). The starting conditions correspond to a gas density $n = 0.1\,\mathrm{cm^{-3}}$, at a temperature of $5 \cdot 10^7$ K, at a reference level which lies 100 pc above the galactic plane in the inner starburst region. These conditions give a gas pressure $nT = 5 \cdot 10^6\,\mathrm{K\,cm^{-3}}$, smaller by a factor of 6 than the value obtained by Schaaf *et al.* (1989). These authors interpreted X-ray data from EXOSAT and obtained $T = 10^8$ K, $n = 0.3\,\mathrm{cm^{-3}}$, with a filling factor of this hot gas $f_h \simeq 0.1$. With a mass loss rate $\dot{M} = 0.8 M_\odot\,\mathrm{yr^{-1}}$ this minimal wind model gives a *ram* pressure of $2 \cdot 10^6\,\mathrm{K\,cm^{-3}}$ at a distance $z = 20$ kpc from the galaxy. Considering that the initial starburst of massive early type galaxies is at least an order of magnitude more intense than the present one of M 82, albeit on a much larger mass and spatial scale, it is immediately convincing that such winds can be driven against an average IC pressure

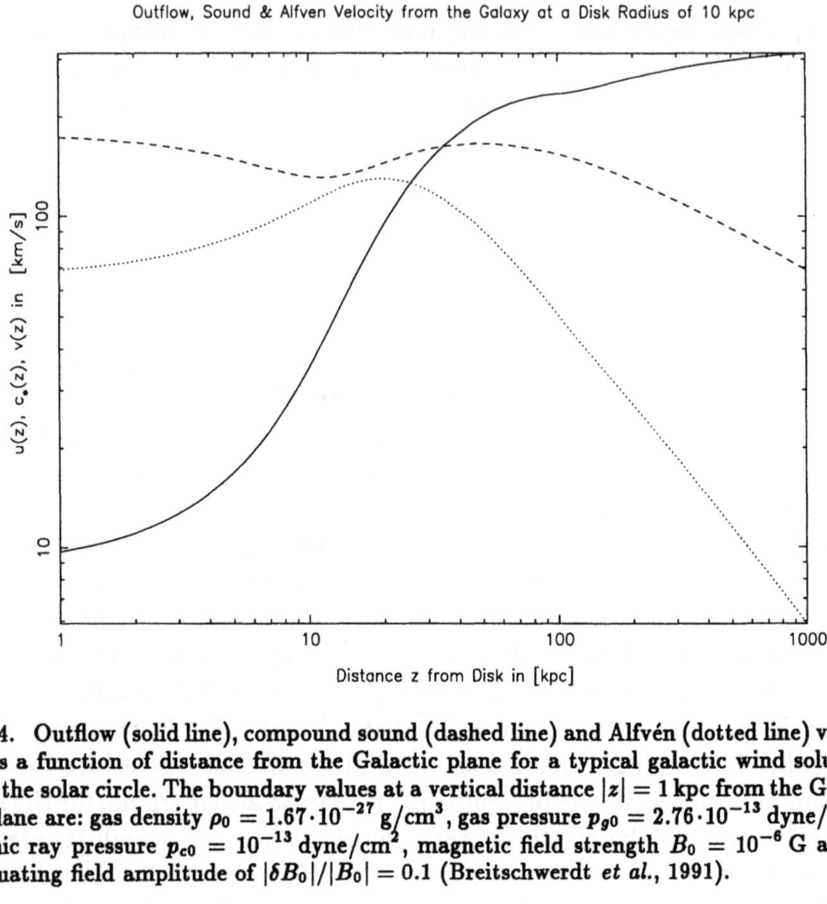

Fig. 4. Outflow (solid line), compound sound (dashed line) and Alfvén (dotted line) velocity as a function of distance from the Galactic plane for a typical galactic wind solution near the solar circle. The boundary values at a vertical distance $|z| = 1$ kpc from the Galactic plane are: gas density $\rho_0 = 1.67 \cdot 10^{-27}$ g/cm^3, gas pressure $p_{g0} = 2.76 \cdot 10^{-13}$ dyne/cm^2, cosmic ray pressure $p_{c0} = 10^{-13}$ dyne/cm^2, magnetic field strength $B_0 = 10^{-6}$ G and a fluctuating field amplitude of $|\delta B_0|/|B_0| = 0.1$ (Breitschwerdt *et al.*, 1991).

$(nT)_{\rm cl} = 10^4\,{\rm K\,cm^{-3}}$ or even a higher one as expected in the central cluster regions (see also David *et al.*, 1990b).

These results may be taken as a dynamical explanation of the finding of Arnaud *et al.* (1992) that IC gas properties are determined by the cluster content of early type galaxies and not of spirals.

The results of Schaaf *et al.* (1989) for M 82 together with the adiabatic character of winds from starburst galaxies can be used to estimate the *efficiency of conversion* of SN energy into wind energy. If, at least within certain limits, the mechanical luminosity $(1/2)\dot{M}u^2$ scales linearly with the pressure at the surface of the disk where the wind commences (our reference level), then we can derive a mechanical luminosity for M 82 through the relation

$$\frac{1}{2}\dot{M}u^2 \simeq 6 f_{\rm h}^{2/3} \left\{ \frac{1}{2} \left(\frac{0.8\ {\rm M_\odot}}{\rm yr} \right) \left(2300\ \frac{\rm km}{\rm s} \right)^2 \right\} = 1.8 \cdot 10^{42}\,{\rm erg\,s^{-1}}\,, \qquad (10)$$

where the quantity in braces on the rhs of eq. (11) is the mechanical luminosity from the model of Fig. 5. The inferred luminosity has to be compared

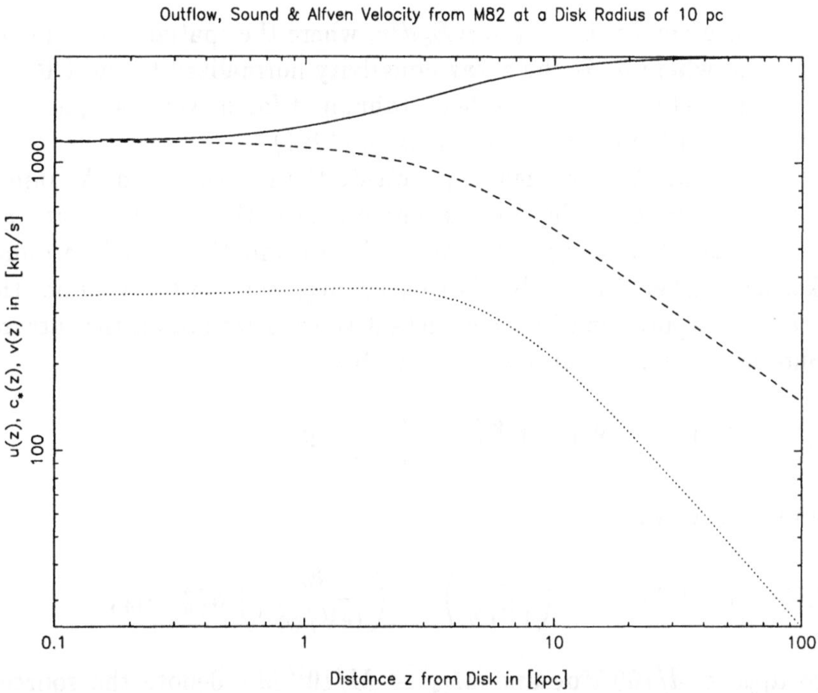

Fig. 5. Outflow (solid line), compound sound (dashed line) and Alfvén (dotted line) velocity as a function of distance from the galactic plane for a typical galactic wind solution from the inner region (R = 200 pc) of the starburst galaxy M 82. The boundary values at a vertical distance $|z|$ = 100 pc from the plane are given in the text. The cosmic ray pressure and the magnetic field strength are $p_{c0} = 4.3 \cdot 10^{-11}$ dyne/cm^2, magnetic field strength $B_0 = 5 \cdot 10^{-5}$ G, respectively, and the fluctuating field amplitude is given by $|\delta B_0|/|B_0| = 0.1$ (Breitschwerdt 1994).

with the energy output $\nu_{SN} E_{SN}$ from M 82 due to SN explosions. Taking $E_{SN} = 10^{51}$ erg, we obtain

$$\frac{1}{2}\dot{M}u^2 = (0.16 - 0.32)\,\nu_{SN} E_{SN} \tag{11}$$

if $\nu_{SN} = 1/3\,\mathrm{yr}^{-1}$ to $1/6\,\mathrm{yr}^{-1}$, as shown by the radio and far infrared observations (Kronberg *et al.*, 1985). Efficiencies Θ_{gal} between 0.2 and 1 had been used in the previous section.

6. Gamma-Ray Fluxes

Eq.(6) permits an estimate of the γ-ray luminosity and fluxes. We shall consider here only pion decay γ-rays from nuclear collisions of CR nuclei with thermal particles. This clearly gives a *lower limit.* Electron fluxes are difficult to quantify even though they might contribute sizeably through inverse Compton γ-rays. The integral γ-ray photon luminosity of a source with gas number density n and CR energy density $E_c = E_{CR}/V$ in the volume V is given by $L_\gamma = \int q_\gamma \, n \, E_c \, d^3r$, where the spatial integral extends over V, and where q_γ is the γ-ray emissivity normalized to the CR energy density. The values of q_γ have been tabulated for power law spectra with various spectral indices α by Drury *et al.* (1994) *.

If the CRs can be assumed to penetrate the source volume V uniformly then one can perform the volume integral over the target density alone. Given that the CR sources, i.e. the galaxies and their wind termination shocks, are rather uniformly distributed over the cluster volume, this is quite a good approximation. For distant sources we obtain the integral γ-ray photon flux for γ-energies $E > 100\,\mathrm{MeV}$

$$F_\gamma(>100\,\mathrm{MeV}) = 8.4 \cdot 10^{-9} \left(\frac{E_c}{1\,\mathrm{eV/cm^3}}\right) d_{100}^{-2} \, M_{14}\,, \tag{12}$$

and in the TeV region

$$F_\gamma(\geq E) = 1.7 \cdot 10^{-12} \left(\frac{E}{1\,\mathrm{TeV}}\right)^{-1.1} \left(\frac{E_c}{1\mathrm{eV/cm^3}}\right) d_{100}^{-2} \, M_{14}\,, \tag{13}$$

where $d_{100} \equiv d/100\,\mathrm{Mpc}$ and $M_{14} \equiv M/10^{14} M_\odot$ denote the source distance and the diffuse source mass , respectively. Here and in the following, integral γ-ray fluxes are given in units of photons $\mathrm{cm^{-2}\,s^{-1}}$. Since we expect a possible deviation from a power law at most at proton energies above $10^{15}\,\mathrm{eV}$ for SNR-produced particles, and at much higher energies for particles accelerated at galactic wind termination shocks, we can ignore any *cutoff* in the momentum spectrum as long as we consider γ-rays of TeV energies and below. In addition, the source spectra from these shocks will both be very hard, and we will use a particle spectral index $\alpha = 4.1$; for galactic wind termination shocks, α might in fact be even smaller. This implies $q_\gamma = 0.46 \cdot 10^{-13}$ for integral emission above 100 MeV, and $q_\gamma = 1.0 \cdot 10^{-17}$ for the integral emission above 1 TeV, in units of photons $\mathrm{s^{-1}\,erg^{-1}\,cm^3\,(H-atom)^{-1}}$.

Let us first calculate the fluxes from M 82.

The total (largely molecular) mass in the central starburst region of volume $V = \pi\,(300\,\mathrm{pc})^2\,(200\,\mathrm{pc})$ is $M = 2 \cdot 10^8 \mathrm{M}_\odot$, corresponding to a mean gas density of $140\,(\mathrm{H-atom})\,\mathrm{cm^{-3}}$ (I. Grenier, private communication). Taking as CR energy density in this region the equipartition value $E_c = E_{\mathrm{CR}}/V = 80\,\mathrm{eV\,cm^{-3}}$, which exceeds the E_c-value for the ISM of our Galaxy by two orders of magnitude (e.g. Völk *et al.*, 1989), and $d = 3.2\,\mathrm{Mpc}$, results in $F_\gamma(>100\,\mathrm{MeV}) = 1.3 \cdot 10^{-9}$, and at $E \gg 100\,\mathrm{MeV}$ $F_\gamma(>E) = 2.6 \cdot 10^{-13} (E/1\,\mathrm{TeV})^{-1.1}$.

These are rather low numbers. Even though the angular size of M 82 is less than an arcminute, the estimated pion decay γ-ray flux is not detectable

* These α's are not to be confused with the slopes of the correlation curves in Fig. 2

by present space instruments like EGRET in the $E > 100\,\mathrm{MeV}$ energy range. Also a TeV detection probably needs an array which must be larger than those arrays which are presently operating or forthcoming, which have typical limiting flux sensitivities of about 10^{-12} photonscm^{-2} s^{-1}. Other nearby starburst galaxies should have comparable fluxes.

For clusters of galaxies the situation is different. For the Perseus cluster, the expected CR energy is given by Eq.(6). Let us distinguish two cases:

In the first we choose, maximally, $\Theta_{\mathrm{gal}} = 1$, i.e. a full transfer of Type II SN-dominated galactic energy production to galactic wind energy, together with $\delta M_{\mathrm{Fe}} = 0.1\mathrm{M}_{\odot}$, $\Theta_{\mathrm{CR}}^{\mathrm{gw}} = 0.3$, and an equal amount of iron in galaxies and in the IC gas, $\epsilon_{\mathrm{cl}} = 0.7$, to obtain $E_{\mathrm{CR}}^{\mathrm{cl}} = 3.7 \cdot 10^{63}\,\mathrm{erg}$, roughly one third of the thermal gas energy. This gives $F_{\gamma}(> 100\,\mathrm{MeV}) = 2.5 \cdot 10^{-8}$ and at much higher energies $F_{\gamma}(> E) = 5.1 \cdot 10^{-12}\,(E/1\,\mathrm{TeV})^{-1.1}$.

As *minimal*, although in our view unrealistically pessimistic, model parameters we choose $\Theta_{\mathrm{gal}} = 0.2$, together with the high value $\delta M_{\mathrm{Fe}} = 0.2\,\mathrm{M}_{\odot}$, advocated by Renzini *et al.* (1993), $\epsilon_{\mathrm{cl}} = 0.7$, and a very low $\Theta_{\mathrm{CR}}^{\mathrm{gw}} = 0.1$, to obtain $F_{\gamma}(\geq 100\,\mathrm{MeV}) = 2.3 \cdot 10^{-9}$ and, in the TeV range, $F_{\gamma}(\geq E) = 4.7 \cdot 10^{-13}(E/1\,\mathrm{TeV})^{-1.1}$.

Taken at face value the maximal TeV fluxes should be detectable by forthcoming systems of imaging atmospheric Cherenkov telescopes.

7. Discussion and Conclusion

The γ-ray fluxes from clusters of galaxies, calculated in the preceding section, do not include any irreversible energization subsequent to the initial acceleration at galactic termination shocks. This is certainly not quite true, since the continued formation of clusters until the present will not only imply adiabatic compressions but also shocks (e.g. Schindler and Müller, 1993) and other entropy generating dynamical processes. Thus, while the dominance of the IC gas does not allow an increase in the total gas energy, it is nevertheless possible to *transfer* energy between different gas components. In this way CR energy can be produced at the expense of energy in the thermal gas. Therefore our estimates for the CR energy are in principle underestimates.

The quantitative implications depend on the detailed history of the system. We mention here the empirical finding by Gavazzi *et al.* (1991), who observe a mean enhancement of the ratio of synchrotron luminosity to far infrared luminosity for cluster spirals relative to spirals in the field, by a factor of five. It has been argued by these authors that this is not due to a decrease of the far infrared emission, but rather due to a synchrotron enhancement through magnetic field compression. An alternative explanation (Völk and Xu, 1994) is that galaxies interact with cold, massive IC gas fragments, which give rise to large scale shocks, accelerating electrons (and by implication also ions). Qualitatively similar effects could also occur in the IC gas.

We have not discussed here the possible consequences of the CR nucleon content for the production of secondary electrons from the decay of charged pions. These electrons constitute part of an expected energetic electron component of the IC medium. In fact, Schlickeiser *et al.* (1987) have argued that rather in-situ accelerated electrons dominate the diffuse radio emission from the COMA cluster, the system, for which they investigated this question. Therefore the energetic electron component may not be a simple indicator for the total nonthermal gas content of clusters, although this question should be reconsidered.

Although even rich clusters are optically thin for TeV photons, there is a distance limitation due to *intergalactic* absorption: Observable sources must be closer than a few hundred Mpc, since otherwise their direct TeV flux will suffer extinction on the optical/infrared intergalactic background radiation fields (Stecker *et al.*, 1992).

Nevertheless we conclude that clusters of galaxies may constitute a *new class* of very high energy γ-ray sources. To the extent that the gas mass increases at least linearly with the galaxy mass, other nearby clusters, like Coma, Pegasus I, UMa I and III, and others could be γ-ray sources with comparable fluxes. Thus the Perseus cluster would not be a unique case. These sources could tell us about the *nonthermal* history of galaxy formation in clusters, in a form that is complementary to the investigation of the *chemical* history through X-ray spectroscopy.

Acknowledgements

We would like to thank A. Mastichiadis, G. E. Morfill, E. Müller, J. Primack, R. Schlickeiser, and M. Steinmetz for critical discussions. Also, we would like to thank M. Malkov and J. Suppanz-Pirsch for their help in preparing the manuscript.

References

Aharonian, F. A., Bhattacharjee, P. and Schramm, D. N.: 1992, *Phys.Rev. D* **46**, 4188.
Aharonian, F. A., Coppi, P. S. and Völk H. J.: 1994, *Astrophys. J.* **423**, L5.
Arnaud, M., Rothenflug, R., Boulade, O., Vigroux, L. and Vangioni-Flam, E.: 1992, *A & A* **254**, 49.
Axford, W. I.: 1981, *Proc. 17th International Cosmic Conference, Paris* **12**, 155.
Bernlöhr, K.: 1992, *A & A* **263**, 54; 1993a, *A & A* **268**, 25; 1993b, *A & A* **270**, 20.
Berezhko, E. G., Yelshin, V. K. and Ksenofontov, L. T.: 1994, *Astroparticle Physics* **2**, 215.
Blumenthal, G. R., Faber, S. M., Primack, J. R. and Rees, M. J.: 1984, *Nature* **311**, 517.
Böhringer, H.: 1994, in *Frontiers of Space and Ground- Based Astronomy*, (eds W. Wamstecker *et al.*), Kluwer Academic Publishers, Dordrecht (NL), p.359.

Breitschwerdt, D. and McKenzie, J. F., Völk H. J.: 1991, *A & A* **245**, 79; 1993, *A & A* **269**, 54.
Breitschwerdt, D.: 1994, Habilitationsschrift, Univ. Heidelberg, 158 p..
Briel, U. G., Henry, J. P. and Böhringer H: 1992, *A & A* **259**, L3.
Chi, X., Dahanayake, C., Wdowczyk, J. W. and Wolfendale, A. W.: 1993, *Astroparticle Physics* **1**, 239.
David, L. P., Arnaud, K. A., Forman, W. and Jones, C.: 1990a, *Astrophys. J.* **356**, 32.
David, L. P., Forman, W. and Jones, C.: 1990b, *Astrophys. J.* **359**, 29; 1991a, *Astrophys. J.* **369**, 121; 1991b, *Astrophys. J.* **380**, 39.
Dressler, A.: 1993, *Sky and Telescope* **4**, 22.
Drury, L. O'C., Aharonian, F. A. and Völk, H. J.: 1994, *A & A* **287**, 959.
Edge, A. C.: 1989, PhD Thesis, Leicester University.
Elbaz, D., Arnaud, M., Cassé, M., Mirabel, I. F., Prantzos, N., Vangioni-Flam, E.: 1992, *A & A* **265**, L29.
Elbaz, D., Arnaud, M., Vangioni-Flam, E.: 1994, to appear in *A & A*.
Fabian, A. C., Hu, E. M., Cowie, L. L. and Grindlay, J.: 1981, *Astrophys. J.* **248**, 47.
Gavazzi, G., Boselli, A., Kennicutt, R.:1991, *Astron. J.* **101**, 1207.
Güsten R. and Mezger, P. G.: 1983, *Vistas Astr.* **26**, 159.
Habe, A., Ikeuchi, S.: 1980, *Prog. Theor. Phys.* **4**, 1995.
Hatsukade, A. C.: 1989, PhD Thesis, Osaka University.
Jaffe, W. J.: 1977, *Astrophys. J.* **212**, 1.
Jokipii, J. R., Morfill, G. E.: 1985, *Astrophys. J.* **290**, L1; 1987, *Astrophys. J.* **312**, 170.
Kerrick A D: 1995, *Astrophys. J.* **438**, L59.
Kronberg, P. P.: 1994, *Rep. Prog. Physics* **57**, 325.
Kronberg, P. P., Biermann, P. L. and Schwab, F. R.: 1985, *Astrophys. J.* **291**, 693.
Larson, R. B.: 1986, *MNRAS* **218**, 409.
Malkov M. A. and Völk H. J.: 1994, to appear in *A & A*.
Montigny, C. von *et al.*: 1995 *Astrophys. J.*, in press.
Ohashi, T.: 1994, Highlight Talk at 17th Texas Symposium on Relativistic Astrophysics, Munich.
Pietsch, W., Böhringer, H. and Morfill, G. E.: 1995, in Cosmic Winds and the Heliosphere, (eds. J. R. Jokipii *et al.*), Univ. Arizona Press, Tucson, in press.
Punch, M. *et al.*: 1992, *Nature* **358**, 477.
Renzini, A., Ciotti, L., D'Ercole, A. and Pellegrini, S.: 1993, *Astrophys. J.* **419**, 52.
Roland, J.: 1981, *Astrophys. J.* **93**, 407.
Roland, J., Sol, H., Paulini-Toth, I. and Witzel, A.: 1981, *A & A* **100**, 7.
Schaaf, R., Pietsch, W., Biermann, P. L., Kronberg, P. P. and Schmutzler, T.: 1989, *Astrophys. J.* **336**, 722.
Schindler, S., Müller, E.: 1993, *A & A* **272**, 137.
Schlickeiser, R., Sievers, A. and Thiemann, H.: 1987, *A & A* **182**, 21.
Schwarz, R. A., Edge, A. C., Voges, W., Böhringer, H., Ebeling, H. and Briel, U. G.: 1992, *A & A* **256**, L11.
Steinmetz, M.: 1993, PhD thesis Technische Universität München.
Stecker, F. W., de Jager, O. C., Salamon, M. H.: 1992, *Astrophys. J.* **390**, L49.
Tinsley, B. M.: 1980, *Fund. Cosm. Phys.*, **5**, 287.
Völk, H. J., Klein, U. and Wielebinski, R.: 1989, *A & A* **213**, L12.
Völk, H. J., Xu, C.: 1994, *Infrared Phys. Tech.* **35**, 527.
White, S. D. M., Frenk, C. S., Davis, M. and Efstathiou,G.: 1987, *Astrophys. J.* **313**, 517.
Zirakashvili, V. N., Breitschwerdt, D., Ptuskin, V. S. and Völk, H. J.: 1993, in Proc. *23rd International Cosmic Ray Conference, Calgary* **2**, 287.

GAMMA-RAYS FROM ACTIVE GALACTIC NUCLEI: EMISSION PROCESSES AND PARTICLE ACCELERATION

REINHARD SCHLICKEISER
Max-Planck-Institut für Radioastronomie, Postfach 2024, 53010 Bonn, Germany

Abstract. It is argued that the high-energy X-ray and γ-ray emission from flaring blazars is beamed radiation from the relativistic jet supporting the relativistic beaming hypothesis and the unified scenario for AGNs. Most probably the high-energy emission results from inverse Compton scattering by relativistic electrons and positrons in the jet of radiation originating *external* to the jet plus pair annihilation radiation from the jet. Future positive TeV detections of EGRET AGN sources will be decisive to identify the prominent target photon radiation field. Direct γ-ray production by energetic hadrons is not important for the flaring phase in γ-ray blazars, but the acceleration of energetic hadrons during the quiescent phase of AGNs is decisive as the source of secondary electrons and positrons through photo-pair and photo-pion production. Injection of ultrahigh energy secondary electrons and positrons into a stochastic quasilinear acceleration scheme during the quiescent AGN phase leads to cooling electron-positron distribution functions with a strong cut-off at low but relativistic energy that under certain local conditions may trigger a plasma instability that gives rise to an explosive event and the flaring γ-ray phase.

Key words: Gamma rays – galaxies: nuclei – galaxies: active – radiation processes – particle acceleration

1. Introduction

Since its launch in April 1991, the experiments on board of the Compton Gamma Ray Observatory have detected many extragalactic active galactic nuclei (AGNs) at photon energies > 20 keV:
(1) 45 AGNs have been reported as ($> 3\sigma$) sources of >100 MeV γ rays (Fichtel et al. 1994, Kanbach 1994, Dermer and Gehrels 1995). Several of them are also seen by the COMPTEL and OSSE instruments (Kurfess 1994). All these AGNs belong to the *blazar category*, i.e. they exhibit high optical polarization, rapid optical variability, flat-spectrum radio emission from a compact core, and very often apparent superluminal ($V_{app} > c$) motion components.
(2) 24 AGNs have been detected by the OSSE experiment in the energy range 50-150 keV (Kurfess 1994): 13 of these are classified as Seyfert 1 galaxies, 5 are Seyfert 2s, 2 are the quasars 3C273 and 3C279 and one is the BL-Lac object PKS 2155-304. Five of the sources are also radio galaxies. No Seyfert AGNs or radio galaxies are detected at photon energies above $E_\gamma > 1$ MeV.

Apparently, there exist two classes of γ-ray emitting AGNs (Dermer and Gehrels 1995):
(I) *γ-ray hard blazars* with power law energy spectra extending to energies $\geq$ 10 GeV, and in the case of Mrk 421 even to TeV energies (Punch et

Space Science Reviews **75**: 299–315, 1996.

al. 1992). The γ-ray emission in these sources most probably is nonthermal radiation from relativistic jets;
(ii) *γ-ray soft Seyferts* with energy spectra that cut-off sharply above ~ 100 keV. The γ-ray emission from these sources most probably is hot thermal emission from the inner accretion disk.

Here I will concentrate on the first class of γ-ray AGNs and discuss the implications of the observations for the nonthermal radiation and particle acceleration processes.

The most remarkable properties of the γ-ray hard blazars are
(1) the high fraction (>17 percent) of superluminal sources in the sample: 0234+285, 0528+134, 0836+710, 3C273, 3C279, 1633+382, 2251+158;
(2) the dramatic peak of the luminosity spectra (=power per $\lg \nu$) at γ-ray frequencies which indicates the importance of nonthermal emission processes in these objects;
(3) the short time variability of the γ-ray emission which is observed in most sources (Montigny et al. 1995). Correcting for the redshift z of the sources the intrinsic variability time scale in the cosmological frame $\delta t_i = \delta t_{obs}/(1+z)$ becomes as short as 0.65 days (0528+134), 0.70 days (1633+382), and 1.3 days (3C279), implying a very compact source from simple light-travel arguments

$$R \leq c\delta t_i \qquad (1).$$

(4) the time-correlated monitoring of the sources at lower frequencies (e.g. Reich et al. 1993, Pohl et al. 1995, Lichti et al. 1995) which very often demonstrates that the γ-ray flare seems to proceed outbursts at lower frequencies, although exceptions from this rule are frequent.

2. Constraints for AGN models from the γ-ray observations

Because of its rapid variability, its high compactness, lack of radio-quiet sources in the sample and high presence of superluminal motion sources it is generally agreed (Blandford 1993, Dermer 1993, Dermer and Schlickeiser 1992) that the γ-ray emission originates in strongly beamed sources, in accord with the relativistic beaming hypothesis and the unified scenario for AGNs that has served well as the baseline model for the central engines of active galactic nuclei (Blandford and Rees 1978, Scheuer and Readhead 1979, Blandford 1990). Simplifying many details, its basic features read as follows: (i) AGNs are powered by accretion of matter onto a supermassive black hole; (ii) in the case of radio-loud AGNs, collimated relativistic jets composed of shocks or plasma blobs move with speed $V_{jet} > 0.7c$ outward along the axes of the accretion disk. The orientation angle θ^* of the jet axis with respect to the observer in individual AGNs is random; and (iii) the line-of-sight of the observer with respect to the jet axis determines the classification of the particular AGN (for review see Antonucci (1993) and Dermer (1993)).

The strongest arguments in favour of the conclusion, that the high-energy X-ray and γ-ray emission is *beamed radiation from the jet*, are both, the absence of intrinsic $\gamma - \gamma$ pair absorption features in the observed γ-ray energy spectra, and the violation of the Elliot-Spapiro relation in at least four γ-ray AGNs. We discuss both in turn.

2.1. Intrinsic compactness

Let us discuss the hypothesis that the high-energy X-ray and γ-ray blazar emission originates as isotropic emission from an uniform spherical (of radius R) source at rest. The optical depth for intrinsic $\gamma + \gamma \rightarrow e^+ + e^-$ attenuation then is (Dermer and Schlickeiser 1994) determined by the source compactness l, i.e. the ratio of the intrinsic source luminosity L to its radius R,

$$\tau \simeq \sigma_T n_\gamma R = \frac{\sigma_T}{4\pi c < \epsilon >} \frac{L}{R} \qquad (2a),$$

where σ_T denotes the Thomson cross section. Using a mean photon energy $< \epsilon >= 1$ MeV and scaling the intrinsic luminosity in units of the solar luminosity $L_{sun} = 2 \cdot 10^{33}$ erg s^{-1} Eq. (2a) becomes

$$\tau = 10^3 \frac{L/L_{sun}}{R/cm} \qquad (2b).$$

The measured γ-ray fluxes above 1 MeV and the known redshifts of the γ-ray blazars imply isotropic intrinsic luminosities of typically $L = 10^{48} L_{48}$ erg s^{-1} (Fichtel et al. 1994) and using the size constraint (1) from the observed time variability in Eq. (2b) yields values of the optical depth

$$\tau \geq 200 \frac{L_{48}}{\delta t_i / 1\ day} \qquad (3)$$

much larger than unity. Such large values for the optical depth for pair attenuation are incompatible with the observed straight power law γ-ray energy spectra over four decades of energy, and rule out isotropic emission at rest.

2.2. Violation of the Elliot-Shapiro relation

A second argument against simple isotropic high-energy emission from spherical accretion is given by applying the argument by Elliot and Shapiro (1974) originally developed for X-ray observations. For spherical black hole accretion the source luminosity is limited by the Eddington luminosity

$$L \leq L_{Edd} < 1.3 \cdot 10^{38} \frac{(M/M_{sun})}{\alpha}\ ergs^{-1} \qquad (4),$$

where M is the mass of the black hole and $\alpha \leq 1$ is a correction factor (Pohl et al. 1995, Dermer and Gehrels 1995) that accounts for the deviation of the Klein-Nishina from the Thomson cross section when calculating the balance between radiation and gravitation forces. Alternatively, the minimum intrinsic time scale for flux variations is

$$\delta t_{i,min} \geq \frac{R_s}{c} = 0.98 \cdot 10^{-5} (\frac{M}{M_{sun}}) \ s \qquad (5),$$

since for the generated high-energy photons to escape from the central source the source size has to be larger than the Schwarzschild radius R_s. Eliminating the mass of the black hole from Eqs. (4) and (5) yields

$$\lg \delta t_{i,min}(s) \geq \lg L(erg \ s^{-1}) - 43.1 + \lg \alpha \qquad (6),$$

which for $\alpha = 1$ agrees with the Elliot-Shapiro relation. Scaling the observed isotropic unbeamed γ-ray luminosity again as $L = 10^{48} L_{48}$ erg s^{-1} we obtain from Eq. (6) that

$$\lg[\frac{\delta t_{obs}/(1+z)}{\text{days}}] \geq \lg L_{48} + \lg \alpha \qquad (7).$$

In Fig. 1 we compare the observed variability time scales and γ-ray luminosities with the Elliot-Shapiro relation, i.e. Eq. (7) calculated with $\alpha = 1$. It can be seen that four sources (0528+134, 1633+382, 3C279. 1406-076) violate the Elliot-Shapiro relation since they lie in the shaded area below the dividing line. At least for these four sources isotropic emission from a simple steady spherical accreting source at rest is ruled out.

Of course, correcting for deviations from the Thomson cross section, i.e. using a value of $\lg \alpha = -1.3$ for a mean photon energy of $< \epsilon >= 10 MeV$, will move the dividing line to the right so that 3C279 and 1406-076 no longer violate this restriction (as has been pointed out by Pohl et al. (1995) and Dermer and Gehrels (1995)); but on the other hand most of this correction is couterbalanced by the fact that the actual source size is certainly larger than the Schwarzschild radius, say $\sim 10 R_s$, so that Eq.(7) with $\alpha = 1$ still serves as a good approximation.

2.3. Relativistic beaming

The discussed absence of photon pair attenuation (section 2.1) and the violation of the Elliot-Shapiro relation (section 2.2) force us to discard the hypothesis of high-energy X-ray and γ-ray emission originating in isotropic sources at rest, and strongly indicate that this emission originates in a source that moves with large bulk speed with respect to us. The natural candidate for this moving source is the relativistic jet hypothesized to exist in AGNs according to the unified scenario. The theory of special relativity

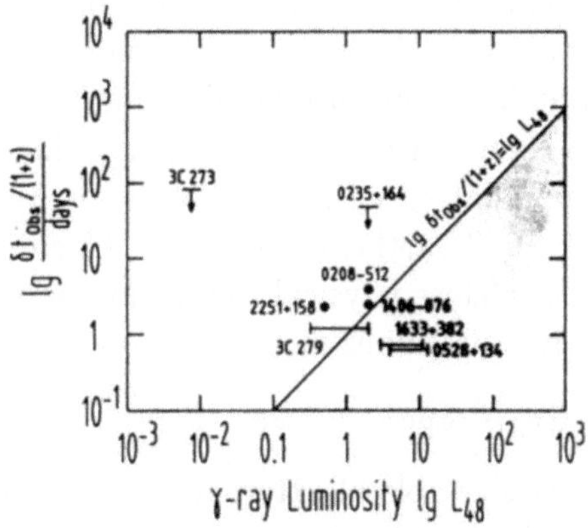

Fig. 1. Comparison of the observed γ-ray variability time scales and luminosities with the Elliot-Shapiro relation for several γ-ray blazars

then provides the relation between observed (denoted by an asterisk *) and intrinsic physical quantities (Rybicki and Lightman 1979, Begelman et al. 1984), as

$$\text{the apparent luminosity } L^*_{app} = \delta^n L, \quad 3 \leq n \leq 4 \qquad (8a),$$

$$\text{the blueshifted frequency } \nu^* = \delta \nu \qquad (8b),$$

$$\text{the time dilatation } t^* = t/\delta \qquad (8c),$$

where the Doppler factor

$$\delta = [\Gamma(1 - \beta \cos\theta^*)]^{-1}, \;\; \beta = V_{jet}/c, \;\; \Gamma = (1 - \beta^2)^{-1/2} \qquad (8d)$$

is determined by the jet's bulk speed and orientation angle. The factor n in the Doppler boosting formula (8a) is model dependent, but for many relevant radiation model lies between 3 and 4. A consequence of the Doppler boosting (8a) is that observers are particularly sensitive to AGNs whose jet axis points towards them ($\theta \to 0$) since the apparent luminosity from these objects is Doppler boosted by a large factor ($\to (2\Gamma)^n$) compared to intrinsically equally bright AGNs with misaligned jet directions.

Since the *intrinsic* luminosity and source size enter the calculation of the compactness in Eqs. (2) we find with Eqs. (8)

$$l = L/R = \delta^{-(1+n)} \frac{L^*_{app}}{ct^*} \qquad (9)$$

a dramatic reduction by the factor $\delta^{(1+n)}$ in the value of the optical depth (3) to values smaller unity provided $\delta \geq 6$. The same reduction factor enters the calculation of the Elliot-Shapiro relation and then implies the non-violation of this relation for the four sources 0528+134, 1633+382, 3C279 and 1406-076. 3C279 is already a well-known superluminal source and the remaining three are prime candidates for superluminal motion. This prediction has been verified by the recent discoveries of superluminal motion components in 0528+134 with apparent speed $\simeq$15c (Pohl et al. 1995) and in 1633+382 with apparent speed $\simeq$6c (Barthel et al. 1995). This discovery also provides conclusive evidence for a close physical connection between γ-ray flaring and the ejection of superluminal blobs in blazars.

2.4. External compactness

In this context the decisive role of external compasctness for γ-ray production has to be emphasized. It has been pointed out by many authors (Carraminana 1992, Bednarek 1993a, Blandford 1993, Dermer and Schlickeiser 1994, Sikora et al. 1994, Blandford and Levinson 1995) that target photon fields external to the jet plasma like the accretion disk photons and/or the reprocesses disk photons in the emission line clouds provide compactness >> 1 inside the γ-ray photosphere. The optical depth for $\gamma - \gamma$ absorption (Dermer and Schlickeiser 1994) produced in the jet at height z above the central black hole due to interactions with

-- with accretion disk photons $$\tau_A = 6 \cdot 10^{-6} \frac{L_{46} M_8}{\Theta_{-4} z^2(pc)} \frac{F(E_\gamma)}{F(E_{\gamma,max})} \quad (10),$$

-- with reprocessed photons $$\tau_s = 0.06 E_\gamma^{0.7}(GeV) L_{46} \tau_{-2} \ln(z/z_i) \quad (11),$$

where L_{46} denotes the disk luminosity in units of 10^{46} erg/s, M_8 the black hole mass in units of 10^8 solar masses, τ_{-2} the optical depth for photon scattering in the emission cloud material in units of 0.01,$\Theta_{-4} = k_b T_A/(m_e c^2)$ the dimensionless accretion disk temperature in units of 10^{-4}. The height z_i denotes the base of the relativistic jet, $E_{\gamma,max} = 3.6/\Theta_{-4}$ GeV is defined by the maximum value of the function

$$F(x) = \left[x^3[\exp(\frac{2}{\Theta x}) - 1]\right]^{-1} \quad (12)$$

characterizing the accretion disk's thermal radiation field. The estimate (10) yields γ-ray transparency ($\tau_A < 1$) at heights greater

$$z(pc) > 2.5 \cdot 10^{-3} L_{46}^{1/2} M_8^{1/2} \Theta_{-4}^{-1/2} \quad (13),$$

which corresponds to $\simeq 500 M_8^{-1/2} R_s$ Schwarzschild radii. At smaller heights the generated γ-ray photons are strongly attenuated. This large size of the γ-ray photosphere is fatal for γ-ray production models close to the central black hole as proposed by Slane and Wagh (1990) and Bednarek (1993b).

The interaction with the reprocessed photons (11) provides transparency ($\tau_s < 1$) at all EGRET energies, in contradiction to Blandford's (1993) conclusion, but would yield strong photon attenuation at TeV energies,

$$\tau_s = 7.5 E_\gamma^{0.7}(TeV) L_{46} \tau_{-2} \ln(z/z_i) \quad (14).$$

As a consequence quasars, which have emission line clouds, should be no sources of TeV γ-ray emission. Nearby lineless BL-Lac objects are more likely to be seen at TeV energies.

3. Nonthermal γ-ray emission processes in jets

We now turn our attention to the relevant γ-ray production processes in relativistic jets under the hypothesis that the jets sometimes are filled with blobs containing relativistic charged particles (e^+,e^-,p?) and thermal particles (p,e^-) which are confined to the jet by suitable magnetic fields of strength B_o. The blobs move with relativistic bulk speed V_{jet}, $\Gamma >> 1$. The origin of these jets and blobs are still a matter of intensive debate. Models include MHD winds ejected from the surface of the rotating accretion disk (for review see Blandford 1990, Camenzind 1990) as well as current-free relativistic e^+-e^- beams (Pelletier and Sol 1992), and it is our hope that the γ-ray observations will provide an important test for these ideas.

Moreover, we assume that at the start of the γ-ray flare, which is related to the release of a new blob, the relativistic particles in the blob have an initial isotropic power law energy spectrum at height z_i:

$$N(\gamma_i, z_i) = N_o \gamma_i^{-s}, \quad \text{for } \gamma_1 \leq \gamma_i \leq \gamma_2 \quad (15).$$

As argued in section 4 this relativistic particle distribution function may result from "slow" acceleration processes during a "quiescent" AGN phase before the γ-ray flare, and there is no need for rapid particle energization during the γ-ray flare rise time. The total density of relativistic electrons and positrons N_o in Eq. (15) is assumed to be so small that the inverse Compton optical depth

$$\tau_{IC} = \sigma_T N_o r_b = 0.6\left(\frac{N_o}{10^{11} cm^{-3}}\right)\left(\frac{r_b}{10^{13} cm}\right) \leq 1 \quad (16)$$

is smaller unity. r_b denotes the size of the blob.

3.1. The role of energetic hadrons

I restrict my discussion to the emission processes of relativistic electrons and positrons and do not discuss the emission from relativistic hadrons because I am convinced that they are less important (but see Mannheim 1995, Mastichiadis 1995 for an alternative view) for the following reasons:

(1) due to their much smaller masses electrons and positrons radiate more efficient than nuclei charge Ze and mass Am_p: for pointlike electromagnetic interactions the "hadron Thomson" cross section

$$\sigma_H = \frac{Z^4}{A^2}(\frac{m_e}{m_p})^2\sigma_T = 1.97 \cdot 10^{-31}\frac{Z^4}{A^2} \quad \mathrm{cm}^2 \tag{17}$$

is several orders of magnitude smaller than the Thomson cross section;
(2) inelastic hadron (A) interactions as $A + p \rightarrow (\pi^o, \pi^+, \pi^-)$, $A + \gamma \rightarrow A + e^+ + e^-$, $A + \gamma \rightarrow A + (\pi^o, \pi^+, \pi^-)$, $A + \gamma \rightarrow (A-1) + N$ (Kazanas and Ellison 1986, Sikora et al. 1987) provide many *secondary* electrons and positrons, and, although hadrons serve as source, the observed radiation is generated in interactions of these secondary electrons and positrons;
(3) the photopair, pion and photopion production processes mentioned under (2) operate most efficiently in regions of high target photon and matter densities, i.e. close to the central black hole. But as shown in section 2.4 such a thick target will be optically thick for the generated γ-rays, and therefore not in accord with the observed straight power law γ-ray energy spectra and the short variability.

To summarize: hadron acceleration may be important during the quiet phase of γ-ray AGNs as source of plenty secondary electrons and positrons, but during the flaring phase the γ-ray radiation is produced by electrons and positrons probably of secondary origin. We will come back to this important issue in section 4.

3.2. Nonthermal γ-ray production by relativistic electrons and positrons

There are three main radiation production processes by relativistic electrons and positrons that are important in relativistic jets:
(1) *the synchrotron-self Compton (SSC) process*: the relativistic electrons and positrons in the jet inverse Compton scatter their own synchrotron radiation to γ-ray energies (Bloom and Marscher 1993, Maraschi et al. 1993);
(2) *inverse Compton scattering of radiation external to the jet*: The relativistic electrons and positrons in the jet inverse Compton scatter

– the accretion disk photons (Dermer et al. 1992, Dermer and Schlickeiser 1993). The anisotropy of the incoming target photons in combination with the angular dependence of the relativistic Doppler boosting favours γ-ray emission at observer's angles where the source exhibits superluminal motion. This γ-ray production process succesfully explains the key observational features as (i) the peak of the bolometric luminosity at γ-ray frequencies (ii) the γ-ray variability time scale $\leq$ few days, (iii) the softening of the spectra between the hard X-ray and medium γ-ray regime, (iv) the high percentage of superluminal sources in the sample, (v) the generation of TeV emission for favourably aligned observers;

– the accretion disk photons reflected from the material through which the jet is passing as e.g. the emission line clouds in quasars (Sikora et al. 1994);

– photons from the universal 2.7 K microwave background and optical photons from the central star cluster;

(3) *pair annihilation radiation*: the relativistic electrons and positrons in the jet annihilate and including all spectral broadening effects due to the energy distribution of the annihilating particles and the jet bulk motion give rise to broad continuum emission in the MeV γ-ray energy range (Henri et al. 1993, Böttcher and Schlickeiser 1995). If the blobs in the jet also contain thermal electrons the relativistic positrons additionally can annihilate with the thermal electrons.

3.3. Comparison of leptonic γ-ray production processes

The observed EGRET γ-ray energy spectra alone are inconclusive to decide on the issue of the most important γ-ray production process in blazars. Advocates of the individual processes have produced remarkable good fits to the EGRET observations. The best tests for the proposed models will be provided by multifrequency time-correlated monitoring of these sources that now are operating (e.g. Lichti et al. 1995).

It is important to realize that probably all three basic radiation mechanisms contribute to the observed γ-ray emission during different phases of the flare with individual contributions that vary as a function of time or height z of the blob above the central object. Moreover, each of these three γ-ray production processes has different Doppler boosting factors and antenna characteristics so the estimate of the relative importance depends strongly on the observer's viewing angle. Another complicating factor is the shortness of energy loss times of the radiating particles implying that the time evolution of the energy spectra of the relativistic electrons and positrons has to be taken into account; not all theoretical calculations include this crucial issue. With all these caveats on the basis of estimates for the total electron loss rates in the various interaction processes (see Dermer and Schlickeiser 1993, 1994 for details), we find that for parameters appropriate to 3C279, the inverse Compton scattering of accretion disk photons dominates the production of γ-rays by

– microwave scattering if $z(\mathrm{pc}) < 360 \frac{L_{46}^{1/2}}{(\Gamma/10)^2(1+z)^2}$,

– Compton scattering of stellar optical photons if $z(pc) < 4.3\ \frac{(L_d/L_{opt})^{1/2}}{(\Gamma/10)}$,

– the synchrotron-self Compton process if $z(pc) < 0.1\ \frac{(L_{46}^{1/2}}{\Gamma(B_o/1\ \mathrm{G})}$,

– Comptonization of scattered radiation if $z(pc) < 0.06\ \frac{M_8^{1/3}\ R_{ELC}^{2/3}(pc)}{\tau_{-2}^{1/3}}$,

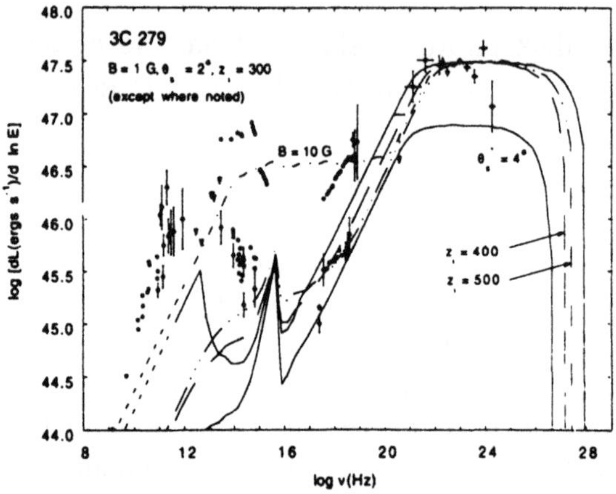

Fig. 2. Model results for the time-integrated luminosity spectrum compared to the multiwavelength observations of 3C279. The radio-through-optical synchrotron emission, which is probably made farther out in the radio jet, is not modelled here. z_i denotes the starting height of the blob at the beginning of the flare in units of half of the Schwarzschild radius of the central object.

where M_8 denotes the central black hole mass in units of 10^8 solar masses, R_{ELC} the distance in units of pc of the emission line clouds from the central black hole, and τ_{-2} the electron scattering optical depth in emission line clouds in units of 10^{-2}. Note that the lack of an observed soft X-ray turnover due to photoelectric absorption in contemporaneous GINGA observations of 3C279 constrains the value $\tau_{-2} \leq 0.25$.

The above constraints together with the size estimate (13) for the γ-ray photosphere place the location of the γ-ray emission site at distances beyond $z \geq 10^{15}$ cm from the central object, which is comparable to the radial extent of the accretion disk and is determined by the obvious absence of $\gamma - \gamma$ pair attenuation in the observed γ-ray spectra.

γ-ray production by inverse Compton scattering of accretion disk photons by outflowing relativistic electrons and positrons in a relativistic jet is a very economical process: for favorable viewing conditions of the observer a total energy in relativistic electrons and positrons in the blob of

$$E_{tot} \geq 10^{50}(r_b/10^{15}\ \text{cm})^2(\gamma_1/10)\ \text{erg} \tag{18}$$

is sufficient to explain the observed intensities. Combined with the observed γ-ray luminosity $L(>100\ \text{MeV}) = 10^{48}L_{48}f\ \text{erg s}^{-1}$, where f is the beaming factor, Eq.(18) implies that the intrinsic relativistic electron and positron loss time

$$t_R = E_{tot}/L(>100\ \text{MeV}) \geq 10^4 L_{48}^{-1}(f/10^{-2})^{-1}\text{sec} \qquad (19).$$

Coupled with the Doppler time-dilation effect, we see from Eq. (19) that very short γ-ray flare durations on the order of a few days can easily be explained in this model. However, the short radiation loss time (19) implies that the time evolution of the electron energy spectrum during the γ-ray flare has to be taken into account consistently. Dermer and Schlickeiser (1993) calculated the time-integrated γ-ray emission from a modified Kardashev-model by instantaneously injecting the power law electron distribution (15) at height z_i at the beginning of the flare, calculating its modification with height in the relativistically outflowing blob due to the operation of the various inverse Compton energy loss processes, and finally integrating the hard X-ray and γ-ray emissivities over height. Comparison with multiwavelength observations of 3C279 is shown in Fig. 2. One notices the good fit to the high-energy emission from this source, including the break at MeV energies which results from incomplete relativistic electron cooling. Moreover the peak of the luminosity spectrum at γ-ray energies is well reproduced by the model.

4. Remarks on the quasilinear particle acceleration in the dense AGN plasma during the quiescent phase

Repeated γ-ray observations indicate a typical duty cycle of γ-ray hard blazars of about 5 percent supporting a "2-phase" kinetic plasma model for the central regions of AGNs (Achatz et al. 1990, Schlickeiser and Achatz 1992). According to the 2-phase model the central powerhouse of AGNs undergoes two repeating phases: in a "quiescent phase" over most of the time (95 percent) relativistic charged particles are efficiently accelerated in the central plasma near the black hole, whereas in a short and violent "flaring phase" the accelerated particles are ejected in the form of plasma blobs along an existing jet structure.

We consider the acceleration of charged particles during the quiescent phase. As sketched in Fig. 3 the central object accretes the surrounding matter through a disk. Associated with the accretion flow is low-frequency magnetohydrodynamic turbulence which is generated by various processes as e.g.:
(a) turbulence generated by the rotating accretion disk at large eddies and cascading to smaller scales (Galeev et al. 1979);
(b) stellar winds from solar-type stars in the central star cluster deliver plasma waves to the accretion flow;
(c) infalling neutral accretion matter becomes ionized by the ultraviolet and soft X-ray radiation of the disk. These pick-up ions in the accretion flow generate plasma waves by virtue of their streaming (Lee and Ip 1987);
(d) if standing shocks form in the neighbourhood of the central object they amplify any incoming upstream turbulence in the downstream accretion shock magnetosheath (McKenzie and Westphal 1969, Campeanu and Schlickeiser 1992).

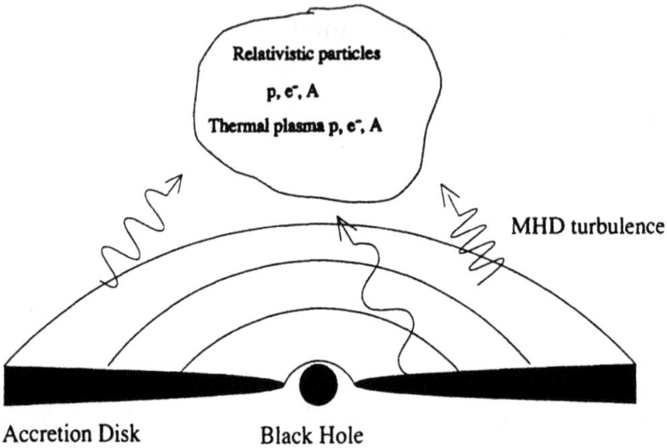

Fig. 3. Sketch of the global structure of the central region of AGNs and particle acceleration during the quiescent phase

These low-frequency MHD plasma waves from the accretion flow are the source of free energy and lead to stochastic acceleration of charged particles out of the thermal plasma.

The dynamics of energetic charged particles (cosmic rays) in cosmic plasmas is determined by their interaction with ambient electromagnetic, photon and matter fields. Among these by far quickest is the particle-wave interaction with electromagnetic fields which very often can be separated into a leading field structure F_o and superposed fluctuating fields δF. Theoretical descriptions of the transport and acceleration of cosmic rays in cosmic plasmas are usually based on transport equations which are derived from the collisionless Boltzmann-Vlasov equation into which the electromagnetic fields of the medium enter by the Lorentz force term. The quasilinear approach to wave-particle interaction is a first order approach in the ratio $q_L \equiv (\frac{\delta F}{F_o})^2$ and requires smallness of this ratio with respect to unity. In most cosmic plasmas this is well satisfied as has been either established by direct in-situ electromagnetic turbulence measurements in interplanetary plasmas, or by saturation effects in the growth of fluctuating fields. Nonlinear wave-wave interaction rates and/or nonlinear Landau damping set in only at appreciable levels of $(\delta F)^2$ and thus limit the value of $q_L \leq 1$. We assume the AGN plasma to have very high conductivity so that any large-scale steady electric fields are absent. We then consider the behaviour of energetic charged particles in a uniform magnetic field with superposed small-amplitude $(\delta B)^2 << B_o^2$ plasma turbulence $(\delta \mathbf{E}, \delta \mathbf{B})$.

We characterize the thermal AGN plasma as a dense plasma, i.e. its electron plasma frequency is large compared to the electron gyrofrequency which is equivalent to values of the ordered magnetic field $B_o << B_c = 3 \cdot 10^{-3}(n_e/cm^{-3})^{1/2}$ G being small with respect to the characteristic magnetic field B_c, and we restrict our analysis to parallel (to the ordered magnetic field) propagating waves. The charged particles then gain energy from cyclotron damping the electric fields $\delta \mathbf{E}$ associated with ten different low-frequency transverse plasma modes (Schlickeiser and Achatz 1993): the four Alfven modes coming as forward and backward moving right-handed or left-handed circularly polarized waves, the two left-handed polarized forward and backward moving ion-cyclotron waves, the two right-handed polarized forward and backward moving Whistler waves, and the two right-handed polarized forward and backward moving electron-cyclotron waves. Determining the relevant quasilinear acceleration rates and transport parameters of non-relativistic charged particles is by no means complete since especially for the interaction with ion- and electron-cyclotron waves thermal resonance broadening effects are particularly important (Schlickeiser and Achatz 1993, Schlickeiser 1994). The acceleration rates and spatial transport parameters can then be used in the kinetic diffusion-convection equation for the isotropic part of the phase space density of charged particles $F(x,p,t)$ which for non-relativistic bulk speed $u << c$ reads

$$\frac{\partial F}{\partial t} - S_o = \frac{\partial}{\partial x}[\kappa \frac{\partial F}{\partial x}] - u\frac{\partial F}{\partial x} + \frac{p}{3}\frac{\partial u}{\partial x}\frac{\partial F}{\partial p} + \frac{1}{p^2}\frac{\partial}{\partial p}[p^2 A \frac{\partial F}{\partial p} - p^2 \dot{p}_{loss} F] \quad (20),$$

where x denotes the spatial coordinate along the ordered magnetic field, p the cosmic ray particle momentum, κ is the spatial diffusion coefficient, A the momentum diffusion coefficient, and S_o denotes the injection source of charged particles.

Stochastic acceleration of particles, characterized by the acceleration time scale $t_A = p^2/A$, competes with continous energy loss processes $\dot{p}_{loss}$, characterized by energy loss time scales $t_L = p/|\dot{p}_{loss}|$, due to Coulomb, ionization and bremsstrahlung intercations with the ambient thermal matter field, synchrotron radiation losses and inverse Compton losses as well as photo-pair and photo-pion losses with the ambient photon radiation fields (see Schlickeiser et al. 1991 and Mannheim and Schlickeiser 1994 for a collection of energy loss rates for electrons and hadrons in AGNs, respectively). Dermer et al. (1995) recently have compared the time scales for escape of particles, continous losses and stochastic acceleration by Alfven and Whistler turbulence, and in Fig. 4 we show their results for energetic protons. At small proton momenta the Coulomb loss time scale is extremely sensitive to the background plasma density and temperature, and for slight changes in the values of these parameters cosmic ray protons may not be accelerated above the Coulomb barrier. Although at small particle momenta the plasma wave's dissipation and the interaction with the cyclotron waves become decisive and might modify the acceleration time significantly, the results of Dermer et al. (1995) shown in Fig. 4 demonstrate that plasma parameter values are possible where the low-frequency turbulence energizes protons to TeV and PeV energies where photo-pair and photo-pion production are effective in halting

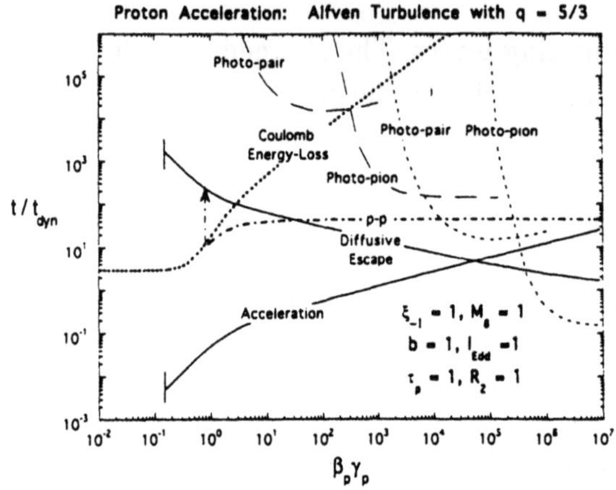

Fig. 4. Acceleration, diffusive escape, and energy-loss time scales for energetic protons through Alfven turbulence with Kolmogorov-type turbulence in a region of $R = 50R_s$ surrounding a supermassive black hole of 10^8 solar masses (from Dermer et al. 1995). A magnetic field strength in equipartition with the radiation energy density and the background plasma energy density, Eddington-limited accretion, and an energy density of wave turbulence equal to 10% of the ambient magnetic field energy density are assumed. The Thomson optical depth and the temperature of the background plasma are chosen to be 1 and 51.1 keV, respectively. The different curves for photo-pair and photo-pion production refer to different assumed temperatures of the background radiation field. The time scales are in units of $t_{dyn} = R/c$.

the acceleration. The corresponding analysis for cosmic ray electrons shows that the external compactness provided by the accretion disk photons leads to heavy inverse Compton losses which suppress the acceleration of low-energy electrons beyond Lorentz factor of $\gamma \approx 10 - 100$. It seems that due to their much smaller radiation loss rate cosmic ray protons are effectively accelerated during the quiescent phase in contrast to low energy electrons.

Now an important point has to be emphasized: *once the accelerated protons reach the thresholds for photo-pair and photo-pion production and the threshold for pion production in inelastic proton-matter collisions they will generate plenty of secondary electrons and positrons of ultrahigh energy* which are now injected at high energies into this acceleration scheme (20). According to the results of Dermer et al. (1995) shown in Fig. 4 it takes about $\simeq 10t_{dyn} = 10M_8$ days for the protons to reach these energies. As solutions of the electron and positron transport equation (20) demonstrate (Schlickeiser 1984, Pohl et al. 1992) a cooling particle distribution ($N(p) \propto p^{-2}$) with a strong cutoff at low (but still relativistic) momentum p_c develops which grows with time as more and more protons hit the photo-pair and photo-pion thresholds. Such bump-on-tail particle distribution functions, which are inverted ($\partial F/\partial p > 0$) below p_c, are collectively unstable with respect to the excitation of electromagnetic waves leading to efficient synchrotron masering in an optically thick environment (Crusius and Schlickeiser 1988), and to the excitation of oblique longitudinal Langmuir waves (Lesch et al. 1989).

As described in detail by Lesch and Schlickeiser (1987) and Achatz et al. (1990), depending on the the local plasma parameters (mainly the electron temperature of the background gas and the density ratio N_o/n_e of relativistic electrons and positrons to thermal electrons) these Langmuir waves either (1) lead to quasilinear plateauing of the inverted distribution function and rapid collective thermalization of the electrons and positrons, or (2) are first damped by nonlinear Landau damping, but ultimately heat the background plasma strongly via the modulation instability once a critical energy density in Langmuir waves $W_c = n_e(k_B T_e/m_e c^2)^2$ has been built up. Both relaxation mechanisms terminate the quiescent phase of the acceleration process.

The almost instantaneous increase of the background gas entropy due to the rapid modulation instability heating leads to an explosive outward motion of the plasma blob carrying the relativistic particles away from the central object. It marks the start of the flaring phase in γ-ray blazars. The initial starting height of the emerging blob z_i entering the calculation of the γ-ray flux in Fig. 2 should be closely related to the size of the acceleration volume in the quiescent phase, both being mainly determined by the size of the γ-ray photosphere. This scenario is supported by the measurements of Babadzanhanyants and Belokon (1985) that in 3C 345 and other quasars optical bursts are in close time correlation with the generation of compact radio jets. After the blast wave has left the central system the central source returns to its quiescent phase, and the same particle acceleration processes described above start again. This sequence of events is in accord with the observed behaviour of γ-ray hard blazars.

5. Summary and conclusions

In this work I argued that the high-energy X-ray and γ-ray emission from flaring blazars is beamed radiation from the relativistic jet supporting the relativistic beaming hypothesis and the unified scenario for AGNs. Most probably the high-energy emission results from inverse Compton scattering by relativistic electrons and positrons in the jet of radiation originating *external* to the jet plus pair annihilation radiation from the jet. Future positive TeV detections of EGRET AGN sources will be decisive to identify the prominent target photon radiation field. The model of Sikora et al. (1994), where reprocessed accretion disk photons in the emission line region are comptonized, apparently does not operate in lineless BL Lac objects and implies significant intrinsic $\gamma - \gamma$ pair attenuation of the generated TeV photons. In the model of Dermer and Schlickeiser (1993), where the accretion disk photons are comptonized, TeV emission is possible from nearby BL Lac objects but triplet pair production and the extreme Klein-Nishina cutoff will give rise to a significant steepening of the γ-ray energy spectra in this range, whereas in quasars the intrinsic $\gamma - \gamma$ attenuation from the reprocessed radiation absorbs the TeV photons. The synchrotron-self Compton models are extremely inefficient at TeV energies due to the extreme Klein-Nishina limit of the second Compton upscattering of MeV photons, and in quasars also suffer from the $\gamma - \gamma$ attenuation.

Direct γ-ray production by energetic hadrons is not important for the flaring phase in γ-ray blazars, but the acceleration of energetic hadrons during the quiescent phase of AGNs is decisive as the source of secondary electrons and positrons through photo-pair and photo-pion production. Injection of ultrahigh energy secondary electrons and positrons into a stochastic quasi-linear acceleration scheme during the quiescent AGN phase leads to cooling electron-positron distribution functions with a strong cut-off at low but relativistic energy that under certain local conditions may trigger a plasma instability that gives rise to an explosive event and the flaring γ-ray phase. Although many details for the stochastic acceleration model in the AGN plasma and for the comparison with time-correlated multifrequency observations still have to be worked out, the present results are very encouraging.

6. Acknowledgements

Discussions and cooperations with C. D. Dermer, H. Mause, M. Böttcher, H. Lesch and M. Pohl are gratefully acknowledged. I also thank the DARA (50 OR 94063) for partial support of my Compton observatory guest investigator programs.

References

Achatz, U., Lesch, H., Schlickeiser, R.: 1990, *AA* **233**, 391
Antonucci, R.: 1993, *ARAA* **31**, 473
Babadzhanyants, M.K., Belokon, E.T.: 1985, *Astrofizika* **23**, 459
Barthel, P.D., et al.: 1995, *ApJ* , in press
Bednarek, W.: 1993a, *AA* **278**, 307
Bednarek, W.: 1993b, *ApJ* **402**, L29
Begelman, M.C., Blandford, R.D., Rees, M.J.: 1984, *Rev. Mod. Phys.* **56**, 255
Blandford, R.D.: 1990, *in Active Galactic Nuclei, ed. T. Courvoisier & M. Mayor, New York, Springer* , 161
Blandford, R.D.: 1993, *Proceedings of the Compton Symp., ed. N. Gehrels & M. Friedlander, New York, AIP* , 533
Blandford, R.D., Rees, M.J.: 1978, *Pittsburg Conference on BL Lac Objects, ed. A. M. Wolfe, Pittsburg, Pittsburg Univ. Press, 328* ,
Blandford, R.D., Levinson, A.: 1995, *ApJ* , in press
Bloom, S.D., Marscher, A.P.: 1993, *Proceedings of the Compton Symp., ed. N. Gehrels & M. Friedlander, New York, AIP* , 578
Böttcher, M., Schlickeiser, R.: 1995, *AA* , in press
Camenzind, M.: 1990, *Rev. Mod. Astron.* **3**, 234
Campeanu, A., Schlickeiser, R.: 1992, *AA* **263**, 413
Carraminana., A.: 1992, *AA* **264**, 127

Dermer, C.D.: 1993, *Proceedings of the Compton Symp., ed. N. Gehrels & M. Friedlander, New York, AIP* , 541

Dermer, C.D., Gehrels, N.: 1995, *ApJ* , in press

Dermer, C.D., Miller, J., Li, H.: 1995, *ApJ* , submitted

Dermer, C.D., Schlickeiser, R.: 1992, *Sci* **257**, 1642

Dermer, C.D., Schlickeiser, R.: 1993, *ApJ* **416**, 484

Dermer, C.D., Schlickeiser, R.: 1994, *ApJS* **90**, 945

Dermer, C.D., Schlickeiser, R., Mastichiadis, A.: 1992, *AA* **256**, L27

Elliot, J.D., Shapiro, S.L.: 1974, *ApJ* **192**, L3

Fichtel, C.E., et al.: 1995, *ApJS* **94**, 551

Galeev, A.A., Rosner, R., Vaiana, G.S.: 1979, *ApJ* **229**, 318

Henri, G., Pelletier, G., Roland, J.: 1993, *ApJ* **404**, L41

Kanbach, G.: 1995, *SSR* , this issue

Kazanas, D., Ellison, D.: 1986, *ApJ* **304**, 178

Kurfess, J.D.: 1994, *in Active Galactic Nuclei across the Electromagnetic Spectrum, ed. T. Courvoisier & A. Blecha, Dordrecht, Kluwer* , 39

Lee, M.A., Ip, W.H.: 1987, *JGR* **92**, 11041

Lichti, G.G., et al.: 1995, *AA* , in press

Lesch, H., Crusius, A., Schlickeiser, R.: 1989, *AA* **209**, 427

Lesch, H., Schlickeiser, R.: 1987, *AA* **179**, 93

Mannheim, K.: 1995, *SSR* , this issue

Mannheim, K., Schlickeiser, R.: 1994, *AA* **286**, 983

Maraschi, L., Ghisellini, G., Celotti, A.: 1993, *ApJ* **397**, L5

Mastichiadis, A.: 1995, *SSR* , this issue

McKenzie, J.K., Westphal, K.O.: 1969, *Planet. Space Sci.* **17**, 1029

Montigny, C. von, et al.: 1995, *ApJ* , in press

Pelletier, G., Sol, H.: 1992, *MNRAS* **254**, 635

Pohl, M., Reich, W., Schlickeiser, R.: 1992, *AA* **262**, 441

Pohl, M., et al.: 1995, *AA* , in press

Punch, M., et al.: 1992, *Nat* **358**, 477

Reich, W., et al.: 1993, *AA* **273**, 65

Rybicki, G.B., Lightman, A.P.: 1979, *Radiative Processes in Astrophysics, New York, John Wiley* ,

Scheuer, P.A.G., Readhead, A.C.S.: 1979, *Nat* **277**, 182

Schlickeiser, R.: 1984, *AA* **136**, 227

Schlickeiser, R.: 1994, *ApJS* **90**, 929

Schlickeiser, R., Achatz, U.: 1992, *in Extragalactic Radio Sources - From Beams to Jets, ed. J. Roland, H. Sol & G. Pelletier, Cambridge. Cambridge University Press* , 214

Schlickeiser, R., Achatz, U.: 1993, *J. Plasma Phys.* **49**, 63

Schlickeiser, R., Biermann, P.L., Crusius-Wätzel, A.: 1991, *AA* **247**, 283

Sikora, M., Begelman, M.C., Rees, M.: 1994, *ApJ* **421**, 153

Sikora, M., Kirk, J.G., Begelman, M.C., Schneider, P.: 1987, *ApJ* **320**, L81

Slane, P., Wagh, S.M.: 1990, *ApJ* **364**, 198

THE HADRONIC MODEL OF ACTIVE GALACTIC NUCLEI

APOSTOLOS MASTICHIADIS
Max-Planck-Institut für Kernphysik
Postfach 10 39 80, D-69029 Heidelberg, Germany

Abstract. We review the hadronic model for Active Galactic Nuclei (AGN). This model, which can be applied to all AGN, advocates the acceleration of protons to ultrarelativistic energies by shock fronts which are formed a few Schwarzschild radii away from the central black hole. The necessary consequences of this hypothesis are discussed. These include the formation of electromagnetic cascades which are initiated by the injection of secondary electrons and photons inside the source, as well as the production and escape of neutrons and neutrinos. As a result of the neutron escape we emphasize that AGN can be sources of TeV radiation.

1. Introduction

The detection of at least 40 Active Galactic Nuclei (AGN) by the Compton Gamma-Ray Observatory at GeV energies (for an updated review see von Montigny et al. (1995) and G. Kanbach, these proceedings) in addition to the detection of Mrk 421 at TeV energies (Punch et al. 1992 and R.C. Lamb, these proceedings) shows that AGN are sites of very energetic phenomena. However all the objects detected so far belong to the category of blazars. These observations are consistent with the idea that the generation of γ-rays and the implied particle acceleration are jet-related phenomena. The hypothesis is that blazars have their jets pointed, within a small angle, towards the observer, therefore the radiation is relativistically boosted and dominates the emission from the core. (For a review of the γ−ray emission from the jets see R. Schlickeiser, these proceedings).

On the other hand EGRET has failed to detect so far even one Seyfert or radio quiet QSO (Lin et al. 1993), objects which by far constitute the majority of all AGN. Furthermore, in the case of the Seyferts one presumably observes phenomena directly associated with the core. This is particularly true in the X-ray regime where the fast variability observed in many cases implies that this radiation is generated within a few Schwarzschild radii of the black hole (Done & Fabian 1989). The high values which X-ray observations imply for the photon compactness $\ell_\gamma = L\sigma_{\rm T}/4\pi R m_e c^3$ (where L is the luminosity, R the radius of the source and $\sigma_{\rm T}$ the Thomson cross-section) mean that any γ-ray of energy > 1 MeV, if generated in this region, would be readily absorbed by the softer photons. Therefore the fact that Seyferts do not emit in the γ-ray regime does not necessarily mean that energetic phenomena do not take place in their cores.

In the present paper we will review the so-called hadronic model of AGN and its important consequences for the TeV and neutrino astronomy. This

Space Science Reviews **75**: 317–329, 1996.

model, developed during the last decade or so, advocates that very efficient proton acceleration is taking place deep inside the cores of all AGN and as a result AGN can be copious producers of neutrinos and high energy γ-rays. The specifics of particle acceleration and its consequences in an AGN environment will be briefly presented in Sections 2 and 3, section 4 will concentrate on the aspects of the predicted TeV emission from these objects, while section 5 will discuss some recent efforts to make the hadronic model self-consistent and at the same time to make it compatible with the latest X-ray/γ-ray observations. We conclude by briefly summarizing our main results in Section 6.

2. Proton acceleration and losses

Although the idea that AGN might be the sites of energetic particle production is certainly not new (Ginzburg & Syrovatskii 1964, Hillas 1984), it was Protheroe & Kazanas (1983) and Kazanas & Ellison (1986) who set the working frame for the hadronic picture of AGN. More specifically, the above authors suggested that a possible mechanism for energetic particle production is that of first order Fermi acceleration at shock fronts. This mechanism is believed to be a very efficient mechanism for proton acceleration from the thermal background. According to the proposed model a standing, collisionless shock in the accretion flow is formed at typical distances 10-100 gravitational radii ($r_{\rm gr}$) around the black hole. Calculations (Ellison & Eichler 1984) have shown that a significant part of the shock energy can be channeled to relativistic particles. At the same time the relativistic particles which are formed provide the necessary back-pressure to support the shock. If the shock compression ratio is greater than 4, then most of the proton energy will be available in the highest energy particles, while exactly the opposite situation occurs for compression ratios less than 4. The particles themselves can be accelerated to ultra high energies ($E_{p,max} \simeq 10^8$ GeV) and their maximum energy is determined by a balance between acceleration and their removal through hadronic collisions on the background thermal plasma. Also, as it was pointed in Kazanas & Ellison, the hadronic collisions between relativistic particles and thermal protons will necessarily inject a population of relativistic secondary electrons the radiation of which would lead to the production of the observed continuum. Therefore, according to the hadronic model, it is the formation of strong shocks in the accretion flow that will ultimately convert part of the gravitational binding energy into electromagnetic radiation.

In a subsequent development Sikora et al. (1987) showed that for a typical AGN environment the most energetic of the protons are more likely to lose energy through collisions with the ambient photons rather than with the background protons. There are two ways with which relativistic protons can interact with photons. The first is electron/positron pair production, an electromagnetic process which has a threshold of $2m_ec^2$. The other is photopion production in which a hadron and various pions are produced. This occurs at higher energies of interaction as the photon energy in the proton rest frame should be at least the rest mass of the pion, i.e. 140 MeV. Despite the fact that proton-photon pair production has a lower threshold and a larger cross section than photopion production, it has at the same time a rather small inelasticity (.001 for collisions close to threshold as compared to an average value of .2 for photopion collisions). It turns out (Begelman et al. 1990) that for soft photon number density slopes of about -2.1 or flatter it is pair production that is the most important energy loss mechanism for the relativistic protons. The situation is reversed for steeper spectra. At any rate the combination of these two processes can provide a more efficient cooling than the proton-proton collisions and their significance becomes more important with increasing proton energy. This happens because as the proton energy increases, it starts to react with lower energy photons and since there are more photons at lower energies, the proton cooling becomes more efficient.

The model therefore requires protons to be continuously accelerated inside a central 'source' of size 10 to 100 r_{gr} around the central black hole up to the energies where the proton gains are balanced by the photon losses as described above. For the gains of protons a diffusive, first order Fermi shock acceleration timescale is assumed (Axford 1981). To calculate the proton loss timescale most authors assume a generic AGN photon spectrum consisting of a black body plus a power law that extends up to $\simeq 1$ MeV. The black body is taken to be the blue bump observed in many AGN while the power law represents the non-thermal AGN continuum. That way a maximum proton energy can be derived. For the products themselves either a simplified appropriate production rate is used (Kirk & Mastichiadis 1989, Begelman et al. 1990) or a more rigorous Monte Carlo approach (Szabo & Protheroe 1994).

The above discussion concerns mainly protons with ultrarelativistic energies. However one has to mention that low energy protons lose energy mainly from Coulomb and ionization losses. Mannheim & Schlickeiser (1994) has shown that these are important for proton kinetic energies below 450 MeV for parameters relevant to the accretion shock. Furthermore, as Dermer et al. (1995) emphasized, the above losses can become so efficient for certain plasma parameters that they may prevent the acceleration of protons above the Coulomb barrier.

3. Consequences of hadronic collisions in AGN

Irrespective of the way ultrarelativistic protons lose energy (i.e. through hadronic collisions with matter or radiation), they are expected to produce γ-rays from neutral pion decay, electrons and positrons from charged pion decay, as well as neutrinos and neutrons. The fate of each species shows interest of its own and so we will examine each one separately.

3.1. Photons

Since photons are produced from neutral pion decay with an energy $\simeq 10\%$ that of the interacting parent proton, one should expect γ-rays to be created within the source with energies up to hundreds of TeV. These γ-rays however have very high optical depths against γ-γ absorption on the softer photons, therefore they will be readily absorbed producing electron-positron pairs. To give an illustrative example, a photon of 1 TeV has an optical depth of a few thousands against absorption on an isotropic photon background field of energy slope of -1 and compactness $\ell_\gamma = 1$. Therefore, even if TeV radiation is produced inside the cores of AGN, it cannot escape the central source. It is interesting here to note that high ℓ_γ sources are optically thick even to MeV radiation. However, it is worth mentioning that the escape probability increases significantly in the case of anisotropic photon fields (Protheroe et al. 1993), therefore if the energetic photons are somehow produced outside the central region, they will have an increased chance of escaping.

3.2. Electrons/positrons

Electrons and positrons are injected at very high energies (again up to hundreds of TeV) either as secondaries from charged pion decay or as a result of γ-γ pair production (see above). Their cooling times against inverse Compton scattering and /or synchrotron radiation are very short compared to the crossing time of the source, therefore they are expected to lose instantaneously their energies producing simultaneously energetic photons. These γ-rays will be absorbed on the background photons producing more electron-positron pairs and that way an electromagnetic cascade will be formed. Generally the above described system of electrons/positrons and photons should be treated simultaneously and this has been done extensively over the last decade with the advent of the pair plasmas (for a review see Svensson 1989 and references therein) in an effort to understand the X-ray properties of AGN. However, one should add here that the pair plasma models were not related to the hadronic picture, i.e. they left the problem of high energy and photon injection totally unspecified.

3.3. Neutrinos

The produced neutrinos will basically escape unattenuated as their cross section for absorption is very small and therefore the corresponding optical depth is negligible for typical AGN parameters. While the estimated flux from individual AGN is probably below current instrumental sensitivities, it is expected that the sum of all AGN can contribute to the diffuse neutrino background and bring it up to observable levels. A detailed account of the expected fluxes is given by F. Stecker in these proceedings.

3.4. Neutrons

Neutrons are produced in about 50% of *pp* collision and 25% of $p\gamma$ collisions. The neutrons, unlike protons which are magnetically confined inside the acceleration region, can escape (Kirk & Mastichiadis 1989, Sikora et al. 1989, Atoyan 1992a). More specifically, Kirk & Mastichiadis found that neutrons with Lorentz factors up to $\gamma_\star \simeq 5 \times 10^7 \ell_\gamma{}^{-1}$ can escape without any attenuation. Neutrons with $\gamma_n > \gamma_\star$ will be subject to photoabsorption, thus they will be absorbed within the source giving rise to hadronic cascades and essentially depositing their energies there.

Neutrons decay back to protons on the average after $\tau_n \simeq 15$ min. Therefore neutrons of Lorentz factors γ_n which do escape the central region will travel an average distance $r_n = \gamma_n c \tau_n \simeq 10^{-5}\gamma_n$ pc before they decay. Upon decay they produce a proton and an electron with Lorentz factors $\gamma_p \simeq \gamma_e \simeq \gamma_n$. Since neutrons suffer no adiabatic losses they can carry a large fraction of the energy generated inside the source at large distances away. This means that the neutrons can play an important role in AGN and various observable or potentially observable effects of the escaping neutrons have been predicted. These include the production of Boron in spallation reactions (Kirk & Mastichiadis 1989), the acceleration of winds in broad absorption line QSOs (Begelman et al. 1991), the onset of relativistic outflows (Kazanas & Mastichiadis 1993, Contopoulos & Kazanas 1994) and the production of cosmic rays around the 'knee' (Protheroe & Szabo 1992). Also, the flat radio spectra of radio loud AGN have been attributed indirectly to escaping neutrons (i.e., after their decay into protons and electrons – Giovanoni & Kazanas 1990). But perhaps the most interesting consequence of the escaping neutrons for the purpose of the present proceedings is related to the production of VHE γ-rays *outside* the central source (Mastichiadis & Protheroe 1990, Atoyan 1992b, Johnson et al 1994). To this we will devote the next section.

4. VHE emission from AGN as a result of neutron escape

As we pointed in the previous section, the leakage of neutrons outside the central regions of AGN is an efficient way of transfering energy from the cores to large distances. An immediate consequence of this picture is the production of high energy γ-rays. These will be created by hadronic collisions both by neutrons on transversing the infalling plasma and from proton-proton collisions that occur further out, after the relativistic neutrons have decayed. Therefore the production rate of γ-rays will depend on the one hand on the grammage the escaping neutrons will cross before they decay and, on the other, on the behavior of the protons which result from the decay of neutrons which avoided absorption en route from the source. These protons are expected to scatter off waves and quickly isotropise. Therefore they are expected to behave as cosmic rays diffusing out of a region up to hundreds of parsecs around the central source. Since the central black hole should dominate the gravitational potential at these radii, it is plausible to assume that this region has an enhanced matter density and that the relativistic protons will undergo frequent *pp* collisions. Mastichiadis and Protheroe (1990) calculated the γ-ray emissivity in two limiting cases, a minimum case where the protons were assumed to move out from the region ballistically and one maximum case where the protons were assumed to diffuse out of the region with a Bohm type diffusion time-scale. As it is to be expected the results between the two cases differ by about an order of magnitude. Atoyan (1992b) and Johnson et al. (1994) on the other hand took an even more optimistic approach by assuming that the protons are effectively trapped in their production sites and lose all their energy in hadronic collisions.

These γ-rays since they are created *outside* the central source will have relatively high chances of escaping photon-photon absorption, in sharp contrast to the γ-rays produced inside the source where the soft photon density is much higher. Assuming that the soft absorbing photons are coming from the central source Mastichiadis & Protheroe and Johnson et al. considered the problem of anisotropic γ-ray escape in some detail by assigning an escape probability which depends (for given parameters of the source photons) on the γ-ray energy, on its distance from the central source and on the direction the propagating photon makes with respect to the source-photon emitting site axis. This way an 'escape' cone can be obtained: γ-rays emitted inside this cone will escape while the ones emitted outside will be absorbed. The absorbed photons are expected to start a pair-synchrotron cascade which however will not contribute significantly to the overall emissivity above 1 TeV. Finally, Johnson et al. extended the results of Mastichiadis & Protheroe and Szabo & Protheroe (1994) and derived the contribution to diffuse γ-ray background from unresolved AGN. They found that at 10 GeV this flux is comparable to the blazar contribution (Stecker et al. 1993). However at higher energies blazars are expected to contribute more (provided that their γ-ray spectrum continues with a -2 slope up to TeV energies).

5. The hadronic model revisited: A self consistent approach

The main uncertainty which comes in the calculation of the TeV and neutrino fluxes of AGN is connected with the way the neutron and neutrino production rates scale with the X-rays. All the models mentioned above specify a priori a photon continuum spectrum on which the protons cool. In this sense, the hadronic models addresses the problem of particle acceleration and injection but lack a self-consistent treatment of the electromagnetic cascades. Furthermore, this overall picture where protons interact with photons producing more photons is highly nonlinear and leaves open many possibilities for feedback effects.

A second related problem which the hadronic models leave unanswered comes from the X-ray observations themselves. Observations with many satellites and baloon experiments have shown that Seyferts are well established sources of X-ray emission starting from the soft regime, going through the 1-10 keV regime and ending to up to hard X-/soft γ-rays. As recent ASCA observations (and earlier observations made with GINGA) have shown there are various components in an AGN X-ray spectrum including a soft excess around .1 keV, a warm absorber, an iron K line, possibly a reflected component and an underlying power law which extends up to hard X-rays (see, for example, Mihara et al. 1994). Especially in the 2-10 keV regime both HEAO-1 and GINGA satellites found that most Seyferts have a characteristic power law of -1.7 (Turner & Pounds 1989). However the origin of this power law still remains a puzzle. This has a deeper implication for the hadronic models of AGN: as stated above these models assume that these X-rays are non-thermal in nature basically coming from the electromagnetic cascades that ensue from the injection of high energy photons and electrons (*cf.* section 3.1 and 3.2). However the recent OSSE observations have already put severe limits to the non-thermal pair plasma models (Maisack et al. 1993). The absence of annihilation lines and the observed spectral turnover at energies less than 100 keV indicate features which are not characteristic of the non-thermal pair plasmas. Interestingly enough the spectrum of the archetypal Seyfert NGC4151 can be fitted by Comptonization of soft photons on thermal electrons in both the GINGA and OSSE energy regimes (Zdziarski et al. 1993, Titarchuk & Mastichiadis 1994). Therefore the new challenge the hadronic models face is to explain the X-ray observations. In other words, the question is whether a self-consistent treatment of the electromagnetic cascades within the framework of the hadronic model, can reproduce the hard X-ray characteristics of AGN.

A first attempt to create a model were the injected secondary photons and electrons produce through electromagnetic cascades a softer radiation which in turn interacts with the relativistic protons was made by Stern et al. (1991) and Stern & Svensson (1991). They found that this system showed limit cycles much like the ones that arise naturally in two-species predator-prey systems (see, for example, May 1975). However, the Monte Carlo approach which they use complicates the interpretation of the results in terms of a specific feedback mechanism. Such a feedback effect was found analytically by Kirk & Mastichiadis (1992 – henceforth "KM92"), using the kinetic equation approach. They showed that relativistic protons become unstable to a combination of proton-photon pair production and electron synchrotron radiation once their number density and energy exceeds a certain critical value. More specifically KM92 found that in order to have feedback protons with energies greater than

$$\gamma_{\rm crit} = (2/b)^{1/3} \tag{1}$$

should be present inside the source. Here $b = B/B_{\rm c}$ denotes the magnetic field in units of the critical field $B_{\rm c} = m_{\rm e}^2c^3/(e\hbar) = 4\cdot 414 \times 10^{13}$G. On the other hand protons above this critical energy should have a density $n_{\rm p} \geq n_{\rm crit}$ as the synchrotron photons must collide with at least one of them before escaping the source. This condition effectively sets the marginal stability criterion which reads

$$n_0 > n_{\rm crit} \equiv \left(\frac{2}{3}\beta - 1\right) b^{1-\beta/3} \left[\int_0^\infty {\rm d}y\, \sigma_{\rm pe}(y) y^{-1-\beta/3}\right]^{-1}, \tag{2}$$

where $\sigma_{\rm pe}$ denotes the cross section of proton-photon pair production in units of the Thomson cross section $\sigma_{\rm T}$ and the protons are supposed to have a distribution $n_{\rm p}(\gamma) = n_0\gamma^{-\beta}$. $n_{\rm p}(\gamma)$ is the differential number density of protons in the volume element $\sigma_{\rm T}R$. The pair-synchrotron instability is not the only loop which could operate on such a system of interacting protons, electrons and photons. KM92 have already discussed a related instability, the photopion-synchrotron loop which, however, operates at a higher threshold than the pair-synchrotron one, i.e. one needs higher concentrations of relativistic protons in the source to set it going. Similarly the presence of an external photon field (as the 'blue bump' which is used in both hadronic and pair plasma models) can play the role the magnetic field played in the loops described above. Therefore we can assume that the produced electron/positron pairs radiate not synchrotron but rather inverse Compton photons and we can find immediately two more loops, namely the proton-photon pair production–Compton loop and the photopion–Compton one. The feedback and marginal stability conditions for the former can be obtained at once from the equations above simply by replacing the quantity

b with $kT_{\rm bb}/m_ec^2$ where $T_{\rm bb}$ is the temperature of the blue bump. However, since the inverse Compton photons have typical energies > 1 MeV, it is possible to set off electromagnetic cascades complicating the picture. This is essentially the instability investigated by Stern et al. (1991).

The feedbacks described above lead eventually to rapid energy losses for the relativistic protons. However the analysis of KM92 employs a stationary proton distribution and uses a linearised set of equations, which exclude electromagnetic cascades, making it impossible to follow the evolution of the system into the saturation phase. Still the above authors were able to show analytically that during the linear growth of the instability, the photons have a number spectral index of $-1-\beta/3$ where β is the proton slope. For $\beta=2$ which is the nominal value of a strong shock the photon slope becomes -1.7 which is the characteristic Seyfert X-ray slope as this was measured by HEAO-1 and GINGA.

To be able to follow the system of protons, electrons and photons in the nonlinear regime Mastichiadis & Kirk (1995) constructed a simplified model which is capable of describing time-dependent effects. In an effort to connect the pair plasma models (by providing them with an injection mechanism) with the hadronic model of AGN (by calculating self-consistently the photon distribution on which the accelerated protons cool) they described the three basic species by a system of three spatially averaged, time-dependent kinetic equations. They included all the relevant physical processes in a self-consistent manner so that the energy lost by one species is gained by another. Table 1 lists all the physical processes used and their relevance to protons, electrons/positrons and photons.

TABLE I

Physical Process	Effects on:	Protons	Electrons/Positrons	Photons
Proton-Photon Pair Production		Energy Loss	Injection	–
Photopion Production		Energy Loss	Injection	Source
Proton-Proton Interactions		Energy Loss	Injection	Source
Synchrotron Radiation		–	Energy Loss	Source
Synchrotron Self Absorption		–	–	Sink
Inverse Compton Scattering		–	Energy Loss	Source
Photon Downscattering		–	Injection	Sink
Photon-Photon Pair Production		–	Injection	Sink
Electron-Positron Annihilation		–	Energy Loss	Source

Energy is supposed to be pumped into the system by a first order acceleration scheme which operates only on protons. The acceleration is a 'slow'

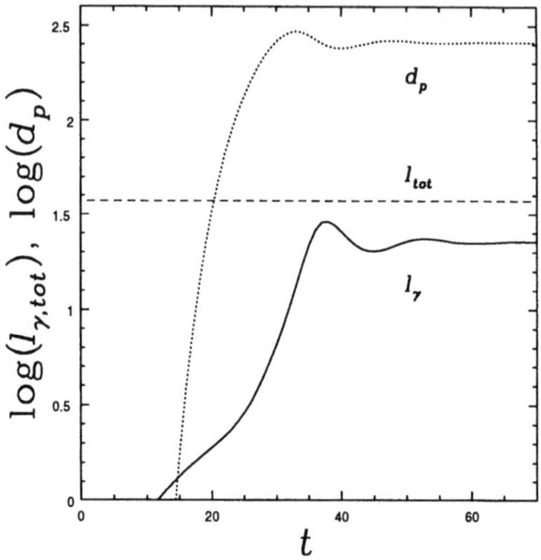

Fig. 1. Photon compactness (full line) and proton depth (dotted line) versus time expressed in units of crossing time in the case where the pair/synchrotron instability operates. Injection parameters are $n_0 = 1.1 n_{\rm crit}$, $t_{\rm acc} = 2t_{\rm cross}$, $b = 10^{-9}$ and $R = 1.5 \times 10^{14}$ cm. The total compactness $\ell_{\rm tot}$ (see Eq. 3) is shown by the dashed line. (Taken from Mastichiadis & Kirk 1995).

process in the sense that the acceleration timescale is taken to be larger than the crossing time of the source (i.e., $t_{\rm acc} \geq t_{\rm cr} = R/c$) and it occurs throughout the whole of the source region. The total amount of energy the accelerator is supposed to put into the products of the hadronic collisions in a steady state can be characterised by a compactness

$$\ell_{\rm tot} = \frac{L_{\rm loss}\sigma_{\rm T}}{4\pi R m_e c^3} \ . \tag{3}$$

Here $L_{\rm loss}$ is the total luminosity given by protons to electrons, photons, neutrons and neutrinos. This luminosity, as shown in Mastichiadis & Kirk, is solely determined by the rate at which protons are injected into the acceleration process. The system is thus totally determined once the details of the proton acceleration are specified inside a source of radius R and magnetic field strength B. The injection of electrons and photons, which are free parameters for the pair plasma models, are internally specified in this case.

Figures 1 and 2 show the evolution of a system which operates above the pair/synchrotron instability. The protons accelerate past $\gamma_{\rm crit}$ ($\simeq 10^3$ in this case) at which point the instability sets and as a consequence there is a fast photon growth and, after some damped oscillations, a saturation of the acceleration and an eventual steady state (Fig. 1). Here $d_{\rm p}$ is a measure of the proton energy density inside the source with $\gamma_p \geq \gamma_{\rm crit}$, while $\ell_{\gamma,\rm tot}$ is the compactness of the self-produced photons. Fig. 2 shows the proton

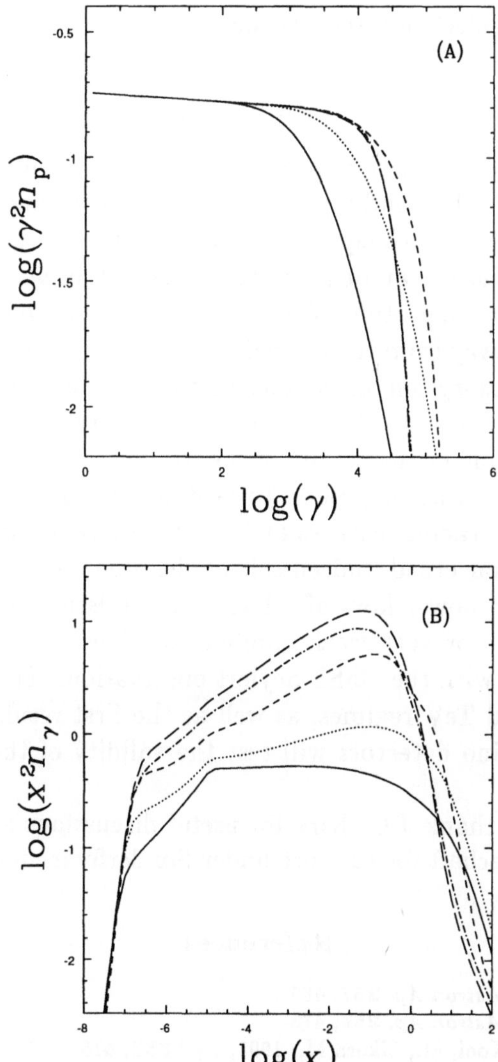

Fig. 2. **(A) The proton distribution n_p for five times t_i ($i = 1, \ldots 5$) after the run was started for the injection parameters specified in Figure 1. The short dashed line is for $t = t_1 = 23$, the dot-dashed line is for $t = t_2 = 28$, the long dashed line is for $t = t_3 = 33$, the full line is for $t = t_4 = 38$ and the dotted line is for $t = t_5 = 43$. All times are expressed in units of t_{cross}. (B) The photon spectrum corresponding to the proton spectrum in (A). Here $x = h\nu/m_e c^2$.**

and photon spectra around the first peak of the photon compactness in Fig. 1. The last curve (dot-dashed line) is close to the final steady state. Here one can clearly see that the photon spectrum breaks above 50 keV while the power law in the energies below the break is about -1.7 in agreement with the linear analysis of KM92. Furthermore there is no evidence for an annihilation line as, for this case, the magnetic energy density exceeds the one in the injected electrons and photons.

6. Summary

The hadronic model of AGN as presented above has some very interesting consequences for the TeV and high energy neutrino astronomy as it predicts that both species can, in principle, be produced from the cores of all AGN. In particular TeV radiation can be produced as a result of *pp* collisions between relativistic and thermal protons. The relativistic protons result from neutron escape and decay away from the central source. The produced TeV radiation can escape $\gamma-\gamma$ absorption on the soft photons as the majority of the latter comes from regions close to the central source, therefore anisotropic effects will play a major role in the escape of the γ-rays.

Furthermore recent developments towards a self consistent hadronic model shows some encouraging features as far as the current hard X-ray observations of AGN are concerned. Indeed this model predicts (as can be seen from Figure 2) 2-10 keV power laws of -1.7, characteristic breaks around some tens of keV and no or very weak annihilation lines. These features are in general agreement with the OSSE Seyfert observations. Future observations both in the X- and TeV regimes, as well as the first results of the soon to be operated neutrino detectors will test the validity of the hadronic AGN picture.

I would like to thank J.G. Kirk for useful discussions and the Deutsche Forschungsgemeinschaft for support under Sonderforschungsbereich 328.

References

Atoyan, A.M.: 1992, *Astron.Ap.* **257**, 465
Atoyan, A.M.: 1992, *Astron.Ap.* **257**, 475
Begelman, M.C., de Kool, M., Sikora M.: 1991, *ApJ* **382**, 416
Begelman, M.C., Rudak, B., Sikora, M.: 1990, *ApJ* **362**, 38
Contopoulos, J., Kazanas, D: 1995, to appear in *ApJ*
Dermer, C.D., Miller, J.A., Li, H.: 1995, to appear in *ApJ*
Done, C., Fabian, A.C.: 1989, *MNRAS* **240**, 81
Ellison, D.C., Eichler, D.: 1984, *ApJ* **286**, 691
Ginzburg, V.L., Syrovatskii, S.I.: 1965, *Ann. Rev. Astron. Astrophys.* **3**, 297
Giovanoni, P.M., Kazanas, D.: 1990, *Nat* **345**, 319
Hillas, A.M: 1984, *Ann. Rev. Astron. Astrophys.* **22**, 425
Johnson, P.A., Mastichiadis, A., Protheroe, R.J., et al.: 1994, *J. Phys. G* **20**, 979
Kazanas, D., Ellison, D.C.: 1986, *ApJ* **304**, 178
Kazanas, D., Mastichiadis, A.: 1995, to appear in the *Proceedings of the 2nd GRO Conference*
Kirk, J.G., Mastichiadis, A.: 1989, *Astron.Ap.* **213**, 75
Kirk, J.G., Mastichiadis, A.: 1992, *Nat* **360**, 135
Lin, Y.C., Bertsch, D.L., Dingus, B.L., et al.: 1992, *ApJLet* **416**, L53
Maisack, M., Johnson, W.N., Kinzer, R.L. et al.: 1993, *ApJLet* **407**, L61
Mannheim, K., Schlickeiser, R.: 1994, *Astron.Ap.* **286**, 983

Mastichiadis, A., Protheroe, R.J.: 1990, *MNRAS* **246**, 279
Mastichiadis, A., Kirk, J.G.: 1995, *Astron.Ap.* **295**, 613
May, R.M.: 1975, *Stability and Complexity in Model Ecosystems*, Princeton University Press, Princeton
Mihara, T., Matsuoka, M., Mushotzky, R.F., et al.: 1994, *Publ.Astron.Soc.Japan* **46**, L137
Protheroe, R.J., Kazanas, D.: 1983, *ApJ* **265**, 620
Protheroe, R.J., Mastichiadis, A., Dermer, C.D.: 1992, *Astroparticle Phys* **1**, 113
Protheroe, R.J., Szabo, A.P.: 1992, *Phys. Rev. Lett.* **69**, 2885
Punch, M., Akerlof, C.W., Cawley, M.F., et al.: 1992, *Nat* **358**, 477
Sikora, M., Begelman, M.C., Rudak, B.: 1989, *ApJLet* **341**, 33
Sikora, M., Kirk, J.G., Begelman, M.C., Schneider, P.: 1987, *ApJLet* **320**, L81
Stecker, F.W., Salamon, M.H., Malkan, M.A.: 1993, *ApJ* **410**, 71
Stern, B.E., Svensson, R.: 1991, in Zdziarski, A.A., Sikora, M. (eds), *Relativistic Hadrons in Cosmic Compact Objects*, Springer-Verlag,
Stern, B.E., Svensson, R., Sikora, M., 1991, in Miller, H.R., Wiita, P.J. (eds), *Variability of Active Galactic Nuclei*, Cambridge University Press, p. 229
Svensson, R., 1989, in Brinkmann, W. (ed), *Physical Processes in Hot Cosmic Plasmas*, Kluwer Academic, p.357
Szabo, A.P., Protheroe, R.J., 1992, in Stenger, V.J. et al. (eds), *High Energy Neutrino Astrophysics*, World Scientific, p. 24
Szabo, A.P., Protheroe, R.J., 1994, *Astroparticle Phys*, **2**, 375
Titarchuk, L., Mastichiadis, A.: 1994, *ApJLet* **433**, L33
Turner, T.J., Pounds, K.A.: 1989, *MNRAS* **240**, 833
von Montigny, C., Bertsch, D.L., Chiang, J., et al.: 1995, *ApJ* in press
Zdziarski, A.A., Lightman, A.P., Maciolec-Niedzwieski, A.: 1993, *ApJLet* **414**, L93

TEV GAMMA-RAYS FROM PROTON BLAZARS

KARL MANNHEIM
Universitäts-Sternwarte, Geismarlandstraße 11, D-37083 Göttingen, Germany
Internet: kmannhe@uni-sw.gwdg.de

Abstract. Proton acceleration in nearby blazars can be diagnosed measuring their intense TeV γ-ray emission. Flux predictions for 1101+384 (Mrk421) and 1219+285 (ON231), both strong EGRET sources (0.1 – 10 GeV), are obtained from model spectra of unsaturated synchrotron pair cascades fitted to publicly available multifrequency data. An experimental effort to confirm the predicted emission in the range 1-10 TeV would be of great importance for the problems of the origin of cosmic rays, the era of galaxy formation and the cosmological distance scale.

1. Introduction

1.1. Relativistic jets

After the initial discovery of 3C273 by COS-B (Swanenburg et al. 1978) it came as a big surprise to most of the community when *numerous* highly variable extragalactic γ-ray sources were discovered with the Energetic Gamma Ray Experiment Telescope EGRET onboard the Compton Gamma-Ray Observatory CGRO (Fichtel et al. 1994, Montigny et al. 1995). The sources were soon identified as blazars, a subclass of active galactic nuclei (AGN), which are kwown to be the most powerful and rapidly variable non-thermal emitters also in other energy bands (Bregman 1990). The γ-ray luminosity of the sources assuming isotropic emission reaches values as high as $\sim 10^{49}$ erg s^{-1}. Intriguingly, at the time of observation the received energy flux from many of these objects appeared to be dominated by γ-rays, similar only to the *pulsed* emission from pulsars, and probably to the emission from the enigmatic γ-ray burst sources and the yet unidentified EGRET sources at high galactic latitude. The TeV detection of Mrk421 by the Whipple group (Punch et al. 1992) now makes it probable that the blazar γ-ray spectra extend to still higher energies.

These new challenging observations can be explained assuming that the γ-rays are emitted in a lighthouse beam directed towards the observer (McBreen 1979). The beaming pattern, and thus the blazar phenomenon, arises from the relativistic bulk motion of plasma jets ejected from the vicinity of a supermassive black hole (Blandford & Königl 1979). Relativistic radio jets are commonly observed in flat-spectrum radio sources; superluminal expansion of the radio structure (Zensus & Pearson 1987), intraday variability (Quirrenbach et al. 1991) and large kinetic energy (Rawlings & Saunders 1991, Celotti & Fabian 1993) indicate bulk motion with Lorentz factors

Space Science Reviews **75**: 331–340, 1996.

$\gamma_j = 10 - 100$ (Begelman et al. 1994), although typical values inferred from observations of jets at VLBI scales yield values ~ 10 or less. Some fraction of the particles in the jet plasma is accelerated to ultra-relativistic energies retaining isotropy in the comoving frame by pitch-angle scattering. Most probably, particles gain energy by repeatedly crossing collisionless shock fronts in the supersonic jet flow (Drury 1983). Anisotropies due to the relativistic bulk motion and oblique shocks complicate the theoretical understanding of the acceleration process considerably (Begelman & Kirk 1990). Nevertheless, the observed power law energy distributions indicate a stochastic nature of the acceleration mechanism.

1.2. The radiation process responsible for γ-ray emission

The high particle energies and the twenty orders of magnitude in frequency of the emitted nonthermal radiation are reminiscent of the properties of cosmic rays in our own Galaxy where electrons produce synchrotron radio emission and protons with Lorentz factors up to 10^{11} generate energetic γ-rays by pion production $p + p \rightarrow \pi^{\circ} + X$ and their subsequent decay. The COS-B detection of the Geminga source, which is now known to be a nearby pulsar, has led to some early speculations whether this source and other unidentified sources could be 'proton quasars' (Moffat et al. 1983, Schlickeiser 1984) where relativistic protons are accelerated. However, proton-matter interactions are ruled out as the radiation mechanism responsible for the variable and highly luminous γ-ray emission from blazars, since this would require unrealistically large amounts of target matter in the jets. It would also imply a significant enhancement of light-elements due to spallation of heavier nuclei which has not been observed (Baldwin et al. 1979). However, it was noticed by Blumenthal (1970) and Colgate (1983) that the compact radiation fields in extragalactic radio sources represent by far the most important target for relativistic protons (cf. also references in Berezinsky et al. 1990). At very high energies protons scatter inelastically by photoproduction of secondaries, i.e. $p + \gamma \rightarrow \pi^{\circ} + p$ and $p + \gamma \rightarrow e^{+} + e^{-} + p$. In contrast to the low-density interstellar medium where γ-rays can escape freely, blazars absorb the energetic γ-rays resulting from the decay of secondaries via pair creation $\gamma + \gamma \rightarrow e^{+} + e^{-}$. As a consequence, unsaturated pair cascades develope which attenuate the γ-ray flux above the energy ϵ^{*} where $\tau_{\gamma\gamma}(\epsilon^{*}) = 1$, but which let γ-rays with energies $\epsilon < \epsilon^{*}$ escape freely (Mannheim et al. 1991).

While it is tempting to explain the observed rapid γ-ray variability in terms of an origin of the emission very close to the central black hole, it is clear that this contradicts the large pair creation opacity in this region. On the other hand, it is not so clear that it is impossible to obtain rapidly varying emission from distant parts of the jet. If the jet starts out as a magnetically confined flow, it could still be collimated to a few light days outside of the central parsec. In this region shocks and other dissipative structures develope as the jet encounters large external pressure gradients. For the radiation process this means that the dominant energy loss process for electrons is synchrotron radiation, not inverse-Compton scattering (i.e. the energy densities obey $a = u_\gamma/u_B < 1$). Observationally, this is found on kpc scales (Harris et al. 1994), as well as for the nuclear jets responsible for beamed X-rays (Biermann et al. 1988). Theoretically, $a \leq 1$ expresses a simple confinement condition: relativistic particles with energy density $u_{\rm rel}$ produce nonthermal radiation with energy density $u_\gamma \leq u_{\rm rel}$. Particles are accelerated by scattering off magnetic field irregularities with energy density $u_{\delta B} \leq u_B$, hence $u_\gamma \leq u_{\rm rel} \leq u_B$. If, on the other hand, the relativistic particle pressure would exceed the magnetic field pressure, the particles would break out of their magnetic confinement turning-off further acceleration, or would generate turbulence leading to $a \sim 1$.

Only if the particles responsible for the γ-ray emission are accelerated deep within the central parsec, so that the energy density of the *external* radiation field exceeds that instrinsic to the jet, inverse-Compton scattering becomes the dominant cooling mechanism. Sikora et al. (1994) have argued that the re-radiation of a central thermal UV spectrum by the broad emission line clouds can produce such a strong external radiation field, if the γ-ray production occurs within the central parsec of the AGN powering the γ-ray blazar. Dermer et al. (1992) and Dermer & Schlickeiser (1993) have argued that direct thermal radiation from an accretion disk represents the majority of target photons. Observed γ-ray spectral breaks greater than $\Delta s = 0.5$ (e.g. Lichti et al. 1995) require further complexity in these models. Moreover, the models require the presence of a thermal radiation field dominating the bolometric luminosity. For BL Lacertae objects with a near-infrared/optical cut-off this emission component has not been found down to levels of the order of $\sim 10^3$ below the nonthermal continuum level (Bregman 1990). Indeed, the emission component can remain undiscovered due to Doppler boosting of the nonthermal emission, but it must show up at *some* level. The central source could also be obscured, but that is inconsistent with the jet pointing towards the observer. Another implication of these models is a rapid electron acceleration process which can overcome the extreme inverse-Compton losses.

1.3. Proton-induced cascades

In the most simple model for blazar emission it is assumed that steady, conical relativistic jets of Lorentz factor $\gamma_{\rm j}$ with opening angle $\Phi \approx \gamma_{\rm j}^{-1}$ are viewed at a small angle θ (Blandford & Königl 1979, Mannheim 1993a). Photons emitted isotropically in the comoving frame are blueshifted with the Doppler factor $\delta = [\gamma(1 - \beta\cos\theta)]^{-1}$ and the apparent luminosity exceeds the comoving frame luminosity by a large margin. The outflow is assumed to advect a tangled magnetic field and relativistic particles in equipartition with the magnetic field. Most probably shock acceleration compensates adiabatic losses during the expansion. Synchrotron-self-absorption frequency, break frequeny (above which the cooling time scale is shorter than the dynamical time scale) and emissivity obey a simple scaling with distance r along the jet, such that the superposition of spectra from scales between $r_{\rm min}$ and $r_{\rm max}$ yields a flat spectrum $S_\nu \propto const.$ up to the break frequency $\nu_{\rm b} \approx 10^{11} - 10^{13}$ Hz above which the spectrum steepens by one power yielding $S_\nu \propto \nu^{-1}$. In the near-infrared/soft X-ray regime the primary electron synchrotron spectrum turns over steeply when the cooling time scale becomes shorter than the acceleration time scale. For protons, the break energy is reached at

$$\gamma_{\rm p,b} = \left(\frac{m_{\rm p}}{m_{\rm e}}\right)^3 \frac{1+a}{1+240a}\gamma_{\rm e,b} \approx 3\cdot 10^9 \left(\frac{\gamma_{\rm e,b}}{100}\right) \tag{1}$$

Note that $a = u_\gamma/u_B$ refers only to the synchrotron photon density in the radio through X-ray range, since higher energy photons are unimportant as a target, for inverse-Compton scattering because of the Klein-Nishina cross-section and for photo-production because of the threshold energy. The proton break energy is much greater than the electron break energy, since the proton energy losses are negligible compared to the electron energy losses at the same energy. The proton-induced luminosity for $\gamma_{\rm p} \leq \gamma_{\rm p,b}$ has the value

$$L_{\rm casc} \simeq \eta \left(\frac{\gamma_{\rm p}}{\gamma_{\rm p,b}}\right) L_{\rm syn} \tag{2}$$

where $\eta = u_{\rm p}/u_{\rm e}$ denotes the energy density ratio of relativistic protons and electrons, respectively. It is very likely that protons do not always reach the extremely high energy $\gamma_{\rm p} = \gamma_{\rm p,b}$. One limitation arises from the condition that the proton Larmor radius must remain smaller than the shock radius which is itself less than the jet radius (Bell 1978). Hence it follows that for a universal cosmic ray proton/electron ratio η it is the proton maximum energy which determines the γ-ray spectrum.

The brightest γ-ray emission is expected from the most compact parts of the jet where most of the nonthermal target photons with $\nu \geq \nu_{\rm b}$ are produced, i.e. at the jet length $r = r_{\rm b}$ (jet radius $r_{{\rm b},\perp} \approx r_{\rm b}\Phi$). At smaller scales $r < r_{\rm b}$ synchrotron-self-absorption and rapid dynamical evolution darken the jet, whereas at larger scales $r > r_{\rm b}$ the radio flux increases and the infrared photon flux decreases. Due to the threshold condition for secondary particle production, e.g. $\nu_{\rm th} \geq 2\cdot 10^{12}(\gamma_{\rm p}/10^{10})^{-1}$ Hz for pions, only the infrared and higher frequency photons are relevant as a target for the protons. Indeed, the jet would be γ-ray bright also on large scales (steep spectrum radio sources), if the proton maximum energy were allowed to reach up to a value $\gamma_{\rm p} \geq \gamma_{\rm p,b}$ at any scale in the jet. However, this is impossible due to the Larmor radius constraint and the finite acceleration time scale. The greatest values of the proton maximum energy seem to be achieved in the hot spots of radio galaxies where $\gamma_{\rm p,max} \approx 3 \cdot 10^{11} \approx .07\gamma_{\rm p,b}$ (Biermann & Strittmatter 1987). Notice, however, that blazars with $L_\gamma \approx 100 L_{\rm syn}$ require $\gamma_{\rm p} = \gamma_{\rm p,b} \approx 10^{10}$ ($\eta = 100$) in the compact parts of the blazar jet. Some of these protons escape after isospin flip $p + \gamma \rightarrow n + \pi^+$ as neutrons . The neutrons then suffer β-decay at a distance of $d_{\rm n} = \tau_{\rm n} c \gamma_{\rm n} \approx$ 100 kpc from the blazar. In this way an extragalactic proton flux without adiabatic losses is injected into the intergalactic medium. In the observer's frame these particles have a Lorentz factor $\gamma \approx \delta \gamma_{\rm p,b} \approx 10^{11}$.

Tacitly it is assumed that the jets become radiative at a rather *large* distance from their origin. Shocks are generated as the jet propagates through the external steep pressure gradient of the elliptical host galaxy. This could happen at a distance as far as 1 kpc from the kinematical center (Sanders 1983).

The spectrum of the proton-induced cascades is complex, producing broken power laws and double-humped spectra between X-ray energies and 100 TeV (Mannheim et al. 1991). The internal absorption of γ-rays in 'proton blazars' (Mannheim 1993a), which determines the precise shape of the cascade spectrum, is difficult to assess, since the photon density is highly inhomogenous, e.g. varying from the limb to the center of the jet and varying with length along the jet. However, the *effective* compactness is strongly constrained once the cascade spectrum is known from observations between keV and GeV photon energies.

1.4. Merits of γ-ray measurements

External absorption of γ-rays by the infrared/optical radiation field produced inside the blazar host galaxy (Mannheim 1993b) and by the cosmic background radiation produced by all galaxies (Gould & Schréder 1966, Stecker et al. 1992, MacMinn & Primack 1994) must be taken into account. This limits the number of possible TeV sources to the few nearest blazars, but it opens up new and challenging possibilities for probing the era of galaxy formation and the cosmic distance scale, since $\tau_{\gamma\gamma}(\epsilon_\gamma) \propto n_{\rm ir}(\epsilon_\gamma) H_\circ^{-1}$ where $n_{\rm ir}(\epsilon_\gamma)$ is the near-infrared photon density at the pair creation threshold $\epsilon_\circ = 2(m_e c^2)^2/\epsilon_\gamma \simeq 0.5/(\epsilon_\gamma/\mathrm{TeV})$ eV. A prerequisite for measuring either $H_\circ$ or $n_{\rm ir}$ via the exponential cut-off produced by external absorption is a theoretical prediction of the unabsorbed γ-ray spectra. To this end I discuss below the γ-ray spectra of 1219+285 and Mrk421 in some detail. The strategy is to find combined electron synchrotron spectra and cascade spectra fitting to the multifrequency data for the same set of parameters. In this way the spectrum is highly constrained yielding firm predictions for the TeV range.

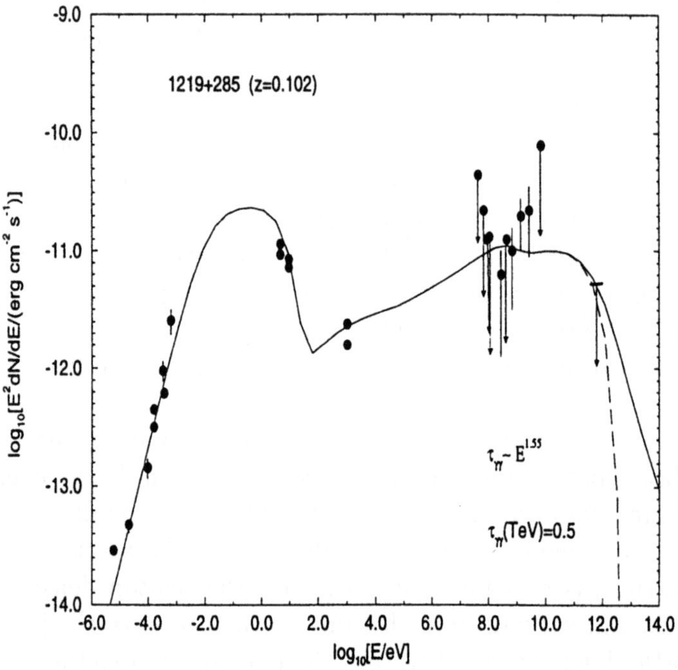

Fig. 1. Multifrequency spectrum of 1219+285. Data were obtained from the NED, Montigny et al. (1995) and Fink (1994). The flux limit at TeV photon energies is a preliminary result obtained by the Whipple group (Rose 1994). The solid line shows the proton blazar model fit for the parameter values in Tab. 1. The dashed line shows the expected flux accounting for external absorption (Stecker et al. 1992)

2. TeV emission from Mrk421 and 1219+285

Internal absorption of γ-rays inside the blazar jets is crucial to the shape of the emerging radiation spectrum. For a steady jet the optical depth in the comoving frame is given by

$$\tau_{\gamma\gamma} = \int_{2(m_e c^2)^2/\epsilon_\gamma}^{\infty} n_{\rm syn}(\epsilon)\sigma_{\gamma\gamma}(\epsilon,\epsilon_\gamma) r_{\rm b,\perp} d\epsilon \simeq \frac{aB^2\epsilon_\gamma}{8\pi \ln[\epsilon_{\rm c}/\epsilon_{\rm b}](m_{\rm e}c^2)^2}\frac{\sigma_{\rm T}}{3} r_{\rm b,\perp} \quad (3)$$

where $a \simeq (1+\eta)^{-1}\beta_{\rm j}\gamma_{\rm j}\Phi \approx (1+\eta)^{-1}$. Above the comoving frame energy ϵ^* where $\tau_{\gamma\gamma}(\epsilon^*) = 1$ the cascade spectrum steepens by one power, since

$$I_\gamma = I_\circ \frac{1-\exp\left[-\tau_{\gamma\gamma}\right]}{\tau_{\gamma\gamma}} \rightarrow I_\circ \epsilon_\gamma^{-1} \quad (\tau_{\gamma\gamma} \gg 1) \quad (4)$$

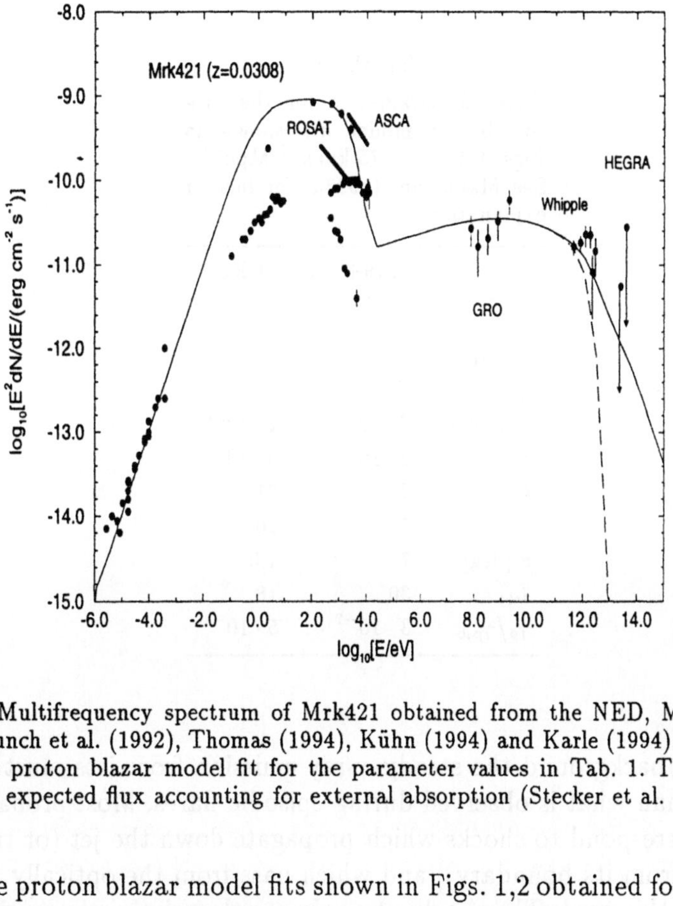

Fig. 2. Multifrequency spectrum of Mrk421 obtained from the NED, Montigny et al. (1995), Punch et al. (1992), Thomas (1994), Kühn (1994) and Karle (1994). The solid line shows the proton blazar model fit for the parameter values in Tab. 1. The dashed line shows the expected flux accounting for external absorption (Stecker et al. 1992)

From the proton blazar model fits shown in Figs. 1,2 obtained for the parameters in Tab. 1 one can infer that the turnover energies in the observer's frame $\delta\epsilon_\gamma^*$ are of the order of TeV, viz.

$$\epsilon_\gamma^*(\mathrm{obs}) \approx \left(\frac{\delta}{10}\right)\left(\frac{\ln[\epsilon_c/\epsilon_b]}{6}\right)\left(\frac{\eta}{100}\right)\left(\frac{B}{10\ \mathrm{G}}\right)^{-2}\left(\frac{r_{b,\perp}}{3\cdot 10^{15}\ \mathrm{cm}}\right)^{-1}\ \mathrm{TeV} \quad (5)$$

It is the length scale $r = r_b$ where the jet becomes optically thin for infrared radiation which regulates the internal absorption to values such that TeV is the typical turnover energy. Another typical energy can be found as a direct consequence: Pairs produced from the optically thick part of the cascade at ϵ_γ^* have Lorentz factors $\gamma_e^* = \epsilon_\gamma^*/2$. The synchrotron photons from these pairs have characteristic energies

$$\epsilon_\gamma^{**}(\mathrm{obs}) = 3\cdot 10^{-14}\delta B\,(\gamma_e^*)^2 \approx 0.02\left(\frac{\delta}{10}\right)^{-1}\left(\frac{B}{10\ \mathrm{G}}\right)\left(\frac{\epsilon_\gamma^*(\mathrm{obs})}{\mathrm{TeV}}\right)^2\ \mathrm{MeV} \quad (6)$$

Below this energy the unsaturated cascade is optically thin, so that the typical X-ray spectral index is $\alpha_X = 0.5 - 0.7$ in marked contrast to saturated cascades with $\alpha_X = 0.9 - 1.0$ (Svensson 1987). To avoid overproducing the

TABLE I

Physical parameters for the proton blazar model fits shown in Figs. 1,2 ($H_o = 75$ km s^{-1} Mpc^{-1}). See Mannheim (1993a) for further explanations

	1219+285	Mrk421
B [G]	4	40
η	30	100
$r_{b\perp}$ [cm]	$7 \cdot 10^{15}$	$2 \cdot 10^{15}$
γ_p	$2 \cdot 10^9$	$6 \cdot 10^7$
δ	7	31
γ_j	5	20
θ [deg]	7	1.5
L_{44}	30	18
$\gamma_p/\gamma_{p,b}$	$3 \cdot 10^{-2}$	$5 \cdot 10^{-4}$

diffuse γ-ray background the steady γ-ray emission from jets must be at a lower level than what is observed during episodic flares. Most probably, the outbursts correspond to shocks which propagate down the jet (or transversally reflect from its boundary), and which pass from the optically thick to the optically thin part. The shocks can enhance the relativistic particle density and increase the particle maximum energies by shock acceleration. No definite predictions for the proton blazar spectra during flux outbursts can be made at present, since this would require to solve non-stationary cascade equations and to couple the radiative transport with a dynamical model for shock propagation and particle acceleration. The optical outbursts (primary electrons) and the hard X-/γ-ray outbursts (cascade pairs) should be correlated, since the cascades develope in the synchrotron photon target produced by the primary electrons. The high-energy cascade emission therefore follows the optical outburst on the very short proton photo-production cooling time scale. This time scale is larger than the optical outburst decay time scale by a factor of $\approx \eta/R$ where $R = L_\gamma/L_o$ denotes the γ/optical luminosity ratio, cf. Eq.(2). If there is a universal value of the proton/electron ratio η, it is predicted that the sources with weakest γ-ray emission compared to the optical emission display the largest delays in γ-rays.

3. Discussion and conclusions

Abundaning geocentrism it is difficult to conveive that baryonic cosmic rays should *not* be present in radio jets. As a matter of fact, there is mounting evidence that extragalactic radio sources produce the observed cosmic ray flux above EeV (Rachen & Biermann 1993). An *in situ* signature of proton acceleration is TeV γ-ray emission from blazars.

However, nature has limited our view of the universe by a very restricted horizon for γ-ray eyes. Due to pair-absorption of γ-rays on cosmic background radiation fields the TeV world ends at a redshift of $z \sim 0.1$. Only the few nearest blazars are therefore potential TeV sources. Nevertheless, the precise value of the TeV horizon depends on two quantities of superior astrophysical interest: The Hubble-constant and the near-infrared photon density which entails the conditions at birth of normal galaxies. We may hope to shed more light on these cosmological cornerstones by measuring and analysing the TeV γ-ray spectra of nearby blazars.

For the two nearby blazars Mrk421 ($z = 0.0308$) and 1219+285 ($z = 0.102$) I obtained flux predictions in the air Cherénkov energy range and above. The predictive power of the calculations is limited by non-simultaneous multifrequency data and by the basic assumptions of the proton blazar model, *viz.* relativistic outflow, isotropy of relativistic protons in the comoving frame, equipartition of relativistic proton and magnetic field pressures, comoving frame maximum Lorentz factors $\gamma_{\mathrm{p,max}} \sim 10^{8-10}$ and a power law index index $p \leq 2$ for the proton energy distribution.

In the TeV range internal absorption of γ-rays competes with external absorption, the former producing a break in the power law spectrum and the latter producing an exponential cut-off. The observed TeV γ-ray emission from Mrk421 is consistent with little external absorption, whereas significant external absorption is likely to be present in the spectrum of 1219+285. If the Hubble-constant had a value $H_\circ \geq 75$ km s^{-1} Mpc^{-1} or if the near-infrared photon density had a value less than the expected value $n_\gamma(0.5\mathrm{eV}) \approx 2 \cdot 10^{-3}$ cm^{-3}, the proton-induced cascade flux would have been above the Whipple upper limit for this source. However, since external absorption does not decrease the observed flux in the air Cherénkov energy range by a large factor, 1219+285 is very likely to be the next blazar detected by a Cherénkov telescope.

Acknowledgements

S. Westerhoff is acknowledged for helpful assistance. This research has made use of the NASA/IPAC Extragalactic Database (NED).

References

Baldwin, J., Boksenberg, A., Burbidge, G., Carswell, R., Cowsik, R., Perry, J., Wolfe, A., 1977, AA, **61**, 165

Begelman, M.C., Kirk, J.G., 1990, ApJ, **353**, 66

Begelman, M.C., Rees, M.J., Sikora, M., 1994, ApJ, **429**, L57

Bell, A.R., 1978, MNRAS, **182**, 443

Berezinsky, V.S., Bulanov, S.V., Dogiel, V.A., Ginzburg, V.L. (ed.), Ptuskin, V.S., Astrophysics of Cosmic Rays, North Holland, Amsterdam, 1990

Biermann, P.L., Strittmatter, P.A., 1987, ApJ., **322**, 643

Biermann, P.L., et al., 1988, AA, **185**, 9

Blandford, R.D., Königl, A., 1979, ApJ, **232**, 34

Blumenthal, G.R., 1970, Phys.Rev. D1, No.6, 1596

Bregman, J.N., 1990, AAR, 125

Celotti, A., Fabian, A.C., 1993, MNRAS, **264**, 228

Colgate, S.A., 1983, 18th Int. Cosmic Ray Conf., Vol.2, p. 230

Dermer, C.D., Schlickeiser, R., Mastichiadis, A., 1992, AA, **256**, L27

Dermer, C.D., Schlickeiser, R., 1993, ApJ, **416**, 458

Drury, L O'C, 1983, Rep.Prog.Phys., **46**, 973

Fichtel, C.E., et al., 1994, ApJS, **90**, 917

Fink, H.H., private communication

Gould, R.J., Schréder, G., 1966, Phys.Rev.Lett., **16**, 252

Harris, D.E., Carilli, C.L., Perley, R.A., 1994, Nat, **367**, 713

Lichti, G.G., et al., 1995, AA, in press

Karle, A., and HEGRA collaboration, 1994, PhD-thesis, University of München

Kühn, M., 1994, PhD-thesis, University of Kiel

Mannheim, K., Krülls, W., Biermann, P.L., 1991, AA, **251**, 723

Mannheim, K., 1993a, AA, **269**, 67

Mannheim, K., 1993b, Phys.Rev.D, **48**, 5270

MacMinn, D., Primack, J., 1994, ApJ, submitted

McBreen, B., 1979, AA, **71**, L17

Moffat, A.F.J., Schlickeiser, R., Shara, M.M., Sieber, W., Tuffs, R., Kühr, H., 1983, ApJ, **271**, L45

Montigny, C. von, et al., 1995, ApJ, in press

Punch, C.W., et al., 1992, Nat, **358**, 477

Quirrenbach, A., et al., 1991, ApJ, **372**, 71

Rachen, J.P., Biermann, P.L., 1993, AA, **272**, 161

Rawlings, S., Saunders, R., 1991, Nat, **349**, 138

Rose, J., private communication

Sanders, R.H., 1983, ApJ, **266**, 73

Sikora, M., Begelman, M.C., Rees, M.J., 1994, ApJ, **421**, 153

Schlickeiser, R., 1984, ApJ, **277**, 485

Stecker, F.W., de Jager, O., Salamon, M.H., 1992, ApJ, **390**, L49

Svensson, R., 1987, MNRAS, **227**, 403

Swanenburg, B.N., et al., 1978, Nat, **275**, 298

Thomas, H.-C., private communication

Zensus, J.A., Pearson, T.J. (eds.), 1987, *Superluminal Radio Sources*, Cambridge University Press

HIGH ENERGY NEUTRINOS FROM QUASARS

F.W. STECKER and M.H. SALAMON*
Laboratory for High Energy Astrophysics
NASA/Goddard Space Flight Center
Greenbelt, Maryland 20771

Abstract. We review and clarify the assumptions of our basic model for neutrino production in the cores of quasars, as well as those modifications to the model made subsequently by other workers. We also present a revised estimate of the neutrino background flux and spectrum obtained using more recent empirical studies of quasars and their evolution. We compare our results with other theoretical calculations and experimental upper limits on the AGN neutrino background flux. We also estimate possible neutrino fluxes from the jets of blazars detected recently by the EGRET experiment on the Compton Gamma Ray Observatory. We discuss the theoretical implications of these estimates.

1. Introduction

Quasars are the most powerful emitters of radiation in the known universe. These remarkable objects are presumably fueled by the gravitational energy of matter falling onto a supermassive black hole at center of the quasar core, though the mechanism responsible for the efficient conversion of gravitational to observed luminous energy is not yet known. If this conversion occurs through the acceleration of high energy cosmic rays, their interactions can lead to the copious production of hadronic and subsequent leptonic high energy byproducts such as neutrinos. It is in this context that quasars have long been considered as potential sources of high energy cosmic ray neutrinos (Berezinsky 1977; Eichler 1979; Silberberg and Shapiro 1979; Berezinsky and Ginzburg 1981).

In related work, Biermann and Strittmatter (1987) suggested that protons could be accelerated to ultrahigh energies in the hot spots found in the jets of radio galaxies (see also Rachen, Stanev and Biermann 1993). Subsequent X–ray production at these sites was considered by Mannheim, Kruells and Biermann (1991). High energy γ-ray and neutrino production in jets was explored by Mannheim, Stanev and Biermann (1992), following the important discovery of strong high energy γ-ray emission from the blazar 3C279 (Hartman, *et al.* 1992). New results on the detection of large fluxes of high energy γ–rays from ~ 40 such sources, obtained using the EGRET detector aboard the Compton Gamma Ray Observatory (Fichtel, *et al.* 1994), have made the study of these high energy sources even more interesting.

An important step in modeling quasar neutrino production process was made when Begelman, Rudak and Sikora (1990) and Stecker, *et al.* (1991)

* NAS-NRC Senior Research Associate, on leave from the Physics Department, University of Utah

Space Science Reviews **75**: 341–355, 1996.

pointed out the importance of photomeson production off thermal UV accretion disk photons as the most significant production channel.

In our work (Stecker, *et al.* 1991), we pointed out that following photopion production, pair cascading off the intense quasar photon fields would both destroy accompanying high energy γ-rays and produce the quasar's X-ray emission (*e.g.*, Done and Fabian 1989). Thus, by normalizing to the X–ray emission, we argued that active galaxies could be by far the strongest sources of ultrahigh energy neutrinos in the universe, producing fluxes detectable on the Earth with present technology. Measurement of these neutrino fluxes and spectra could thus provide key information as to the fundamental astrophysical nature of quasars.

We further pointed out that by integrating over all quasars in the Universe, a large and measurable isotropic neutrino background would be produced. Because of the degradation of γ-rays to X–ray energies, the energy in quasar background neutrinos could be comparable to that of the cosmic X–ray background rather than the much smaller energy of the cosmic γ-ray background. We also noted that radio quiet quasars are a factor of 10 more abundant than their radio loud counterparts (Sramek and Weedman 1978; Kellerman, *et al.* 1989). They thus take on corresponding significance when considering their contribution to the high energy cosmic neutrino background.

Szabo and Protheroe (1994) have followed up on this basic model with independent calculations. They have examined a series of models and assumed that nucleons are trapped indefinitely in the quasar core. In such models, the nucleons eventually lose energy and interact with the X–rays in the core as well as the UV photons, even though the X–ray field of the core is optically thin to nucleons. The result is a much higher flux of lower energy neutrinos than we calculated. The difference between the predicted event rates in neutrino detectors between the two calculations is not as pronounced. This is because the detectability of high energy neutrinos ($\propto \sigma_{\nu N} \times$ range of produced muon) is $\propto E_\nu \ln(E_\nu)$ in this energy range.

An extensive discussion of quasar neutrino production took place at an historic workshop in Hawaii, with almost all of the above-mentioned workers taking part. The reader is referred to the proceedings of this workshop (Stenger, *et al.* 1992) for more background into this subject. In that workshop, Stecker, *et al.* , Szabo and Protheroe, Biermann, and Sikora and Begelman all gave estimates of the neutrino background from AGN.

In this paper, we will update our model using new information on quasar evolution and will also consider neutrino production in blazar jets. We will then compare our results with other work. We will also review the present experimental upper limits on cosmic high energy neutrino fluxes and their implications.

2. The AGN Phenomenon

The standard "AGN paradigm" assumes that infalling matter forms an accretion disk around a black hole. It has been suggested by Kazanas and Ellison (1986) that an accretion disk shock will form at a distance of several Schwartzschild radii ($R_s = 2GM/c^2$) from the black hole. First–order Fermi acceleration of charged particles at this shock then converts a significant fraction of the gravitational energy into highly relativistic particles which have the characteristic hard power law spectrum typical of shock acceleration. Other possible energy generation mechanisms involve electromagnetic acceleration by black hole accretion disk dynamo models. In any model where high energy cosmic rays are produced at the core of the quasar, such cosmic rays (even neutrons) will be trapped by the intense radiation fields. Thus, a necessary consequence of *any* acceleration mechanism in which protons reach high energies is the production of high energy neutrinos through the collisions of these particles with the intense photon fields in quasars.

Recent observational studies bearing on quasar evolution and theoretical models of quasar formation and evolution can be used to guide one in choosing the basic physical parameters needed to calculate neutrino fluxes from individual quasars and also to calculate the neutrino background from all "unresolved" quasars. There is empirical justification for assuming that all quasars run at some constant high efficiency. Quasar evolution appears to be characterised a phase of emission near the maximum allowed rate. This is quantified in terms of the the Eddington limit ($L_{\rm edd} = 4\pi GMm_pc/\sigma_T$ where M is the mass of the black hole), the maximum steady state luminosity that can be produced before radiation pressure disrupts the accretion flow. Studies have shown that quasars and Seyfert galaxies are emitting at a substantial fraction of the Eddington luminosity at all observed redshifts. For quasars, this fraction, which we will call the "Eddington fraction", appears to be usually above 10^{-1}, and for the less luminous Seyfert galaxies, above 10^{-2} with a substantial number above 10^{-1} (Padovanni and Rafanelli 1988).

These results are in accord with theoretical beliefs that as long as there is an adequate supply of gas available, the luminosity produced by the "central engine" of the quasar should approach $L_{\rm edd}$. This also implies that the bright phase of a quasar *should not be substantially longer than the Eddington time*, $t_{\rm edd} = Mc^2/L_{\rm edd} = 4.4 \times 10^8$ yr. Thus, it appears that continuous models of quasar evolution are ruled out (Cavaliere and Padovani 1988).

Haehnelt and Rees (1993) have pointed out that the similarity between the timescale $t_{\rm edd}$ and the evolutionary timescale for the hierarchial buildup of galaxy size masses at high redshifts allows for a natural explanation of the strong evolution of the quasar population. In their work, they point out that if the characteristicly large Eddington fraction of quasars is insensitive to redshift, the only way to understand the rapid increase of quasar luminosity with redshift is to allow that the black holes which are active at low redshifts are different sources with typically smaller masses than those which were emitting at higher redshifts.

We thus arrive at a picture where all black holes in their active phase had an adequate supply of gas to accrete near their Eddington limit so that the Eddington fraction is both large and roughly constant with redshift, as we assumed in our original paper (Stecker, *et al.* 1991). In this scenario, typically the larger mass black holes turned on and used up their supply of gas early, to be replaced in later generations by smaller mass black holes, which, in turn, used up their gas. Such a picture would naturally explain why the quasar X–ray spectral index is no different at $z = 3$ than at $z = 0$, even though their mean luminosity at $z = 3$ is $\sim$30 to 40 times larger than that at $z = 0$ (Elvis 1994). The short lifetime scenario allows for both density and luminosity evolution of the quasar population. However, the ROSAT results (Boyle, *et al.* 1993) appear to be consistent with the simple assumption of pure luminosity evolution, and we will use that analysis accordingly in updating our calculation of the neutrino background from quasars.

3. Cosmic Ray Interactions in Quasar Cores

Let us assume that cosmic rays are accelerated in the centers of active galaxies (see section 4) and examine their expected interactions and energy losses, including those which lead to γ-ray and neutrino production. First, we consider the photon environment.

A characteristic quasar spectrum has comparable amounts of power in each logarithmic energy interval, but the spectrum is not completely flat. Typically, the most prominant feature of the spectrum is the "blue bump", which has a pronounced peak in the hard UV and is thought to be thermal emission from the black hole accretion disk. This feature is thought to arise from the release of energy through viscous processes in the accretion disk (*e.g.*, Laor 1990).

There is also usually a thermal feature in the IR region, thought to be due to dust intercepting and reradiating the UV emission. This hypothesis is supported by the fact that the observable "blue bump" is not as pronounced in AGN which have prominant thermal IR features. Dust can only exist outside of the nucleus, as it is destroyed by sputtering, so the IR spectrum must arise at a large distance from the "central engine" (Barvainis 1987). The lack of variability in the IR spectrum also implies that the emission is from a large region (Edelson and Malkan 1987).

Between the IR and UV emission features there may also be a further continuum component (Loska and Czerny 1990; Laor 1990), but this component appears to be not associated with the very central core (Done *et al.* 1990).

The most rapidly variable part of the spectrum is the X–ray region. Its emission is similar in luminosity to that of UV bump but its rapid time variability indicates that the X–rays are produced in a smaller region than the UV flux (Lawrence, *et al.* 1987; McHardy and Czerny 1987; McHardy 1988). Thus, *in the very central regions, only the UV and X–ray photon fields are of importance.*

The lack of strong X-ray absorption features in quasar spectra (Mushotzky 1982; Turner and Pounds 1989) implies that the secondary X–rays are produced in regions of low column density. This puts strict limits on the amount of target gas for *pp* interactions, and one finds that the very large photon density in the quasar core makes $p\gamma$ interactions the dominant energy loss mechanism. This is a threshhold interaction, and the dominant ν production channel at energies just above the threshold is $\gamma p \rightarrow \Delta \rightarrow n\pi^+$ (Stecker 1968).

For protons traveling through a photon field of characteristic size R, the optical depth to photomeson production for protons of energy E_p can be approximated by $\tau(E_p) \sim \epsilon n(\epsilon)\sigma_o R$ where σ_o at the Δ resonance peak is $\approx 5 \times 10^{-28}$ cm^2 and $n(\epsilon)$ is the differential number density of photons at the resonance energy in cm^{-3} eV^{-1}. An estimate for $n(\epsilon)$ from the accretion disk can be found by assuming that all sources run at some constant efficiency (see Section 2). Fitting accretion disk models to the UV spectrum (Laor 1990) gives a typical luminosity of $(0.03 - 0.1)L_{\rm edd}$ ergs s^{-1}. We have assumed a "blue bump" UV luminosity of $\sim 0.05L_{\rm edd}$ is emitted isotropically. Given the weak dependence of the characteristic accretion disk temperature on black hole mass and the very similar UV bump shapes seen in the observational data, we can assume a luminosity independent 'generic' UV quasar spectrum. The quasar X–ray spectrum is also fairly universal, typically being an $\epsilon^{-1.7}$ power law above 2 keV (Mushotzky 1982; Turner and Pounds 1989), with a cutoff in the MeV energy range (Rothschild, *et al.* 1983). The total X-ray luminosity is roughly the same as that in the UV bump *i.e.* $L_{\rm x} \approx L_{\rm UV} \approx 0.05L_{\rm edd}$. Normalizing so that $L \sim 4\pi R^2 c \int \epsilon n(\epsilon) d\epsilon = 0.1L_{\rm edd}$, we can calculate the mean UV and X–ray photon density within the UV emission region of radius R. We thus obtain

$$n(\epsilon) \approx \frac{5\times 10^{14}}{L_{45}} {\rm cm}^{-3}{\rm eV}^{-1} \begin{cases} 0.165\epsilon & \epsilon < 1 \text{ eV} \\ 0.165\epsilon^{-0.9} & 1 < \epsilon < 40 \text{ eV} \\ 5.35\times 10^{-5}\epsilon^2 e^{-\epsilon/15} & 40 < \epsilon < 192 \text{ eV} \\ 4.15\times 10^{-2}\epsilon^{-1.7} & 192 < \epsilon < 10^6 \text{ eV} \end{cases} \quad (1)$$

where L_{45} is the total UV luminosity in units of 10^{45} ergs s^{-1}. The $1/L$ scaling of $n(\epsilon)$ follows from the linear scaling of both R and L with M (*i.e.*with $L_{\rm edd}$ and R_s above.)

The optical depth from the "blue bump" UV photons is $\gg 1$. The optical depth is independent of the luminosity since $\tau(E_p) \propto n(\epsilon)R$ and $n(\epsilon) \propto 1/L$

while $R \propto L$. The high optical depth of the UV photon field implies that the secondary neutrons also will not generally escape the luminous core region, so that comparable amounts of power are generated via $n\gamma$ collisions. With roughly half of the energy loss going into π^0's and the other half into $\pi^{\pm}$'s, the luminosity for $\nu_\mu + \overline{\nu}_\mu$ is $\sim 0.4L_x$, and for $\nu_e + \overline{\nu}_e$ is $\sim 0.2L_x$. As the quasar photon spectrum drops exponentially between 40 and 200 eV, it is overwhelmingly the UV bump photons which interact with the high-energy nucleons. We note that for this typical quasar spectrum, the X–rays are not optically thick.

4. Proton Acceleration in Quasar Accretion Disk Shocks

It has been suggested by Kazanas and Ellison (1986) that protons can be accelerated to very high energies by diffusion across the strong shock generated at the inner edge of the black hole accretion disk. First-order Fermi acceleration of protons in strong (non-relativistic) shocks produces a power-law proton energy spectrum $\propto E_p^{-2}$ up to a maximum energy $E_{\rm max}$. This maximum proton energy is determined by equating the proton acceleration time, $t_{\rm acc}(E_p)$, with the $p\gamma$ lifetime, $t_{p\gamma}(E_p)$. The proton acceleration time is given by $t_{\rm acc}(E_p) \approx 2.2 \times 10^{-4}(R_{\rm shock}/R_s)(E_p/m_p)B^{-1}$ (Kazanas and Ellison (1986). The $p\gamma$ lifetime is given by $t_{p\gamma}(E_p) \approx (N_\gamma \sigma_{p\gamma} c\kappa)^{-1}$, where κ is the mean elasticity and N_γ is the number density of photons above the photoproduction threshhold. The shock radius, $R_{\rm shock}$, is fixed at $\sim 10R_s$ by our assumption that the X–ray luminosity is $0.05L_{\rm edd}$. The magnetic field B is taken to be 10^3 $L_{45}^{-1/2}$ G, which assumes approximate equipartition of the magnetic energy density with the quasar accretion disk "UV bump" photon density. (Note that this is a change from our earlier work (Stecker, *et al.* 1991) where we just assumed a luminosity independent value of B = 10^3 G.) Setting the loss time equal to the acceleration time, and noting that $N_\gamma \propto 1/L$ (see section 3), we obtain $E_{\rm max}$ given by equation (2). For $L_{45} > 0.1$, $E_{\rm max} \propto L^{1/2}$, a result similar to that obtained by Szabo and Protheroe (1994) using this same assumption about the B field. We note that at all relevant luminosities the loss rate at $E_{\rm max}$ due to $p\gamma \rightarrow e^+e^-$ interactions is negligible compared to photopion production (see discussion in the next section).

$$E_{\rm max} = 10^9 {\rm GeV} \cdot \begin{cases} 0.25L_{45}^{0.375}, & L_{45} \leq 1 \\ 0.25L_{45}^{0.5}, & 1 < L_{45} \leq 10^2 \\ 25, & L_{45} > 10^4 \end{cases} \qquad (2)$$

For all luminosities, the maximum proton energy is limited by $p\gamma$ interactions rather than by the scale size of the shock region.

5. The Production of Ultrahigh Energy Neutrinos

Given a power law (roughly $\propto E^{-2}$) proton spectrum, the pγ pion producing interactions occur mostly close to the threshhold energy. The dominant ν production channel at energies just above the threshold is $\gamma p \rightarrow \Delta \rightarrow n\pi^+$ (Stecker 1968). The cross-section for this reaction peaks at photon energies of $\epsilon' = 0.35 m_p c^2$ in the proton rest frame. In the observers frame $\epsilon E_p = 0.35$ eV-EeV with E_p in EeV and ϵ in eV. For UV bump photons, with a mean energy of ~30 eV, this translates to a characteristic proton energy of $E_c \sim 10^7$ GeV.

It is possible to estimate the isotropic X–ray flux from the core of the closest and brightest quasar, 3C273. Its X–ray spectrum is observed to exhibit an iron line at 6.4 keV of equivalent width $\sim$ 50 eV (Turner, *et al.* 1990). The currently accepted explanation for this line in quasars is fluorescence from the X–ray illuminated accretion disk , so that much of the X–radiation is probably unbeamed. The isotropic X–ray emission in the 2–10 keV band is then $\sim 10^{46}$ ergs s^{-1}, giving a total X–ray flux of $\sim 10^{47}$ ergs s^{-1}.

It now appears that, in the case of the *Seyfert galaxy* NGC 4151, at least some of the X-ray flux and perhaps most or all of the X-ray flux may be of thermal origin with Comptonization playing a major role (Zdziarski, Lightman and Maciolek-Niedzwiecki 1993; Titarchuk 1994). Such a contribution to a *quasar* X-ray spectrum may correspondingly lower our estimate of its neutrino flux, since the neutrino flux in our model is proportional only to the non-thermal AGN component of the X–ray emission. (See, however, Zdziarski *et al.*, 1994; 1995.) To take account of this eventuality, we will assume that, on average, only 30% of quasar X-ray emission is non-thermal in normalizing our calculations. Using this estimate, we calculated the neutrino spectrum of 3C273. The result shown in Figure 1 was calculated with the assumptions given in Stecker *et al.*(1991) using Monte Carlo techniques.

The mean pion energy is $< E_\pi > \sim 0.2 E_p$, while the mean neutrino energy is $\sim E_\pi/4 \sim 0.05 E_p$ (Stecker 1979). Thus the neutrino spectrum from proton interactions with the UV bump extends from $E_c/20$ to $E_{\max}/20$ *i.e.* from $\sim 10^6$ to $\sim 10^8$ GeV, with a spectral index of ~ 2, mirroring that of the proton spectrum. Above $\sim 10^9$ GeV, the cutoff in the proton spectrum leads to a cutoff in the neutrino spectrum. Below $E_c/20$ relativistic kinematics give rise to a flat ν energy distribution (Stecker 1979).

It is probable that there is a range of X–ray and γ-ray compactness in quasars, and hence a range of optical depths for $p\gamma$ processes. For more compact quasars, the maximum energy of the protons would be lower, with $p\gamma$ interactions with X-rays determining the cutoff. In models with higher compactness and indefinitely large nucleon containment times, interactions with X–rays will result in the more copious production of lower energy neutrinos with a softer low energy spectrum (see, *e.g.*, Sikora, *et al.* 1987; Szabo and

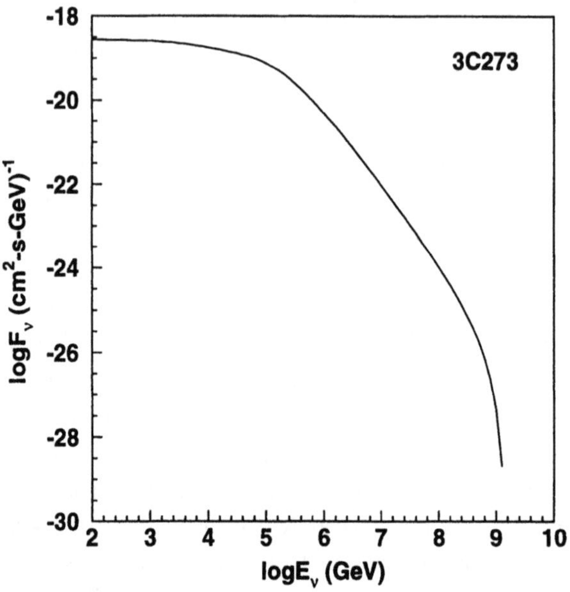

Fig. 1. The predicted $\nu_\mu + \overline{\nu}_\mu$ flux from 3C273, with $L_x = 10^{47}$ ergs s^{-1} (Piccinotti, *et al.* 1982) and redshift $z = 0.158$. (A Hubble constant of 50 km-s^{-1}-Mpc^{-1} has been assumed in our calculations.) Note that the $\nu_e + \overline{\nu}_e$ flux is half that of the $\nu_\mu + \overline{\nu}_\mu$ flux.

Protheroe 1994). In this scenario, $p\gamma$ pair-production can compete with photomeson production as an energy loss process at the lower energies, which can reduce some of the low energy neutrino production (Sikora and Begelman 1992). The physics of quasar neutrino production at the "lower energies" is probably intermediate between the various models discussed in the literature.

For the photon densities in our model, the effective optical depth for the $p\gamma$ pair-production process is ≤ 1, owing to the fact that the energy loss rate for this process is two orders of magnitude smaller than that for photopion production (see, *e.g.*, Puget, Stecker and Bredekamp 1976; Sikora and Begelman 1992). It should be noted that even though the loss rate for pair-production is more important than for photopion production at energies significantly below E_{max} (owing to the lower pair production threshold energy), it is always negligible compared to the acceleration rate (see, *e.g.*, Figure 1 of Sikora and Begelman 1992).

A soft spectrum of low energy neutrinos can also arise from the decay of pions produced in *pp* interactions. Using the X–ray data, we have argued that such interactions should not account for the bulk of the neutrino production (see section 3). In our initial work (Stecker, *et al.* 1991), we estimated that, at most, 10% of AGN neutrinos would be from *pp* interactions. This conclusion is consistent with the suggested accretion disk production estimate of Nellen, *et al.* (1993).

6. The Extragalactic Neutrino Background Spectrum

The diffuse ν flux can be found by integrating over the luminosity function (LF) for quasars. The most recent determinations of the X-ray LF, as well as high redshift luminosity evolution, come from the ROSAT satellite (Boyle, *et al.* 1993). We will adopt model I of Boyle, *et al.* (1993). For this model, the LF at $z = 0$ is given by

$$\rho_o(L_{44})\Delta L_{44} = \begin{cases} 7.6 \times 10^{-7}\mathrm{Mpc}^{-3} L_{44}^{-1.7}, & L_{\min} < L < L_1 \\ 3.2 \times 10^{-7}\mathrm{Mpc}^{-3} L_{44}^{-3.4}, & L_1 < L < L_{\max} \end{cases} \tag{3}$$

where the luminosity function ρ is per luminosity interval in units of $(10^{44}$ erg s$^{-1})^{-1}$, L_{44} is the 0.3 to 3.5 keV X-ray luminosity in units of 10^{44} erg s^{-1}, $L_{\min} = 10^{39}$ erg s^{-1} , $L_1 = 10^{43.8}$ ergs s^{-1} and $L_{\max} = 10^{47}$ ergs s^{-1}. These luminosities refer to the X-ray flux in the ROSAT range and must be multiplied by a factor 17 to transform them into a total X-ray flux. The source luminosity as a function of redshift, z, is taken to be (Boyle, *et al.* 1993)

$$L(z) = \begin{cases} L(0)(1+z)^{2.73}, & z \leq 2 \\ L(0)3^{2.73}, & z > 2 \end{cases} \tag{4}$$

We calculate the diffuse neutrino background assuming an epoch of formation of quasar activity corresponding to $z_{max} = 4$ and note that our results are not very sensitive to z_{max} in the range between 2 and 4.

Figure 2 shows our new background result, obtained assuming that ~30 % of the quasar X-ray flux is from the $p\gamma$ interactions and subsequent electromagnetic cascading (see section 5). Our results indicate that the diffuse background neutrino flux from quasars could dominate over the atmospheric background in the range 10^5 to 10^{10} GeV.

7. Possible Origin of the Broad Line Emission Region

The broad line region (BLR), a characteristic feature of AGN, is thought to be made up of a large number of dense, illuminated clouds (Davidson and Netzer 1979). The large line widths of $\sim$ 5000 km s^{-1} can be explained by doppler shifts due to orbital motions. If the clouds have predominantly circular orbits, then the velocity can be calculated as $v = \sqrt{GM/r}$, or $\beta = \sqrt{R_s/2r}$. For these high velocities, the clouds have to be within $\sim 1700R_s$, or 5×10^{16} cm for a 10^8 solar mass AGN black hole.

An outstanding problem in this scenario is the formation and stability of these clouds. While the two phase instability, in which dense condensations can exist in pressure equilibrium with a hot, rarified gas is seen as the best and currently accepted model (Krolik, McKee and Tarter 1981), there are many problems with this explanation (Mathews 1986; Elizur and Ferland 1986).

However, we have previously showed that the high energy neutrino luminosities predicted from quasars will have a profound effect on stars close to the center of the host galaxy, producing stellar winds, swelling the atmospheres and even causing their total disruption (Stecker, *et al.* 1985; Gaisser, *et al.* 1986; Stecker, *et al.* 1991). The column density for absorption of a $\sim 10^6$ GeV neutrino is $X = m_H/\sigma_{\nu N} \sim 2 \times 10^9$ g cm^{-2}, while total column density for a solar mass star $< \rho > R \sim 10^{12}$ g cm^{-2}. Using a conservative disruption criterion that the neutrino energy deposited in the star be greater than its nuclear energy generation, we find a sphere of stellar disruption, $R_{\rm SSD}$, within a radius of

$$R_{\rm SSD} \approx 30 L_{45}^{1/2} \left(\frac{M_*}{M_O}\right)^{-1.1} \text{ light days} \tag{5}$$

where M_* and M_O are the masses of the irradiated star and the Sun respectively. This radius is the same as that inferred above for the BLR, and matches well with the current best observational determinations (Netzer 1989; Clavel, *et al.* 1991) and previous calculations have shown that outflowing material from stellar disruption provides, at least qualitatively, an environment similar to that in which the broad lines must arise Kazanas 1989). Thus, high energy neutrinos from an AGN core could play a key role in producing the BLR.

8. Neutrinos from the Cores and Jets of Blazars

The evidence for relativistic bulk motions and beaming in blazars ("optically violent variable" quasars and BL Lac objects) implies that the intrinsic luminosities and sizes of the emission regions are more uncertain than in radio quiet objects. However, if the above discussion of the origin of the BLR is correct, then the cores of these objects should similarly produce neutrino fluxes. These objects would be expected to contain the same sort of nucleus as the radio quiet objects, but with the addition of jet emission from further out.

With the observation of γ-ray emission above 100 MeV from at least 40 blazars out to redshifts greater than 2 by the EGRET group (Fichtel 1994), it has now become possible to estimate the corresponding high energy neutrino emission from the jets of these sources. Let us assume that the observed γ-rays are produced in the relativistic jets of these objects and are beamed because (a) as discussed before, we do not expect γ-rays above a few MeV to escape from the core of a typical quasar, and (b) EGRET observed time variability on the order of a day from 3C279, implying relativistic beaming. Let us further assume the the observed γ-rays are produced by high energy protons, leading to pion production and the subsequent decay of neutral pions into γ-rays and accompanying charged pions into electrons, positrons and neutrinos. We further note that the observed differential spectrum of 3C279 is consistent with a roughly E^{-2} power law up to energies in excess of 5 GeV (Hartman, *et al.* 1992), as would be expected for shock accelerated proton primaries leading to pion decay γ-rays. Also, the detection of γ-rays above 0.5 TeV from Mrk 421 by the Whipple Observatory group (Punch, *et al.* 1992), in combination with the EGRET detection, implies a roughly E^{-2} spectrum for this source up to TeV energies and possibly beyond. Such high energies are hard to account for from electron acceleration alone, since radiative energy losses of electrons at such energies are quite severe. Mannhiem, *et al.* (1992) have proposed a model for proton acceleration in blazar jets.

If we make the above assumptions, we expect that the ratio of γ-rays to $\nu_\mu + \overline{\nu}_\mu$ should be 1:1. Assuming an E^{-2} differential spectrum, the ratio above a given energy should be 2:1, since the mean energy of a γ-ray from pion decay is roughly twice the energy of a neutrino from the decay of a pion of equal energy. We can then predict TeV neutrino fluxes from the EGRET sources. For Mrk 421, with a reported γ-ray flux above 0.5 TeV of $\sim 1.2 \times 10^{-11}$ cm^{-2} s^{-1}, we obtain an immediate predicted neutrino flux above 1 TeV of $\sim 6 \times 10^{-12}$ in these same units. This source is at a low enough redshift (0.031) so that no significant γ-ray absorption at TeV energies is expected, however, the other EGRET sources are at higher redshifts where pair production interactions with intergalactic infrared photons can cut off the observed γ-ray spectrum (Stecker, De Jager and Salamon 1992). Indeed, although Mrk 421 is the weakest of the EGRET sources yet reported, none of the other sources searched for were detected by the Whipple team above 0.5 TeV (Kerrick, *et al.* 1993). However, correspondingly high energy neutrinos can reach us unattenuated, so that we may use an extrapolation of the EGRET spectrum to predict TeV neutrino fluxes.

For the jet in 3C279, in its more quiescent state, corresponding to the second EGRET observing period, extrapolation of the E^{-2} γ-ray spectrum leads to a predicted jet muon neutrino flux of $\sim 4 \times 10^{-11}$ above 1 TeV. For the flaring phase the corresponding neutrino flux would be $\sim 2 \times 10^{-10}$, twice the minimum detectable flux for DUMAND II. Our predicted neutrino flux above 1 TeV for the core of this source would be $\sim 10^{-13}$.

For the 3C273 jet, extrapolating the EGRET spectrum (C. von Montigny, *et al.* 1993) leads to a predicted ν_μ flux above 1 TeV of $\sim 10^{-13}$. The corresponding prediction for the core flux is $\sim 4 \times 10^{-14}$.

It is interesting to note that for superluminal jet sources, with a typical bulk Lorentz factor of ~ 10, the beamed flux would appear to be $\sim 10^3$ times brighter than that of an isotropic source with the same intrinsic production rate.

Using the 1:1 correspondence of γ-rays to ν_μ's (see above), combined with an extrapolated estimate of the blazar γ-ray background which we published recently (Salamon and Stecker 1994), we immediately obtain the estimate of the neutrino background flux from blazar jets shown in Figure 2, assuming the spectrum continues as a power law well above TeV energies. Figure 2 also shows the AGN neutrino background fluxes calculated by us here and by other workers.

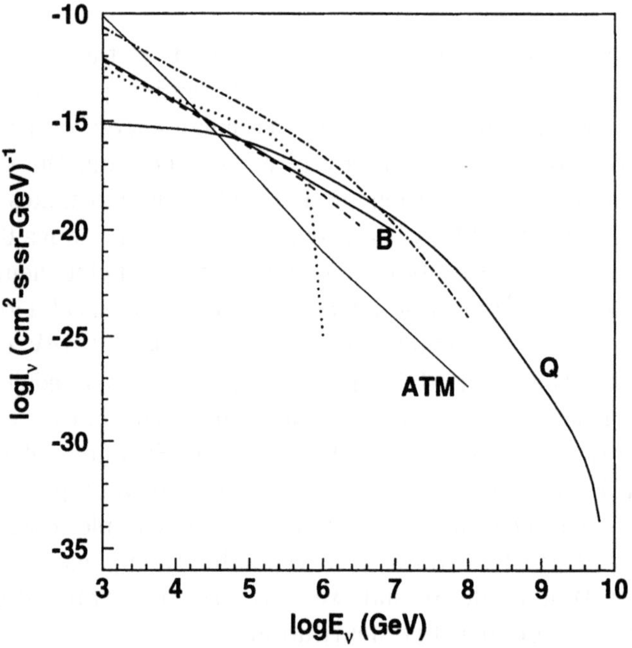

Fig. 2. Our revised integrated high energy $\nu_\mu + \overline{\nu}_\mu$ neutrino background from quasars (Q) and blazars (B) (thick solid lines), as well as the fluxes calculated by Sikora and Begelman (1992) (dotted line), Biermann (1992) (dashed line), and the geometric mean for the $b = 1$ model of Szabo and Protheroe (1994) (dash-dot line). Also shown is the horizontal $\nu_\mu + \overline{\nu}_\mu$ flux from high energy cosmic rays interacting with the Earth's atmosphere (ATM, solid line) (Stecker 1979).

9. Discussion

As can be seen from Figure 2, the high energy neutrino background fluxes from AGN calculated by various workers are comparable. This is not too surprising, since they are mainly based on the assumptions of shock acceleration of protons to high energies followed by photomeson production and subsequent charged pion decay. All results are normalized to AGN X-ray or X-ray background energy fluxes in some way.

The results shown in Figure 2 lead to predictions of observable high energy neutrino background fluxes. Indeed, presently existing data are beginning to test the different predictions of the models (Zas, Halzen and Vazquez 1993; Rhode 1994; Aglietta, *et al.* 1994; Gaisser, Halzen and Stanev 1995). In particular, the analyses of upper limits on muon fluxes from the Frejus experiment (Rhode 1994; Gaisser, *et al.* 1995) are consistent with our quasar background flux prediction, but appear to rule out part of the range of model parameters considered by Szabo and Protheroe (1994). Our prediction is also consistent with bounds on giant horizontal air showers established with the AKENO array (Zas, Halzen and Vazquez 1993) and the EAS-TOP collaboration (Aglietta, *et al.* 1994). However, the predicted flux is not very far below the presently established bounds.

The event rates for various ν detectors can be calculated from our background spectrum given in Figure 2. The Fly's Eye detector, with its energy threshhold of 10^8 GeV would expect to see $\ll 1$ event per year, consistent with their present non-detection. Their next generation instrument, the HiRes Eye would see < 1 *downward* events per year. Above the HiRes energy threshold of $\sim 10^8$ GeV, *upward* neutrinos are severely attenuated by absorption in the Earth. However, the upward event rates at energies greater than 10^5 GeV for DUMAND II, AMANDA and NESTOR are of order 10^2 yr^{-1}.

We note that $n\gamma$ interactions produce a significant flux of $\overline{\nu}_e$'s at 6.3 PeV, the Glashow resonance energy for the $\overline{\nu}_e e^- \to W^-$ process. An observation of an enhanced event rate at this energy would indicate that AGN are optically thick to neutrons. We estimate that DUMAND II would see on the order of 10 events per year from W^- resonance production. (The reader is referred to the articles in Stenger, *et al.* (1992) for a discussion of various neutrino detector experiments.)

We conclude that results from high energy neutrino detectors either planned or presently under construction should be able to test the viability of the idea that significant proton acceleration in AGN cores ultimately produces a substantial fraction of their observed X-ray luminosity.

References

Aglietta, M., *et al.* : 1994, preprint.
Barvainis, R.E. : 1987, *Astrophys. J.* **320**, 537.
Berezinsky, V.S. : 1977, *Proc. Neutrino 77*, (Moscow: Nauka Pub.), p. 177.
Berezinsky, V.S. and Ginzburg, V.L. : 1981, *Mon. Not. R. Astron. Soc.* **194**, 3.
Begelman, M.C., Rudak, B. and Sikora, M. : 1990, *Astrophys. J.* **362**, 38.
Biermann, P. : 1992, in *High Energy Neutrino Astrophysics*, ed. V.J. Stenger, J.G. Learned, S. Pakvasa and X. Tata, (Singapore: World Scientific), p. 86.
Biermann, P. and Strittmatter, P.L. : 1987, *Astrophys. J.* **322**, 643.
Boyle, B.J., *et al.* : 1993, *Mon. Not. R. Astr. Soc.* **260**, 49.
Cavaliere, A. and Padovani, P. : 1988, *Astrophys. J.* **333**, L33.
Clavel, J., *et al.* : 1991, *Astrophys. J.* **366**, 64.
Davidson, K. and Netzer, H. : 1979, *Rev. Mod. Phys* **51**, 715.
Done, C. and Fabian, A.C. : 1989, *Mon. Not. R. Astr. Soc.* **240**, 81.
Done, C., *et al.* : 1990, *Mon. Not. R. Astr. Soc.* **243**, 713.
Edelson, R.A., and Malkan, M.A. : 1987, *Astrophys. J.* **323**, 516.
Eichler, D. 1979 : *Astrophys. J.* **232**, 106.
Elizur M. and Ferland, G.J. 1986 : *Astrophys. J.* **305**, 35.
Elvis, M.S. : 1994, in *Multi-Wavelength Continuum Emission of AGN* ed. T.J.-L. Courviosier and A. Blecha (Dordrecht: Kleuwer), p. 25.
Fichtel, C.E., *et al.* : 1994 *Astrophys. J. Suppl.* **94**, 551.
Gaisser, T.K., Halzen,F. and Stanev, T. 1995 : *Phys. Rpts.*, in press.
Gaisser, T.K. Stecker, F.W., Harding A.K., and Barnard, J.J. : 1986, *Astrophys. J.* **309**, 674.
Haehnelt, M. and Rees, M.J. : 1993, *Mon. Not. R. Astr. Soc.* **263**, 168.
Hartman, R.C., *et al.* : 1992, *Astrophys. J. Lett.* **385**, L1.
Kazanas, D. : 1989, *Astrophys. J.* **347**, 74.
Kazanas, D. and Ellison, D.C. : 1986, *Astrophys. J.* **304**, 178.
Kellerman, K.I., *et al.* : 1989, *Astron. Jour.* **98**, 1195.
Kerrick, A.D., *et al.* : 1993, *Proc. 23rd Intl. Cosmic Ray Conf.* (Calgary: U. Calgary Press), **1**, 405.
Krolik, J.H., McKee, C.F. and Tarter, C.B. : 1981, *Astrophys. J.* **249**, 422.
Laor, A. : 1990, *Mon. Not. R. Astron. Soc.* **246**, 369.
Lawrence, A., Pounds, K.A., Watson, M.G. and Elvis, M.S. : 1987, *Nature* **325**, 694.
Loska, Z. and Czerny, B : 1990, *Mon. Not. R. Astron. Soc.* **244**, 43.
Mannheim, K., Kruells, W.M. and Biermann, P.L. : 1989, *Astron. Astrophys.* **251**, 723.
Mannheim, K., Stanev, T., and Biermann, P.L. : 1992, *Astron. Astrophys.* **260**, L1.
Mathews, W.G. 1986, *Astrophys. J.* **305**, 187.
McHardy, I.M. : 1988, *Mem. Soc. Ast. It.* **59**, 239.
McHardy, I.M. and Czerny, B. : 1987, *Nature* **325**, 696.
Montigny, C. von *et al.* : 1993, *Astron. and Astrophys. Suppl.* **97**, 101.
Mushotzky, R.F. : 1982, *Astrophys. J.* **256**, 92.
Nellen, L., Mannheim, K. and Biermann, P.L. : 1993, *Phys. Rev.* **D47**, 5270.
Netzer, H. : 1989, *Comments Astrophys.* **14**, 137.
Padovani, P. and Rafanelli, P. : 1988, *Astron. Astrophys.* **205**, 53.
Piccinotti, G., *et al.* : 1982, *Astrophys. J.* **253**, 485.
Puget, J.L., Stecker, F.W. and Bredekamp, J.H. : 1976, *Astrophys. J.* **205**, 638.
Punch, M, *et al.* : 1992, *Nature* **358**, 477.
Rachen, J.P., Stanev, T. and Biermann, P.L. : 1993, *Astron. Astrophys.* **273**, 377.
Rhode, W. : 1994, in *Proc. Neutrino '94*, in press.
Rothschild, R. E., *et al.* : 1983, *Astrophys. J.* **269**, 423.
Salamon, M.H. and Stecker, F.W. : 1994, *Astrophys. J. Lett.* **430**, L21.
Silberberg, R. and Shapiro, M.M. : 1979, *Proc. 16th Intl. Cosmic Ray Conf. (Kyoto)*, **10**, 357.

Sikora, M. and Begelman, M.C. : 1992, in *High Energy Neutrino Astrophysics*, ed. V.J. Stenger, J.G. Learned, S. Pakvasa and X. Tata, (Singapore: World Scientific), p. 114.
Sikora, M., Kirk, J.G., Begelman, M.C. and Schneider, P. : 1987, *Astrophys. J. Lett.* **320**, L81.
Sramek, R.A. and Weedman, D.W. : 1978, *Astrophys. J.* **221**, 468.
Stecker, F.W. : 1968, *Phys Rev. Lett* **21**, 1016.
Stecker, F.W. : 1979, *Astrophys. J.* **228**, 919.
Stecker, F.W., De Jager, O.C. and Salamon, M.H. : 1992, *Astrophys. J. Lett.* **390**, L49.
Stecker, F.W., Done, C., Salamon, M.H. and Sommers, P. : 1991, *Phys. Rev. Letters* **66**, 2697; E (1992): *ibid.*, **69**, 2738.
Stecker, F.W., Harding, A.K. and Barnard, J.J. : 1985, *Nature* **316**, 418.
Stenger, V.J., Learned, J.G., Pakvasa, S. and Tata, X., ed. : 1992, *Proc. High Energy Neutrino Astrophysics Workshop (U. Hawaii)*, World Scientific, Singapore.
Szabo, A.P., and Protheroe, R.J. : 1994, *Astroparticle Physics* **2**, 375.
Titarchuk, L. : 1994, *Astrophys. J.* **434**, 570.
Turner, M.J. *et al.*: 1990, *Mon. Not. R. Astron. Soc.* **244**, 310.
Turner, T.J. and Pounds, K.A. : 1989, *Mon. Not. R. Astron. Soc.* **240**, 833.
Zas, E., Halzen, F., and Vazquez, R.A. : 1993, *Astroparticle Phys.* **1**, 297.
Zdziarski, A., Lightman, A.P. and Maciolek-Niedzwiecki, A. : 1993, *Astrophys. J.* **414**, L93.
Zdziarski, A., *et al.*: 1994, *Mon. Not. R. Astron. Soc.* **269**, L55.
Zdziarksi, A., *et al.*: 1995, *Astrophys. J. Lett.* **438**, L63.

POSSIBLE EPISODIC GAMMA-RAY SOURCES

Moving Fragile Target Crosses Relativistic Particle Beam

F.A.AHARONIAN and A.M.ATOYAN *

Max-Planck Institut für Kernphysik, Postfach 103980, D-69029 Heidelberg, Germany

Abstract. A model for production of episodic γ-ray event at interaction of a moving gas target with a beam of relativistic particles is proposed. The typical duration of γ-ray emission is limited by the flight time of the target across the beam as well as by the time of destruction and/or expulsion of the target by luminous beam. The time-dependent radiation spectra of the expanding and moving gas cloud irradiated by the beam are calculated for the galactic binary systems Her X-1 and AE Aquarii which are reported as episodic γ-ray emitters at very high energies. Some predictions and observational tests for the model are discussed.

1. Introduction

In his classification of VHE γ-ray emitters T.Weekes (1992) has referred X-ray binaries to the category of serendipitous sources, i.e. the objects which "... do not have any firm *a priori* basis for their selection as candidate VHE γ-ray sources". Nevertheless, this very class of sources, in particular Cygnus X-3, has played perhaps the most important role in the 80s in the renewed interest in ground based γ-ray observations. But ironically, the same sources have created, to some extent, a certain doubt on the credibility of most results of ground-based γ-ray astronomy in the past. Many experts treat these data with a healthy degree of scepticism (e.g. Weekes, 1992; Lamb, 1995) due to both the low confidence level of experimental results and the reported unusual characteristics of the primary radiation. As a result, X-ray binaries are no more considered as primary objectives for future VHE γ-ray observations (see, e.g. Goret, 1993). However, the community seems to be far from drawing a final conclusion on the reported episodic γ-ray events from several sources, in particular from the X-ray binaries Her X-1, Vela X-1, and from the cataclysmic variable AE Aquarii. Moreover, there are groups strongly believing in the reality of this phenomenon (see e.g. Chadwick et al., 1990; Raubenheimer, 1993).

If we admit that the γ-ray episodic events from X-ray binaries do exist, then we are facing serious problems of interpretation of many extraordinary features which are common for most of reported events from different objects, namely: (a) the irregular shifts of periods of the γ-ray pulsations relative to the canonical (spin) period; (b) the absence of the orbital phase modulation, and especially, observations of pulsed γ-rays in the phase of

* On leave from Yerevan Physics Institute, Armenia

Space Science Reviews **75**: 357–373, 1996.

deep eclipse of the X-ray source; (c) the episodic character of γ-radiation with duration $\Delta t \leq 1\,\mathrm{h}$.

The standard "atmospheric target" model of γ-ray production in X-ray binaries (bombardment of the normal star's atmosphere by the relativistic particle beam accelerated by the compact object) is not able to explain the above mentioned peculiarities. Since the beam obviously originates from the compact object (the signature of the accelerator is seen in the form of the signal modulation with the spin period), the only possibility left for explanation of these peculiarities is to assume some other target(s). In principle, the material for the target can be provided by the normal companion star in the form of accretion gas, stellar wind, massive clouds ejected from the star, etc. In order to explain the frequency shifts of γ-ray pulsations with characteristic scale $\Delta\nu/\nu_0 \geq 0.1\%$ one has to assume γ-ray production in the fast moving targets with $v/c \geq 10^{-3}$. The γ-ray production site(s) cannot be very close to the compact object, at least for those events which are observed in the phase of deep eclipse. From this point of view the Cheng-Ruderman (1989) and Slane-Fry (1989) models proposed to explain the frequency shifts by production of γ-rays in the periphery of the fast rotating accretion disk, are not able to account for the episodes observed during the eclipse. A proper model should allow for the γ-ray production at distances comparable with the lengthscale of the binary system.

The massive gas outflows from the normal star induced, for example, by the X-ray heating of the outermost regions of the star (Harpaz and Rappaport, 1994), could serve as effective targets for the relativistic beam accelerated by the compact object. In this model (Aharonian and Atoyan, 1991) the episodic γ-radiation is expected due to the chance probability of intersection of the trajectory of ejected mass with the propagation cone of the accelerated particles. Under the intense beam the target (gas cloud) becomes fragile. Therefore, the duration of the γ-radiation is limited not only with the flight time of the cloud across the beam, but also with the lifetime (destruction and/or expulsion) of the cloud under the beam. If the beam of relativistic particles is modulated with the spin period of the compact object, one may expect a similar modulation of the secondary γ-radiation. However, the period of γ-pulsations may be shifted due to the double Doppler effect (i.e. reemission of the pulsed signal from the moving target). Since γ-ray emission arises at the collision of accelerated particles with clouds irregularly ejected from the surface of the normal star there is no reason to expect a correlation between the orbital phase of the binary and γ-ray episodes. Moreover, at high enough trajectories of clouds, γ-ray events may be expected even in the eclipse of the X-ray source. Thus, all the peculiarities of γ-ray episodes reported from X-ray binaries could be in principle understood within the "moving target crosses beam" scenario.

2. The Scenario

The basic assumption of the proposed model of production of γ-ray episodes in the binary systems is the hypothesis that the normal star *spontaneously* ejects massive gas clouds which may occasionally, depending on their trajectories, intersect the propagation cone of the beam of relativistic particles which are *continuously* accelerated by the compact companion star (e.g. pulsar). The cartoon of the scenario is shown in Fig.1. The relativistic particles entering the cloud produce γ-rays. The production and escape of γ-rays from the cloud take place under extreme conditions of strong impact of the beam on the cloud's parameters (size, density, temperature, magnetic field) and its trajectory. The heating and expansion of the cloud as well as its expulsion from the system result in variation in time of principal parameters defining the γ-ray luminosity of the irradiated cloud: (1) the gas density n, the size $R_{\rm cl}$, and the distance l of the cloud from the accelerator which define the γ-ray production rate; (2) the optical depths of the cloud which define the absorption (and regeneration) rates of γ-rays interacting with the ambient gas and low-frequency photons. Therefore, the γ-ray spectrum is characterized by strongly nonlinear dependence on the parameters of the beam and the cloud.

2.1. General Requirements to the Basic Parameters

In the model proposed, we assume that the beam consists of relativistic protons; they reach the target without drastic energy losses which is not the case for relativistic electrons. Admitting the episodes observed from some X-ray binaries, in particular from Her X-1, at energies $E \geq 100$ TeV (Baltrusaitis *et al.*, 1985; Dingus *et al.*, 1988; Gupta *et al.*, 1990; Roberts *et al.*, 1993), we should assume that the energy spectrum of protons extends up to at least 1 PeV. Commonly the power-law spectrum $dN/dE \propto E^{-\alpha}$ is suggested, though the monoenergetic particle beam with $E_0 \geq 1$ PeV also could be considered (Hillas, 1984). Note that the difference between power-law and monoenergetic proton spectra is significant not only for the formation of the spectrum of the first-generation γ-rays, but also for essentially different efficiencies of heating of the cloud.

The interpretation of the observed offset (in many γ-ray episodes) of the period by the *double* Doppler effect requires different directions of the primary and reflected signals (see Fig.1). This implies that the pulsed γ-ray production occurs when the moving target *is out* of the line of sight between observer and the beam accelerator. This requirement opposes the conventional belief that γ-rays are seen when a target intervenes between the pulsar and observer. Therefore, to see the radiation one has to require the protons of the beam to be isotropized in the magnetic field of the cloud,

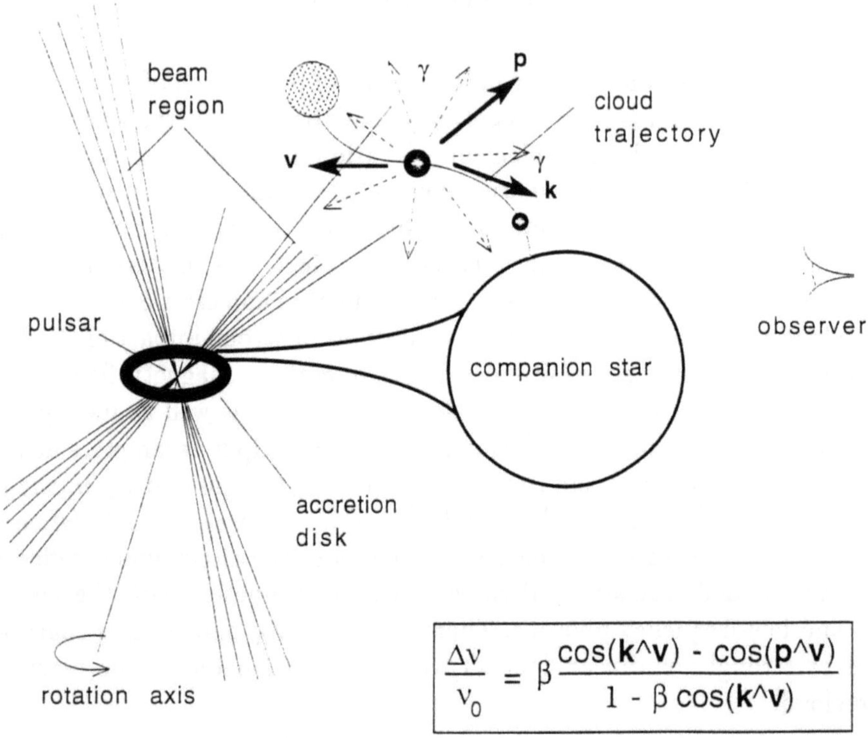

Fig. 1. The cartoon of the scenario "moving target crosses beam". The shift of the frequency of pulsations relative to the spin frequency of the pulsar, $\Delta\nu/\nu_0$, due to the double Doppler effect is shown, where **p** and **v** are the unit vectors in directions of the proton beam and the cloud velocity, respectively, and **k** is the unit vector in the direction of the photon pointed from the cloud to observer. The frequency shift, depending on the orientation of **p** and **k** relative to **v** , is of order of $\beta = v/c$; $\Delta\nu = 0$ if $\mathbf{p} \parallel \mathbf{k}$.

i.e. the gyroradius should not exceed the radius of the cloud. This leads to the lower limit to the magnetic field

$$B \geq 300\,(E/10^3\,\mathrm{TeV})\,(\mathrm{R_{cl}}/10^{10}\,\mathrm{cm})^{-1}\,\mathrm{G}\,. \tag{1}$$

The pulsed component with a period $P \approx P_0$ can be observed if

$$R_{\mathrm{cl}} \leq R_{\mathrm{max}} = cP_0 = 3\cdot 10^{10}\,(P_0/1\,\mathrm{s})\,\mathrm{cm}\,. \tag{2}$$

The second requirement for effective production of the pulsed component is that the interaction time of protons τ_{pp} with production of secondary π^0 mesons should not exceed the period of pulsations, i.e. the gas density

$$n \geq n_{\mathrm{min}} \approx (cP_0\sigma_{\mathrm{pp}})^{-1} \approx 7\cdot 10^{14}\,(P_0/1\,\mathrm{s})^{-1}\,\mathrm{cm}^{-3}\,, \tag{3}$$

where $\sigma_{\mathrm{pp}} \approx 5\cdot 10^{-26}\,\mathrm{cm}^2$ is the cross-section of the inelastic $p-p$ interactions at TeV energies.

Eqs.(2)–(3) relate only to the stage of *pulsed* radiation production. However, the efficiency of the overall (pulsed+unpulsed) γ-ray production does not correlate with these conditions; the γ-ray luminosity can be even higher at later stages of the cloud evolution when the cloud becomes larger.

The power of the beam can be estimated from the observed γ-ray fluxes. The typical fluxes reported for episodic events from different binaries are $F_\gamma(\geq 1\,\mathrm{TeV}) \geq 10^{-10}\,\mathrm{cm^{-2}s^{-1}}$. Then, the γ-ray production rate is

$$\dot{N}_\gamma(\geq E) \approx 1.1 \cdot 10^{34}\, F_{-10}\, d_{\mathrm{kpc}}^2\, (E/1\,\mathrm{TeV})^{-\alpha+1}\, \mathrm{s}^{-1}\,, \tag{4}$$

where $F_{-10} \equiv F_\gamma(\geq 1\,\mathrm{TeV})/10^{-10}\,\mathrm{cm^{-2}s^{-1}}$, and $d_{\mathrm{kpc}} \equiv d/1\,\mathrm{kpc}$ is the distance to the source. The maximum γ-ray production rate is reached when the time of the relativistic proton escape from the cloud is much larger than the proton interaction time $\tau_{\mathrm{pp}} = (\sigma_{\mathrm{pp}} n c)^{-1}$. In this case $\dot{N}_\gamma$ depends only on the proton injection rate:

$$\dot{N}_\gamma(\geq E) = \frac{< mx >_\alpha \dot{N}_{\mathrm{p}}(\geq E)}{(\alpha-1)f}\,. \tag{5}$$

Here $f \approx 0.5$ is the inelasticity at $p-p$ interactions, and $< mx >_\alpha$ is the so-called spectrum-weighted moment; $< mx >_\alpha \approx 0.2$ for $\alpha = 2$ (e.g. Gaisser, 1990). Since the cloud at a distance l from the accelerator covers a solid angle $\Delta\Omega = \pi(R_{\mathrm{cl}}/l)^2$, one obtains the minimum luminosity of the beam per unit solid angle:

$$L_{\mathrm{p}} \approx 2.2 \cdot 10^{35}\, (l/R_{\mathrm{cl}})^2\, F_{10}\, d_{\mathrm{kpc}}^2\ \mathrm{erg\,s^{-1}\,sr^{-1}}\,. \tag{6}$$

At $R_{\mathrm{cl}} \ll l$ only a small fraction ($\leq 1\%$) of the total power of the beam propagating within the solid angle $\Omega_0 \sim 1\,\mathrm{sr}$ is absorbed in the cloud. However, at some stages of its evolution (expansion), the cloud may reach sizes comparable with the distance to the pulsar, thus much larger fraction of the beam power can be transformed into γ-rays.

The lifetime of the cloud effectively emitting γ-rays is defined by thermal expansion of the cloud. The characteristic temperature of the cloud is determined from the thermal balance (*heating rate* = *radiative cooling rate*). The energy of the beam absorbed in the cloud is $\Delta\dot{W}_{\mathrm{p}} = \pi(R_{\mathrm{cl}}/l)^2 L_{\mathrm{p}}$. Some part of this energy is thermalized due to ionization losses of protons. Since the Coulomb losses of protons are important only at energies below few GeV, the fraction η of the total energy $\Delta\dot{W}_{\mathrm{p}}$ which goes to the heating of the cloud is very sensitive to the spectrum of the relativistic protons, especially at low energies. For example, in the case of the power-law spectrum in the total energy of protons with $\alpha \sim 2$ the heating efficiency η is about a few percents, whereas the power-law spectrum in the kinetic energy gives $\eta > 0.1$. The ionization heating efficiency is very small ($\eta \leq 10^{-6}$) for the monoenergetic proton beam with $E_0 \geq 1\,\mathrm{PeV}$. However, even in this case we cannot neglect the heating of the cloud since the heating is provided also by the secondary particles, in particular by the low energy electrons and X-rays. In addition, intense heating may be caused by the turbulence developed in the plasma by the relativistic beam. Thus, η should be considered as a model parameter to be chosen in reasonable limits, e.g. within $10^{-3} - 10^{-1}$.

The minimum temperature of the cloud can be estimated from the condition of the maximum (blackbody) cooling rate, $T_{\min} = \eta \Delta \dot{W}_{\rm p}/(4\pi R^2 \sigma_{\rm SB})$. Therefore,

$$T \geq 2.5 \cdot 10^4 \mathrm{K} \left(\frac{L_{\rm p}}{10^{37}\,\mathrm{erg\,s^{-1}\,sr^{-1}}}\right)^{1/4} \left(\frac{\eta}{0.1}\right)^{1/4} \left(\frac{l}{10^{11}\,\mathrm{cm}^2}\right)^{-1/2} \qquad (7)$$

The characteristic timescale of thermal expansion of the cloud (in vacuum) can be estimated as $\Delta t \sim R_{\rm cl}/v_{\rm s}$, where $v_{\rm s} \simeq \sqrt{kT/m_{\rm p}}$ is the sound speed. For the cloud size $R_{\rm cl} \sim 10^{10}\,\mathrm{cm}$ it would correspond to the scales of $\Delta t \simeq 1\,\mathrm{h}$.

Another limitation on the duration of episodic pulsed emission is imposed by the flight time of the moving cloud across the beam. For typical distances $l \sim 10^{11}$ cm and velocities of order of Keplerian, $v \sim 10^7 - 10^8$ cm/s, the flight time $\Delta t \sim l/v \leq$ few hours. Actually, this time may be even less for relatively light clouds which can be accelerated under the pressure of the beam to velocities essentially exceeding the Keplerian ones. The beam may affect motion of the cloud if the beam pressure exceeds the gravitational force. If the beam particles lose their initial direction in the cloud's magnetic field, an essential if not all momentum of the beam, $p = \Delta \dot{W}_{\rm p}/c$, transfers to the cloud, i.e. $F_{\rm beam} = dp/dt \sim L_{\rm p}\pi(R/l)^2/c$. Then, the critical radius R_* after reaching which $F_{\rm beam}$ exceeds gravitational attraction, $F_{\rm g} = GM_{\rm cl}M/l^2$, to the compact object:

$$R_* \simeq 3.6 \cdot 10^{10} \left(\frac{M_{\rm cl}}{10^{22}\,\mathrm{g}}\right)^{1/2} \left(\frac{M}{1\,M_\odot}\right)^{1/2} \left(\frac{L_{\rm p}}{10^{37}\,\mathrm{erg\,s^{-1}\,sr^{-1}}}\right)^{-1/2} \mathrm{cm}. \qquad (8)$$

Thus, during its expansion the cloud absorbs more and more energy and momentum of the beam. For clouds of relatively small masses (typically, $M_{\rm cl} \leq 10^{22}$ g, the stage when $F_{\rm beam}$ exceeds $F_{\rm g}$ can be reached, and the cloud may be quickly accelerated and be thrown out of the system.

The intense beam of charged particles not only heats the cloud but can also generate turbulent magnetic fields. The maximum field which can be induced is determined from the condition of equipartition with thermal gas: $B_{\max} \sim \sqrt{8\pi\, n\, kT}$. For example, for the cloud density $n = 10^{15}\,\mathrm{cm}^{-3}$ and temperature $T = 3 \cdot 10^4$K (compare with Eq.(3) and Eq.(7)), one obtains $B \approx 300$ G. Thus, the field required by Eq.(1) for the effective opening of the beam in the cloud, can be provided by the beam itself, almost independently on the initial magnetic field in the cloud.

2.2. Application to Her X-1 and AE Aquarii

More than 10 X-ray binaries and cataclysmic variables are reported as episodic γ-ray sources at TeV and PeV energies (Chadwick *et al.*, 1990; Weekes, 1992). Here we assume that the phenomenon of VHE γ-ray episodes is a real effect, and discuss the principal model requirements for 2 sources, Her X-1 and AE Aquarii.

1. *Her X-1/HZ Herculis.* This low mass binary system has a spin period of the pulsar $P_0 = 1.24$ s, and the orbital period $P_{\rm orb} \approx 1.7$ days. The distance between the pulsar with the mass $\simeq 1.3\,M_\odot$ and the center of the normal companion star with $\simeq 2.2\,M_\odot$ is $a = 6.3 \cdot 10^{11}$ cm. The radius of the normal star which fills up its Roche lobe is $\approx 2.6 \cdot 10^{11}$ cm. This determines the scale of distances l of a cloud from the pulsar (at least, in the cases of γ-ray episodes observed in the eclipse of the X-ray source). The fluxes of pulsed γ-rays during episodes lasting $\Delta t \leq 1$ h are typically $F_\gamma(\geq 1\,{\rm TeV}) \geq 10^{-10}\,{\rm cm^{-2}s^{-1}}$ (Dowthwaite *et al.*, 1984; Lamb *et al.*, 1988; Resvanis *et al.*, 1988). For the estimated distance to the source within $2 - 6$ kpc (Hutchings, 1978), the luminosity of the beam, in accordance with Eq.(6), is of about $L_{\rm p} \geq 10^{39}\,{\rm erg\,s^{-1}sr^{-1}}$. Gamma-radiation modulated with spin period of the pulsar $P_0 = 1.24$ s can be observed only at those stages of the cloud evolution, when the density exceeds $5 \cdot 10^{15}\,{\rm cm^{-3}}$, and the size of the cloud is less than $4 \cdot 10^{10}$cm. On the other hand, for the reasonable luminosity of the beam, the size of the cloud cannot be much less than 10^{10} cm. Also, the product $n\,R$ should not exceed $10^{26}\,{\rm cm^{-2}}$ to avoid the drastic absorption of γ-rays in the cloud. Thus, the range of the cloud mass favorable for the production and escape of pulsed γ-ray signal should be about of $M_{\rm cl} \sim 10^{22} - 10^{23}$ g.

2. *AE Aquarii.* This cataclysmic variable consists of $0.94 M_\odot$ white dwarf and $0.74 M_\odot$ K-type star orbiting with period 9.88 h at the distance $a = 3 \cdot 10^{11}$ cm from each other. The spin period of the white dwarf is about $P_0 \simeq 33$ s. The source shows flaring activity in the radio, optical, and γ-rays (see e.g. Meintjes *et al.*, 1992 and references therein). TeV pulsed γ-ray episodes from this source have been observed by the Potchefstroom and Durham groups (Meintjes et al., 1992; Bowden et al., 1992). The pulsed γ-ray signals detected by the Potchefstroom group have, as a rule, periods shifted with respect to P_0, and presumably are correlated with optical emission during flares (Meintjes et al., 1992). In the Durham dataset the most of the power appears in the second harmonic (Bowden et al., 1992). The TeV fluxes observed at the level $10^{-10} - 10^{-9}\,{\rm cm^{-2}s^{-1}}$, are comparable with the fluxes observed from Her X-1. However, since the source is located at the distance $d \leq 100$ pc, the required luminosity of the beam does not exceed $5 \cdot 10^{35}\,{\rm erg\,s^{-1}\,sr^{-1}}$. Therefore, clouds are heated to lower temperatures, $T \leq 10^4$ K, resulting in smooth γ-ray spectra without profound absorption feature which appears in the γ-ray spectra of Her X-1. Besides,

since the spin period of the compact object accelerating particles in the AE Aquarii is much larger than in Her X-1/Hz Herculis, it allows for significantly larger range of the cloud parameters for production of pulsed radiation, and perhaps more frequent occurrence of detectable γ-ray events.

3. Gamma-Ray Spectra

In the framework of the proposed scenario the γ-radiation is initiated by interactions of relativistic proton beam with the gas cloud. In the case of an optically thin cloud γ-rays produced at decays of secondary π^0-mesons repeat the spectrum of the parent protons. In fact, at the initial stages the dense cloud may be optically thick, i.e. $\Lambda = m_p n R_{cl} \geq 100\,\mathrm{g/cm^2}$. Since $\Lambda \propto R_{cl}^{-2}$, the cloud during its expansion eventually becomes optically thin with respect to the absorption of γ-rays in the gas. However, even at $\Lambda \ll 100\,\mathrm{g/cm^2}$, the cloud may still be not transparent for VHE γ-rays interacting with the optical/UV radiation of the thermal gas heated by the beam up to temperatures $T \sim 10^5 K$. The secondary (e^+, e^-) pairs from photon-photon interactions produce broad spectrum of synchrotron and inverse Compton γ-rays which in their turn interact with ambient gas and radiation field producing new particles, etc. Thus, unusual conditions can be created in the cloud, where relativistic electron-photon cascades are effectively developing in all three substances (gas, radiation, and magnetic field) in the cloud. The beam has a strong impact on the density of all these substances which leads to strong nonlinear dependence of the resulting broad-band radiation spectrum on the beam luminosity.

In the optical and UV bands the thermal radiation of the cloud masks the nonthermal component. However, in radio and X-rays the nonthermal radiation could be observable. Since the cloud typically remains opaque with respect to Thomson scattering and to the photoelectric absorption for a long time, the radiation transfer in the gas plays principal role in the formation of the spectrum of X-rays. In the radio domain the synchrotron radiation is produced by low energy electrons ($E \leq 100\,\mathrm{MeV}$) in strong magnetic fields, $B \sim 10^2 - 10^3\,\mathrm{G}$. At these conditions the radiation at GHz frequences may be strongly modified due to synchrotron reabsorption, resulting in flat or inverted radio spectra.

In the calculations following processes are taken into account:

(a) *in the gas*: production and decay of π-mesons produced at $p-p$ interactions; ionization losses and bremsstrahlung of relativistic electrons; (e^+, e^-) pair production, Compton scattering and photoelectric absorption of γ-rays;
(b) *in the photon field*: (e^+, e^-) pair production at $\gamma - \gamma$ collisions; inverse Compton scattering of relativistic electrons on the thermal photons;
(c) *in the magnetic field*: synchrotron radiation and self-absorption.

The relativistic cascade processes in the expanding cloud have been calculated under following assumptions:

(a) it is assumed that relativistic charge particles are partially captured in the cloud with energy-dependent escape time $\tau_{esc} = (R_{cl}/c) \times (1 + R_{cl}/R_L)$, where $R_{cl} = R_{cl}(t)$ is the radius of the cloud at instant t, and $R_L = E/eB$ is the Larmor radius. This implies that energetic particles with $R_L > R_{cl}$ cross end escape the cloud without deflections in the magnetic field;

(b) the magnetic field induced by the beam in the cloud is in equipartition with the thermal gas;

(c) the temperature of the cloud is determined from the thermal balance; the fraction η of the injected beam power which goes to the heating of the cloud is taken as a free parameter within $0.1 - 10\,\%$;

(d) the expansion of the spherical cloud occurs with the sound speed.

The detailed description of the method and approximations used in calculations of the time-dependent cascade development in the expanding target will be published elsewhere. Here we discuss briefly the typical spectra expected during γ-ray episodes from Her X-1 and AE Aquarii.

1. **Her X-1.** In calculations we assume a luminosity of the proton beam accelerated by the pulsar $L_p = 2 \cdot 10^{39}\,\mathrm{erg\,s^{-1}\,sr^{-1}}$. The time evolution of the spectrum of radiation emerging from the cloud with mass $M_{cl} = 2 \cdot 10^{22}\,\mathrm{g}$ entering into the propagation cone of the beam at the distance $l = 10^{11}\,\mathrm{cm}$ from the pulsar is shown in Fig.2. The partial contributions from different radiation processes are also shown. Note that for a given energy E the curves correspond to the fluxes integrated in the range from E to $10 \times E$. Since the predicted differential spectra, with prominent emission and absorption features, are very unusual, this kind of presentation of fluxes illustrates properly the expected counting rates of detectors having typically dynamical ranges of order of 10.

At the first stages, $t \sim$ several minutes, after the entrance of the cloud into the propagation cone of the beam, the flux is too low to be observed, because the cloud is rather small, thus it absorbs only a small fraction of the beam luminosity. Later the cloud, during its thermal expansion, covers more solid angle, absorbing and respectively reemitting more energy which releases in the form of both thermal (optical/UV) and nonthermal radiations. For the chosen parameters the observable fluxes of TeV γ-rays ($F_\gamma \sim 10^{-10}\,\mathrm{cm^{-2}s^{-1}}$) are reached after $\sim 20\,\mathrm{min}$. However, due to expansion of the cloud (and decrease of the gas density) the characteristic time of $p-p$ interactions at $t \geq 1\,\mathrm{h}$ exceeds the beam modulation period $P_0 = 1.24\,\mathrm{s}$ which leads, as discussed above, to the disappearance of pulsed radiation. Thus, if the sensitivity of a detector is enough to see only the pulsed component of radiation, one might conclude that the γ-ray emission has lasted $\leq 1\,\mathrm{h}$. In fact, detectors with

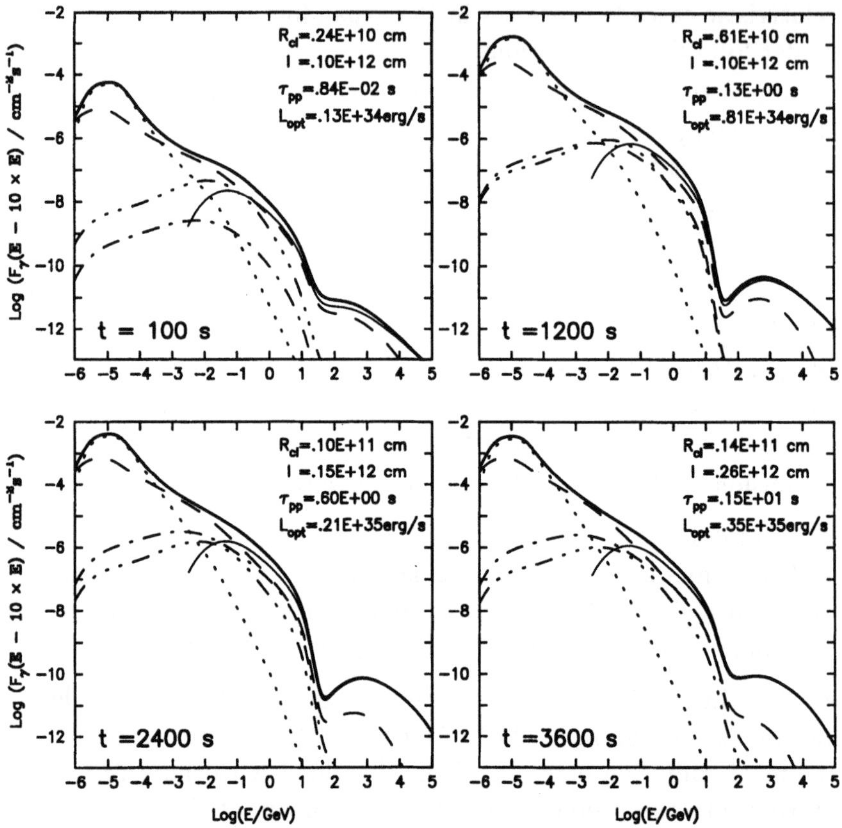

Fig. 2. The evolution of the radiation spectra expected during a γ-ray episode from Her X-1. The fluxes and principle parameters of the cloud (the radius R_{cl}, the distance l from the pulsar, the $p-p$ interaction time τ_{pp}, and the thermal optical luminosity within $300-600$ nm are presented at 4 different instances t after entrance of the gas cloud into the propagation cone of the beam. The contributions of different processes are shown by solid line (π^0 decays), dashed line (synchrotron radiation), dot-dashed line (inverse Compton), fancy line (electron bremsstrahlung), dotted line (multiple Compton scattered radiation in the cold gas). The heavy solid line is the total spectrum emerging from the cloud. The mass of the cloud, its initial radius and distance from the pulsar are $M_{cl} = 2 \cdot 10^{22}$ g, $R_{cl} = 2 \cdot 10^9$ cm, and $l = 10^{11}$ cm, respectively. The luminosity of the proton beam is $L_p = 2 \cdot 10^{39}$ erg s^{-1}sr^{-1}. The distance to Her X-1 is taken $d = 3$ kpc.

enhanced sensitivities *would be able to detect DC γ-ray signals even after the disappearance of pulsations*, since the total flux may remain on the same level up to several hours, provided that the cloud does not leave the region wherein the beam propagates. However, for typical trajectories clouds would cross the beam during $t \leq 1$ h. In fact, at large distances, $l \geq 3 \cdot 10^{11}$ cm, the episodes could last up to several hours, though in this cases the TeV fluxes cannot exceed 10^{-11} cm^{-2}s^{-1}.

On the other hand, if the trajectory of the cloud intersects the beam close to the compact object, $l \ll 10^{11}$ cm, the fluxes of TeV γ-rays may

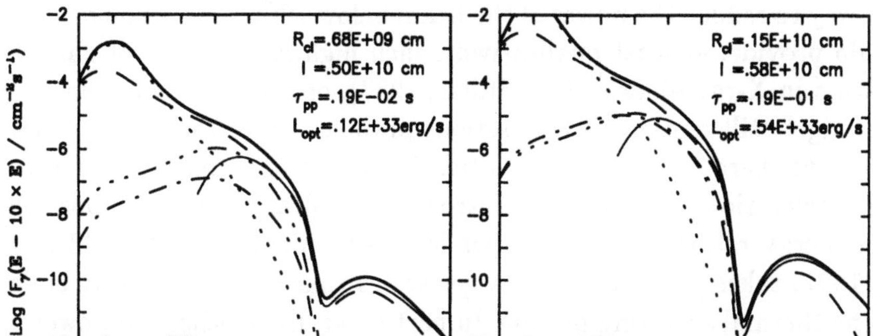

Fig. 3. Same as Fig.2, but for the cloud mass $M_{\rm cl} = 2 \cdot 10^{21}$ g, its initial radius $R_{\rm cl} = 5 \cdot 10^{8}$ cm, and initial distance from the pulsar $l = 5 \cdot 10^{9}$ cm.

reach $10^{-9}\,{\rm cm}^{-2}{\rm s}^{-1}$. The duration of such events is defined not only by the crossing time, but also by the possible quick change of the trajectory of the cloud and its expulsion from the system under the strong pressure of the beam. The time evolution of the γ-ray spectra for a very strong and short episode is demonstrated in Fig.3. Note, that for eclipsing binaries like Her X-1, this kind of episode occurring very close to the compact object is impossible to detect during the deep eclipse of the X-ray source.

One of the disturbing peculiarities of VHE γ-radiation from X-ray binaries is their high luminosities, leading to hard energy budget problem. In some cases this implies the beam power $\dot{W}_{\rm p} \geq 10^{39}$ erg/s, substantially higher than the Eddington luminosity of the neutron star (e.g. Hillas, 1987). In particular, in our calculations for Her X-1 we have to assume the beam luminosity $L_{\rm p} = \dot{W}_{\rm p}/\Omega_0 \sim 2 \cdot 10^{39}\,{\rm erg\,s}^{-1}\,{\rm sr}^{-1}$. This value may be reduced by a factor of few assuming monoenergetic beam of VHE particles as supposed by Hillas (1984) for Cyg X-3. Another possibility is to assume very narrow beam of particles (Johnson, 1994), $\Omega_0 \ll 1\,{\rm sr}$. Nevertheless, we do not see a realistic way to reduce the power of the beam below 10^{38} erg/s. Though some models do provide, at least qualitatively, such high luminosities of particle acceleration in X-ray binaries (e.g. Hillas, 1987), one has to admit that the energy budget still remains a very serious problem. Nevertheless, it should be noted that there is no obvious relation between the accretion luminosity and the acceleration power of the neutron star (pulsar). It is quite possible that the energy release in the relativistic binary systems like SS 443 (e.g. Kundt, 1991) takes place in much larger scales than we see in X-ray emission of the thermal accreting plasma. In particular, the mechanical power of ejected matter in the recently observed superluminal motion in the galactic object GRS1915+10, which originally was discovered as a transient soft γ-ray source, exceeds 10^{41} erg/s (Mirabel and Rodriguez, 1994)

2. **AE Aquarii.** Another possible episodic γ-ray source at TeV energies reported is AE Aquarii (Meintjes et al., 1992; Bowden et al., 1992). The γ-ray production in this source occurs at much smaller luminosity scales, however due to proximity of the source, the γ-ray fluxes could be detectable. The bolometric luminosity of the white dwarf (blue star), which peaks most probably in the ultraviolet, is rather uncertain and is estimated within 10^{34}–10^{35} erg/s (Patterson, 1979). (Thus, the spin-down luminosity of the white dwarf recently estimated as $\approx 10^{34}$ erg/s (De Jager *et al.*, 1994) does not exceed the gain of kinetic energy due to the gas accretion). This implies that the luminosity of the beam may reach $L_{\rm p} = \dot{W}_{\rm p}/\Omega_0 \sim 10^{34}-10^{35}\,{\rm erg\,s^{-1}sr^{-1}}$.

Much lower luminosity of the beam in AE Aquarii, if compared with Her X-1, is essentially compensated by its proximity, leading to the same level of the expected fluxes from both sources. However, there is noticeable difference in the temporal and spectral characteristics of the flaring radiation from these two sources. In particular, in AE Aquarii clouds are heated to lower temperatures, $T \leq 10^4$K. The slower thermal expansion of the cloud implies slower evolution of cloud parameters, resulting in less prominent temporal and spectral evolution of the radiation from radio to γ-rays.

In Fig.4a the energy fluxes of the radiation emitted by the cloud at different instances of its motion in the elliptic Keplerian orbit with the perihelion (i.e. the closest distance from the white dwarf) $l_{\rm min} = 5\cdot 10^{10}$ cm are shown*. It is seen that for the beam luminosity $L_{\rm p} = 10^{34}\,{\rm erg\,s^{-1}sr^{-1}}$ the calculated fluxes are at the level of the reported fluxes (e.g. Bowden, 1992). Two features of the γ-ray spectra should be noted: (i) due to relatively low temperature of the cloud, the absorption of VHE γ-rays in the field of thermal (blackbody) photons is almost negligible, resulting in smooth γ-ray spectra; (ii) though the primary spectra of the protons in the beam is supposed to

* Note that in AE Aquarii the beam luminosity is insufficient to affect the trajectory of the cloud (see Eq.(8), thus the assumption of the Keplerian trajectories appears a good approximation.

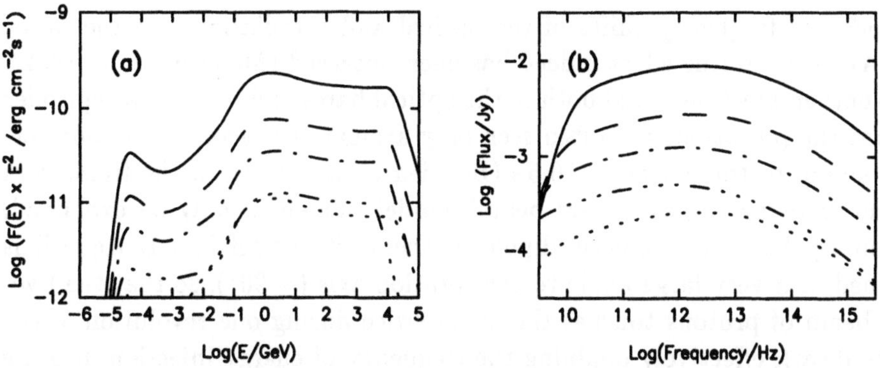

Fig. 4. The evolution of fluxes of the broad-band radiation from AE Aquarii: a) the energy fluxes ($F(E)\times E^2$) from X-ray to γ-ray energies; b) the flux densities ($J_\nu = F(\nu)\times\nu$) of the nonthermal (synchrotron) radiation from radio to optical frequences. It is assumed that the cloud is moving around the white dwarf in an elliptic trajectory with the perihelion $l_{min} = 5\cdot 10^{10}$ cm, and enters the propagation cone of the beam 1 h before perihelion. The fluxes are shown at different instances before and after perihelion: $t = 0$ (at perihelion) – solid line; $t = -1$ h – dotted line; $t = +0.5$ h – dashed line; $t = +1$ h – dot-dashed line; $t = +3$ h – fancy line. The mass of the cloud is $M_{cl} = 2\cdot 10^{22}$ g, the initial size of the cloud $R_{cl} = 10^{10}$ cm, and the luminosity of the beam $L_p = 10^{34}$ erg s^{-1} sr^{-1}.

extend up to PeV energies (as in the case of Her X-1) the expected fluxes of γ-rays at $E \geq 10$ TeV are significantly lower than for Her X-1. This feature is due to inability of the relatively low magnetic field (induced by the moderately intense beam) to effectively open the beam of PeV protons.

The expected fluxes of synchrotron radiation from radio to optical frequencies are shown in Fig.4b. At GHz frequences the cloud is optically thick with respect to synchrotron self-absorption, resulting in the flux density increasing with frequency ($J_\nu \propto \nu^\alpha$ with $\alpha > 0$). This spectral feature as well as the absolute fluxes agree with often detected radio flares with typical duration $t \sim 1$ h (Bastian et al., 1989). At frequences $\nu \geq 100$ GHz the cloud become thin* , and standard optically thin synchrotron spectra are produced. This radiation extends up to optical frequences, where, however, it is substantially masked by the thermal optical radiation of the cloud heated by the beam. For the cloud's parameters used, the calculated optical luminosity within 300 – 600 nm reaches $L_{\rm opt} \approx 10^{32}$ erg/s. Since both the time of $p-p$ interactions, τ_{pp}, and the light crossing time of the cloud, $t \sim R_{cl}/c$, are significantly less than the relatively large spin period of the white dwarf $P_0 \approx 33$ s, the pulsed optical radiation of the cloud is expected. As explained above, the reflection of the pulsed signal from the moving target shifts the frequency of pulsations seen by an observer, and obviously during the same

* In Her X-1, the clouds, due to larger magnetic fields induced, are optically thick below 10^{12} Hz, therefore the radio flux at GHz frequences is strongly suppressed.

episode the frequency shifts of the optical and γ-radiations are the same. For AE Aquarii this effect indeed has been observed (Meintjes *et al.*, 1992).

Another effect observed both in the optical flares and VHE γ-ray episodes is that the pulsations are often seen only around the second harmonic corresponding to the period $\simeq 16.5$ s (Bowden *et al.*, 1992). In the framework of the "moving target crosses beam" model this effect may be explained, assuming that the magnetic dipole in the white dwarf in AE Aquarii is inclined at a very large angle to the rotation axis ($\sim 90°$), so that the two-side beam of protons touches the cloud twice during one revolution of the white dwarf, effectively doubling the frequency of energy injection into the cloud.

At last, the drastic absorption of X-rays below 10 keV results in negligible flaring soft X-ray luminosity of the cloud, much less than the observed pulsed X-radiation of AE Aquarii with $L_x(0.1 - 4\,\text{keV}) \approx 10^{31}\,\text{erg/s}$ (Patterson *et al.*, 1984) which is attributed to the radiation of the accretion disk.

4. Discussion

The current pessimistic attitude to the X-ray binaries as VHE γ-ray sources is due to not only the low credibility of most of the reported detections of episodic γ-ray events, but also to the unusual features of γ-ray signals. The main purpose of this paper is to study, if these features are really so extraordinary to raise the question of the reality of the phenomenon itself. We believe that they are not.

An observation of intense VHE γ-radiation implies the existence of not only a powerful beam of relativistic particles, but also of a target nearby. An attractive feature of X-ray binaries is that they contain both counterparts, the pulsar as an obvious accelerator and the normal star as a possible target. However, observations of γ-ray episodes from X-ray binaries (e.g from Cyg X-3, Vela X-3, Her X-1) at orbital phases around 0.5 excludes the normal star atmosphere as a target (Hillas, 1987). Another potential target could be accretion gas near the compact object (see e.g. Hillas, 1987; Protheroe and Stanev, 1987). However, in this case it is not obvious how one could account for those episodic events which are observed in the eclipse of Her X-1 and Vela X-1. Thus, a model which would be able to reproduce γ-ray episodes at all orbital phases should allow a target to be far from both compact object and the normal star.

The outflow of large amount of gas from the companion star, e.g. in the form of dense clouds, can serve as an appropriate targets for production of episodic γ-ray events. To leave the surface of the star the clouds should have velocities exceeding hundreds of km/s. The hypothesis of fast moving target naturally explains two features of reported γ-ray episodes, namely the irregular offset of the period of pulsations relative to the spin period of the compact object (due to the double Doppler effect), and timescales of the reported signals $t \sim 1\,\mathrm{h}$. The duration of the signal is defined by the flight time of the cloud across the beam as well as by possible destruction and/or expulsion of the cloud.

The episodic character of the γ-ray emission is obviously connected with the chance probability for the spontaneously ejected clouds to intersect the propagation cone of the beam. Therefore, the frequency of appearance of such episodic events may differ from source to source, depending on the geometry and parameters of the binary system. Furthermore, one could expect a long-term variation of the γ-ray activity of one and the same source, associated with the long-term variation of the mass-loss rate of the normal star. It is interesting to note that the mass loss from X-ray heated normal star, in particular in the low-mass X-ray binaries, could be not a continuous process but rather consisting of series of spikes with typical duration of $1 - 100\,\mathrm{yr}$. Very high rates of mass loss up to $10^{-5} M_{\odot}\,\mathrm{yr}^{-1}$, significantly exceeding the Eddington limit, implies that most of the mass ejected from the secondary star is not accreted onto the compact object, but is lost from the system (Harpaz and Rappaport, 1994). It seems very intriguing to attribute the presumable decline of the γ-ray flux from Cygnus X-3 (and perhaps also from Her X-1) in the last decade (Weekes, 1992; Protheroe, 1994: Lamb, 1995) to this effect of variation of the rate of mass ejection from the companion star.

The "moving target crosses beam" model predicts not only drastic variation of the γ-ray fluxes, but also significant change of the spectral and time structures of signals. The energy spectra expected in the TeV domain, in particular for powerful sources, are untypically hard having, at some stages, profound $\gamma - \gamma$ absorption feature at $E \sim 100\,\mathrm{GeV}$ (see Figs. 2,3). This effect could in principle explain the nondetection of a signal by the Durham Cherenkov telescope during the γ-ray episode claimed to be detected by the Fly's Eye (Baltrusaitis, 1985) at much higher energies.

At energies below 1 GeV the cloud is transparent for γ-rays even during very intensive episodes. For Her X-1 the peak fluxes above 100 MeV could reach $10^{-5}\,\mathrm{cm}^{-2}\mathrm{s}^{-1}$. These fluxes lasting only $\sim 1\,\mathrm{h}$ could be marginally detected by the EGRET. With the effective detection area of the EGRET instrument $\approx 10^3\,\mathrm{cm}^2$ at 100 MeV, this flux results in tens of detected events only. Being, however, supported by simultaneous detections at TeV energies, where the number of expected γ-ray events are even higher, the significance of the detection would be much increased. Note that in the case of VHE γ-ray episodes, when the fluxes are expected at least 1 order of magnitude higher than the detectable quasi-steady flux, the main disadvantage of the ground-based γ-ray detectors (that they are dominated by the cosmic ray background) essentially disappears. Furthermore, the main advantage associated with high rate capabilities (due to large detection areas) becomes pronounced.

Here we would like to emphasize that the strategy of searching for episodic signals with strongly evolving radiation spectrum perhaps should be different from the procedure optimized for the quasi-steady sources. In particular, the imaging Cherenkov technique for rejection of the cosmic ray background should be implemented very cautiously. Since the shape parameters of images depend on the energy of primaries, the application of the cut optimized for standard power-law spectra could lead to sweeping out of γ-ray induced events as well. Therefore, for the rejection of the cosmic ray background perhaps it is worthwhile to exploit mainly the high ($\delta\psi \approx 0.1^\circ$) angular resolution of the stereoscopic imaging systems.

Another important expectation following from the strong evolution of the cloud parameters is that the pulsed component disappears at stages when either the radius of the cloud becomes larger than the light crossing time and/or when the density of the gas becomes low enough so that $\tau_{\rm pp} \geq P_0$. In fact, the flux intensity could remain at high level during some time even after this stage. However if the instrument sensitivity is not high enough for DC detection, one might conclude that the signal has disappeared at all. From this point of view the binaries containing large spin period compact objects (like Vela X-1 or AE Aquarii) seem to be favorable for production of detectable episodic events. For the same reason, it could be expected that with enhanced sensitivities of detectors the number of flaring γ-ray events would drastically increase. Since the detection of short and week DC signals is rather difficult problem, the multiwavelength simultaneous observations of synchrotron radio and thermal optical/UV flaring radiations unavoidably produced during the same episodes, will significantly increase the credibility of γ-ray detections. In addition, simultaneous detection of pulsed signals with *the same period shifts* in different energy bands would unambiguously indicate that these signals originate in *one and the same* moving target.

Finally, we note that the "moving target crosses beam" scenario can be invoked also for explanation of other short time-scale high energy phenomena, in both galactic and extragalactic scales.

Acknowledgements

We thank R.C.Lamb and H.J.Völk for valuable discussions. AMA is grateful to the Max-Planck-Institute für Kernphysik for hospitality during his visit to Heidelberg; his work was supported through the Verbund-forschung Astronomie/Astrophysik of the German BMBF.

References

Aharonian, F.A. and Atoyan, A.M.: 1991, *Astrophys.J.* **381**, 220.
Baltrusaitis, R.M. *et al.*: 1985, *Astrophys.J.* **293**, L69.
Bastian, T.S., Dulk, G.A., and Chanmugam, G.: 1989, *Astrophys.J.* **324**, 431.
Bowden, C.C.G. *et al.*: 1992, *Astroparticle Physics* **1**, 47.
Chadwick, P.M., McComb, T.J.L., and Turver, K.E.: 1990, *J.Phys.G* **16**, 1773.
Cheng, K.S. and Ruderman, M.: 1989, *Astrophys.J.* **337**, L77.
De Jager, O.C. *et al.*: 1994, Mont. Not. R. Astron Soc. **267**, 577.
Dingus, B.L. *et al.*: 1988, *Phys. Rev. Letters* **61**, 1906.
Dowthwaite, J.C. *et al.*: 1984 *Astrophys.J.* **342**, L35.
Gaisser, T.K.: 1990, Cosmic Rays and Particle Physics, Cambridge University Press.
Goret, P.: 1993, *Proc. Towards a Major Atmospheric Cherenkov Detector II* (edt. R.C.Lamb), Iowa State University, 25.
Gupta, S.K. *et al.*: 1990, *Astrophys.J.* **354**, L13.
Harpaz, A. and Rappaport, S.: 1994, *Astrophys.J.* **434**, 283.
Hillas, A.M.: 1984, *Nature* **312**, 50.
Hillas, A.M.: 1987, *Proc. Very High Energy Gamma-Ray Astronomy* (edt. K.E.Turver), D.Reidel Publ.Co., 71.
Hutchings, B: 1979, Proc. Physics and Astrophysics of Neutron Stars and Black Holes, North-Holland, 202.
Johnson, P.A.: 1994, *Astroparticle Physics* **3**, 53.
Kundt, W.: 1991, *Comments Astrophys.* **15**, 255.
Lamb, R.C.: 1995, *Space Sci. Rev.*, this volume.
Lamb, R.C et al.: 1988, *Astrophys.J.* **328**, L13.
Meintjes, P.J. *et al.*: 1992, *Astrophys.J.* **401**, 325.
Mirabel, I.F. and Rodriguez, L.F.: 1994, *Nature* **371**, 46.
Patterson, J.: 1979, *Astrophys.J.* **234**, 978.
Patterson, J. *et al.*: 1984, *Astrophys.J.* **240**, L133.
Protheroe, R.J. and Stanev, T.: 1987, *Nature* **328**, 136.
Protheroe, R.J.: 1994 *Astrophys.J. Suppl.* **90**, 883.
Raubenheimer, B.C.: 1993, *Proc. Towards a Major Atmospheric Cherenkov Detector II* (edt. R.C.Lamb), Iowa State University, 3.
Resvanis, L.K. *et al.*: 1988, *Astrophys.J.* **328**, L9.
Roberts, M.D. *et al.*: 1993, *Proc. 23rd Int. Cosmic Ray Conf., Calgary* **1**, 322.
Slane, P. and Fry, W.F.: 1989, *Astrophys.J.*, **342**, 1129.
Weekes, T.C.: 1992, *Space Sci. Rev.* **59**, 315.

TOPOLOGICAL DEFECTS AND HIGHEST ENERGY COSMIC AND GAMMA RAYS

G. SIGL
Department of Astronomy & Astrophysics
Enrico Fermi Institute, The University of Chicago, Chicago, IL 60637-1433

Abstract. In this paper we review the hypothesis that a substantial part of the cosmic ray flux observed above about 10^{19} eV may be produced by decaying or annihilating topological defects left over from phase transitions in the early universe at grand unification energy scales ($\approx 10^{16}$ GeV). Possible signatures of cosmic ray producing defect models are discussed which could be tested experimentally in the near future. We thereby focus on model independent universal spectral properties of the predicted particle fluxes.

Key words:

highest energy γ-rays – topological defects – cosmic strings

1. Introduction

It is commonly believed that cosmic rays are mainly produced due to first order Fermi acceleration (see, e.g., Gaisser 1990) at astrophysical shocks. A potential source of cosmic rays (CR) of ultrahigh energies (UHE) (i.e. above $\approx 10^{18}$ eV) are relativistic shocks contained in radiogalaxies and active galactic nuclei (see, e.g., Cesarsky 1992; Rachen & Biermann 1993). The recent observation of cosmic rays above 10^{20} eV by the Fly's Eye (Bird, *et al.* 1993, 1994, 1995), and AGASA (Hayashida, *et al.* 1994; Yoshida, *et al.* 1995) experiments, and the experiment at Yakutsk (Efimov, *et al.* 1991; Egorov 1993) may, however, not be easily explained by this mechanism (Sommers 1993; Sigl, Schramm, & Bhattacharjee 1994; Elbert & Sommers 1995).

Therefore, it has been suggested that such superhigh energetic cosmic rays could have a non-acceleration origin (Hill 1983; Hill, Schramm, & Walker 1987; Bhattacharjee & Rana 1990; Bhattacharjee 1991; Bhattacharjee, Hill, & Schramm 1992; Sigl, Schramm, & Bhattacharjee 1994; Bhattacharjee & Sigl 1995) as, for example, the decay of supermassive elementary "X" particles associated with Grand Unified Theories (GUTs). These particles could be radiated from topological defects (TDs) formed in the early universe during phase transitions caused by spontaneous breaking of symmetries implemented in these GUTs [for a review on TDs, see Vilenkin (1985)]. This is because TDs, like ordinary and superconducting cosmic strings, domain walls and magnetic monopoles, are topologically stable but nevertheless can release part of their energy in the form of these X particles due to physical processes like collapse or annihilation. The X particles with typical GUT scale masses of the order of 10^{16} GeV decay subsequently into leptons and quarks. The strongly interacting quarks fragment into a jet of hadrons which

Space Science Reviews **75**: 375–385, 1996.

results in typically of the order of $10^4 - 10^5$ mesons and baryons. It is assumed that these hadrons then give rise to a substantial fraction of the highest energy cosmic ray (HECR) flux as well as a considerable neutrino flux.

The shapes of the nucleon and γ-ray spectra predicted within such TD models are thus expected to be universal (i.e., independent of the specific process involving any specific kind of TD) and to be only dependent on the physics of X particle decay. This is because at HECR energies nucleons and γ-rays have attenuation lengths in the cosmic microwave background (CMB) which are small compared to the Hubble scale. Cosmological evolutionary effects which depend on the specific TD model are therefore negligible. In contrast, the predicted neutrino flux especially at lower energies would depend on the TD model.

Independently of the spectral shapes of the predicted nucleon and γ-ray fluxes the question arises whether a specific TD model is capable of producing a HECR flux at an observable level. In this paper we give an overview of the status of topological defect models for cosmic ray production with emphasis on spectral shapes and absolute flux levels. We organize the rest of the paper as follows: In section 2 we discuss the universal injection spectra of nucleons and γ-rays and evolve it to the spectra actually observed. Based on this we present in section 3 distinct signatures of these spectra which could be tested experimentally against other models of HECR production. The capability of various TD models of producing observable HECR fluxes is discussed in section 4.

2. The Universal Particle Spectra in Defect Models

In this section we discuss the universal nucleon and γ-ray spectral shapes within defect models. These are exclusively determined by the physics of X particle decay and the subsequent propagation of the decay products through intergalactic space.

2.1. Injection Spectra of Nucleons, γ-Rays and Neutrinos

We assume that each X-particle decays into a lepton and a quark each of an energy approximately half of the X particle mass m_X. The quark hadronizes by jet fragmentation and produces nucleons, γ-rays and neutrinos, the latter two from the decay of neutral and charged pions in the hadronic jets. The hadronic route is expected to produce the largest number of particles and we will focus on this channel.

The spectra of the hadrons in a jet produced by the quark are, in principle, given by Quantum Chromodynamics (QCD). Suitably parametrized QCD motivated hadronic spectra that fit well the data in collider experiments in the GeV–TeV energies have been suggested in the literature (Hill 1983). However, there is a great deal of uncertainty involved in extrapolating from the "low" energy data to the extremely high energies arising in the present situation. To study the sensitivity of results to the assumed hadronization spectrum, we will also use different injection spectra suggested by Chi, *et al.* (1993) on certain phenomenological grounds.

The injection spectrum, i.e., the number density of particles produced per unit time per unit energy interval, for the species a, where a can denote nucleons (N), γ-rays (γ) or neutrinos (ν), can be written as

$$\Phi_a(E_i, t_i) = \frac{dn_X(t_i)}{dt_i}\frac{2}{m_X}\frac{dN_a(x)}{dx}, \tag{1}$$

where $x \equiv 2E_i/m_X$, E_i being the energy at injection and dN_a/dx the effective fragmentation function describing the production of the particles of species a from the original quark. We will consider the following two cases for the fragmentation function:

(1) *QCD-motivated injection spectra*

In this case, the *total* hadronic fragmentation spectrum dN_h/dx is taken to be of the form (Hill 1983)

$$\frac{dN_h(x)}{dx} = \begin{cases} \frac{15}{16}x^{-1.5}(1-x)^2 & \text{if } x_0 \le x \le 1 \\ 0 & \text{otherwise} \end{cases}, \tag{2}$$

where the lower cutoff x_0 is typically taken to correspond to a cut-off energy $\approx 1\,\text{GeV}$. Assuming a nucleon content of $\approx 3\%$ and the rest pions, we can write the fragmentation spectra as (Bhattacharjee, Hill, & Schramm 1992; Aharonian, Bhattacharjee, & Schramm 1992)

$$\begin{aligned} \frac{dN_N(x)}{dx} &= (0.03)\frac{dN_h(x)}{dx}, \\ \frac{dN_\gamma(x)}{dx} &= \left(\frac{0.97}{3}\right)2\int_x^1 \frac{1}{x'}\frac{dN_h(x')}{dx'}dx', \end{aligned} \tag{3}$$

where we omit the neutrino spectrum which is similar to the photon spectrum.

(2) *Phenomenological injection spectra*

Recently, somewhat different injection spectra have been suggested by Chi, *et al.* (1993) on the following phenomenological grounds. UHE γ-rays as well as protons generate lower energy γ-rays by $\gamma - \gamma_b$ and $p - \gamma_b$ collisions with background photons (γ_b). The electromagnetic component of the energy lost in these collisions develops cascades in the universal radio background (URB), the cosmic microwave background (CMB) and in the infrared background (IRB) (in order of decreasing energy of the propagating photon). The measured flux of extragalactic γ-rays in the 100 MeV energy region (Fichtel, Simpson, & Thompson 1978) provides constraints on the form of the nucleon and gamma ray injection spectra at energies above

$\approx 5 \times 10^{19}$ eV. Based on these considerations, Chi, *et al.* (1993) have suggested injection spectra which we describe by the following fragmentation functions:

$$\frac{dN_{\rm N}(x)}{dx} = A_{\rm N}\, x^{-1.5}\,, \quad \frac{dN_\nu(x)}{dx} = \frac{dN_\gamma(x)}{dx} = A_\gamma\, x^{-2.4}\,, \tag{4}$$

with

$$\frac{A_\gamma}{A_{\rm N}} \approx 0.028 \left(\frac{10^{15}\,{\rm GeV}}{m_{\rm X}}\right)^{0.9}. \tag{5}$$

The condition (5) comes from the requirement (Chi, *et al.* 1993) that the photon-to-nucleon (γ/N) ratio at injection at energy $E = 10^{20}$ eV, i.e., at $x = 2 \times 10^{20}\,{\rm eV}/m_{\rm X}$ be ≈ 60. The spectra (4) are absolutely normalized such that the total quark energy $m_{\rm X}/2$ is injected above 5×10^{19} eV (which we will use as our normalization point to the observed HECR flux)

2.2. The Evolved Spectra

The evolution of the spectra is governed by energy-loss and/or absorption of the particles as they propagate through the extragalactic medium. Assuming uniform energy injection the diffuse flux (in units of number of particles per area, time, solid angle and energy) today ($t = t_0$), at energy E_0, is calculated by (see, e.g., Bhattacharjee & Rana 1990)

$$j(E_0) = \frac{3}{8\pi}\, ct_0 \int_0^{z_{\rm i,max}(E_0)} dz_{\rm i} (1 + z_{\rm i})^{-5.5} \frac{dE_{\rm i}(E_0, z_{\rm i})}{dE_0} \Phi(E_{\rm i}, z_{\rm i})\,. \tag{6}$$

Here, $z_{\rm i}$ is the injection redshift corresponding to the injection time $t_{\rm i}$, $E_{\rm i}(E_0, z_{\rm i})$ is the necessary injection energy, and the maximum injection redshift $z_{\rm i,max}(E_0)$ is determined from the condition $E_{\rm i}(z_{\rm i,max}, E_0) \leq m_{\rm X}/2$.

In the continuous energy-loss approximation (Berezinsky & Grigor'eva 1988; Aharonian, Kanevsky, & Vardanian 1990) the general energy-loss equation in terms of redshift z can be written as

$$\frac{1}{E}\frac{dE}{dz} = \frac{1}{1+z} + \frac{(1+z)^{1/2}}{H_0}\beta[(1+z)E]\,. \tag{7}$$

The first term is due to redshift of the (relativistic) particle energy and the second term describes losses due to interactions with the background medium in terms of the energy-loss rate $\beta(E) = -(dE/dt)/E$ (H_0 is the Hubble constant). For a particle observed at the earth with an energy E_0 the necessary injection energy $E_{\rm i}(z_{\rm i})$ at redshift $z_{\rm i}$ can be found by integrating Eq. (7) back from $z = 0$ to $z = z_{\rm i}$ with $E_{\rm i}(z = 0) = E_0$.

For nucleons one can use the expression for $\beta(E)$ derived in Berezinsky and Grigor'eva (1988). The important process here is the Greisen-Zatsepin-Kuz'min (GZK) (Greisen 1966; Zatsepin & Kuz'min 1966) effect in which

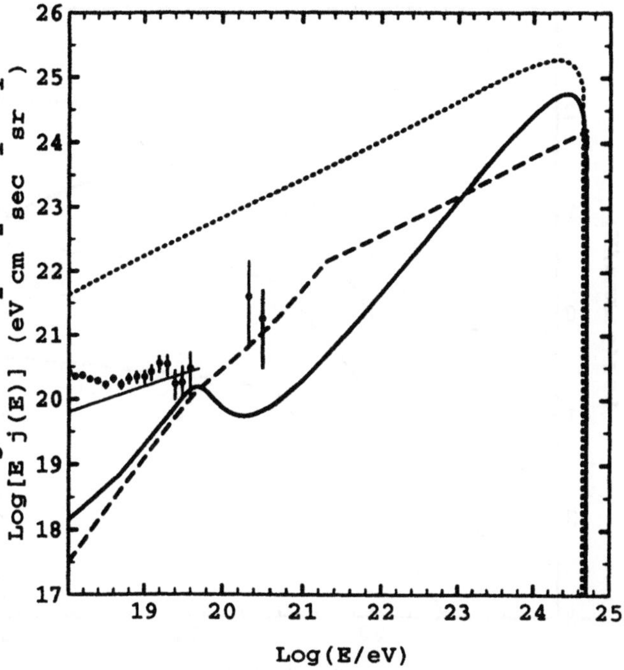

Fig. 1. Evolved neutrino (dotted line), γ-ray (dashed line) and proton (solid line) spectra typically predicted in defect models. An X particle mass of $m_X = 10^{16}$ GeV and the phenomenologically motivated injection spectra discussed in section 2.1 were used. The combined proton and gamma ray flux was normalized at 5×10^{19} eV to the light (supposedly extragalactic) flux component (thin solid line) fitted (Bird, *et al.* 1993, 1994) to the high energy Fly's Eye data (dots with error bars, except the second highest energy event which was measured by the AGASA experiment). The two highest energy events observed would thus very likely have been γ-rays within the TD scenario.

UHE nucleons above about 6×10^{19} eV (today) lose energy drastically due to photopion production off the CMB photons. This gives rise to the onset of a sharp fall ("cutoff") of the extragalactic UHE nucleon flux. We neglect the secondary γ-ray and neutrino contribution caused by the interactions of UHE nucleons with the CMB.

The evolution of the UHE γ-ray spectrum is mainly governed by *absorption* of the UHE photons through e^+e^- pair production on the CMB and on the URB. The developing electromagnetic cascades cause an increase of the effective penetration length of the UHE γ-rays. This effect, however, depends rather strongly on the poorly known strength of the intergalactic magnetic field. We therefore consider here only the absorption of the UHE γ-rays on the CMB and URB photons and use the absorption lengths given in Aharonian, Bhattacharjee, & Schramm (1992).

Similarly, UHE neutrinos can be absorbed by the thermal neutrino background but we are not so much interested in the neutrino fluxes here.

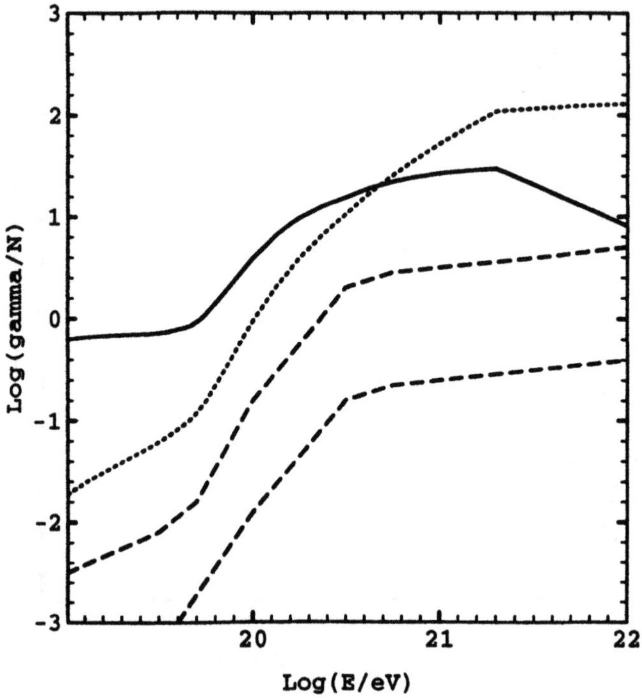

Fig. 2. The γ/N ratios predicted within the QCD- and the phenomenologically motivated injection scenario in TD models (dotted and solid line, respectively) compared to the ones resulting from acceleration spectra with injection slopes 2 (upper dashed) and 3 (lower dashed).

Figure 1 shows an example for the spectra typically resulting. Only the neutrino spectrum depends on the cosmological history of X particle injection for which we assumed a t^{-3} dependence which is typical for the more interesting TD scenarios (see section 4 below). Note that TDs alone are not able to explain the CR flux below a few 10^{19} eV without overproducing CRs at higher energies. More detailed calculations would have to solve the full Boltzmann equations for propagation and take non-uniform injection (discrete sources) into account. The latter effect can give rise to anisotropies in the HECR flux.

3. Signatures for Defect Sources of Highest Energy Cosmic Rays

As can be seen from Figure 1 one expects a γ/N flux ratio which steeply increases with increasing energy around 10^{20} eV. This is mainly due to the fact that the nucleon attenuation length drops drastically above about 6×10^{19} eV because of the GZK effect whereas the γ-ray attenuation length is slowly rising within this same energy range. A γ/N ratio strongly increasing with energy is therefore expected not only within TD models. There is,

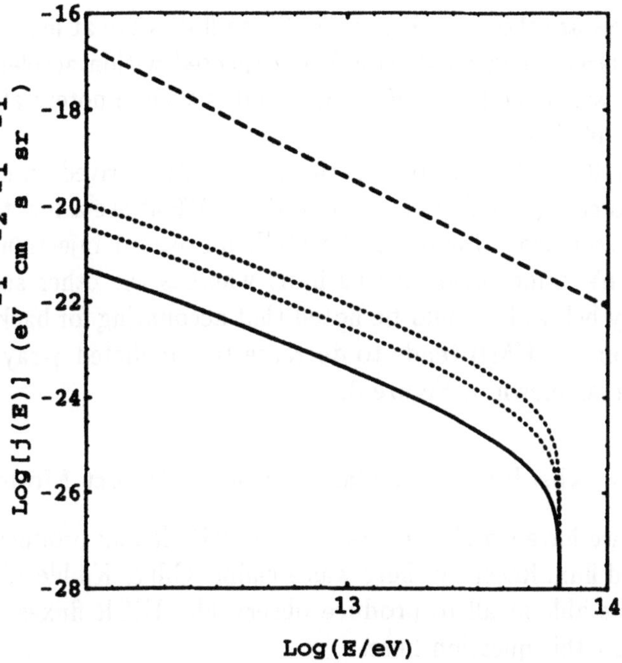

Fig. 3. Analytical approximation to the γ-ray fluxes below 100TeV resulting from cascading in the CMB of γ-rays injected within defect models. The solid curve corresponds to the phenomenologically motivated injection scenario (independent of m_X). The upper and lower dotted curves are for QCD motivated injection for $m_X = 10^{16}$ GeV and 10^{15} GeV, respectively. Also shown is the approximate charged cosmic ray flux (dashed curve).

however, a crucial difference between these and standard acceleration models which was first pointed out by Aharonian, Bhattacharjee, & Schramm (1992): Within TD models γ-rays dominate over nucleons already at injection whereas in case of shock acceleration γ-rays are only produced as secondaries of protons which are the primary accelerated particles. TD scenarios therefore predict a considerably higher γ/N ratio. This is demonstrated in Figure 2 where the two injection scenarios from section 2.1 are compared with two acceleration scenarios which yield particularly high γ/N values. With future instruments like the giant air shower array (Cronin 1995) it should be possible to distinguish γ-rays from nucleons at least on a statistical basis.

The high amount of high energy γ-ray injection in TD models leads to an enhanced electromagnetic cascade in the various photon backgrounds as compared to acceleration models. The injected energy is recycled below a certain threshold energy which today is of the order of 100TeV. For cascading in the CMB alone the resulting cascade spectrum can be approximated analytically (Svensson & Zdziarski 1990). The spectra predicted for the defect models are shown in Figure 3. These fluxes are at least two orders of magnitude larger than what would be expected within acceleration models (Wdowczyk & Wolfendale 1990) which makes them a potential signature for a TD origin of HECR.

The HEGRA collaboration (Karle, *et al.* 1995) derived an upper limit on the γ-ray to charged CR flux ratio of about 1% above a few times 10^{13} eV. As can be seen from Figure 3 the QCD-motivated injection scenario for $m_X = 10^{16}$ eV comes near to this limit whereas the other scenario is still considerably below. It should be noted that accounting for background fields different from the CMB tends to decrease the predicted γ-ray fluxes in the energy interval shown in Figure 3.

4. Absolute Flux Calculation in Defect Models

Up to now we have simply normalized the HECR flux produced by TDs to the observed flux. Recently there was a claim (Gill & Kibble 1994) that TDs might not be able at all to produce observable HECR fluxes. We therefore briefly review this question here.

4.1. Ordinary Cosmic Strings

A cosmic string network can be formed in the early universe if for example a global or gauge symmetry group U(1) is spontaneously broken. The network is expected to reach the so called scaling regime where the length scale describing its large scale structure increases proportional to cosmic time t. In this regime closed string loops are continuously chopped off the network at a rate per volume $\approx \beta/t^4$ and with a typical length αt. Here, α and β are dimensionless constants given by string simulations (Austin, Copeland, & Kibble 1993). It has been shown (Bhattacharjee 1989; Bhattacharjee & Rana 1990) that cosmic strings can measurably contribute to the HECR flux only if a fraction $f \gtrsim 10^{-3}$ of the total energy in these loops is dissipated in the form of X particles on a time scale much smaller than the evolution time scale of the strings. In that case the X particle injection history is given by

$$\frac{dn_X}{dt} = f\alpha\beta\mu m_X^{-1}t^{-3}, \tag{8}$$

where μ is the square of the symmetry breaking scale. Gill & Kibble (1994) argue that this criterion is unlikely to be fulfilled under reasonable circumstances. However, we feel that this criterion could be satisfied, for example, in the following two cases: First, a considerable fraction of the loops forms in lowest frequency modes which lead to loop collapse within one oscillation period (Kibble & Turok 1982). This could happen if gravitational backreaction effects smooth out higher frequency wiggles on the loop. Or high-harmonic loops self-intersect and decay into smaller and smaller loops, finally releasing their energy in relativistic particles (Siemens & Kibble 1994). Second, hybrid systems of TDs could be formed such as light domain walls bounded by cosmic string loops. The domain walls in these systems could then aid the complete collapse of the string loops which form the boundaries of these walls.

4.2. Superconducting Cosmic Strings

Superconducting cosmic strings (SCS) (Witten 1985) as potential HECR sources were first proposed by Ostriker, Thomson, & Witten (1986). This was further pursued by Hill, Schramm, & Walker (1987). The relevant process is the ejection of charge carriers (which are the X particles in this case) carrying the electric current on the string once this current reaches a certain saturation value. The absolute X particle injection rate is in principle given by Eq. (8) with f replaced by $L_s/(100G\mu t)$ where G is the Newtonian constant and L_s is the saturation length of the string loop.

The SCS scenario has become less promising because of several problems: First, it is possible that the SCS loops may never achieve the saturation current at all (Davis & Shellard 1989). Second, if the X particles decay within the high magnetic field region around the string the produced HECR are expected to be quickly degraded in energy due to synchrotron losses and other processes (Berezinsky & Rubinstein 1989). Third, it can be shown (Sigl, *et al.* 1995) that SCS which would substantially contribute to the HECR flux today would likely have produced spectral CMB distortions as well as light element abundances by ^{4}He photodisintegration which violate experimental bounds. This is mainly because the X particle injection rate is proportional to t^{-4} rather than t^{-3} as for ordinary cosmic strings.

4.3. Magnetic Monopoles

Magnetic monopoles come from one particular type of TD solutions in spontaneously broken non-abelian gauge theories which are allowed in essentially all GUT models. Their formation in the early universe is inevitable in most models and leads to the well known "monopole problem" (for a review, see Kolb & Turner 1990), since their abundance is severely constrained. Besides the cosmological bound which limits the energy density in monopoles, the Parker bound limits their mean galactic abundance from their interactions with the galactic magnetic field (for an updated Parker bound, see Adams *et al.* 1993).

Monopoles (M) can be formed in metastable bound states with antimonopoles ($\bar{M}$) ("monopolonium") which spiral in and finally collapse resulting in M-$\bar{M}$ annihilation. Monopolonium formed around the period of nucleosynthesis can thus give rise to HECR today (Hill 1983). The HECR flux can in principle be calculated from the monopole abundance and the relative monopolonium abundance at formation, ξ_f (Bhattacharjee & Sigl

1995). It turns out that, for example, for a monopole energy density $\approx 10^{-2}$ of the critical density, a symmetry breaking scale $m_X \approx 10^{16}$ GeV (which satisfies the Parker bound) and $\xi_f \approx 10^{-8}$ an observable HECR flux could be produced.

In addition, this scenario would not violate bounds on light element abundances and CMB distortions (Sigl, *et al.* 1995). Since the average distance between monopoles in this scenario is much smaller than average defect separations in other TD models one would expect comparatively isotropic fluxes.

In summary, SCS are highly unlikely to produce a substantial part of the observed HECR flux. In contrast, in our feeling ordinary cosmic strings and other TDs cannot be ruled out yet as HECR sources due to an incomplete understanding of the relevant physical processes involved. In particular, HECR production by magnetic monopole annihilation still seems to be an interesting option.

Acknowledgements

This work was supported, in part, by the DoE, NSF and NASA at the University of Chicago, by the DoE and by NASA through grant NAGW-2381 at Fermilab, and by the Alexander-von-Humboldt Foundation.

References

Adams, F.C., *et al.*: 1993, *Phys. Rev. Lett.* **70**, 2511
Aharonian, F.A., Bhattacharjee, P., & Schramm, D.N.: 1992, *Phys. Rev.* **D46**, 4188
Aharonian, F.A., Kanevsky, B.L., & Vardanian, V.V.: 1990, *Astrophys. Space Sci.* **167**, 93
Austin, D., Copeland, E.J., & Kibble, T.W.B.: 1993, *Phys. Rev.* **D48**, 5594
Berezinsky, V.S., & Grigor'eva, S.I.: 1988, *Astron. Astrophys.* **199**, 1
Berezinsky, V.S., & Rubinstein, H.R.: 1989, *Nucl. Phys.* **B323**, 95
Bhattacharjee, P.: 1989, *Phys. Rev.* **D40**, 3968
Bhattacharjee, P.: 1991, in Nagano, M., & Takahara, F., ed(s)., *Astrophysical Aspects of the Most Energetic Cosmic Rays*, World Scientific: Singapore, 382
Bhattacharjee, P., Hill, C.T., & Schramm, D.N.: 1992, *Phys. Rev. Lett.* **69**, 567
Bhattacharjee, P., & Rana, N.C.: 1990, *Phys. Lett.* **B246**, 365
Bhattachajee, P., & Sigl, G.: 1995, Report *FERMILAB-Pub-94/387-A*, to be published in *Phys. Rev. D.*
Bird, D.J., *et al.*: 1993, *Phys. Rev. Lett.* **71**, 3401
Bird, D.J., *et al.*: 1994, *Astrophys. J.* **424**, 491
Bird, D.J., *et al.*: 1995, Preprint *astro-ph/9410067*, to appear in *Astrophys. J.* (March 1, 1995)
Cesarsky, C.T.: 1992, *Nucl. Phys. B (Proc. Suppl.)* **28B**, 51
Chi, X., *et al.*: 1993, *Astropart. Phys.* **1**, 239
Cronin, J.: 1995, 'Design Considerations for a Ground Array', to appear in the Snowmass Report on Cosmic Rays
Davis, R.L., & Shellard, E.P.S.: 1989, *Nucl. Phys.* **B323**, 209
Efimov, *et al.*: 1991, in Nagano, M., & Takahara, F., ed(s)., *Astrophysical Aspects of the Most Energetic Cosmic Rays*, World Scientific: Singapore, 20

Egorov. T.A.: 1993, in *Proceedings of Tokyo Workshop on Techniques for the Study of Extremely High Energy Cosmic Rays*. ICRR, Univ. of Tokyo, 35
Elbert. J.W., & Sommers, P.: 1995, Preprint *astro-ph/9410069*, to appear in *Astrophys. J.* (March 1, 1995)
Fichtel. C.E., Simpson, G.A., & Thompson, D.J.: 1978, *Astrophys. J.* **222**, 833
Gaisser. T.K.: 1990, *Cosmic Rays and Particle Physics*, Cambridge University Press: Cambridge
Gill, A.J., & Kibble, T.W.B.: 1994, *Phys. Rev.* **D50**, 3660
Greisen. K.: 1966, *Phys. Rev. Lett.* **16**, 748
Hayashida, N., *et al.*: 1994, *ICRR-Report-324-94-19*, to appear in *Phys. Rev. Lett.* (1995)
Hill, C.T.: 1983, *Nucl. Phys.* **B224**, 469
Hill, C.T., Schramm, D.N., & Walker, T.P.: 1987, *Phys. Rev.* **D36**, 1007
Karle, *et al.*: 1995, Report *MPI-PhE/94-28*, to appear in *Phys. Lett. B* (1995)
Kibble. T.W.B., & Turok, N.: 1982, *Phys. Lett.* **B116**, 141
Kolb, E.W., & Turner, M.S.: 1990, *The Early Universe*, Addison-Wesley: New York
Ostriker, J.P., Thompson, C., & Witten. E.: 1986, *Phys. Lett.* **180**, 231
Rachen, J.P., & Biermann, P.L.: 1993, *Astron. Astrophys.* **272**, 161
Siemens, X.A., & Kibble, T.W.B.: 1994, Preprint *hep-ph/9412216*
Sigl, G., Schramm, D.N., & Bhattacharjee, P.: 1994, *Astropart. Phys.* **2**, 401
Sigl, G., *et al.*: 1995, in preparation.
Sommers, P.: 1993, in *Proceedings of Tokyo Workshop on Techniques for the Study of Extremely High Energy Cosmic Rays*, ICRR, Univ. of Tokyo, 23
Svensson, R., & Zdziarski, A.A.: 1990, *Astrophys. J.* **349**, 415
Vilenkin, A.: 1985, *Phys. Rep.* **121**, 263
Wdowczyk, J., & Wolfendale, A.W.: 1990. *Astrophys. J.* **349**, 35
Witten. E.: 1985, *Nucl. Phys.* **B249**, 557
Yoshida, S., *et al.*: 1995, *ICRR-Report-325-94-20*, to appear in *Astropart. Phys.* (1995)
Zatsepin, G.T., & Kuz'min, V.A.: 1966, *JETP. Lett.* **4**, 78

INTERGALACTIC MAGNETIC FIELDS, AND SOME CONNECTIONS WITH COSMIC RAYS

PHILIPP P. KRONBERG
University of Toronto, Dept. of Astronomy, Toronto M5S 1A7, Canada

Abstract. An overview is presented of the methods of probing for the geometry, and strength of intergalactic magnetic fields. Recent results are briefly surveyed for galaxy halos, galaxy clusters, and the intergalactic medium on various scales, and some relevant physical processes and radiation processes are mentioned, as well as the "coupling" between intergalactic magnetic fields and cosmic rays.
The general trend of recent results indicates that, wherever we detect intergalactic hot gas and galaxies, we also find magnetic fields at levels of $\sim 10^{-7}$ G, or higher. The hitherto undetected, weaker fields in the rarified i.g.m. and in large intergalactic voids could be probed by both Faraday rotation, and possibly using very energetic CR nuclei ($> 10^{20}$eV), and/or transient extragalactic γ ray bursts.

1. Introduction

Observations over the last ten years have produced magnetic field detections not only in galaxy disks, but also in galaxy halos, clusters of galaxies, and in some very distant galaxy systems which produce both absorption lines and Faraday rotation of the radiation from background quasars. Generally, the more we look for extragalactic magnetic fields, the more ubiquitous we find them to be.

The presence of magnetic fields in diffuse astrophysical plasmas can be observationally verified by four commomly used methods:

(1) Through the detection of synchrotron radiation, which requires both cosmic ray electrons and a magnetic field. The emissivity, in units of radiated power per unit volume, can be expressed as

$$\varepsilon(\nu) \propto n_{er0}\, \Xi(\gamma) B sin\phi^{(\gamma+1)/2} \nu^{(1-\gamma)/2} \tag{1}$$

where the energy distribution of the cosmic ray electrons is given by $n_{er}(E) = n_{er0}E^{-\gamma}$. ϕ is the average electron pitch angle with respect to the local field direction, B is the magnetic field strength, ν the frequency of observation, and Ξ (γ) is a function of γ.

(2) Through the detection of Zeeman Effect splitting of a radio transition in the interstellar gas.

$$\nu = \nu_{nm} \pm eB(4\pi mc)^{-1} \tag{2}$$

Space Science Reviews **75**: 387–399, 1996.

(3) An induced polarization of optical starlight (intrinsically unpolarized) due to intervening dust grains which align themselves with the interstellar magnetic field.

(4) Faraday rotation due to a magnetic field in the presence of non-relativistic electrons. The rotation measure from a source with an electron *column density* N_e ($\equiv 10^{19}\ N_{19}\mathrm{cm}^{-2}$) and average line-of-sight magnetic field strength $\langle B_{||}\rangle$ μG, which is at a redshift z, is

$$RM = 2.7(1+z)^{-2} N_{19} \langle B_{||}\rangle \text{ rad m}^{-2}. \tag{3}$$

Unfortunately, Zeeman splitting (2), which gives a reasonably direct measurement of field *strength*, is only just detectable in our Galaxy at the level of a few μG. It is virtually undetectable in external galaxies, much less in hot intergalactic gas, because of the Doppler smearing of the radio lines due to turbulence and thermal line broadening, combined with the normally weak field strengths. Optical polarization of starlight (3) is, up to now only practical for nearby, relatively face-on galaxies, and in any case cannot detect the field *sign*, nor the field strength.

Thus, for our purposes, methods (1) and (4) are the most practical for detecting the geometrical ordering of magnetic fields in most galaxy systems and intergalactic space. Measurements of the actual field strength, however, are not so easy: For synchrotron radiation (1), which is relatively easy to detect, especially at the lower radio frequencies, we see that the measured emissivity $\varepsilon(\nu)$ is weighted by the relativistic electron density (eq.(1)), which is normally impossible to measure directly (except by direct cosmic ray measurements in the vicinity of the Sun)). Likewise, in the case of Faraday rotation (4), the observed RM (eq. (2)) is scaled by the co-spatial non-relativistic electron density (n_e), and also inversely by the square root of the number of reversals along the line of sight. In this case, however, companion measurements of n_e *are* sometimes possible: X-ray observations of thermal bremsstrahlung has, as we describe in §4, recently provided n_e distributions in the intracluster medium (i.c.m.) of clusters of galaxies. This, combined with an RM "signal", has led to relatively assumption-free estimates of the magnetic field strength in the intergalactic space within clusters of galaxies.

Evidence for a widespread interstellar magnetic field at some strength level has existed since the early days of radio astronomy, when it was discovered that the galactic disk is pervaded by diffuse synchrotron emission – the dominant signal in low resolution maps of the radio sky at frequencies below 500 MHz.

Beginning in the 1960's, Faraday RM's from linear polarization measurements of extragalactic radio sources confirmed that the galactic magnetic field had an organized component over the disk of our Galaxy (Davies

1968). Subsequent measurements of the RM's of ever larger numbers of (polarized) extragalactic radio sources has led to a more refined modelling of the large scale Galactic magnetic field structure (Simard-Normandin and Kronberg 1980, Vallée et al. 1988, Clegg et al. 1992, Rand and Lyne 1994). The Galactic magnetic field appears organized on a grand scale, and also has some large scale field reversal(s). Recent studies of external galaxies (cf. Wielebinski 1990) indicate that all disk galaxies are permeated by large scale magnetic fields. This suggests that all galaxies with interstellar cosmic rays and thermal gas probably possess an accompanying interstellar magnetic field.

2. Faraday Rotation Measures (RM) and the RM Sky

A polarized extragalactic radio source can undergo Faraday rotation in the following places: (i) In the Earth's ionosphere, which, depending on location, time of day, and level of solar activity produces less than $1 - 2$ rad/m^2. This amount is usually compared with (ii) that produced in the Milky Way, which varies from ~ 2 at high galactic latitudes to ~ 500 rad/m^2 and more close to the galactic plane. (iii) An intervening galaxy cluster can Faraday rotate a background radio source by typically 20 to 100 rad/m^2, and some dense, "cooling flow" galaxy clusters have been known to produce up to 10000 rad/m^2 in large (≥ 10 kpc) zones near the cluster core. Such large RM's appear, however to be rare cases, in that most radio galaxies and quasars have integrated RM's less than ~ 20 rad/m^2. (iv) *within* the large, galaxy -- size outer radio lobes of extended radio galaxies, the Faraday rotation is ususally small, $\leq$ 20rad/m^2. (v) Finally, on the largest known scales, any widespread, all-pervading intergalactic magnetic field could exist at some level, but it has not yet been detected; The cumulative RM, up to a redshift of about 2.5 is ≤ 3 rad/m^2.

Figure 1 shows a recent plot of the integrated Faraday rotation measures on an Aitoff equal area projection in galactic coordinates (l, b). It can be seen that higher RM's are prevalent at $b \leq 28°$. In contrast, the majority of sources at high b are small, with a few clear exceptions. These likely indicate an unusually dense cluster environment around the radio galaxy or quasar, an intervening galaxy cluster, or (statistically more likely) an intervening galaxy system in front of a distant quasar (*cf.* Kronberg and Perry, 1982).

In cases where the system causing the exess RM can be further studied by other observations -- using X-ray data, or absorption lines produced by the associated thermal gas -- there is some prospect of estimating the actual magnetic field strength associated with the gaseous component of the system in question (*e.g.* galaxy cluster, group, or individual galaxy halo/disk).

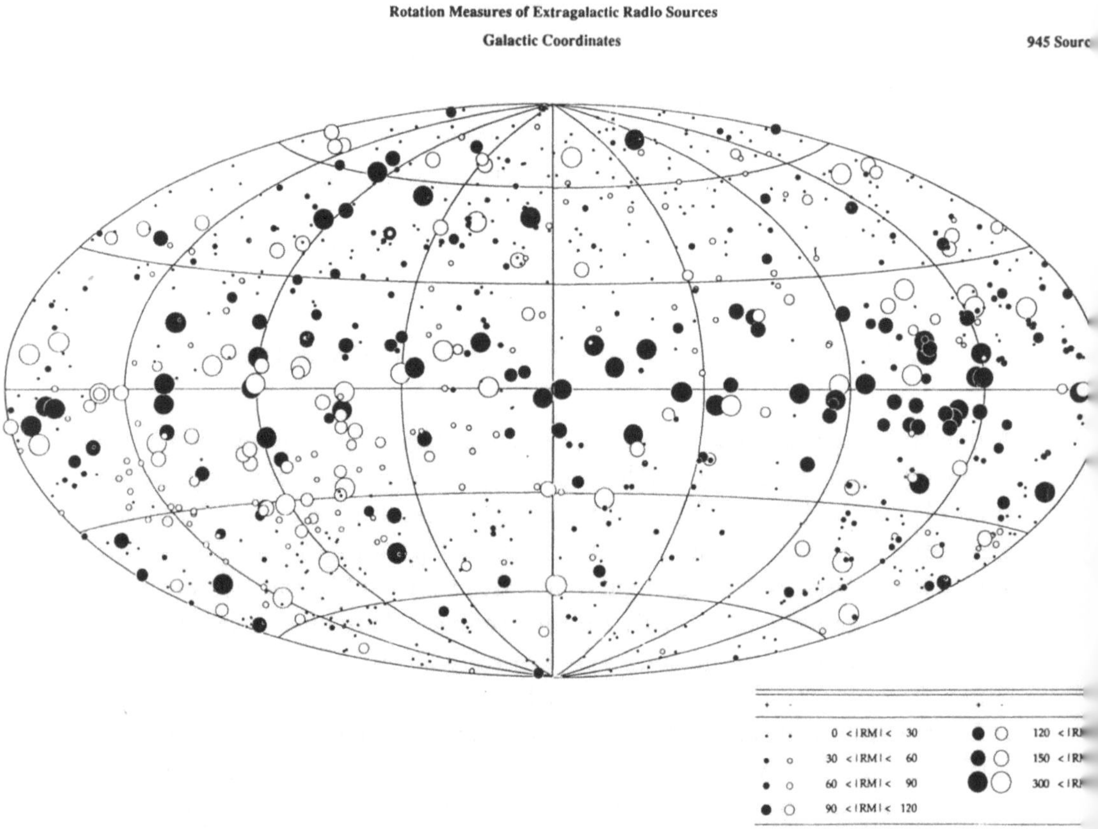

Fig. 1. The Faraday Rotation measures of 945 extraglactic radio sources projected onto galactic coordinates (l, b). The galactic center (0,0) is at the center of the plot. Solid circles denote positive RM's, and open circles indicate negative values. The diameters increase with increasing |RM|.

3. Widespread Magnetic Fields in the Milky Way, and other Nearby Galaxies, Galaxy Halos, and Extended Radio Galaxies

I shall omit, for lack of space, a description of the large organized magnetic fields in the disks of spiral galaxies like the Milky Way. For recent reviews, the reader is referred to Kronberg (1994) and Wielebinski and Krause (1993). Field strengths of from $1 - 4\mu$G are found to be typical of diffuse interstellar medium in such systems. These values come from various lines of evidence. Examples are Zeeman Splitting and pulsar RM/Dispersion measure ratio

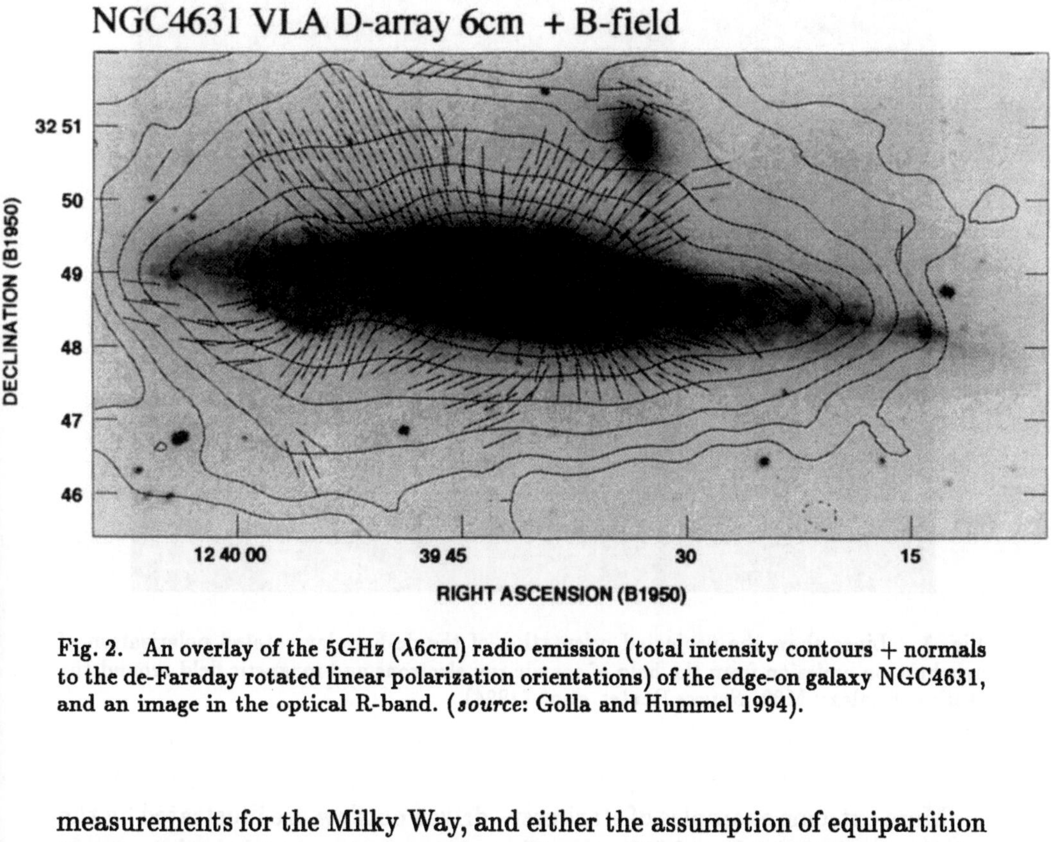

Fig. 2. An overlay of the 5GHz (λ6cm) radio emission (total intensity contours + normals to the de-Faraday rotated linear polarization orientations) of the edge-on galaxy NGC4631, and an image in the optical R-band. (*source*: Golla and Hummel 1994).

measurements for the Milky Way, and either the assumption of equipartition for the disk synchrotron emission, or the combination of RM data with X-ray, or recombination line data for the external galaxy in question.

I call attention here to recent evidence for magnetoplasma, including synchrotron-emitting cosmic rays, which have been more recently found to be ejected from galaxies. Low level radio polarimetry of some edge-on spiral galaxies with moderate star forming activity has recently revealed that cosmic ray gas and magnetic fields are ejected far into the galaxy's halo. The inferred magnetic field strengths in these out–of–plane halos is a few microgauss, *i.e.* comparable with that in the denser disk. As the recent example of NGC 4631 from Golla and Hummel (1994) shows (*cf.* Fig. 2), such out-of plane fields are remarkably organized, and appear to be directed out of the plane if the star formation activity (hence outflow wind) is sufficiently intense. What is unknown at present is how much of this magnetized halo gas, over cosmic time, has leaked into the i.g.m. from such "mild outflow" galaxy systems such as NGC4631.

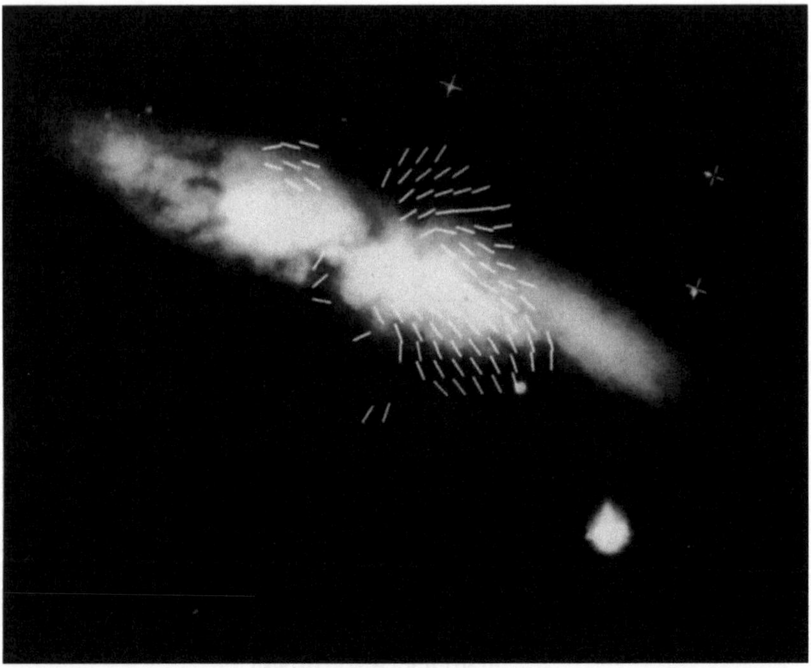

Fig. 3. Lines show the projected orientation of the de-Faraday rotated polarization of synchrotron emission from the halo of cosmic ray electrons and magnetic field around the starburst galaxy M82. *Source* Reuter *et al.* (1994).

More extreme examples of starburst–driven outflow are illustrated by the nearby starburst galaxy M82, whose halo has recently revealed a large-scale organized magnetic field, as illustrated in Fig. 3 below. M82's halo has also been detected in X-rays, H_{alpha}emission, and in other emission lines. The origin of M82's outflow appears localized to the $\sim$ 700pc–wide starburst region, and the outflow velocity is significantly above M82's escape velocity. Since the origin of the ouflow is driven by supernovae, such systems are clearly able, in principle, to eject both cosmic rays and magnetic field into the i.g.m.

Another source of large, magnetized cosmic ray gas are the well-known diffuse, strongly emitting radio lobes of some radio galaxies and quasars. The supra-galaxy scale morphology of double radio sources, their associated large energies (Burbidge 1956), and the high collimation of their radiating jets make them a remarkable phemonenon of the extragalactic universe. Large scale jets and associated radio source structure (reviewed recently by Bridle 1991) also have their analogues on stellar scales, *e.g.* in binary

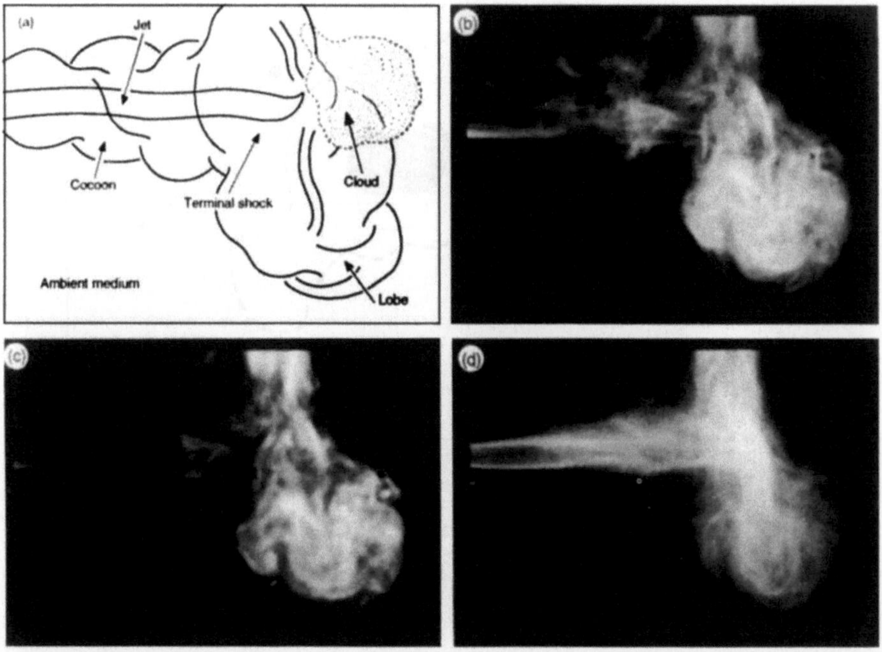

Fig. 4. Simulations of an extragalactic radio source jet and lobe by D.A. Clarke (1993). The panels show: (a) a sketch of the principal features, (b) the total synchrotron emissivity (*cf.* eq.(1)), (c) the magnetic pressure, and (d) the velocity shear modulus integrated along the line of sight. (*cf.* Kronberg (1994) for a related discussion of the magnetization of the i.g.m. by extended radio source lobes).

star systems and protostellar objects, where a magnetized accretion disk produces a highly collimated outflow (cf. Mestel 1972, Lovelace, Berk and Contopoulos 1991). Figure 4 shows, from simulations by D.A. Clark (1993), how we now are beginning to understand the field amplification mechanisms which can "inflate" significant volumes of intergalactic space with magnetic fields – which produce the (easily detectable) synchrotron radiation (panel (b) of the figure). Of particular note is the filamentery structure, not so easily detectable, but recently confirmed in observations (*e.g.* Perley, Dreher and Cowan 1994). These represent zones of amplified magnetic field, and are an important constraint on the amplification mechanisms for these large-scale intergalactic fields.

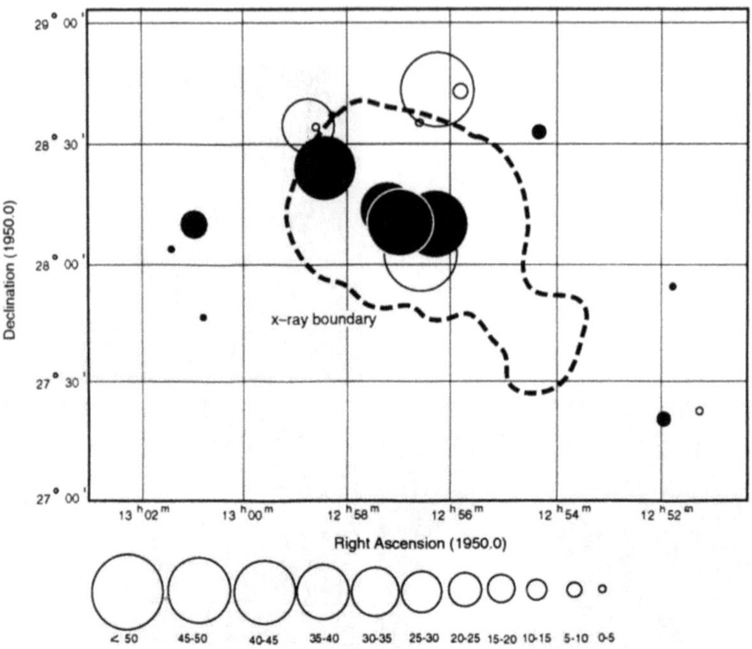

Fig. 5. RM distribution of 18 background radio sources relative to the X-ray boundary of the Coma Cluster of galaxies, which shows a clear excess RM added by the intracluster medium (source Kim et al. 1990). Filled and open circles show positive, and negative RM's respectively. The dashed line shows the ROSAT X-ray boundary at approx. 0.12 counts/(400arcsec^2) for the ROSAT PSPC detector, in the range 0.5 – 2.4 keV (from Briel, Henry, and Böhringer 1992).

4. Recent Detections and Measurements of Magnetic Fields in Galaxy Clusters

Intergalactic gas in galaxy clusters has a typical electron density of 10^{-4} - 10^{-2}cm^{-3}, temperature in the range $10^7 - 10^8$K, and an extent of 1 Mpc for the cluster core. At this temperature range, which makes clusters significant X-ray sources ($L_x \approx 10^{43} - 10^{45}$erg s^{-1}), the ion sound speed is comparable to the galaxies' velocity dispersion in the cluster, which is $400 - 1200$km s^{-1}. A minority of clusters, such as the Coma cluster, contain an enhanced population of cosmic ray electrons which, over dimensions comparable to that of the hot gas, emit a diffuse "halo" of synchrotron radiation, thus revealing an intracluster magnetic field (Fig. 5). For a more extensive review of magnetic fields in galaxy clusters the reader is referred to Kronberg (1994).

5. Magnetic Fields in the Wider Intergalactic Medium

Cosmic rays serve as particularly effective "illuminators" of intergalactic magnetic fields at low radio frequencies. This is because of the relatively high spectral density of synchrotron radiation at low frequencies, reflecting the high value of γ , typically $2.4-3$, which defines the power law distribution of cosmic ray electron energies: $N(E) \propto E^{-\gamma}$ (cf. eq. 1). The critical frequency (near which most of the synchrotron radiation is emitted) is related to E by $\nu_C = (3eB\sin\phi/4\pi m^3c^5)E^2$ (Pacholczyk 1970) . An advantage of observing at the lowest possible radio frequencies is that we preferentially detect the lowest energy CR electrons, which survive the longest to "keep illuminating" the associated magnetic field.

To calculate the longest possible loss time for a CR electron, we define a "cosmic background- equivalent" magnetic field, B_{bge}, for which a CR electron's energy loss rate by synchrotron radiation ($dE/dt \propto |B|^2E^2$) equals that due to inverse Compton scattering off the microwave background radiation $(dE/dt) \propto \varepsilon_{bg}E^2$, where $\varepsilon_{bg} = 4.8.10^{-13}(1+z)^4$erg cm^{-3}. Thus, when $B = B_{bge}$ the energy density of the magnetic field $(B^2/8\pi)$ equals that in the cosmic background radiation (ε_{bg}). For $z \ll 1$, to which we restrict the present discussion, $B_{bge} \approx 3.10^{-6}$ G. Larger magnetic field strengths will cause shorter, synchrotron radiation-dominated lifetimes, whereas for $B < B_{bge}$ a CR electron's loss rate, being dominated by inverse Compton scattering, will depend only on its energy and ε_{bg} (cf. Rees 1967, Pacholczyk 1970 for a discussion of the basic radiation processes).

The inverse Compton-dominated lifetime, τ_{max}, for synchrotron-radiating CR electrons which are observed at a frequency ν, expressing the electron energy in terms of ν and B, can be written as

$$\tau_{max} \sim 3.10^8(\nu/327\text{MHz})^{-0.5}(|B|/|B_{bge}|)^{0.5} \text{ years} \qquad (4)$$

where $(|B|/|B_{bge}| \leq 1)$. The factor $(|B|/|B_{bge}|)^{0.5}$, is probably not far below unity as suggested by recently measured i.c.m. field values. Thus we see that the maximum "fossil lifetime" of relativistic electrons scales to first order by $\simeq \nu^{-1/2}$. Observing frequencies of interest are between approximately 20MHz and 400MHz: Below 20MHz, refraction and absorption by the ionized component of the interstellar medium and/or the ionosphere become important. Above 400 MHz, the emissivity becomes too low for the most sensitive detection of diffuse intergalactic synchrotron emission. In addition, the higher energy electrons, which radiate on average at higher frequencies, are less likely to have propagated to large distances from their acceleration sites because of their shorter loss times. Observing frequencies below $\sim$ 20MHz are conceivable from space, if one is restricted to high galactic latitudes.

Until recently both the low resolution, and difficulties with earthbound and solar interference and ionospheric phase fluctuations, have prevented accurate imaging at frequencies below $\sim$ 300 MHz. To distinguish true diffuse intergalactic synchrotron emission from a blend of discrete extragalactic radio sources, an angular resolution of $1'$ or better is required. This, at $\nu = 75$ MHz for example, requires a well-filled antenna array (to achieve the required image quality), which extends to $\approx$ 30 km or more (to achieve the required resolution). Furthermore, interference suppression and phase irregularities in the ionosphere need to be eliminated, which require compute-intensive data processing techniques. These requirements are being approximately fulfilled for the first time (by the NRAO VLA recently outfitted to low frequencies), the Giant Metre Wave Telescope (GMRT) at Pune, India, and the Westerbork Synthesis Radio Telescope (WSRT) in the Netherlands). These instrumantal developments open up new possibilities for the detection of large scale i.g. magnetic fields. Relation (4) shows that radio images at $\nu \sim 30$ MHz could image diffuse, magnetized intergalactic gas in the local universe which was produced $\sim 10^9$ years ago, on the assumption that $|B|$ is not much different from $|B_{bge}|$. Such observations could give clues from the local Universe to the evolution of magnetic field strength over a significant period of cosmic time in such systems, e.g. in old, extended radio source lobes. An important complement observation will be that of sensitive imaging of diffuse X-ray emission which is produced by either inverse Compton or thermal bremsstrahlung.

Figure 6 shows a WSRT image published by Kim et al. (1989) of the 326 MHz intergalactic synchrotron emission surrounding the Coma cluster of galaxies, which may be prototypical of the type of radiation which could be imaged in future at much lower frequencies. They found emission extending beyond the Coma cluster, indicating an extended, magnetized region on a supracluster scale. Measurements of this type do not yield the field strength directly, but the assumption of equipartition between particle and magnetic energy density gives field strengths of ca. $2.10^{-7}(1+k)^{2/7}$ G in the region in Kim et al.'s observations, which extend well outside of the cluster core. (k is the ratio of CR proton to electron energies, usually estimated at between 10 and 100.)

These first sensitive low frequency images suggest that future lower frequency images of high sensitivity and resolution will yield a great deal of new information on the distribution of intergalactic magnetic fields on large scales. Low frequency radio astronomy will play an important role in the detection of intergalactic magnetic fields. Precision low frequency images at high resolution of the extragalactic sky may prove in future to be as important for studies of galaxy evolution and cosmology as the recent microwave background surveys.

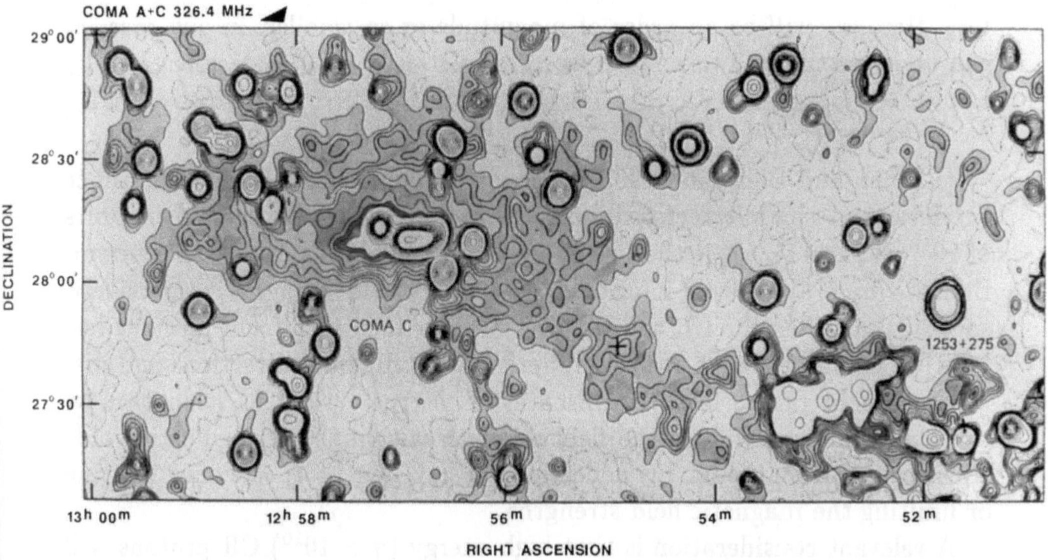

Fig. 6. Radio image showing 326 MHz emission along in intergalactic "bridge", which appears to connect the Coma cluster of galaxies (Coma C) with the Coma A complex of radio emission elsewhere in the Coma supercluster (Source: Kim K.T., Kronberg, P.P., Giovannini G, and Venturi T., 1989).

6. Possible Future Use of Cosmic Rays to infer the Strength and Scaling of an Intergalactic Magnetic Field

The observed cosmic ray (CR) spectrum of light baryons, i.e. mainly protons, is found to extend to *ca.*10^{20} eV. Recent data from the Fly's Eye detector indicate that a chemical "switchover" occurs near $10^{18} - 10^{19}$eV, in the sense that CR's above this range are mostly protons and Helium whereas heavy nuclei appear more predominant below 10^{18}eV (Gaisser et al. 1993). These most energetic CR protons are almost certainly extragalactic in origin. If, as this suggests, the Universe is pervaded with these most energetic CR particles, then it is instructive to compare the average distance between the acceleration sites (= 100 Mpc if luminous FRII radio sources are the origin of the CR's) with the gyroradius of 10^{20} eV protons in the presence of a widespread intergalactic magnetic field. An upper limit to the latter is currently $\sim 10^{-9}$ G. The gyroradius, r_g, of a 10^{20} eV proton in a 10^{-9} gauss field is, interestingly, also of order 100 Mpc.

In the nearby Universe, 10^{20} eV protons will travel in nearly straight lines for $B_{i.g.} \ll 10^{-9}$ gauss, which may be characteristic of the interiors (voids) of the large, $\approx$ 100 Mpc bubbles which define the large scale distribution of optically visible matter. Around the periphery of the voids, i.e. the bubble "surfaces" where the galaxies lie, they will be substantially deflected if $B_{i.g} = 10^{-7 \to -6}$ G, and possibly diffused in galaxy clusters and superclusters. Here, r_g will be an order of magnitude or so smaller. In any case, r_g can be approximated to a mean free path CR$(E, \tau, B_{i.g.})$, which is given by

$$\lambda_{CR_{max}} \simeq 30(E_{18}/B_{-9})(\tau_{17}/d_{bubble})\mathrm{Mpc} \tag{5}$$

(Rachen and Biermann 1993) where τ is the CR particle's lifetime since acceleration to its initial (maximum) energy in units of 10^{17} sec, E is in units of 10^{18} eV, and d_{bubble} is the size of the cosmic bubble in Mpc; The assumptions we made about void interior and periphery field strengths, respectively, suggest that when the CR acceleration sites are identified (thus far none has been), a combination of position differences for different directions and distances can in future lead to estimates of intergalactic field strengths. Of particular interest will be the field strength within the intergalactic voids where, apart from cosmic ray deflection, we have few prospects of estimating or limiting the magnetic field strengths.

A relevant consideration is that high energy ($\gamma > 10^{10}$) CR protons will lose energy with cosmic time, principally due to high energy $\gamma - p$ interactions (which produce pions) with the cosmic microwave background radiation (Greisen 1966, Zatsepin and Kuzmin 1966, Stecker 1968). Since the MBR photon density decreases with time in a cosmology-dependent way, the proton CR propagation (related to τ) is complex (cf. Hill and Schramm 1985, Berezinsky and Grigor'eva 1988). The essence is that, for $\gamma > 10^{11}$, the attenuation length is very small, only $\simeq$ 10 Mpc. This results in a cutoff in the energy spectrum of cosmic rays from cosmologically distant sources with high CR energies.

The distribution in *time* of secondary γ rays due to e^+e^-–photon cascades from very high energy primary γ ray events is another, potentially powerful tracer of intergalactic magnetic fields, as Plaga has proposed elsewhere in this volume (see also Plaga 1995). Plaga's technique, if it can be uniquely associated with *i.g. fields*, is of great interest in this context since it would be many orders of magnitude more sensitive than Faraday rotation probes.

The fact that the Universe is opaque to high energy proton CR's beyond a few tens of Mpc is of interest for this discussion for the following reasons: (i) The CR energy spectrum becomes modified in ways which can be observationally tested, (ii) Ii implies that direct sources of such CR's can only be identified in the local universe, and (iii) given this "convenient" isolation of local extragalactic CR sources, the deflection and diffusion of high energy proton CR's in the local universe will depend on the intergalactic magnetic field, as discussed above.

Future observations of the locations and isotropy of proton CR's will be influenced by the strength and morphology of the intergalactic field, so that discrepancies between the arrival directions of primary CR sources and their production sites might be established in future. Field strengths $\gg 10^{-9}$ G will cause significant deflection, whereas the converse is true if $B_{i.g.} \ll 10^{-9}$ G. A very much lower level régime of field strengths, which might obtain in cosmic voids, might be measurable in future from e^+e^-−photon cascades.

Acknowledgements

I thank David Clarke and Götz Golla, for assistance in providing the illustrations, Peter Biermann for useful discussions, the Killam Foundation for a Fellowship, and the Natural Sciences and Engineering Research Council of Canada for a research grant.

References

Berezinsky, V.S., and Grigor'eva, S., 1988 *Astron. Astrophys.* **199**, 1
Bridle, A.H., *Am. Inst. of Phys. Conf. Series*, **254**, 386
Briel, U.G., Henry,J.P., and Böhringer, H. 1992 *Astron. Astrophys.* **259**, L31
Burbidge, G.R., 1956, *Astrophys J.* **124**, 416
Clarke, D.A. 1993, *Jets in Extragalactic Radio Sources* (Ringberg Workshop Proc.) ed H.-J. Röser and K. Meisenheimer (Berlin:Springer) p.243
Clegg A.W., Cordes, J.M., Simonetti, J.H., and Kulkarni, S.R. 1992, *Astrophys J.* **386**, 143
Davies, R.D. 1968, *Nature***218**, 435
Gaisser T.K.,, Stanev T., Tilav S., Corbato, S. C., Dai, H. Y., Dawson, B. R., Elbert, J. W., Emerson, B., Kieda, D.B., Luo, M., Ko, S., Larson, C., Loh, E.C., Salamon, M. H., Smith, J.D., Sokolsky, P., Sommers, P., Tang, J., Thomas, S.B., and Bird, D.J. 1993 *Phys. Rev. D*, **47** 1919
Golla, G., and Hummel, K. 1994 *Astron. Astrophys.*, , xxx
Greisen, K. 1966 *Phys. Rev. Lett.*, **16**, 748
Hill C.T., and Schramm, D.N., 1985 *Phys. Rev. D*, **31**, 564
Kim, K.-T., Kronberg, P.P., Dewdney, P.E., and Landecker, T.L. 1990, *Astrophys J.* **355**, 29
Kim, K.-T., Kronberg, P.P., Giovannini, G., and Venturi, T. 1989, *Nature* **341**, 720
Kronberg, P.P. 1994, *Rep. Prog. Phys.* **57**, 325
Kronberg, P.P., and Perry, J.J. 1982, *Astrophys J.* **263**, 653
Lovelace, R.V.E., Berk,H.L., and Contopoulos J. 1991 *Astrophys J.* **379**, 696
Mestel, L.,*Stellar Evolution* ed. H-Y Chiu and A Muriel (Cambridge MA: MIT Press), p.643
Pacholczyk, A., 1970 *Radio Astrophysics* (San Francisco: Freeman)
Perley, R.A., Dreher,J.W., and Cowan, J.J. 1984, *Astrophys J.* **285**, L35
Plaga, R. 1995 *Nature* (submitted)
Rand, R.J., and Lyne, A.G., 1994 *Mon. Not. Roy. Astron. Soc.* **268**, 497
Rees, M.J., 1967 *Mon. Not. Roy. Astron. Soc.* **137**, 429
Reuter, H.-P., Klein, U., Lesch, H., Wielebinski,R. and Kronberg, P.P. 1994 *Astron. Astrophys.* **282**, 724
Simard-Normandin M., and Kronberg, P.P. 1980, *Astrophys J.* **242**, 74
Stecker, F. W., 1968 *Phys. Rev. Lett.*, **21**, 1016
Vallée, J.P., Simard-Normandin M., and Bignell, J.P. 1980 *Astrophys J.* **331**, 321
Wielebinski R., 1990 *The Interstellar Medium in Galaxies*, (Dordrecht:Kluwer), p.349
Zatsepin, G.T,. and Kuz'min, V.A., 1966 *JETP Lett.*, **4**, 78

ABSORPTION OF HIGH ENERGY GAMMA-RAYS BY LOW ENERGY INTERGALACTIC PHOTONS

F.W. STECKER
Lab. for High Energy Astrophysics
NASA Goddard Space Flight Center
Greenbelt, MD 20771

and

O.C. DE JAGER
Dept. of Physics
Potchefstroom University
Potchefstroom 2520, South Africa

Abstract. Following our previously proposed technique, we have used the recent γ-ray observations of Mkr421 to place theoretically significant constraints on the magnitude of the intergalactic infrared radiation field (IIRF). Our 2σ upper limits are consistent with normal IR production by stars and dust in galaxies. They rule out exotic mechanisms proposed to produce a larger IIRF. Although they are still subject to revision and are unconfirmed, the data on the spectrum of Mkr421 hint at a possible absorption cutoff which could be produced by an IIRF of the magnitude expected from stellar emission and reprocessing in galaxies. Using models for the low energy intergalactic photon spectrum from microwave to ultraviolet energies, we calculate the opacity of intergalactic space to γ-rays as a function of energy and redshift. These calculations indicate that the GeV γ-ray burst recently observed by the CGRO EGRET detector originates at a redshift less than ~ 1.5.

1. Introduction

The potential importance of the photon-photon pair production process in high energy astrophysics was first pointed out by Nikishov (1961). Unfortunately, his early paper overestimated the energy density of the low energy IIRF by about three orders of magnitude. However, with the discovery of the cosmic microwave background radiation, Gould and Schreder (1966) and Jelley (1966) were quick to point out the opacity of the universe to photons of energy greater than 100 TeV. Stecker (1969) and Fazio and Stecker (1970) generalized these calculations to high redshifts, showing that photons originating at a redshift z will be absorbed above an energy of $\sim 100(1+z)^{-2}$ TeV. Following the detection of the blazar 3C279 at energies of 0.1 to 5 GeV by the EGRET team (Hartman, *et al.* 1992), we (Stecker, De Jager and Salamon 1992, Paper I) proposed that one look for pair-production absorption features in blazar spectra at TeV energies to determine the intensity of the IIRF. *Direct* observational determinations of the extragalactic infrared background radiation are fraught with difficulties (see section 3).

As we discussed in Paper I, an EGRET- strength "grazar" (observed γ-ray blazar) with a hard spectrum extending to multi-TeV energies is potentially detectable with ground-based telescopes. Pair-production interactions

Space Science Reviews **75**: 401–412, 1996.

with the cosmic microwave background radiation will not cut off the γ-ray spectrum of a nearby (low redshift) source below an energy of $\sim$ 100 TeV (Fazio and Stecker 1970). The techniques involved in making ground-based observations of TeV γ-ray sources have been reviewed by Weekes (1988).

Shortly after Paper I was published, the Whipple Observatory group reported the discovery of $\sim$ TeV γ-rays coming from the nearby blazar Mkr 421, a BL Lac object (Punch *et al.* 1992). This source has a hard, roughly E^{-2} γ-ray spectrum extending to the TeV energy range (Lin, *et al.* 1992; Mohanty, *et al.* 1993) where the pair production process involving infrared photons becomes relevant. Following this discovery, we made use of the TeV observations of Mrk 421 to carry out our suggestion of using such sources to probe the IIRF (Stecker and De Jager 1993; De Jager, Stecker and Salamon 1994).

There are now over 40 grazars which have been detected by the EGRET team (Fichtel, *et al.* 1994). These sources, optically violent variable quasars and BL Lac objects, have been detected with a redshift range extending beyond 2. Of these objects, only the low reshift BL Lac, Mkr421 was seen by the Whipple telescope. The fact that the Whipple team did not detect the much brighter EGRET source 3C279 at TeV energies (Vacanti, *et al.* 1990, Kerrick, *et al.* 1993) is consistent with our previous predictions (Paper I; Stecker 1993) because this source is at a much higher redshift (z=0.54). However, the interpretation is somewhat complicated by the significant time-variability of this source at GeV energies and the fact that simultaneous measurements were not made.

The TeV spectrum of Mkr421 obtained by the Whipple Observatory group is consistent with a perfect $E^{-2.06}$ power-law, showing no absorption out to an energy of at least 3 TeV (Mohanty, *et al.* 1993). It is this lack of absorption below 3 TeV which we (Stecker and De Jager 1993) used to put upper limits on the extragalactic infrared energy density in the $\sim$ 0.2 to 1 eV energy range ($\sim$ 1 μm to 5 μm) (See also Dwek and Slavin 1994). However, there *is* possible evidence of absorption in the Mkr421 spectrum *above* $\sim$ 3 TeV, which we have interpreted as possible intergalactic absorption allowing an estimate the IIRF at wavelengths longer than $\sim$ 5 μm (De Jager, Stecker and Salamon 1994).

In this paper, we will review those results. We will also use semiempirical models of the low energy intergalactic radiation fields from microwave to ultraviolet energies to calculate the opacity of the universe as a function of redshift to γ-rays extending from the EGRET range (GeV energies) to energies in the PeV region. Using these calculations, we can place a limit on the redshift from which the multi-GeV γ-ray burst observed by EGRET can have originated.

2. The Absorption of TeV Photons by Infrared Photons

Consider the interaction of a TeV γ-ray of energy $E(z) = (1+z)E$ with a soft photon of energy $\epsilon(z) = (1+z)\epsilon$ at a redshift z, where E and ϵ are the presently observed (z=0) photon energies. Pair production is expected above the threshold energy condition $E\epsilon(1+z)^2 x > 2(mc^2)^2$, where $x = (1-\cos\theta)$ and θ is the angle between the photon directions. The cross section is given by (Heitler 1960)

$$\sigma[E(z),\epsilon(z),x] = 1.25\times10^{-25}(1-\beta^2)\left[2\beta(\beta^2-2)+(3-\beta^4)\ln\left(\frac{1+\beta}{1-\beta}\right)\right] \ \mathrm{cm}^2 \tag{1}$$

where $\beta = (1-2(mc^2)^2/(E\epsilon x(1+z)^2))^{1/2}$. For γ-rays in the TeV energy range, this cross section is maximized when the soft photon energy is in the infrared range:

$$\epsilon(E) \simeq \frac{2(mc^2)^2}{E} \simeq 0.5\left(\frac{1\ \mathrm{TeV}}{E}\right)\ \mathrm{eV} \tag{2}$$

For a ~ 1 TeV γ-ray, this corresponds to a soft photon wavelength near the K-band ($\sim 2\mu$m). Thus, infrared photons with wavelengths around 2μm will contribute most to the absorption of TeV γ-rays. If the photons with energy ϵ have a number density $n(\epsilon)d\epsilon$ cm^{-3}, the optical depth for γ-rays will be

$$\tau(E) = \frac{c}{H_o}\int_0^z dz\,(1+z)(1+\Omega z)^{-1/2}\int_0^2 dx\frac{x}{2}\int_{\epsilon_1}^{\infty} d\epsilon\; n(\epsilon)\sigma\left[2x\epsilon E(1+z)^2\right] \tag{3}$$

where $\epsilon_1 = 2m^2c^4/Ex(1+z)^2$ (Stecker 1971).

We (Stecker and De Jager 1993) used the Whipple data on the TeV energy spectrum from Mkr421 to put limits on τ, and therefore to put limits on the infrared radiation density. For a low redshift source ($z \ll 1$) such as Mk421, the calculation of the optical depth, τ, simplifies considerably. Taking the soft photon spectrum as $n(\epsilon) = \kappa\epsilon^{-\alpha}$, for small redshifts , z, (with $\Omega = 1$)

$$\tau(E) = \frac{2^\alpha cz}{(\alpha+1)H_o}\left(\frac{2(mc^2)^2}{E}\right)n\left[\frac{2(mc^2)^2}{E}\right]\int_1^{\infty} u^{-\alpha}\,du\,\sigma[\beta=\sqrt{1-1/u}] \tag{4}$$

For normal galaxies (Stecker *et al.* 1977) with $n(\epsilon) = \kappa_1\epsilon^{-2.55}$, in units of eV^{-1}cm^{-3} with ϵ in eV, we find

$$\tau(E) = 143\kappa_1 E_{\mathrm{TeV}}^{1.55} \tag{5}$$

and, for ϵu_ϵ as constant (or $n(\epsilon) = \kappa_2\epsilon^{-2}$), we find that

$$\tau(E) = 120\kappa_2 E_{\mathrm{TeV}}. \tag{6}$$

Expressions (5) and (6) for $\tau(E)$ should roughly hold up to $E \sim 10$ TeV.

Using the 11 spectral points then available, Mohanty *et al.* (1993) were able to fit a single power law between their data set below 5.1 TeV and the EGRET data set at $\sim$ GeV energies (Lin, *et al.* 1992). The spectrum that they obtained was

$$dN/dE = (1.02 \pm 0.14) \times 10^{-11} E^{-2.06 \pm 0.04} \text{ photons/cm}^2\text{/s/TeV} \tag{7}$$

We generalized this fit to include absorption by assuming spectra of the form

$$dN/dE = KE^{-\Gamma} \exp\left[-\tau(E)\right] \tag{8}$$

with K, Γ and κ as free parameters. Then the best χ^2 fit is for

$$dN/dE = (1.11 \pm 0.21) \times 10^{-11} E_{\mathrm{TeV}}^{-2.04 \pm 0.04} \exp\left(-\tau(E)\right) \tag{9}$$

where (assuming $H_o = 75$ km/s/Mpc) $\tau(E) = 0.048 E_{\mathrm{TeV}}$ (for $\alpha = 2$) and $\tau(E) = 0.043 E_{\mathrm{TeV}}^{1.55}$ (for $\alpha = 2.55$) (again valid only up to $E \sim 10$ TeV) with a minimum $\chi^2_8 = 4.9$ for $11 - 3 = 8$ degrees of freedom for both models for $n(\epsilon)$. The upper limits on τ are obtained by allowing all free parameters to vary simultaneously, which allows us to construct confidence ellipsoids. The 1σ and 2σ upper limits on κ_1 in Eq. (5) (i.e. for $\alpha = 2.55$) are $\kappa_1 < 9 \times 10^{-4}$ and $\kappa_1 < 1.9 \times 10^{-3}$ respectively, whereas the corresponding numbers for κ_2 in Eq. (6) (i.e. for $\alpha = 2$) are $\kappa_2 < 2.1 \times 10^{-3}$ and $\kappa_2 < 4.0 \times 10^{-3}$ respectively. The 2σ or 95% upper limits on $\tau(E)$ are then $\tau(E) < 0.27 E_{\mathrm{TeV}}^{1.55}$, which implies $n(\epsilon) < 1.9 \times 10^{-3} \epsilon^{-2.55}$ $\mathrm{eV}^{-1}\mathrm{cm}^{-3}$ for $\alpha = 2.55$. For $\alpha = 2$ we find $\tau(E) < 0.48 E_{\mathrm{TeV}}$, which implies $n(\epsilon) < 4.0 \times 10^{-3} \epsilon^{-2}$ $\mathrm{eV}^{-1}\mathrm{cm}^{-3}$. The latter upper limit (for $H_o = 50$ km/s/Mpc) is shown on Fig. 1.

Even though the EGRET and Whipple detections were not simultaneous, they can be cautiously combined. Although the intensity of Mkr421 may occasionally vary by large amounts on a timescale of days (Kerrick, *et al.* 1995), it was shown by the Whipple group that averaging on a timescale of several weeks results in a 20% variation at most. Such a timescale is the one relevant to the above results.

In the above analysis, the spectral information *above* 3.4 TeV was neglected since the highest energy points were only upper limits. However, we note that using the data from Mohanty *et al.* (1993) and extrapolating the $E^{-2.06}$ differential spectrum to energies above 3 TeV, we expect ~ 58 ON-source events (source γ-rays and cosmic-ray background) between 3.4 TeV and 5.1 TeV, whereas only 26 *on*-source events were observed, which is consistent with the 27 *off*-source cosmic ray background events collected with similar exposure. Furthermore, above 5.1 TeV we expect ~ 32 *on*-source events, whereas only 10 were observed against the 12 background events. These data may be interpreted as indicating the presence of an absorption feature in the *integral* spectrum of Mkr421 manifested at the highest energies. Such an interpretation allows us to obtain an estimate of the IIRF, rather than an upper limit (De Jager, Stecker and Salamon 1994). This empirical estimate is consistent with theoretical estimates of the IIRF (Stecker, *et al.* 1977).

There are three main caveats to consider with regard to the possible cutoff in the high energy end of the observed spectrum of Mkr421 being considered as an absorption effect of the IIRF: (1) The source spectrum may have an *intrinsic emission* cutoff above $\sim$ 3 TeV. However, one may then ask why the intrinsic spectrum would turn over at the energy naturally expected for an IIRF absorption feature. We also note that emission models involving Compton scattering of high energy electrons in the source would not produce such an apparently sharp cutoff (Ghisellini, Maraschi and Treves 1985; Dermer and Schlickheiser 1993; Bloom and Marscher 1993). (2) *Intrinsic absorption* in the source may produce a high–energy cutoff. However, we consider it highly improbable that the infrared characteristics of the source would conspire to exactly mimic that which would be produced by a reasonably estimated IIRF. Also, if the observed IR flux from Mkr421 (Makino, *et al.* 1987) is from the galaxy, the implied optical depth would not be large enough to produce an observable absorption effect, while if it originates in the jet and is beamed, the intrinsic flux may be considerably reduced as well as the interaction rate. We note, however, that if the absorption *is* intrinsic, then our estimate becomes the best upper limit extant on the IIRF. (3) Perhaps, *most importantly*, there is a systematic uncertainty in the event rate derived by Mohanty, *et al.* (1993) in the energy range above 5 TeV, which is as yet poorly determined. A possible *oversubtraction* of assumed cosmic ray produced events, which can become more confused with γ-ray produced events at the higher energies, might have artificially supressed the highest energy event rate reported by Mohanty, *et al.* (1993) (R. Lamb, priv. comm.). We note, however, that the Crab nebula, a much stronger source, was observed by the Whipple group to have a power–law spectrum with *no* such turnover up to $\sim$ 10 TeV (Vacanti, *et al.* 1991).

We also note that intergalactic electron-photon cascading (Protheroe and Stanev 1993; Aharonian, Coppi and Volk 1993) is not expected to significantly influence the TeV spectrum of Mkr421 since the source is observed within a point-spread function with HWHM of $\sim$ 0.2 degrees, and it can be shown that an intergalactic magnetic field of 10^{-20} G or greater would deflect the cascade TeV electrons away from the source direction, creating a γ-ray halo much larger than the point-spread function. Since the measured magnetic field strengths in superclusters are of the order of a μG (Kronberg 1994 and references therein), it is highly unlikely that the intergalactic field would be as weak as 10^{-20} G.

Thus, our interpretation of the Mkr 421 spectrum (De Jager, *et al.* 1994) may indeed be the first determination of the IIRF at a level which is expected from stellar processes in galaxies (Stecker, *et al.* 1977).

3. The Intergalactic Infrared Background Radiation Field

A theoretical upper limit for the infrared energy density from starlight in normal galaxies was estimated by Stecker, *et al.* (1977), assuming that all of the energy from nucleosynthesis in stars is emitted in the infrared. Estimates of the intergalactic and optical radiation densities were obtained by Yoshii and Takahara (1988), Takahara (1990), and Tyson (1990), based on galaxy counts and other modelling considerations. Cowie, *et al.* (1990) have estimated the IIRF using a K-band deep galaxy survey.

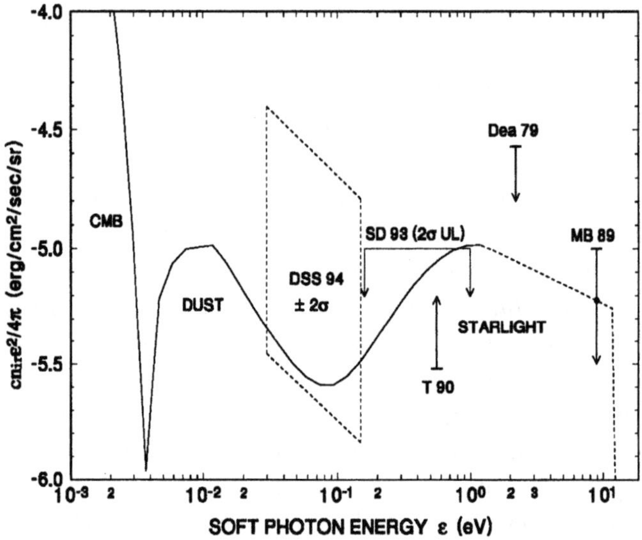

Fig. 1. The low energy intergalactic photon spectrum from the microwave to the Lyman limit in the ultraviolet modeled by including the cosmic microwave background (CMB) spectrum and dust and starlight spectra modeled by Franceschini, *et al.* (1994) extrapolated into the UV using the results of Martin and Bowyer (1989), indicated as 'MB 89'. The lower limit given by Tyson (1990) is indicated as 'T 90' and the upper limit of Dube, *et al.* (1979) is indicated as 'Dea 79'. We also show our previous upper limit (Stecker and De Jager 1993, 'SD 93') and our estimate of the IIRF (De Jager, Stecker and Salamon 1994, 'DSS 94'), obtained from the TeV data on Mkr421 (see text).

Direct measurements of the extragalactic infrared and optical radiation are plagued by the difficulty of separating the true extragalactic component from galactic radiation, zodiacal light and instrumental background (e.g. radiation from rocket exhaust). The observational upper limits obtained in the near infrared by Matsumoto, *et al.* (1988) and the optical (Dube, *et al.* 1979, see Fig. 1) all lie well above the theoretical starlight spectrum shown in Fig. 1, as well as the other estimates mentioned above. Thus, it is clear that the present directly obtained upper limits are not theoretically significant. However, as we discussed in the previous section, already the TeV data from Mkr421 have provided both estimates and upper limits which *are* theoretically significant. In fact, the estimates of De Jager, Stecker and Salamon (1994) indicate that at least 30% of the energy emitted by stars and dust in galaxies winds up in the infrared.

The intergalactic background radiation from galactic starlight and dust reradiation has recently been modeled in some spectral detail (Franceschini, *et al.* 1994; MacMinn and Primack 1994). As discussed by Franceschini, *et al.* (1994), our results are consistent with their theoretical models. Using their models for stellar and dust emission, together with the COBE results on the cosmic microwave-FIR background radiation and extrapolating to the Lyman limit in the UV, we obtain a full spectrum low energy intergalactic photon background as shown in Fig. 1. To obtain the UV estimate shown in Fig. 1, we have used the results of Martin and Bowyer (1989). It should be kept in mind that this UV estimate is very uncertain and represents an upper limit on the UV flux. This flux may actually drop off significantly at energies below the Lyman limit. (To allow for this possibility in our calculations below, we have used two other model estimates.)

Fig. 1 also shows our previously obtained upper limit (Stecker and De Jager 1993) and IIRF estimate (De Jager, Stecker and Salamon 1994). It can be seen that our estimate from the Mkr421 data lies somewhat above the model. This can be from either of three causes: (1) uncertainty in the interpretation of the Mkr421 event rate (R. Lamb, private communication), (2) some intrinsic absorption in Mkr421 (see previous discussion), or an underestimate in the modeled IIRF owing to the omission of contributions from infrared bright extragalactic sources (De Zotti, *et al.* 1994).

4. The Opacity of the Universe to Gamma-Rays

By making use of the modeled intergalactic spectrum shown in Fig. 1, one can calculate the opacity of the universe at various redshifts over a wide range of γ-ray energies. Fig. 2 shows the optical depth of the universe to γ-rays originating at the redshift corresponding to the grazars designated. We can also use the low energy intergalactic photon spectrum shown in Fig. 1 to obtain a redshift limit on the 17 February 1994 high energy γ-ray burst observed by EGRET (Hurley, *et al.* 1994). The highest energy photon in this burst had an energy of 18 ± 5 GeV. The corresponding opacity to photons in this energy range is shown in Fig. 3. In all of the models of the intergalactic UV background which we used (see figure caption), the universe becomes opaque to ~ 20 GeV photons at a redshift of ~ 1.5. (These models will be elaborated on in a more detailed treatment elsewhere by Stecker, Salamon and De Jager.) In all of the models considered here, we assumed that there are no intergalactic ionizing photons above the Lyman limit of 13.6 eV. The probable existence of such photons would only decrease our estimate of the absorption redshift. Therefore, one obtains the interesting result that the source of this burst lies at a redshift which is not uncommon for high energy astrophysical sources. One should be cautioned, however, that the existence of one burst in this redshift range does not necessarily rule out exotic cosmological models for γ-ray bursts, such as those involving topological defects; a string located at a "low" redshift would be brighter and more observable at high energies than one at very high redshifts.

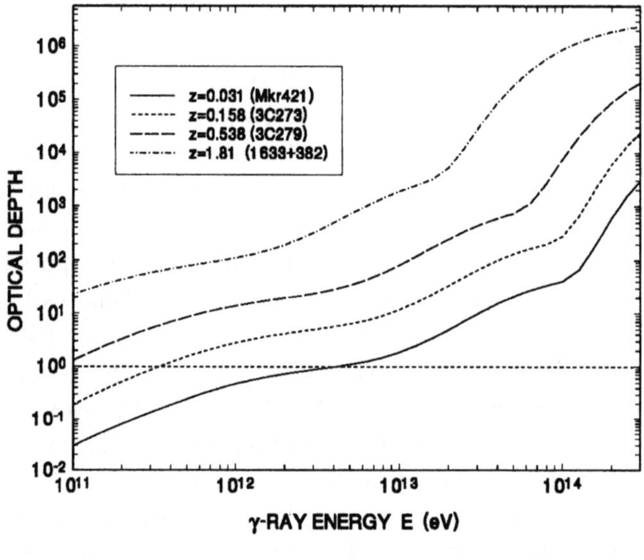

Fig. 2. Optical depth versus energy for photons from four EGRET grazars obtained using the photon spectrum shown in Fig. 1 and taking $H_o = 50$ km/s/Mpc.

We have used the opacity curve shown in Fig. 2 and $H_o = 50$ km/s/Mpc to fit the Whipple data on Mkr421 as shown in Fig. 4. The Whipple spectrum is well fitted by a simple E^{-2} source spectrum with absorption included ($\chi^2 = 7.7$ for $8 - 2 = 6$ degrees of freedom). Fig. 4 also shows the latest totaled EGRET spectra (Lin, *et al.* 1994). It can be seen that an extrapolation of our Mkr421 fit on the TeV data will adequately agree with the EGRET results.

Again, using the opacity curves shown in Fig. 2, we have calculated predicted turnovers in other EGRET observed grazars at various redshifts, assuming power-law spectra with the indices fitted by the EGRET team (Fichtel, *et al.* 1994). Our results are shown in Fig. 5. It can be seen that we predict turnovers in such spectra which should be observable with sensitive

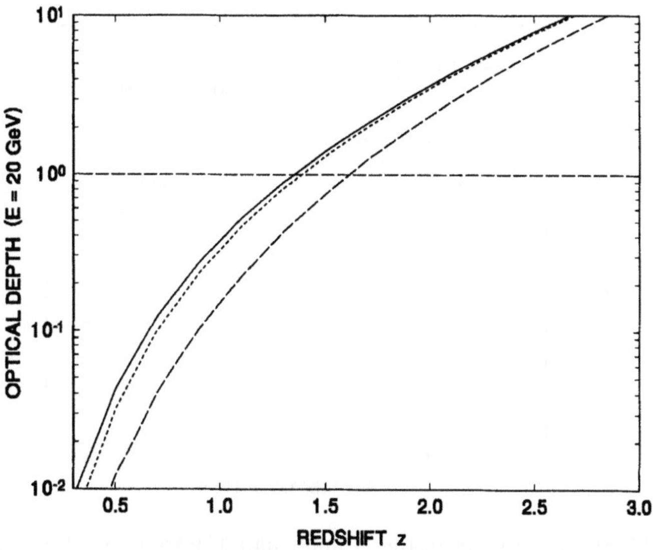

Fig. 3. Optical depth versus redshift for photons of energy 20 GeV (assuming $H_o = 50$ km/s/Mpc). The *solid line* was obtained using the IIRF model shown in Fig. 1. The *short dashed line* was obtained by modifying the UV model with an exponential cutoff of the form $\exp(-\epsilon/7.7\ \mathrm{eV})$ above 7.7 eV. For the *long dashed line* the starlight component between 0.1 eV and 1.2 eV was mirrored into the range 1.2 to 13.6 eV, resulting in symmetric peak around 1.2 eV. In all cases, we took an absolute cutoff in the UV at the Lyman limit of 13.6 eV. The results shown can be used to put an upper limit on the redshift of the 17 February 1994 burst observed by EGRET (see text).

enough multi-GeV and sub-TeV detectors and used in concert to test our model of intergalactic absorption.

We have also pointed out that by combining a measurement of the opacity of the universe to high energy γ-rays with a direct measurement of the IIRF, one can determine the value of the Hubble parameter H_0 at cosmologically significant distances, independent of the distance ladder (Salamon, Stecker and De Jager 1994). Such a determination of H_0 has become even more critical with the recent, relatively "local" measurement of H_0 using Cepheids in the Virgo cluster (Pierce, *et al.* 1994; Freedman, *et al.* 1994). These determinations using observations form the Hubble Space Telescope give values of $H_0 \sim 80-90$ km s^{-1} Mpc^{-1}, resulting in a *prima facie* conflict with age determinations of old stars in globular clusters. A determination of a smaller value for H_0 sampled at truly cosmological redshifts could resolve this conflict. Indeed, it has very recently been suggested by Wu, *et al.* (1994) that the "local" value of H_0 could be 1.2 to 1.4 times larger than its true cosmological value, owing to gravitational effects.

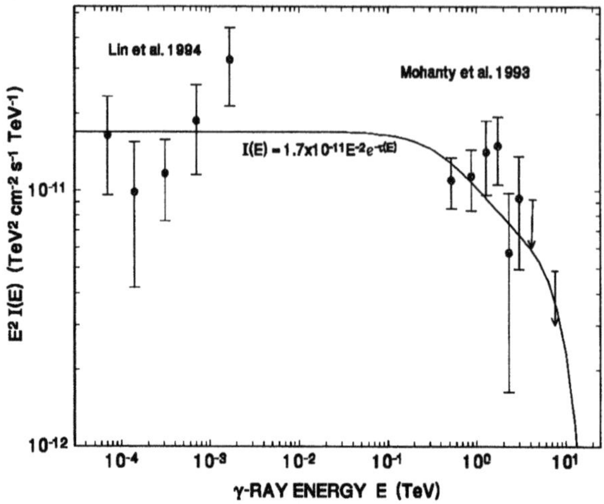

Fig. 4. Mkr421 spectrum fitted to the Whipple data (Mohanty, *et al.* 1993) with absorption calculated using the model spectrum in Fig. 1 and $H_o = 50$ km/s/Mpc. The upper limits above 3 TeV are at the 2σ level. The last point follows from the integral flux above 5 TeV, which was converted to a differential point given the expected absorption spectrum. Also shown are the most recent EGRET data points (Lin, *et al.* 1994).

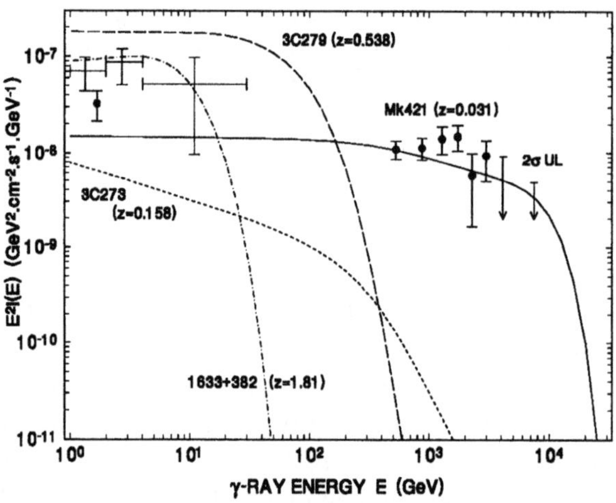

Fig. 5. Power-law spectra extapolated from EGRET data for four observed grazars. The absorption calculations assumed $H_o = 50$ km/s/Mpc. The solid circles are data points for Mkr421. The plain crosses are data points for the high redshift grazar 1633+382 (Mattox, *et al.* 1993).

5. Conclusions

The observational upper limits on the extragalactic infrared background radiation field are not very useful in constraining the various theoretical models proposed. We (Stecker, *et al.* 1992) have suggested that using sub-

TeV and multi-TeV observations of grazars, one can derive useful information about the extragalactic infrared background provided that the intrinsic source spectrum does not cut off in this energy range. This appears to be the case with Mkr421. If the spectrum of an extragalactic source extends beyond 1 TeV, then the extragalactic infrared field should cut off the observed spectrum between $\sim$ 10 GeV and a few TeV, depending on the redshift of the source. There is possible evidence for such a sharp drop in the spectrum of Mkr421 above 3 TeV where intergalactic absorption is expected to occur. We have used this evidence to estimate the intensity of the IIRF.

Our conservative upper limits on the IIRF, obtained from the Whipple data on Mkr421 below 3 TeV, rule out various exotic mechanisms for producing larger fluxes, such as some exploding star, decaying particle, massive object and black hole models (Carr 1988). They are consistent with the extragalactic near infrared background originating from ordinary stellar processes in galaxies (Stecker, *et al.* 1977; Franceschini *et al.* 1994).

The background flux in the 1 to 5 μm range is both estimated and theoretically expected to be in the range $\sim 10^{-9}$ to $\sim 10^{-8}$ w m^{-2} sr^{-1}. Such a flux may be eventually detectable with the DIRBE experiment on COBE, provided that the foreground radiation from zodiacal light can be modeled well enough to be subtracted out (Mather, 1982; Hauser 1990; Mather, *et al.* 1990). If this flux can be directly detected, it may be used in combination with high energy γ-ray absorption measurements to determine the value of the Hubble parameter at cosmologically significant redshifts (Salamon, Stecker and De Jager 1994). Our estimates of the opacity of the universe to γ-rays can also be used to constrain the redshifts of high energy γ-ray burst sources. The highest energy burst observed by EGRET is constrained to originate at a redshift less than $\sim$ 1.5.

It would be desirable to extend γ-ray observations to sub-TeV energies and to grazars at differing redshifts with higher sensitivity instruments in order to accurately determine the low energy intergalactic background radiation fields (see Fig. 5).

Acknowledgements

We thank M.H. Salamon for discussions of the manuscript.

References

Aharonian, F.A., Coppi, P. and Volk, H.J.: 1993, *Proc. 23rd Int'l. Cosmic Ray Conf.*, Univ. of Calgary Press, Calgary, **1**, 451.

Bloom, S.D. and Marscher, A.: 1993, in M. Friedlander, N. Gehrels, and D.J. Macomb (eds.), *Compton Gamma Ray Observatory*, AIP, New York, **280**, p. 578.

Carr, B.J.: 1988, in A. Lawrence (ed.), *Comets to Cosmology*, Springer-Verlag: Berlin, p. 265.

Cowie, L.L., Gardner, J.P., Lilly, S.J. and McLean, I.: 1990, *Astrophys. J. Lett.* **360**, L1.

De Jager, O.C., Stecker, F.W. and Salamon, M.H.: 1994, *Nature* **369**, 294.
Dermer, C.D. and Schlickheiser, R.: 1993, *Astrophys. J.* **416**, 458.
De Zotti, G. *et al.* : 1994, preprint.
Dube R.R., Wickes, W.C. and Wilkinson, D.T.: 1979, *Astrophys. J.* **232**, 333.
Dwek, E. and Slavin, J.: 1994, *Astrophys. J.* , in press.
Fazio, G.G. and Stecker, F.W.: 1970, *Nature* **226**, 135.
Fichtel C.E. *et al.* : 1994, *Astrophys. J. Suppl.* **94**, 551.
Franceschini, *et al.* : 1994, *Astrophys. J.* **427**, 140.
Freedman, W. *et al.* : 1994, *Nature* **371**, 757.
Gould, R. J. and Schreder, G.P.: 1966, *Phys. Rev. Lett* **16**, 252.
Ghisellini, G. and Maraschi, L.: 1989, *Astrophys. J.* **340**, 181.
Hartman, R.C. *et al.* : 1992, *Astrophys. J. Lett.* **385**, L1.
Hauser, M.G. *et al.* : 1990, in *After the First Three Minutes*, AIP, New York, **222**, p. 161.
Heitler, W.: 1960, *The Quantum Theory of Radiation*, London, Oxford Press.
Hurley, K. *et al.* : 1994, *Nature* **372**, 652.
Jelley, J.V.: 1966, *Phys. Rev. Lett.* **16**, 479.
Kerrick, A.D. *et al.* : 1993, *Proc. 23rd Int'l. Cosmic Ray Conf.*, Univ. of Calgary Press, Calgary, **1**, 405.
Kerrick, A.D. *et al.* : 1995, *Astrophys. J. Lett.* **438**, L63.
Kronberg, P.P.: 1994, *Rep. Prog. Phys.* 325.
Lin, Y.C. *et al.* : 1992, *Astrophys. J. Lett.* **401**, L61.
Lin, Y.C. *et al.* : 1994, in C.E. Fichtel, N. Gehrels, and J.P. Norris (eds.), *The 2nd Compton Symp.*, AIP, New York, **304**, p. 582.
MacMinn, D. and Primack, J.R.: 1994, preprint.
Makino, F. *et al.* : 1987, *Astrophys. J.* **313**, 662.
Martin, C, and Bowyer, S.: 1989, *Astrophys. J.* **338**, 677.
Mather, J.C.: 1982, *Optical Engineering* **21**, 769.
Mather, J.C. *et al.* : 1990, *Astrophys. J. Lett.* **354**, L37.
Mohanty, G. *et al.* : 1993, *Proc. 23rd Int'l. Cosmic Ray Conf.*, Univ. of Calgary Press, Calgary, **1**, 440.
Matsumoto, T., Akiba, M. and Murakami, H.: 1988, *Astrophys. J.* **332**, 575.
Mattox, J.R., *et al.* : 1993, *Astrophys. J.* **410**, 609.
Nikishov, A.I.: 1962, *Sov. Phys. JETP* **14**, 393.
Pierce, M.J. *et al.* : 1994, *Nature* **371**, 385.
Protheroe, R.J. and Stanev, T.: 1993, *Mon. Not. R. Astron. Soc.* **264**, 191.
Salamon, M.H., Stecker, F.W. and De Jager, O.C.: 1994, *Astrophys. J. Lett.* **423**, L1.
Stecker, F.W.: 1969, *Astrophys. J.* **157**, 507.
Stecker, F.W.: 1971, *Cosmic Gamma Rays*, Baltimore, Mono Book Co.
Stecker, F.W.: 1993, in J. Tran Thanh Van (ed.), in *Particle Astrophysics*, Proc. Rencontres de Blois, Gif-sur-Yvette: Editions Frontiers, p. 111.
Stecker, F.W. and De Jager, O.C.: 1993, *Astrophys. J. Lett.* **415**, L71.
Stecker, F.W., De Jager, O.C. and Salamon, M.H.: 1992, *Astrophys. J. Lett.* **390**, L49 (Paper I).
Stecker, F.W., Puget, J.L. and Fazio, G.G.: 1977, *Astrophys. J. Lett.* **214**, L51.
Takahara, F.: 1990, *Prog. Theor. Phys.* **83**, 1071.
Tyson, J.A.: 1990, in S. Bowyer and C. Leinert (eds.), in *The Galactic and Extragalactic Background Radiation*, IAU, Netherlands, p. 245.
Vacanti, G. *et al.* : 1990, in *Proc. 21st Int'l. Cos. Ray Conf.*, Univ. of Adelaide, Adelaide, **2**, 329.
Vacanti, G. *et al.* : 1991, *Astrophys. J.* **337**, 467.
Wu, X.-P., Deng, Z., Zou, Zh., Fang, L.-Zh., and Qin, B.: 1994, submitted to *Astrophys. J. Lett.* .
Weekes, T.C.: 1988, *Physics Reports* **160**, 1.
Yoshii, Y. and Takahara, F.: 1988, *Astrophys. J.* **326**, 1.

PROBING THE ERA OF GALAXY FORMATION VIA TEV GAMMA RAY ABSORPTION BY THE NEAR INFRARED EXTRAGALACTIC BACKGROUND

D. MACMINN* and J. R. PRIMACK
Santa Cruz Institute for Particle Physics
University of California, Santa Cruz, CA 95064

February 20, 1995

Abstract. We present models of the extragalactic background light (EBL) based on several scenarios of galaxy formation and evolution. We have treated galaxy formation with the Press-Schecter approximation for both cold dark matter (CDM) and cold+hot dark matter (CHDM) models, representing a moderate ($z_f \sim 3$) and a late ($z_f \sim 1$) era of galaxy formation respectively. Galaxy evolution has been treated by considering a variety of stellar types, different initial mass functions and star formation histories, and with an accounting of dust absorption and emission. We find that the dominant factor influencing the EBL is the epoch of galaxy formation. A recently proposed method for observing the EBL utilizing the absorption of ~ 0.1 to 10 TeV gamma-rays from active galactic nuclei (AGN) is shown to be capable of discriminating between different galaxy formation epochs. The one AGN viewed in TeV light, Mrk 421, does show some evidence for a cutoff above 3 TeV; based on the EBL models presented here, we suggest that this is due to extinction in the source. The large absorption predicted at energies > 200 GeV for sources at $z > 0.5$ indicates that observations of TeV gamma-ray bursts (GRB) would constrain or eliminate models in which the GRB sources lie at cosmological distances.
Key words: diffuse radiation–galaxies: evolution–gamma rays:sources

1. Introduction

We have developed an indirect test of the era of galaxy formation by modeling the extragalactic background light (EBL) in the near infrared portion of the spectrum—a domain difficult to observe directly due to galactic and zodiacal contamination (Matsuura et al. 1994, Mattila 1990). A new method for probing this spectrum has been suggested (Stecker et al. 1992) which involves the absorption of TeV γ-rays by the EBL. TeV γ-rays have recently been detected from Markarian 421 (Punch et al. 1992), an AGN at a redshift of $z = 0.031$. If several more distant TeV sources exist, it should be possible to observe for the first time the EBL, and thus constrain the timescale over which galaxies form. The potential of this technique is illustrated by the constraints it places upon an earlier claim for a detection of the EBL (Matsumoto et al. 1988).

It has been known for many years that high energy γ-rays from sources at cosmological distances will be absorbed by a diffuse background of long wavelength photons (Gould & Schréder 1967) through electron-positron pair

* Now at the University of Chicago, Dept. of Astronomy & Astrophysics.

production. This point had been of little interest, however, until the discovery of a class of high energy γ-ray emitters by the EGRET detector aboard the Compton Gamma Ray Observatory (Lin et al. 1992). These sources, all AGNs of various subclasses, were found to emit a substantial amount of energy in the GeV energy domain with a roughly E^{-2} spectrum. While the physical description of the emission process is uncertain, it is possible that this spectrum could continue into the TeV range for some of these sources, a hypothesis given support by the detection of Mrk 421 at TeV energies. Mrk 421 is the nearest of the EGRET sources and should suffer from only limited EBL absorption of its γ-rays as we will show, but sources at greater distances should show evidence of absorption in the TeV domain which can serve as an indicator of the era of galaxy formtion.

2. Extragalactic Background Models

Models of the EBL require several inputs: a library of stellar spectra and a function describing the number distribution by mass of a stellar population (the initial mass function, IMF); a characterization of galactic classes (elliptical, spiral) and the evolution of their respective stellar populations (the star formation rate, SFR) and dust content; a function describing the formation of galaxies as a function of redshift; and assumptions about the geometry of the Universe—the density parameter (Ω_0), the cosmological constant (Λ) and the Hubble constant (H_0). Such models have been created before (Partridge & Peebles 1967; Yoshii & Takahara 1988; Franceschini et al. 1994) with a variety of different assumptions, but each has a perhaps oversimplified model of galaxy formation. The conclusion reached by these earlier modelers is that the EBL proves to be a poor test of cosmology, meaning geometry, as the uncertainties in the IMF and SFR overshadow differences due to Ω_0 and H_0. We concur that the EBL provides a poor test of geometry. But this by no means limits its usefulness as a probe of cosmology, for the dominant factor influencing the EBL is the era of galaxy formation, which in modern theories depends strongly on the nature of dark matter. For simplicity, we have chosen to consider here only those models with a flat ($k = 0$) geometry and $\Lambda = 0$, i.e. $\Omega = 1$; the arguments that the age of the Universe $t_0 \sim 13$ Gyr then require $H_0 \approx 50$ km s^{-1} Mpc^{-1}. Considering also $\Omega_0 < 1$, especially with $\Lambda > 0$, would tend to increase differences due to the galaxy formation epoch as such models form galaxies earlier than $\Omega = 1$ CDM and much earlier than CHDM. A larger value of H_0, as has been suggested by Freedman et al. (1994), would not change the central conclusion of this work as the effects of the era of galaxy formation on the EBL are largely independent of H_0. The specific dark matter models used to characterize this era would likely be supplanted by a new generation of models; likely too, new models would differ in their predictions for the era of formation, and it is here that this proposed test acquires its value as a model independent probe.

Observational work has suggested several models for the IMF; we have considered the full range predicted by various authors (Scalo 1986; Basu & Rana 1992). We have assumed the SFR to have the functional form of a decaying exponential in time (Bruzual & Charlot 1993), with two classes of galaxies considered: ellipticals, composing 28% of the total and with a characteristic star formation time scale of 0.5 Gyr, and spirals, 72%, with a characteristic time scale of 6 Gyr. The instantaneous recycling approximation has been used to model the evolution of the gas content and metallicity within each galaxy (Tinsley 1980). Dust absorption has been treated in a fashion similar to Guiderdoni & Rocca-Volmerange (1987); the dust emission spectrum is taken from Désert et al. (1990), and is based on a three component model corresponding to the PAH molecules ($\sim$ 10 to 30μm), warm dust from active star forming regions (30 to 70μm), and cold "cirrus" dust (70 to 1000μm). Galaxy formation has been treated with the Press-Schecter approximation, which expresses the number density of galaxy halos of a given mass as a function of redshift based on the results of N-body simulations (Klypin et al. 1993). Two different cosmological models for structure formation have been considered: the cold dark matter (CDM) model ($\Omega_{CDM} = 0.9$, $\Omega_B = 0.1$, bias of 1.5), representing a moderately early era of galaxy formation ($1 < z_f < 3$), and cold+hot dark matter (CHDM) ($\Omega_{CDM} = 0.6$, $\Omega_\nu = 0.3$, $\Omega_B = 0.1$), representing a late era of galaxy formation ($0.2 < z_f < 1$). The latter is normalized to COBE and both are based on the simulations of Klypin et al. (1993). There is a difficultly in describing a galactic population in this manner in that the Press-Schecter function merely predicts the abundance of dark matter halos and makes no distinction between a cluster of galaxies and a single galaxy. One must therefore also assume a maximum galactic mass; we set the upper limit at $M_{\mathbf{max}} = 5 \times 10^{12}$ $M_\odot$ based on a comparison between the galactic luminosities obtained for a power law mass to light ratio and the normalization process described below, and we have checked that the results do not depend sensitively on $M_{\mathbf{max}}$.

The final parameter to be set is the normalization of the SFR, or the efficiency of star formation for a given galactic type. This can be directly tied to the local galaxy population through a comparison of model galactic luminosities with those of the nearby population. A number of studies have been made of the galaxy local luminosity function (LLF). Most recent are the work of Loveday et al. (1992) based on the Stromlo-APM southern sky survey (LEPM LLF), the LLF of Efstathiou et al. (1988) based on an average of several surveys (EEP LLF), and the LLF of DeLapparent et al. (1989) based on the CfA survey (LGH LLF). Each has characterized the

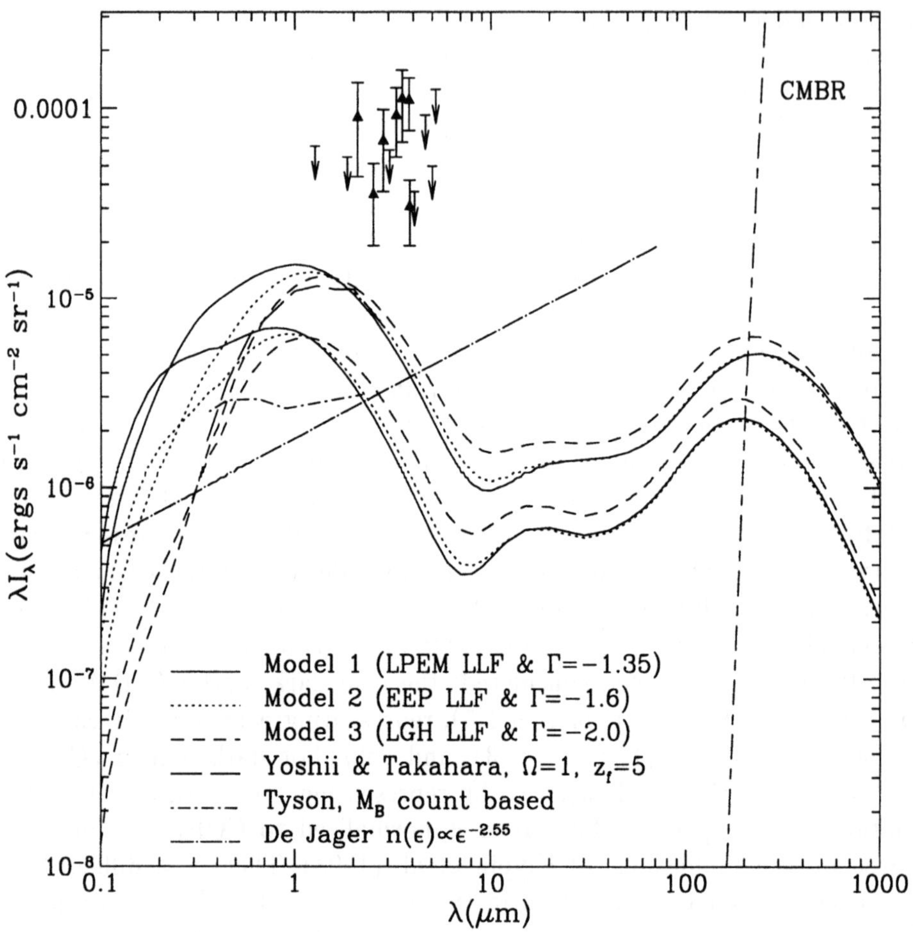

Fig. 1. The EBL at $z = 0$ for a variety of models. The upper set of curves is for the CDM model of galaxy formation, the lower set for CHDM. The number distribution of stars by mass was taken to be a power law of the form $N \propto m^{\Gamma-1}$; three values for Γ were considered for each galaxy formation model. Two of those presented here represent the extreme values for both CDM and CHDM galaxy formation models, with Model 1 representing the bluest spectra and Model 3 the reddest; Model 2 is midway between these extremes. The near vertical line at the right represents the CMBR. Data of Yoshii & Takahara (1988) is taken from their model with cosmological parameters similar to those used herein; the Tyson (1990) data is a direct estimate of the EBL based solely on number counts of galaxies. The DeJager et al. (1994) model is based on a power law EBL fit to the possible absorption of Mrk 421 γ-rays shown in Fig. 4. Observations (triangles) and upper limits (arrows) are from Matsumoto et al. (1988); a similar experiment failed to confirm this measurement (Matsuura et al. 1994).

luminosity distributions of the galaxies in the blue magnitude band as a Schecter function,

$$\phi(y)\,dy = \phi_\star y^\alpha e^{-y}\,dy \quad \text{where } y = L/L_{\star B}. \tag{1}$$

If one makes the assumption that the most massive galaxies are also the most luminous, one can compare the number densities of the LLF and the Press-Schecter function (at $z = 0$) and require that the average age model galaxies, at the present epoch, have the expected luminosities by deriving a function which expresses the SFR normalization as a function of galactic mass. The luminosity of the galaxies is thus normalized by forcing the model galaxies to duplicate the local population.

The model results are shown in Figure 1. Note the large separation between the galaxy formation model predictions, this will be seen for any combination of model parameters given that we include both early and late galaxy formation models and that we require that the present day model galaxies match the local population. The larger EBL flux predicted by the CDM model arises for several reasons: (1) stars have been contributing their light to the EBL for ~ 2 Gyr longer than in the CHDM case, (2) the initial burst of star formation in early-type galaxies has redshifted from the optical to the near IR, and (3) the galaxies are older at a given redshift and hence are composed of more evolved stars, producing a brighter flux in the red and near IR–see Figure 2. In addition to the parameter choices discussed above, we have also varied the other quantities not fixed by observational data to ensure that no large deviations were found.

There is some question as to the presence of a significant contribution from other sources than those considered here, particularly in the far IR ($\sim 100\mu$m) (although, as noted below, our primary interest is in the 1 to 10 μm region). The general nature of the galaxy models used herein is meant to represent both normal galaxies and the average properties of starburst and infrared luminous galaxies; the far IR flux predicted here is consistent with that of both Beichman & Helou (1991) and Hacking & Soifer (1991), each based on a luminosity function of extragalactic objects derived from the IRAS point source catalog and including quasar/AGN-like objects as well as infrared bright galaxies. Other sources not included in this catalog would have to be at high redshift or intrinsically dim; the former has no known population with which to be identified, and the latter would be of little effect. Given that we have not considered exotic IR producing mechanisms, however, the lowest curve in Figure 1 should be taken as a lower limit to the EBL.

3. TeV Gamma-Ray Absorption

The cross section for the absorption of γ-rays through e^+e^- pair production is given by Gould & Schréder (1967); the total absorption is found by assum-

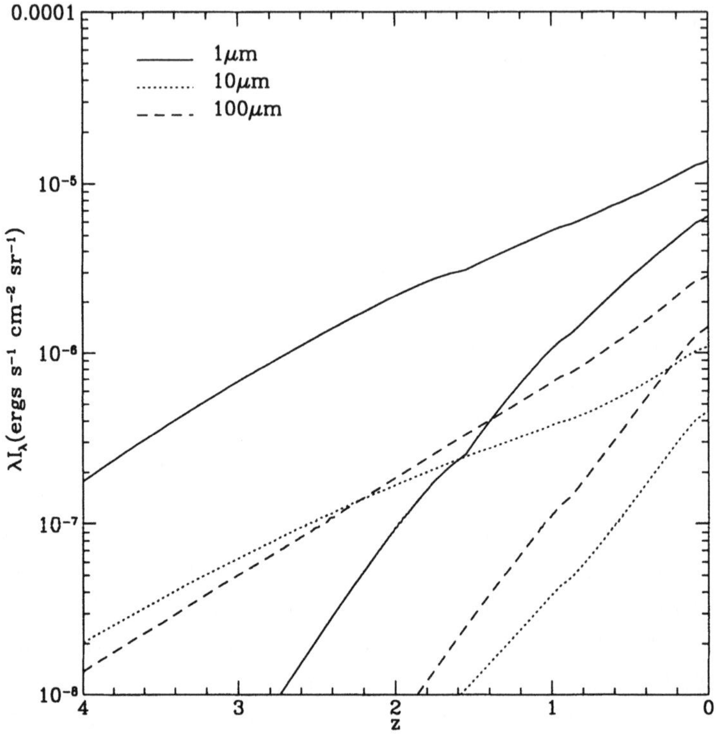

Fig. 2. The cumulative EBL flux at several different wavelengths plotted as a function of redshift of contributing sources. The upper three curves are for Model 2, CDM; the lower three for Model 2 CHDM. The flux at 1μm is dominated by starlight, that at 10μm by a mix of starlight and dust and at 100μm by dust. Note that $\sim$ 30% of the CDM model flux comes from sources at $z > 2$, and $\sim$ 50% from $z > 1.0$. CHDM predicts only $\sim$ 1% from $z > 2$, and $\sim$ 30% from $z > 0.7$.

ing an exponential attenuation of a source's flux where the optical depth is given by

$$\tau(E) = \frac{c}{H_0} \int_0^{z_s} \int_0^2 \int_{2m_e^2c^4/Ex(1+z)^2}^{\infty} \frac{x}{2}\sqrt{1+z}\, n(\epsilon, z)\sigma(xE'\epsilon')\, dz\, dx\, d\epsilon \quad (2)$$

where z_s is the redshift of the source; $E' = E(1+z)$ and $\epsilon' = \epsilon(1+z)$; $x = 1 - \cos\theta$, where θ is the encounter angle of the photons; $n(\epsilon, z)$ is given by the EBL models in Fig. 1; and σ is the pair production cross section. This cross section is maximized when

$$\epsilon \approx \frac{1}{3}\left(\frac{1\ \mathrm{TeV}}{E}\right)\ \mathrm{eV} \quad (3)$$

where ϵ is the EBL photon energy and E is the γ-ray energy. The photon number density for the EBL models drops off above 1 eV ($\sim$ 1 μm) and

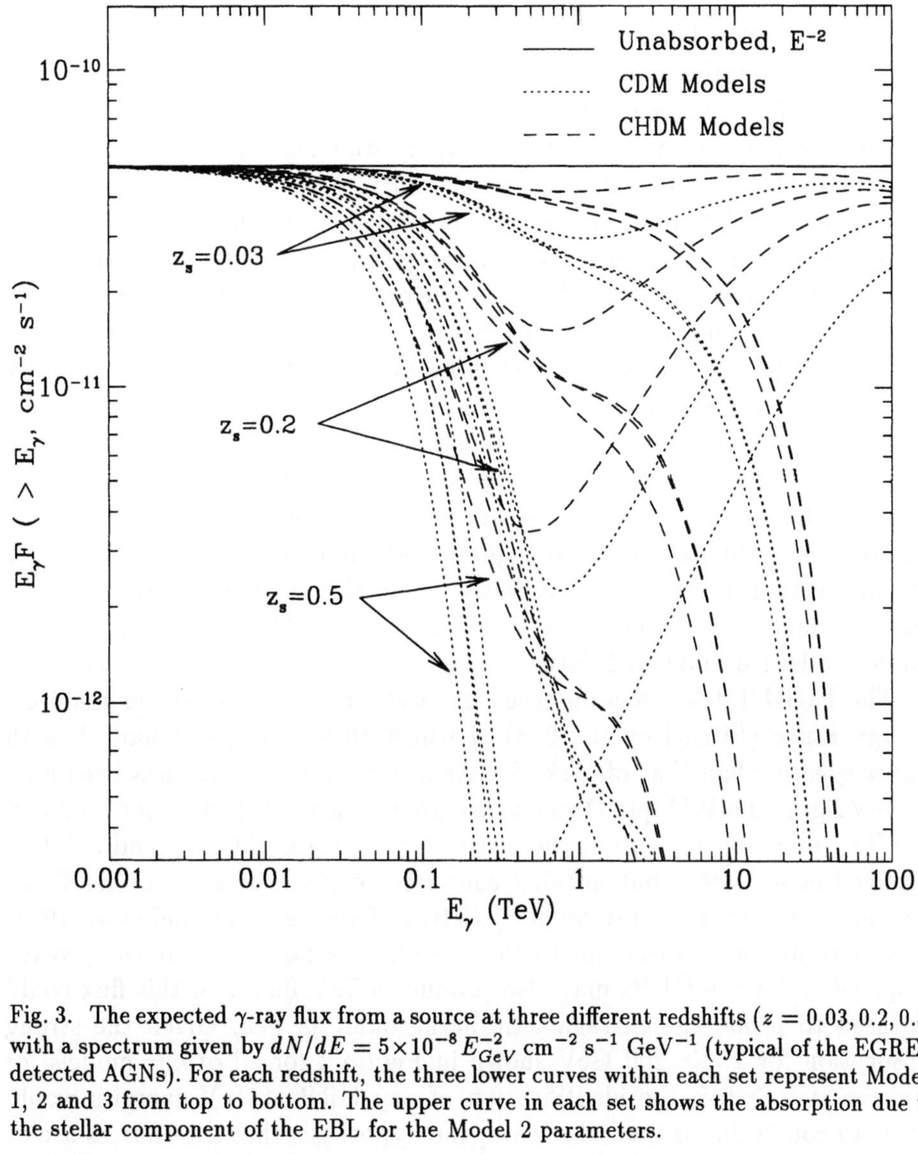

Fig. 3. The expected γ-ray flux from a source at three different redshifts ($z = 0.03, 0.2, 0.5$) with a spectrum given by $dN/dE = 5\times10^{-8} E_{GeV}^{-2}$ cm^{-2} s^{-1} GeV^{-1} (typical of the EGRET detected AGNs). For each redshift, the three lower curves within each set represent Models 1, 2 and 3 from top to bottom. The upper curve in each set shows the absorption due to the stellar component of the EBL for the Model 2 parameters.

below 0.1 eV; the absorption effect due to the stellar component of the EBL will be most pronounced for γ-ray energies of 300 GeV to 3 TeV—see Figure 3. Above 10 TeV, it is likely that the dust-emitted component of the EBL makes the Universe mostly opaque, and above 100 TeV the cosmic microwave background completely absorbs any γ-rays from sources at cosmological distances. The predicted absorption at $E > 10$ TeV is similar to that suggested by Stecker et al. (1992) based on their EBL models, particularly for sources at $z > 0.1$. There is also an effect due to the redshift of the source. The absorption at some redshift along the path from the source will occur between a background photon of energy $\epsilon_0(1+z)$ and a γ-ray of energy $E_0(1+z)$, hence the lower cutoff in a source's spectrum will scale with $\sim (1+z)^{-2}$. For sources at $z \geq 0.5$ there is an additional effect from the galaxy formation models in that the magnitude of the EBL at large z is quite small for cosmologies with late galaxy formation such as CHDM.

This last point, combined with the exponential dependence of γ-ray absorption upon the model predictions of the EBL, gives this test its potential power—provided that there are several sources which do possess a E^{-2} γ-ray spectrum stretching into the TeV range, and that there are instruments with the sensitivity to detect fluxes on the order of a few events per cm^2 day. The one source known to emit TeV γ-rays, Mrk 421, may already constrain an earlier measurement of a diffuse infrared emission component of the sky which was attributed to the EBL—see Figure 4. The Mrk 421 spectrum also shows some evidence for a cutoff above 2 TeV; the size and sharpness of this cutoff is too large to explain by absorption with our physically motivated EBL models and may instead represent absorption intrinsic to the source. This explanation would suggest that the mechanism responsible for the production of the γ-rays was based on an electron Comptonization process (Dermer & Schlickeiser 1993; Sikora et al. 1994) rather than a proton based model (Mannheim 1993).

The EGRET team has observed at least 31 AGNs to date in the GeV energy range (Fitchel et al. 1994) of which 10 are at $z < 1$ and all with fluxes greater than that of Mrk 421. As of yet, only Mrk 421 has been seen in TeV light; the Whipple Observatory group has published upper limits on the TeV emission for two of the nearer GeV sources (3C 273 and 3C 279) (Fennell et al. 1993), but nothing conclusive can yet be said. EGRET has also observed several γ-ray bursts (GRB) in GeV light (Fitchel et al. 1994) with a comparable spectrum to the AGNs. It is possible that the process responsible for the GRBs may also produce a TeV flux, and this flux could be used to probe the extragalactic background as well. Given the strong absorption for $E_\gamma \geq 200$ GeV shown in Figure 3 for all of our models for sources at $z \geq 0.5$, the identification of any GRB at TeV energies would serve to constrain, or even eliminate, the hypothesis that the sources are at cosmological distances (see also the paper by Stecker and DeJager in these proceedings). Note also that the smallest absorption predicted here for any source should be taken as an upper limit to the TeV flux.

Observations at 100 GeV, below the cutoff expected for any model of the EBL, should help clarify whether the spectrum of a given source shows intrinsic absorption. The Whipple telescope is currently sensitive down to ~ 500 GeV. There are a number of instruments planned (and some funded) which can perhaps achieve lower thresholds as well as several other detectors to complement Whipple, chiefly the GLAST satellite, sensitive from 10 MeV up to $\sim$100 GeV (GLAST workshop, 1994); the MILAGRO detector,

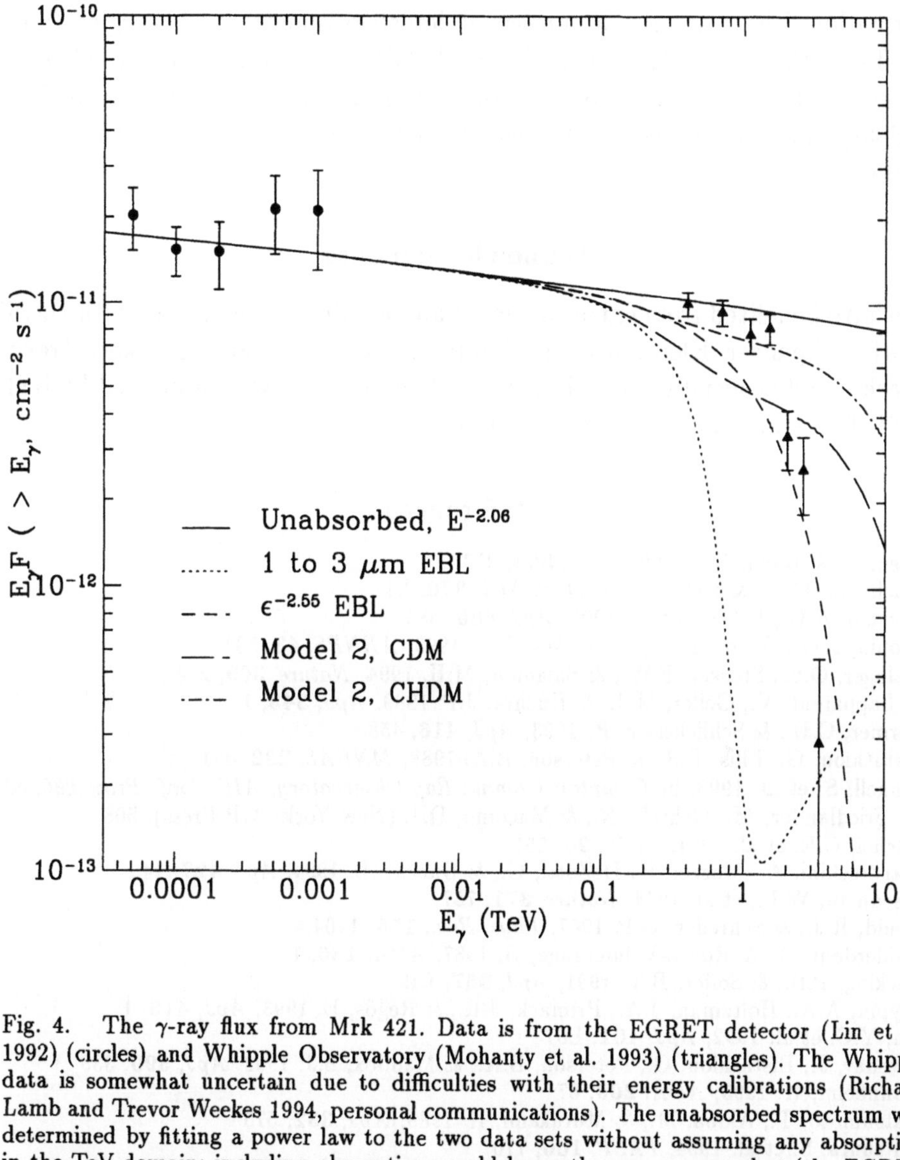

Fig. 4. The γ-ray flux from Mrk 421. Data is from the EGRET detector (Lin et al. 1992) (circles) and Whipple Observatory (Mohanty et al. 1993) (triangles). The Whipple data is somewhat uncertain due to difficulties with their energy calibrations (Richard Lamb and Trevor Weekes 1994, personal communications). The unabsorbed spectrum was determined by fitting a power law to the two data sets without assuming any absorption in the TeV domain; including absorption would lower the power somewhat (the EGRET data alone suggests an $E^{-1.96}$ spectrum) and bring both Model 2 curves into agreement with the first four Whipple points. The $\epsilon^{-2.55}$ is from the De Jager et al. (1994) model in Fig. 1. The 1 to 3 μm curve shows the absorption which would be expected if one interpets the measurement of Matsumoto et al. (1988) as a background of extragalactic origin; due to the sharp cutoff below 1 TeV, reasonable fits of the observed spectrum to this background are not possible.

operating above 300 GeV; and the Themis and Solar One arrays, operating at energies > 100 GeV (Rene Ong 1994, personal communication). If further sources are found to emit TeV γ-rays, the new detectors, combined with existing facilities, should be able to constrain many theories of galaxy formation. There are also a number of detectors operating at energies above 10 TeV—the CASA-MIA, CYGNUS, HEGRA, and TIBET arrays (Cronin et al. 1993); given the absorption of such γ-rays predicted in this work, it unlikely that such detectors will see *any* γ-rays from sources at cosmological distances.

Acknowledgements

We have benefited from encouragement and helpful conversations with David Dorfan, Spencer Klein, Richard Lamb, Rene Ong, Floyd Stecker, Trevor Weekes and especially Anatoly Klypin. This research was supported by NSF REU funds and a grant from the DOE at UCSC.

References

Basu, S., & Rana, N.C. 1992, *ApJ*, **393**, 373
Beichman, C.A., & Helou, G. 1991, it ApJ, **370**, L1
Bruzual, A.G., & Charlot, S. 1993, *ApJ*, **405**, 538
Cronin, J.W., Gibbs, K.G., & Weekes, T.C. 1993, *ARNPS*, **43**, 883
DeJager, O.C., Stecker, F.W., & Salamon, M.H. 1994, *Nature*, **369**, 294
de Lapparent, V., Geller, M.J., & Huchra, J.P. 1989, *ApJ*, **343**, 1
Dermer, C.D., & Schlickeiser, R. 1993, *ApJ*, **416**, 458
Efstathiou, G., Ellis, R.S., & Peterson, B.A. 1988, *MNRAS*, **232**, 431
Fennell, S. et al. 1993, in *Compton Gamma Ray Observatory, AIP Conf. Proc. 280*, ed. Friedlander, M., Gehrels, N., & Macomb, D.J. (New York: AIP Press), 508
Fitchel, C.E. et al. 1994, *ApJS*, **94**, 551
Franceschini, A., Mazzei, P., DeZotti, G., & Danese, L. 1994, *ApJ*, **427**, 140
Freedman, W.L., et al. 1994, *Nature*, **371**, 757
Gould, R.J., & Schréder, G.P. 1967, *Phys. Rev.*, **155**, 1404
Guiderdoni, B., & Rocca-Volmerange, B. 1987, *A&A*, **186**, 1
Hacking, P.B., & Soifer, B.T. 1991, *ApJ*, **367**, L49
Klypin, A.A., Holtzman, J.A., Primack, J.R., & Regös, E. 1993, *ApJ*, **416**, 1
Lin, Y.C. et al. 1992, *ApJ*, **401**, L61
Loveday, J., Efstathiou, G., Peterson, B.A., & Maddox, S.J. 1992, *ApJ*, **390**, 338
Mannheim, K. 1993, *A&A*, **269**, 67
Matsumoto, T., Akiba, M., & Murakami, H. 1988, *ApJ*, **332**, 575
Matsuura, S. et al. 1994, *PASP*, **106**, 770
Mattila, K. 1990, in *Galactic and Extragalactic Background Radiation, IAU Symposium Vol. 139*, ed. Bowyer, S., & Leinert, C. (Dordrecht:Kluwer Academic), 245
Mazzei, P., Xu, C., & DeZotti, G. 1992, *A&A*, **256**, 45
Mohanty, G. et al.1993, in *Proc. 23rd Int'l. Cosmic Ray Conf.*, 1, 440
Partridge, R.B., & Peebles, P.J.E. 1967, *ApJ*, **148**, 377
Punch, M. et al. 1992, *Nature*, **358**, 477
Scalo, J.M. 1986, *Fund. Cosmic. Phys.*, **11**, 1
Sikora, M., Begelman, M.C., & Rees, M.J. 1994, *ApJ*, **421**, 153
Stecker, F.W., DeJager, O.C., & Salamon, M.H. 1992, *ApJ*, **390**, L49
Tinsley, B. 1980, *Fund. Cosmic. Phys.*, **5**, 287
Tyson, J.A. 1990, in *Galactic and Extragalactic Background Radiation, IAU Symposium Vol. 139*, ed. Bowyer, S., & Leinert, C. (Dordrecht:Kluwer Academic), 245
Yoshii, Y., & Takahara, F. 1988, *ApJ*, **326**, 1

RECENT RESULTS FROM COMPTEL

HANS BLOEMEN
SRON-Utrecht *
Sorbonnelaan 2, 3584 CA Utrecht, The Netherlands

February 6, 1995

Abstract. The COMPTEL telescope aboard the Compton Gamma Ray Observatory has put MeV γ-rays into the midst of astronomy. Among recent highlights are the discovery of intense MeV emission from blazar-type active galactic nuclei, the surprising non-detection of any Seyfert galaxy at MeV energies, the first image of the Milky Way in the light of the ^{26}Al line at 1.809 MeV (possibly including a detection of the Vela supernova remnant), the discovery of 3–7 MeV emission from the Orion complex, which can be identified with nuclear interaction lines of ^{12}C and ^{16}O at 4.44 and 6.13 MeV, the detection of the ^{44}Ti line at 1.15 MeV from the supernova remnant Cas A, and the first results on the spectrum and propagation of low-energy (1–100 MeV) cosmic-ray electrons.

1. Introduction

COMPTEL is one of the four scientific instruments aboard NASA's Compton Gamma Ray Observatory, which has been in orbit since April 1991. It provides for the first time extensive imaging possibilities in the MeV range (about 0.75–30 MeV), which was largely unexplored hitherto. A description of the instrument characteristics is given by Schönfelder et al. (1993). COMPTEL has a location accuracy of typically 1° in a wide field-of-view of one steradian. With an energy resolution of 5–8%, it is capable of imaging in narrow energy bands, which has great potentials for γ-ray line studies. A broad range of research topics is addressed by COMPTEL. This paper summarizes recent highlights, of which some are merely first steps with exciting prospects. COMPTEL is a collaborative effort of four institutes: the Max-Planck Institut für extraterrestrische Physik in Munich (Germany), SRON-Utrecht (The Netherlands), the Astrophysics Division at ESTEC in Noordwijk (The Netherlands), and the Space Science Center at UNH (Durham, USA). The OSSE instrument on board the Compton Observatory operates at the low-energy side of COMPTEL. EGRET is operative at the high-energy side. The fourth instrument is the BATSE all-sky (burst) monitor.

2. Active Galactic Nuclei

Among the most surprising results of the Compton Observatory is the non-detection so far of any Seyfert galaxy above a few hundred keV (Maisack et al. 1995), which is in sharp contrast to expectations based on the intense

* also Leiden Observatory

H. Völk & F.A. Aharonian (eds.), TeV Gamma-Ray Astrophysics, 423–434.

MeV emission reported earlier for NGC 4151 and MCG 8-11-11 (Perotti et al. 1981ab). Equally fascinating is the discovery of intense, but highly variable, γ-ray emission from many blazar-type active galactic nuclei (AGNs) up to redshifts as large as $z \approx 2$; prior to the launch of the Compton Observatory, only the low-redshift quasar 3C273 was seen (by COS-B in the 100-MeV regime; Swanenburg et al. 1978). The fourth AGN previously detected at (MeV) γ-ray energies is the radio galaxy Cen-A, for which first COMPTEL results have been presented by Steinle et al. (1995).

Blazars are believed to be AGNs for which the outflowing plasma jets are closely aligned with our line-of-sight, often showing superluminal motion. This class of AGNs includes compact, radio loud, flat-spectrum radio sources, BL Lac objects, and optically violent variable (OVV) quasars. Over 40 high-latitude ($|b| > 10°$) sources seen by EGRET above 100 MeV can be identified with blazars (Hartman et al. 1994, Fichtel et al. 1994), of which a dozen is listed as marginal detections in the latest EGRET catalogue presented by Fichtel et al. Only a subset is seen by COMPTEL, as expected on the basis of a simple extrapolation of the EGRET (power-law) spectra (Hermsen et al. 1993a, Collmar et al. 1993, Blom et al. 1995, Williams et al. 1995). But the COMPTEL observations add a valuable finding, namely that the power-law spectra (ranging from $E^{-1.5}$ to $E^{-2.6}$) show a break (or possibly a bump) in the 1-10 MeV range. An example is shown in Figure 1. Because of the large field-of-views of EGRET and COMPTEL many blazar sources have been observed frequently, which has shown that several sources are variable down to a time scale of weeks, in a few cases even days. Another ~40 high-latitude sources seen by EGRET, of which most (~80%) are listed as marginal detections by Fichtel et al. (1994), has so far not been identified.

An interesting new development is that COMPTEL is now finding evidence for sources at high Galactic latitudes that can be associated with EGRET's marginal blazar detections or have no EGRET counterpart at all. Bloemen et al. (1995) have presented first evidence for a blazar-type AGN that is exceptionally bright at MeV energies. It is firmly detected by COMPTEL in the 1–10 MeV range, but only a weak ~3σ source is seen in the EGRET data, being the weakest source in the EGRET catalogue (see Figure 2). Within the 99% error region of this new COMPTEL source (GRO J0516-609), two potential AGN counterparts can be identified. These are the quasars PKS 0506-612 and PKS 0522-611, separated by ~ 2°, which are both compact, radio loud, flat-spectrum radio sources, like all AGNs detected at high-energy gamma rays. The former is the preferred counterpart by the weak EGRET detection, which has a smaller error region.

The radiated power per decade of energy (νF_ν) of GRO J0516-609 strongly peaks at ~3 MeV (Figure 3). If it has indeed a blazar origin, this strongly peaked νF_ν spectrum not only rules out a synchrotron self-Compton model for the γ-ray emission (no break in the γ-ray spectrum is expected), but

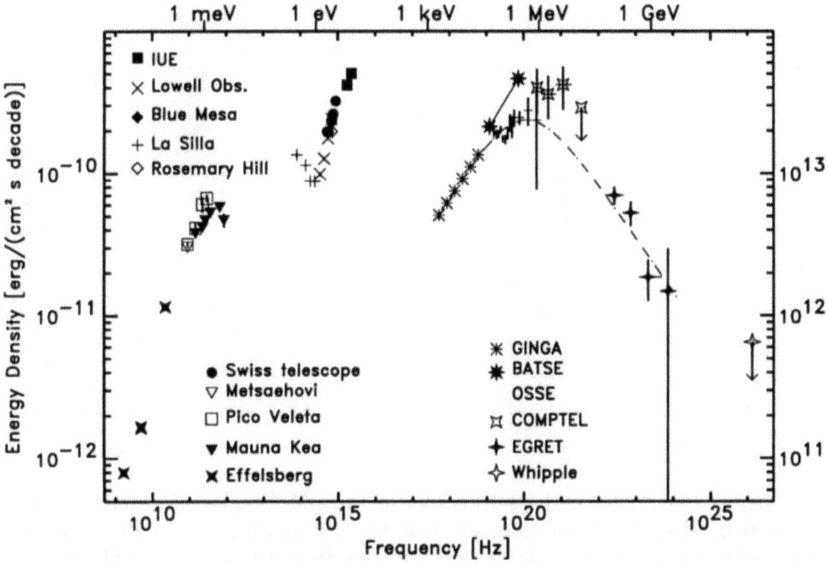

Fig. 1. Compilation of 3C273 observations, illustrating that the radiated power per decade of energy (νF_ν or differential flux spectrum multiplied by E^2) shows a peak in the COMPTEL MeV energy range. All data points were obtained during a two-month period (June-July 1991). *From Lichti et al. 1995.*

also tends to exclude a subset of the models that attribute the (MeV) γ-rays to Comptonization of an external radiation field. These are the models that ascribe the steepening of the spectrum in the MeV range to cooling of the high-energy portion of the relativistic electron distribution, implying a change in slope across the break of (only) ~0.5 (Dermer et al. 1992, Dermer & Schlickeiser 1993, Sikora, Begelman, and Rees 1994). The Comptonization model proposed by Blandford (1993) cannot be excluded, because the change in slope is not determined. Also, it should be emphasized that a Comptonization model may well explain the EGRET part of the spectrum. Alternative models that could account for the intense MeV emission of GRO J0516-609 are the proton initiated cascade model of Mannheim & Biermann (1992) and the $e^\pm$ annihilation model of Henri, Pelletier, and Roland (1993) (both in addition to inverse-Compton emission as considered above). In view of the stringent EGRET 30–100 MeV upper limit (which suggests an MeV bump rather than an MeV break), the model of Henri et al. seems to be preferable because it predicts a distinct spectral bump due to a broad blue shifted $e^\pm$ annihilation line that is Doppler boosted in a relativistic beam.

The break in blazar spectra at MeV energies approximately coincides with the well known MeV bump in the diffuse cosmic γ-ray background radiation, of which the origin is still unkown. "MeV blazars" like GRO J0516-609 would be particularly good candidates for this MeV bump (Bloemen et al. 1995).

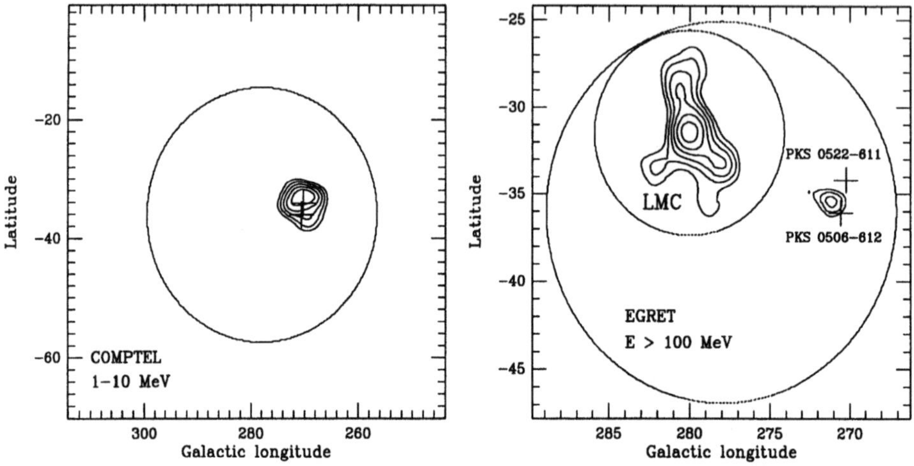

Fig. 2. *Left:* COMPTEL map (1-10 MeV) of the LMC region showing the newly discovered source GRO J0516-609 located near the blazar-type AGNs PKS 0506-612 and PKS 0522-611. There is no evidence for emission from the LMC. *Right:* EGRET map (> 100 MeV) of the LMC and PKS 0506-612/0522-611 region. The weak excess seen near the two quasars is at the 3σ level. The small circle includes additional data in order to show the LMC with best sensitivity. The map was smoothed for pictorial purposes. *From Bloemen et al. 1995.*

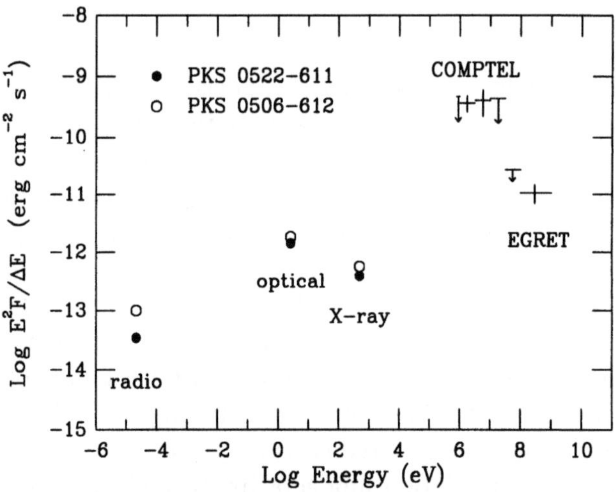

Fig. 3. Differential flux spectrum of GRO J0516-609 (see Fig. 2) multiplied by E^2 (i.e., radiated power per decade of energy). The radio, optical, and X-ray observations were not obtained simultaneously with the COMPTEL and EGRET observations. The upper limits are at the 2σ level. *From Bloemen et al. 1995.*

Padovani et al. (1993) and Salamon & Stecker (1994) have already shown that blazars may be responsible for at least a substantial fraction of the diffuse background in the 100-MeV regime.

3. Gamma-ray lines from recent nucleosynthesis

With a spectral resolution of 5–8% and a γ-ray line sensitivity of $\sim 10^{-5}$ photon cm^{-2} s^{-1}, COMPTEL has good potentials for studies of de-excitation lines from radionuclei, which provide direct constraints for nucleosynthesis modelling. Of particular interest are the lines from ^{22}Na (1.275 MeV), ^{26}Al (1.809 MeV), ^{44}Ti (1.156 MeV), and ^{56}Co (0.847, 1.238 MeV), for which the available observations are summarized below. Nuclear excitation due to energetic particle interactions is addressed in Section 4.

The 1.275 MeV line from the decay of ^{22}Na into ^{22}Ne (lifetime $\tau \simeq 3.8$ years) has so far not been detected from any astrophysical source. Explosive hydrogen burning in (neon-rich) novae (Clayton and Hoyle 1974) is considered to be the most prominent scenario for the production of ^{22}Na, but upper limits from HEAO-3, SMM, and OSSE set stringent limits on the ejected ^{22}Na mass (typically $< (1-5) \times 10^{-7} M_{\odot}$; e.g. Leising et al. 1993 and references therein). A COMPTEL study of recent novae is in progress.

The 1.809 MeV line emission from ^{26}Al ($\tau \simeq 10^6$ years), first observed from the central region of the Galaxy by HEAO-3 (Mahoney et al. 1984), has now for the first time been mapped along the Milky Way by COMPTEL (Diehl et al. 1994, 1995a). ^{26}Al can be produced at various sites (early predictions were already given by Ramaty and Lingenfelter 1977). Although the structured appearance of the ^{26}Al emission along the Milky Way seen by COMPTEL (Figure 4) clearly excludes a dominant contribution from (bulge) novae (Diehl et al. 1994, 1995a), alternative scenarios that have been proposed could so far not be distinguished. These scenarios largely invoke massive stars, as reviewed by Prantzos (1991). Apart from a possible detection of the nearby Vela supernova remnant (Diehl et al. 1994, 1995b), essentially all other localized emission regions that can be seen in Figure 4 must be due to an ensemble of sources. The intermittent nature of massive-star formation and the 1.809 MeV line, both on time scales of typically 10^6 years, may indeed explain the structured appearance of the 1.809-MeV Milky Way, but it remains to be seen whether Wolf-Rayet stars, supernova explosions, or possibly other sources are prime contributors.

The 1.156 MeV line from the decay of ^{44}Ti into ^{44}Ca ($\tau \simeq 80$ years) is not only of importance for nucleosynthesis studies, but also provides an interesting possibility to search for the remnants of obscured supernovae that have occured over the past few hundred years. The prime known candidate, Cas A, was indeed detected by COMPTEL (Iyudin et al. 1994), as illustrated in Figure 5. The observed flux of approximately 7×10^{-5} photon cm^{-2} s^{-1} translates into an ejected ^{44}Ti mass of $(1-4) \times 10^{-4}$ M$_{\odot}$, which is in the range of expected values. A search for other ^{44}Ti sources in the COMPTEL survey of the Milky Way is in progress.

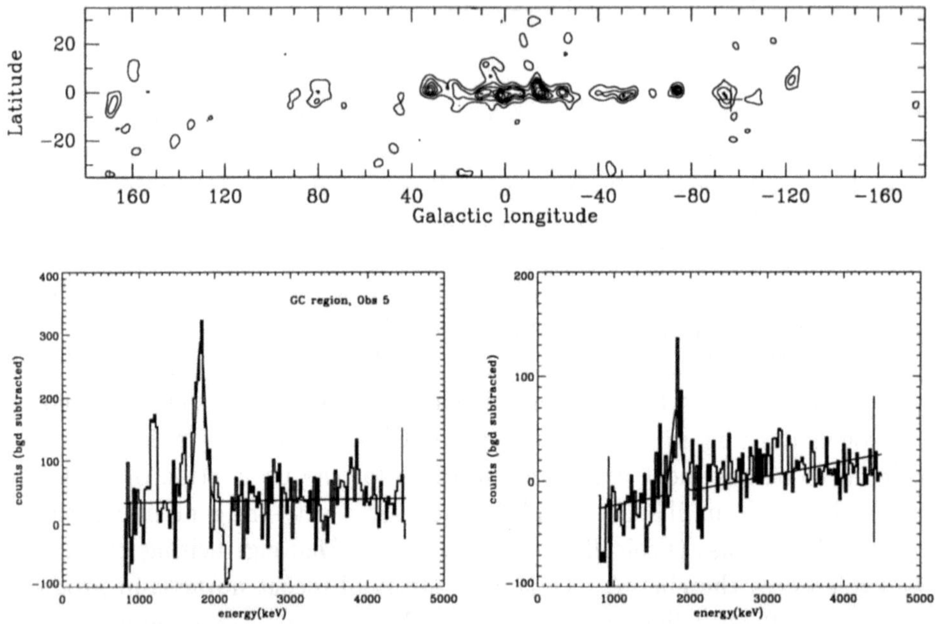

Fig. 4. *Top:* COMPTEL map of the Galactic ^{26}Al 1.809 MeV emission, showing evidence for localized emission regions along the Milky Way (including a possible detection of the Vela supernova remnant at $\ell \simeq -97°$). The continuum radiation is removed using a model derived from the observed emission at adjacent energies. *Bottom:* Examples of background-subtracted counts spectra, obtained from one single observation of the inner Galaxy (left) and a combination of available observations of the Vela region (right). Typical statistical error bars are shown. *From Diehl et al. 1995ab.*

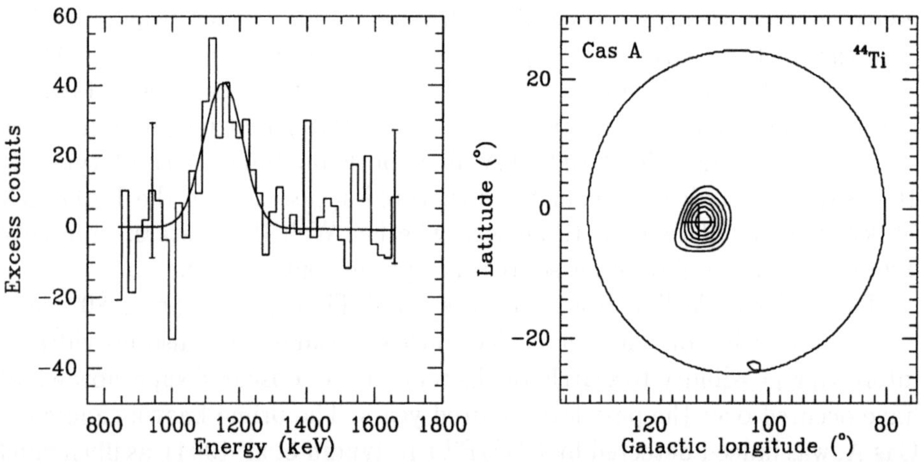

Fig. 5. Background-subtracted counts spectrum of the Cas-A supernova remnant, showing evidence for the 1.157 MeV line of ^{44}Ti, together with a map of the Cas-A region in the light of this line (the position of Cas-A is indicated). Typical statistical error bars are shown in the spectrum. *From Iyudin et al. 1994.*

The observation of γ-ray lines from the decay chain $^{56}Ni \rightarrow {}^{56}Co \rightarrow {}^{56}Fe$ ($\tau \simeq 0.31$ years) provides direct diagnostics of nucleosynthesis in supernova explosions (Clayton, Colgate, and Fishman 1969). So far, there have been two realistic opportunities to study the early phase of explosive nucleosynthesis, namely SN 1987A (Type II) in the LMC and SN 1991T (Type Ia) in the Virgo cluster. The ^{56}Co lines were seen from SN 1987A by several balloon experiments and extensively studied with SMM (Leising and Share 1990 and references therein). Four years after the explosion, when the Compton Observatory was launched, OSSE detected the subsequently expected 0.122 MeV line from $^{57}Co \rightarrow {}^{57}Fe$, with $\tau \simeq 1.1$ years (Kurfess et al. 1992). The line fluxes from supernovae of Type Ia, like SN 1991T, are expected to be much higher than from SN Type II. So far, however, no evidence for line emission from SN 1991T was reported by OSSE and COMPTEL. In view of the distance uncertainty this may not be surprising, but the lack of γ-ray lines is hard to reconcile with the high optical brightness (driven by ^{56}Co decay) at the time of the COMPTEL and OSSE observations.

4. Gamma-ray lines from nuclear interactions

Until recently, γ-ray lines resulting from accelerated particle interactions had been observed only from solar flares (e.g. Chupp 1990), namely the lines of ^{12}C at 4.44 MeV, of ^{16}O at 6.13 MeV, and of ^{20}Ne, ^{24}Mg, ^{28}Si, ^{32}S and ^{56}Fe in the 1–2 MeV range. COMPTEL has now detected emission in the 3–7 MeV range from the Orion complex, which was identified with the ^{12}C and ^{16}O de-excitation lines at 4.44 and 6.13 MeV (Bloemen et al. 1994a), as illustrated in Figure 6. In view of early theoretical predictions (Meneguzzi and Reeves 1975, Ramaty et al. 1979), this discovery came as a pleasant surprise, the observed flux (a total of $\sim 10^{-4}$ photon cm^{-2} s^{-1} in the two lines) being much larger than anticipated. So far, Orion has revealed only upper limits in the 1–3 MeV range.

Bykov and Bloemen (1994) have shown that these results may find a novel explanation in the strong overabundance of oxygen in the supernova ejecta from massive stars (Weaver and Woosley 1993, Woosley et al. 1993), in combination with the efficient acceleration of low-energy cosmic-ray nuclei (up to tens of MeV/nucleon) in regions with strong stellar winds and multiple supernova explosions (Bykov and Toptygin 1990, Bykov and Fleishman 1992). In this scenario, the γ-ray lines originate from energetic nuclei that are extracted from the enriched hot medium in regions of massive star formation and bombarding the ambient dense gas in clouds and shells. The width of nuclear interaction lines provides crucial information on the processes that occur, but the available COMPTEL observations of Orion are inconclusive. If produced predominantly by accelerated heavy nuclei interacting with ambient gas (as in the above scenario), the lines are significant-

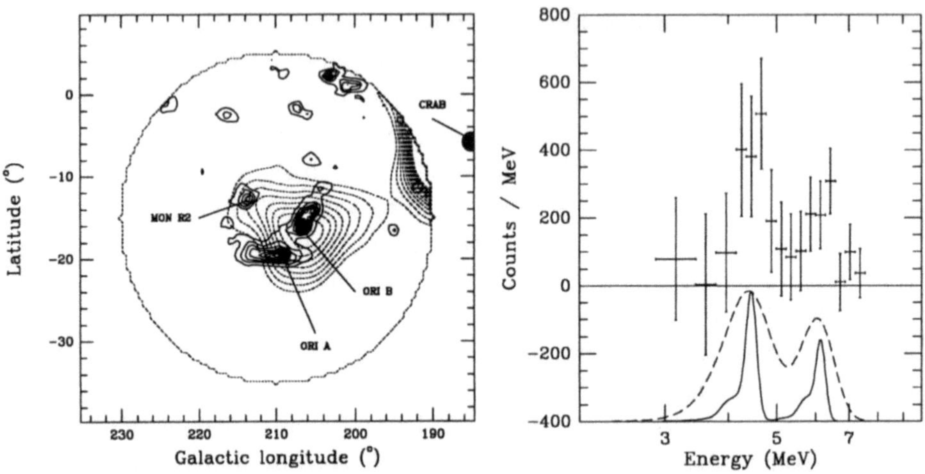

Fig. 6. COMPTEL map (dotted contours) and spectrum of the Orion Complex of molecular clouds and star-forming regions (the nearest birthplace of massive stars a a distance of $\sim$ 400 pc), showing evidence for 3–7 MeV emission that can be attributed to the 4.44 and 6.13 MeV de-excitation lines of carbon and oxygen following energetic nuclear interactions. The Columbia CO observations are superimposed for comparison (solid contours). The excess in the upper right originates from the Crab pulsar. *From Bloemen et al. 1994a.*

ly broader than the same lines produced by the inverse reactions involving accelerated protons and α-particles. Already scheduled further COMPTEL observations of the Orion complex may provide this important piece of information. Detailed model spectra of the Orion region for various compositions of the accelerated particles are presented by Ramaty et al. (1994).

The study of nuclear interaction lines has broad applications: it provides direct insight into particle acceleration (in fact the very origin of cosmic rays), into the composition and enrichment of the ISM in regions of massive star formation (with important implications for nucleosynthesis studies), and into the role of cosmic rays in the ISM in general (such as cosmic-ray ionization and heating). In addition, nuclear interactions may play an important role in the production of light elements, not resulting from ordinary nucleosynthesis (see Bykov, these proceedings).

5. Diffuse Galactic continuum emission

Preliminary broad-band images of the Milky Way obtained with COMPTEL (Bloemen et al. 1994b, Strong et al. 1995) show a structured ridge of emission along the Galactic equator. Although a considerable contribution from unresolved point sources cannot be excluded, a significant fraction of the observed emission is presumably of diffuse origin, due to interactions of cosmic-ray (CR) electrons with the interstellar gas (bremsstrahlung) and radiation fields

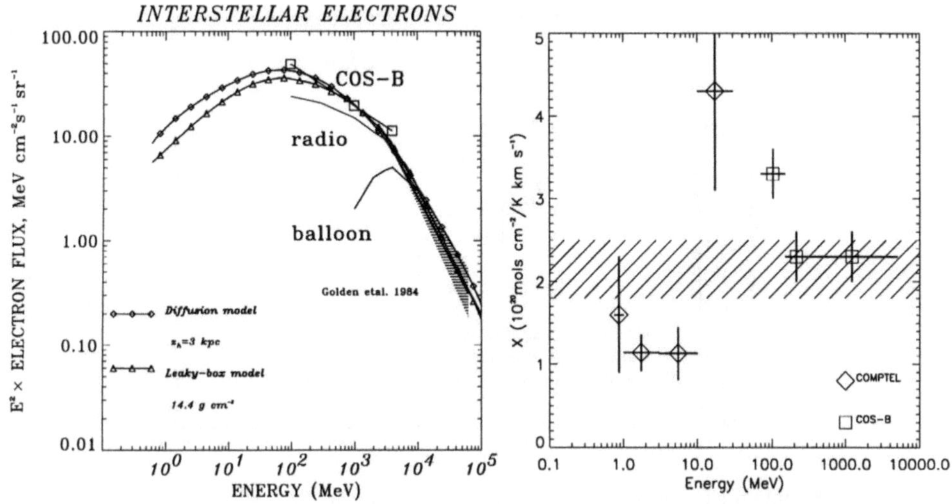

Fig. 7. *Left:* Interstellar electron spectrum, illustrating the unique information that is provided by COMPTEL at low energies ($\lesssim$ 100 MeV). Two propagation/confinement models were considered in extracting the electon spectrum from the COMPTEL (and COS-B) γ-ray emissivities, namely a leaky-box model with a grammage of 14.4 g cm^{-2} and a diffusion model including energy losses, with a diffusion coefficient of 3×10^{28} cm^2 s^{-1} ($\lesssim$ 4 GeV) and a halo half thickness of 3 kpc. *Right:* Values of the "CO-to-H_2 conversion factor" derived from COMPTEL and COS-B studies, showing evidence for an energy dependence of the relative contribution from molecular clouds to the total γ-ray emission from the disk, with a peak contribution at 10-100 MeV. *From Strong et al. 1995.*

(inverse Compton). Starting from the assumption that the emission seen by COMPTEL *is* of diffuse origin and due to electron bremsstrahlung, Strong et al. (1994, 1995) fitted a model based on HI and CO surveys of the interstellar gas distribution in order to derive the γ-ray emissivity spectrum. This provides unique information on the low-energy CR electron spectrum ($\lesssim$ 100 MeV), as shown in Figure 7.

It should be emphasized, however, that the observed emission and model show distinct differences (Bloemen et al. 1994b), unlike what is seen at higher γ-ray energies ($>$ 50 MeV, as reviewed by Bloemen 1989 and Strong 1994). This may be due to an important contribution from unresolved point sources, but it is certainly plausible that the structured appearance of the MeV Milky Way results from a very inhomogeneous distribution of the low-energy CR electrons, which is largely determined by the distribution of the CR sources and the characteristics of the ambient interstellar medium. A closely related and very interesting preliminary finding from the analysis by Strong et al. is the indication for an energy dependence of the relative contribution from molecular clouds to the total γ-ray emission from the disk, with a peak contribution at 10-100 MeV (Figure 7). If confirmed in further analyses, this may have important implications for our understanding of the acceleration and propagation of CR electrons in/near molecular clouds.

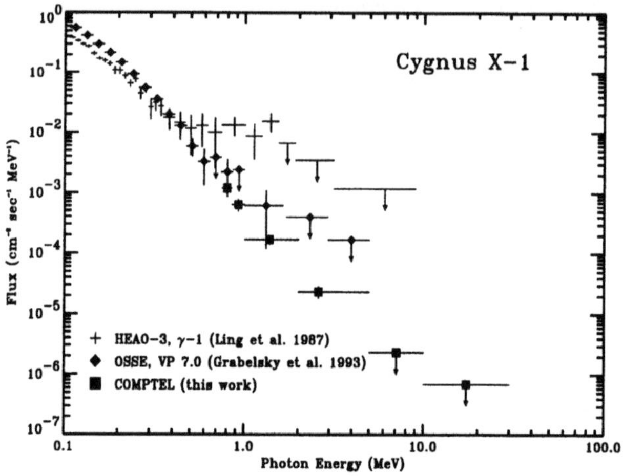

Fig. 8. COMPTEL spectrum of Cygnus X-1 incorporating all available observations (obtained in the period June 1991 to February 1994), together with representative spectra from HEAO-3 and OSSE. A drastic upturn in the spectrum at MeV energies (as seen in the HEAO-3 spectrum shown here) has so far not been seen by COMPTEL nor OSSE for any black-hole candidate. *From McConnell et al. 1995.*

6. Gamma-ray emission from Galactic compact objects

Only very few distinct γ-ray point sources in the Galactic disk have so far been recognized by COMPTEL. Apart from pulsars, uniquely identified through their timing signatures, the study of most other source candidates is hampered by our limited knowledge of the diffuse emission from the disk, in particular in the inner Galaxy.

Pulsed γ-ray emission of six pulsars has been reported now. In addition to the previously known γ-ray emission from the Crab and Vela pulsars, four others are seen by the Compton Observatory, namely Geminga, PSR B1509-58, PSR B1706-44, and PSR B1055-52. Very different properties in terms of energy spectra and light curves are found. Only Crab is seen by all four GRO instruments (see e.g. Ulmer 1994 and references therein). PSR B1706-44, and PSR B1055-52 have very hard spectra and are only observed by EGRET at energies above a few hundred MeV. The spectra of Vela and Geminga (the strongest sources in the EGRET energy regime, in addition to Crab) break around 100 MeV, with evidence for interesting phase dependencies. Nevertheless, Vela is clearly detected by COMPTEL but only a hint of Geminga is seen (Hermsen et al. 1993b, 1994). In contrast, PSR B1509-58 is mainly visible in the hard X-ray and low-energy γ-ray domain and not seen by COMPTEL beyond about 1 MeV (Bennett et al. 1993, 1994). A search for other pulsars in the COMPTEL data, using $\dot{E}/d^2$ as a measure of detectability, is negative so far (Bennett et al. 1993, 1994). Spectral analyses of the Crab and Vela pulsars as a function of phase are in progress. Such analyses (including the study of time variability) are a must to make progress in our understanding of the γ-ray emission from pulsars.

Black-hole candidates form another category of potential (MeV) γ-ray sources. So far, however, there is no indication from the OSSE and COMPTEL data for a drastic upturn in the spectra of black-hole candidates around 1 MeV, as found sporadically in previous observations. Two X-ray binaries have so far been seen by COMPTEL, Cygnus X-1 and Nova Persei. Cygnus X-1 is steadily detected up to several MeV (McConnell et al. 1994, 1995), but the flux is far below the HEAO-3 observations presented by Ling et al. (1987), as shown in Figure 8. Nova Persei was seen up to $\sim$ 2 MeV during a flare in the Summer of 1992 (van Dijk et al. 1994). Although both Cygnus X-1 and Nova Persei do not show evidence for excessive MeV emission, single temperature Wien spectra that fit the hard X-ray emisson quite well fall far below the MeV measurements. Further spectral studies and model fitting are in progress.

In conclusion, it may be clear from the above that MeV γ-ray astronomy has become an extremely rich field of research. This paper has only given a flavour of the ongoing work and results, with some findings not addressed at all (in particular for solar flares and γ-ray bursts). With an expected mission lifetime of 6–10 years, the COMPTEL telescope aboard the Compton Observatory will provide further insight into the MeV γ-ray sky for several years to come.

Acknowledgements

I am grateful to all colleagues who have made the COMPTEL mission such a great success and made this work possible. The COMPTEL project is supported by NASA under contract NAS5-26645, by the German goverment through DARA grant 50 QV 90968, and by the Netherlands Organisation for Scientific Research NWO.

References

Bennett, K., et al.: 1993, *Proc. 23rd Int. Cosmic Ray Conf. (Calgary)*, Vol. 1, p. 172
Bennett, K., et al.: 1994, *Astrophys. J. Suppl. Ser.* **90**, 823
Blandford, R.D.: 1993, *Proc. First Compton Symposium*, eds. Friedlander et al., AIP Con. Proc. 280, p. 533
Bloemen, H.: 1989, *Ann. Rev. Astron. Astrophys.* **27**, 469
Bloemen, H., et al.: 1994a, *Astron. Astrophys.* **281**, L5
Bloemen, H., et al.: 1994b, *Astrophys. J. Suppl. Ser.* **92**, 419
Bloemen, H., et al.: 1995, *Astron. Astrophys.* **293**, L1
Blom, J.J., et al.: 1995, *Astron. Astophys.*, in press
Bykov, A.M., Bloemen, H.: 1994, *Astron. Astrophys.* **283**, L1
Bykov, A.M., and Toptygin, I.N.: 1990, *Sov. Phys. JETP* **71**, 702
Bykov, A.M., Fleishman, G.D.: 1992, *Mon. Notices Royal Astron. Soc.* **255**, 269

Chupp, E.L.: 1990, *Physica Scripta* **T18**, 15
Clayton, D.D., Colgate, S.A., and Fishman, G.J.: 1969, *Astrophys. J.* **155**, 75
Clayton, D.D., Hoyle, F.: 1974, *Astrophys. J.* **187**, L101
Collmar, W., et al.: 1993, *Proc. 23rd Int. Cosmic Ray Conf. (Calgary)*, Vol. 1, p. 168
Dermer, C.D., Schlickeiser, R., Mastichiadis, A.: 1992, *Astron. Astrophys.* **256**, L27
Dermer, C.D, Schlickeiser, R.: 1993, *Astrophys. J.* **416**, 458
Diehl, R., et al.: 1994, *Proc. Second Compton Symposium*, eds. Fichtel et al., AIP Conf. Proc. 304, p. 147
Diehl, R., et al.: 1995a, *Astron. Astrophys.* , in press
Diehl, R., et al.: 1995b, *Astron. Astrophys.* , submitted
van Dijk, R., et al.: 1994, *Proc. Second Compton Symposium*, eds. Fichtel et al., AIP Conf. Proc. 304, p. 197
Fichtel, C.E., et al.: 1994, *Astrophys. J. Suppl. Ser.* **94**, 551
Hartman, R.C. et al.: 1994, *Proc. Second Compton Symposium*, eds. Fichtel et al., AIP Conf. Proc. 304, p. 563
Henri, G., Pelletier, G., Roland, J.: 1993, *Astrophys. J.* **404**, L41
Hermsen, W., et al.: 1993a, *Astron. Astrophys. Suppl. Ser.* **97**, 97
Hermsen, W., et al.: 1993b, *Proc. 23rd Int. Cosmic Ray Conf. (Calgary)*, Vol. 1, p. 176
Hermsen, W., et al.: 1994, *Astrophys. J. Suppl. Ser.* **92**, 559
Iyudin, A., et al.: 1994, *Astron. Astrophys.* **284**, L1
Kurfess, J.D., et al.: 1992, *Astrophys. J.* **399**, L137
Leising, M.D., et al.: 1993, *Proc. First Compton Symposium*, eds. Friedlander et al., AIP Con. Proc. 280, p. 137
Leising, M.D., and Share, G.H.: 1990, *Astrophys. J.* **357**, 638
Lichti, G.G., et al.: 1995, *Astron. Astrophys.* , in press
Ling, J.C., et al.: 1987, *Astrophys. J.* **321**, L117
Mahoney, W.A., et al.: 1984, *Astrophys. J.* **286**, 578
Maisack, M., et al.: 1995, *Astron. Astrophys.* , in press
Mannheim, K., Biermann, P.: 1992, *Astron. Astrophys.* **53**, L21
McConnell, M., et al.: 1994, *Astrophys. J.* **424**, 933
McConnell, M., et al.: 1995, in *Gamma-Ray Astronomy* (COSPAR Symp. Hamburg), *Adv. Space Res.*, in press
Meneguzzi, M., Reeves, H.: 1975, *Astron. Astrophys.* **40**, 91
Padovani, P., et al.: 1993, *Monthly Notices Roy. Astron. Soc.* **260**, L21
Perotti, F., et al.: 1981a, *Astrophys. J.* **247**, L63
Perotti, F., et al.: 1981b, *Nature* **292**, 133
Prantzos, N.: 1991, in *Gamma-Ray Line Astrophysics*, AIP Conf. Proc. 232, p. 129
Ramaty, R., Lingenfelter, R.E.: 1977, *Astrophys. J.* **213**, L5
Ramaty, R., Kozlovsky, B., Lingenfelter, R.E: 1979, *Astrophys. J. Suppl. Ser.* **40**, 487
Ramaty, R., Kozlovsky, B., Lingenfelter, R.E: 1994, *Astrophys. J. Letters* , in press
Salamon, M.H., Stecker, F.W.: 1994, *Astrophys. J.* **430**, L21
Sikora, M., Begelman, M.C., Rees, M.J.: 1994, *Astrophys. J.* **421**, 153
Schönfelder, V., et al.: 1993, *Astrophys. J. Suppl. Ser.* **86**, 657
Steinle, H., et al.: 1995, in *Gamma-Ray Astronomy* (COSPAR Symp. Hamburg), *Adv. Space Res.*, in press
Strong, A.W.: 1994, *Proc. Second Compton Symposium*, eds. Fichtel et al., AIP Conf. Proc. 304, p. 463
Strong, A.W., et al.: 1994, *Astrophys. J. Suppl. Ser.* **92**, 425
Strong, A.W., et al.: 1995, *Astron. Astrophys.* , in press
Swanenburg, B.N., et al.: 1978, *Nature* **275**, 298
Ulmer, M.: 1994, *Astrophys. J. Suppl. Ser.* **90**, 789
Weaver, T.A., Woosley, S.E.: 1993, *Physics Reports* **227**, 65
Williams, O.R., et al.: 1995, *Astron. Astrophys.* , in press
Woosley, S.E., Langer, N., Weaver, T.A.: 1993, *Astrophys. J.* **411**, 823

The cosmic microwave background

M. Lachièze-Rey
Service d'Astrophysique, Centre d'études de saclay,
F-91191 Gif sur Yvette Cedex, France

April 16, 1995

Abstract

I briefly recall the main properties of the Cosmic Microwave Background. In this very short presentation, I do not go into the details which may be found in the recent and excellent review papers by White, Scott & Silk (1994) and Bond (1995). At the light of the most recent observational results, the CMB appears to confirm very well the big bang models. I also introduce some characteristics which go beyond the global, homogeneous and isotropic, cosmology and concern the formation of large scales structures in the Universe.

1 Introduction

In july 1965, the radioastronomers Arno Penzias et Robert Wilson discovered an electromagnetic radiation which was diffuse, isotropic, unpolarized, and apparently suffered no seasonal (or any detected) variation. This discovery was reported in the Astrophysical Journal.

H. Völk & F.A. Aharonian (eds.), TeV Gamma-Ray Astrophysics, 435-450.

In the same issue, Robert Dicke, Jim Peebles, Peter Roll et Donald Wilkinson, from the Institute of Advanced Studies at Princeton, suggested that this was the radiation emitted by the universe as a whole, during a primordial very dense and hot phase. This had been in fact predicted about 15 years before by Gamov, Alpher et Herman, in the frame of the big–bang models. This discovery was considered, and still remains now, a very strong argument in favor of the big–bang models, initially introduced by Georges Lemaître under the name "Atome Primitif". Penzias et Wilson got the physics Nobel price in 1978.

JThe discovery was partly due to chance. In 1959, the NASA launched the ECHO satellite, the weak expected signal of which required a very sensitive receptor. The Bell Company, for the purpose of communication with this satellite, had constructed a horn antenna, 3 m diameter, at Holmdel, 50 km from Princeton. Penzias and Wilson intended to use it to observe astronomical sources with low angular diameter, for which it was the most sensitive radiotelescope at the time. Its small dimensions and high directivity allowed a precise measurement of the gain, and the identification of all contributions to the noise, the conditions for absolute measurements. Penzias and Wilson used the antenna in 1963, with a MASER receptor, tuned to the wavelength 7,3 cm (4.08 GHz). They tried to calibrate this instrument to use it as a radiotelescope. The first observations revealed fluxes higher than expected, that could not be explained by the estimated contributions from sky, antenna, wave guide, or any piece of the instrument, nor by atmospheric, human, or astronomical sources.

During the same period, Robert Dicke (director of the IAS, Princeton) and Jim Peebles had calculated that the Universe should be filled of such radiation (also predicted by Georges Gamov in 1949, and by Doroshkevich and Novikov in 1964). In the same time, Roll and Wilkinson had began to conceive a radiometer for its detection. Finally, the Princeton group recognized soon the nature of the detected radiation. After these first observations, the spectral properties of this radiation were explored : further measurements confirmed soon its thermal nature. In the middle of 1966, the spectrum was established to be thermal from 2.6 mm to 21 cm. This is now confirmed with a very good precision by the FIRAS instrument aboard the *COBE* satellite. I will present here a short overview of the CMB. A recent and excellent review has been written by White, Scott & Silk (*Annu. Rev. Astron. Astrophys. 1994. 32: 319–70*).

2 Cosmological origin of the CMB

Observations strongly suggest a remote and universal origin for the CMB :

- Diffuse and extremely isotropic, it cannot be due to the contributions of unresolved sources, which would imply some granularity, which is not observed. It must be due to a diffuse medium.

- The observation of the JSunyaev–Zeldovich effect, i.e., interactions with the gas of clusters of galaxies, implies an origin of the CMB further away than these clusters, extragalactic and very remote. CMB shows a global anisotropy with a very characteristic dipolar dependence. It is interpreted as due to the Doppler effect caused by our motion with respect to the last scattering surface, the source of the CMB. The good fit with the convergence direction of galactic proper velocities implies an origin further than a few 10 Mpc for the CMB.

- The thermal spectrum, now confirmed with a very good precision, reveals a global past equilibrium state. No plausible source other than the whole universe in its distant past has been convincingly proposed.

- Optical observations of molecular transitions have suggested that the universe was filled with hotter radiation in the past (Songaila et al., 1994). The big–bang models predict a temperature decrease $T(z) \propto 1 + z$, as observed.

Predictions from the big–bang models The big–bang models are characterized by an Jexpansion law $R(t)$, approximatively proportional to $t^{2/3}$ in the last 10 billion years (matter dominated Universe). Cosmic distances grow $\propto R(t)$, and cosmic volumes $\propto R^3(t)$. The energy density of Jmatter dilutes $\propto R^{-3}(t)$, that of radiation $\propto R^{-4}(t)$. The radiation temperature decreases $\propto R^{-1}(t)$.

Table 1: Chronology for the CMB

$z > 3\ 10^6$	Bremsstrahlung & double Compton diffusion create the CMB photons with a black–body spectrum.
$3\ 10^6 > z > z_y \approx 10^5$	No distortion expected in big–bang models. Kinetic equilibrium is insured : any photon excess leads to Bose–Einstein spectrum with $\mu \neq 0$.
$z_y > z > z_{rec} \approx 10^4$	No distortion expected in big–bang models. Compton diffusion onto hot electrons would distort the spectrum, with a characteristic shape.
$z > z_{rec}$	No more matter–radiation interaction : the black-body spectrum is preserved with $T \propto 1 + z$.

Table 2: The average CMB characteristics

Temperature	T=2.726K	$\propto 1+z$
Energy density	$\rho \approx 4\ 10^{-13}$erg cm^{-3} $\approx 4.4\ 10^{-34}$g cm^{-3}	$\propto (1+z)^4$
Photon density	$n \approx\ 400$cm^{-3}	$\propto (1+z)^3$
Average photon energy	$\varepsilon \approx 10^{-15}$ erg $\approx 6.2\ 10^{-4}$ eV	$\propto 1+z$

The recombination is defined as the moment when hydrogen ceased to be ionized, at a temperature $T_{rec} \approx 4\,000$ K. Before it, electrons and photons where thermally and dynamically coupled, with their two populations in mutual thermodynamical equilibrium. The table 1 details some pre–recombination processes : the equilibrium was created before z_{BE} and not destroyed after. After the recombination, the CMB propagated freely in the Universe : it remained isotropic, conserved its thermal spectrum with an adiabatic decrease of its temperature $T \propto 1+z$ J; this dependence has been recently confirmed by molecular observations. The table 2 gives the main (present) average characteristics of the CMB. At recombination, matter and radiation decoupled *dynamically* : matter became free to condense and to initiate the process of galaxy and large scale structures formation. On the other hand the *thermalJ* decoupling occurred later. The baryonic matter conserved a thermal spectrum for some time, with the same temperature than radiation. Finally the thermal decoupling became total, and matter became cooler than radiation, with $T \propto (1+z)^2$.

The Universe has however suffered reionisation later after the recombination. A sufficiently important optical depth for photon scattering (as predicted in some specific models for galaxy formation) would have generated a second recombination, so that the last scattering surface, from which originates the CMB would be at $z < 1000$.

3 Beyond isothermality and isotropy

The CMB is thermal and isotropic with a very high precision. But the details of the cosmic models predict that it may have suffered some distortions, at the moment of the recombination, or later. These distortions can be global or local. *Global* distortions do not break the isotropy of the CMB and modify only its spectral shape, independently of the direction in the sky. They are characterized by their spectral shape. On the other hand, *local* distortions, i.e., anisotropies, are imprinted in specific directions in the sky, by local processes. They are characterized by their angular scale θ. They can be thermal – temperature variations ΔT – or not.

It is also usual to distinguish *intrinsic* and *secondary* distortions. Intrinsic distortions were imprinted at the recombination or before.

Secondary ones were imprinted after. The table 3 gives an overview of the main types of possible distortions. Different kinds are expected, depending on the cosmological model adopted, and on the cosmogonical scenario describing galaxy and large scale structure formation.

CMB anisotropies from cosmogonic models

It is generally believed that galaxies and large scale structures form through the process of gravitational instability. Very small density fluctuations $\delta \equiv \delta\rho / < \rho > \approx 10^{-6}$ were created very early in the primordial universe, and then have condensed under their own gravitational attraction. Different models have been constructed according to this general scheme. They differ by

- The moment and process of appearance of these primordial fluctuations: already present at the Planck era (as initial conditions for the big–bang models), quantum fluctuations occurring during an inflationary phase, topological defects (cosmic strings) occurring during a cosmic phase transition, ... have been considered.
- The nature of these initial fluctuations : *isocurvature* (or isothermal) fluctuations concerning (almost) only matter ; or *adiabatic* ones, concerning equally both matter and radiation.
- The nature of the matter present in the universe : *baryonic* only ; or with hypothetic *non baryonic* component(s), classified as *cold* (cdm), *hot*, or *warm*.
- The values of the cosmological parameters: the Hubble constant

Table 3: CMB distortions

INTRINSIC (imprinted at recombination)	global	non thermal	Compton (y) or chemical potential μ (not predicted in big–bang models)
	local	all thermal	• At largest angular scales : primordial gravitational waves & potential fluct. (Sachs–Wolfe) • Doppler effect (matter velocities): (at medium angular scales) • Adiabatic fluctuations (dominant at small scales)
		But :	Depth of last scattering surface washes out fluctuations at smaller angular scales
DIPOLE	concerns whole sky	purely thermal	Earth motion with respect to last scattering surface
AFTER DECOUPLING	global	thermal	not recognizable
		only non thermal	Compton distortions (y)
	local	thermal	gravitational (not detectable)
		non thermal	Sunyaev–Zeldovich or others (y)
other contributions			Unresolved extragalactic sources Discrete sources (galaxies,...) The Galaxy Solar System, Earth, ...

H_0, the deceleration parameter q_0 and the cosmological constant Λ.

- The statistical properties of the initial fluctuations : gaussian or not, and their distribution as a function of the spatial scale, expressed by the *power spectrum*.

In principle, these characteristics specify entirely the fluctuations and allow to define a cosmogonical scenario. According to their nature, fluctuations may experience a first relativistic amplification, then freeze out or suffer damping, up to the recombination. At this moment, they imprint their mark on the CMB, under the form of intrinsic thermal distortions (see table 3). Then a new phase of amplification begins.

Distortions after recombination

Global distortions may have been imprinted onto the CMB by encounters with hot electrons, more or less uniformly distributed in a diffuse intergalactic medium. The resulting comptonisation would leave a specific non thermal signature in the spectrum, expressed by a parameter y. The FIRAS instrument aboard *COBE* provides the observational limit $y < 2.5\ 10^{-5}$, which constrains severely the amount of diffuse intergalactic ionized gas, and rules out the possibility that such diffuse gas would be the source of the observed X-ray background, as it was proposed earlier.

Electromagnetic waves, which follow the geodesics of a spacetime, are affected by the fluctuations of the gravitational potential, resulting from the non linear formation of large scale structures, much after the recombination. The calculations show however that the resulting effects are presently below the detectability limits.

We know for certain that clouds of intergalactic ionized gas exist. Thus the CMB photons *locally* suffer Compton diffusions onto the free electrons, when they propagate through such zones. The well known Sunyaev–Zeldovich effect corresponds to the crossing of the hot X–ray emitting gas in clusters of galaxies : the thermal electrons communicate part of their energy to the photons and modify their spectrum. The *kinetic* Sunyaev–Zeldovich effect is due to the bulk velocity of the gas cloud, i.e., the proper velocity of the cluster. Also, the thermal velocities of the electrons imprint a *thermal* Sunyaev–Zeldovich effect, already detected in some clusters. The search for both effects will be very active in the next years. It has been largely pointed out that measurements of this effect, together with X–ray data, may help to determine the distance of the cluster, and thus bring useful contributions to the problem of the distance scale in cosmology, and the estimation of H_0. The measurement of the kinetic Sunyaev–Zeldovich effect would give access to the proper velocities of clusters. Comparable distortions, although generally at lower levels, may be expected from any kind of ionized gas clouds between the last scattering surface and us, like for instance ionized bubbles around quasars or gas clouds in the process of radiative collapse.

4 Observations of the CMB

Many different experiments have been proposed to explore and measure the properties of the CMB. "Absolute" measurements intend to establish *global* properties of the CMB, and thus measure its intensity (temperature), at some given wavelength(s), regardless the direction on the sky. *Relative* measurements only detect flux differences between 2 or more directions onto the sky, separated by θ, i.e., anisotropies. The observation wavelength must be adapted to test the desired physical effect, but also to minimize other contributions and optimize the sensitivity. It is advantageous to use several wavelengths : the resulting spectral information provides some knowledge of the physical process indeed observed, and allows to subtract other unwanted components (from our Galaxy or other non cosmological sources). Peculiar observations are also made in precise directions, like those of some clusters of galaxies to detect the Sunyaev–Zeldovich effect.

Absolute measurements Absolute measurements are difficult to perform because of limited sensitivity, necessity of precise calibration, and unwanted contributions. The FIRAS instrument aboard *COBE* provided the spectrum and the temperature of the CMB with high precision, which will be quite difficult to improve (Gush et al., 1990, Mather et al., 1994). At wavelengths from 0.1 to 10 mm, stringent upper limits onto possible distortions were established as $y < 2.5\ 10^{-5}$ (see previous section). This constitutes a very important confirmation of the big–bang models and provides a fundamental information on the post–recombination universe.

Anisotropies The *dipole anisotropy* was detected very soon after the discovery of the CMB. It is now very well observed and interpreted. Its precise position dependence now offers a possible calibration for the instruments.

Anisotropy measurements at different angular scales do not use the same technic and concern distinct physical effects. They are deduced from one-beam mapping, or from 2-, 3-, or 4-beams differential experiments. Interferometry has also shown to be very efficient to explore the very small angular scales. One of the main goal of these observations is to determine the statistics (gaussian or not, spectral shape) of the intrinsic CMB fluctuations.

The DMR instrument aboard *COBE* has provided the first widely accepted detection of cosmological anisotropies of the CMB (Smoot et al., 1992) : the quadrupole level ($\theta = 90$deg) $\frac{\Delta T}{T} \approx 5\ 10^{-6}$; and the modes between 10deg and 90deg, for which $\frac{\Delta T}{T} \approx 10^{-5}$. Preliminary analyses suggested a "scale invariant" spectral shape, corresponding to a spectral index $n \approx 1$. Anisotropies have also been detected by other missions (the soviet satellite *RELICT*, balloon borne and polar experiments), with detected levels comparable to those from the DMR.

Cross–correlations give a mutual confidence in the results. They also constrain more the spectral shape, now estimated to values $n \approx 1.2 - 1.6$.

All these results are presently being discussed and interpreted. Concerning very large angular scales, they are not directly related to the formation of structures like galaxies, clusters or even superclusters. Thus they do not discriminate between different models for large scale structure formation. However extremely specific scenarii – like cdm for instance – assume a relation between the spatial scales relevant for structure formation (up to few 10 Mpc) and those corresponding to the DMR observations (around 1 000 Mpc). For such models, the DMR results provide a convenient normalization for the power spectrum of fluctuations. They concern however mainly the physics in the primordial universe. Testing structure formation would require an angular resolution at least one order of magnitude better.

5 Conclusion and perspectives

The CMB is now very well observed, and its average properties established with a good precision. All predictions of the big–bang models have been confirmed without alternative explanations. Much additional informations are also provided. The dipole anisotropy indicates the motion of our galaxy with respect to the last scattering surface, i.e., the best approximation for an "absolute" frame. Beyond the dipole anisotropy, *COBE* has provided the first – highly desired – anisotropy detections. The very low level of the quadrupole anisotropy excludes that our Universe could be rotating, or be in anisotropic expansion. The detected anisotropies will help to set constraints onto the physics of the primordial Universe.

Observations of the CMB also provide information onto the post–recombination Universe and its ionisation history : they already exclude the presence of large amounts of hot ionized intergalactic gas. Observations of the Sunyaev–Zeldovich effect initiate new and complementary methods to determine the cosmological parameters, and also offer a new tool to explore the physics of the galaxy clusters.

JThe astronomy of the CMB is only beginning. Different projects with polar stations, balloons and satellites are under preparation and study. The developments of HEMT and bolometric detectors allow the planning of very ambitious missions, to examine anisotropies at a deeper level, at smaller angular scales, and at different wavelengths. Spectrophotometric information will allow to discriminate between different possible origins for the detected anisotropies : intrinsic vs. post–recombination, thermal vs. non–thermal, kinetical vs. thermal Sunyaev-Zeldovich effect, etc. This will unveil some aspects of the primordial universe (inflation, primordial gravitational waves, topological defects, large scale fluctuations in gravitational potential,...). The main expected output will be however to provide the first direct tests of cosmogonic models, for galaxy and large scale structure formation. We will learn about the statistics of the fluctuations to the Friedmann–Lemaître models : its gaussian or non gaussian character, its power spectrum (in particular the spectral index). This will open, in any case, a new era in cosmology.

References

[1] Bond J. R. 1995, in *Cosmology and Large Scale Structure*, ed. R. Schaeffer, Elsevier Science Publishers, Netherlands (1995)

[2] Gush H.P, Halpern M., & Wishnow E. 1990, Phys. Rev., 65, 537

[3] Mather J.C. 1994 Ap. J. (Lett.), 420, 439

[4] Songaila A. et al., 1994, Nature 371, p.43

[5] Songaila A. et al., 1994, Nature 371 p.43

[6] Smoot G.F. et al. 1992, Ap. J. (Lett.), 396, L1

[7] White M., Scott D. & Silk J. 1994 ARAA, 32, 319

MIX
Papier aus verantwortungsvollen Quellen
Paper from responsible sources
FSC® C105338

If you have any concerns about our products,
you can contact us on
ProductSafety@springernature.com

In case Publisher is established outside the EU,
the EU authorized representative is:
Springer Nature Customer Service Center GmbH
Europaplatz 3, 69115 Heidelberg, Germany

Printed by Libri Plureos GmbH
in Hamburg, Germany